Proceedings of the
Third International Conference on

B Physics and CP Violation

Proceedings of the
Third International Conference on

B Physics and CP Violation

Taipei, Taiwan December 3–7 1999

Editors

H. -Y. Cheng
Academia Sinica

W. -S. Hou
National Taiwan University

World Scientific
Singapore • New Jersey • London • Hong Kong

Published by

World Scientific Publishing Co. Pte. Ltd.

P O Box 128, Farrer Road, Singapore 912805

USA office: Suite 1B, 1060 Main Street, River Edge, NJ 07661

UK office: 57 Shelton Street, Covent Garden, London WC2H 9HE

British Library Cataloguing-in-Publication Data
A catalogue record for this book is available from the British Library.

B PHYSICS AND CP VIOLATION
Proceedings of the Third International Conference

ISBN 981-02-4376-6

Printed in Singapore by Uto-Print

Preface

1999 was a tremendous year. Direct CP violation finally showed up in the K system, while the holy grail of B factories — mixing dependent CP violation in $B \to J/\psi K_S$ mode — turned up at the Tevatron, although just short of "measurement". Also in February, the CLEO-II experiment ended its operation. With 10 million recorded $B\overline{B}$ pairs, it remains unchallenged for B physics output. But this will not be for long, since the two asymmetric e^+e^- B factories, namely BABAR at SLAC and BELLE at KEK, turned on successfully in early summer, followed by CLEO-III which turned on in November. It seems that everyone is jockeying for the great competition on: "*B Physics and CP Violation*". At the change of Millennium and dawn of new revelations, excitement is in the air.

With great anticipation, 140 high energy physicists from all over the world descended on Taipei in December, undeterred by the strong September earthquake, to attend the Third Conference of the series started by Professor A.I. Sanda in Nagoya in October 1994. They met with 60 or so locals to review and discuss the state of the art and future prospects of the field. The Conference was augmented strongly by physics from K, charm and hyperon systems. It seems that there is not only competition in the B arena, but between B and K as well. The race is to unravel the Kobayashi-Maskawa matrix, especially the size and origin of CP phases, but the higher aim is to uncover signs of New Physics.

The Conference provided a forum for intensive discussions at the technical level. The strong attendance at an off the beaten path location attested to the worldwide interest on the subject. Organized around physics topics to maximize interaction between experimenters and theorists, the Conference indeed enjoyed many spirited discussions.

Another indication of elevated interests is in the bidding for the sequels. A special Advisory Committee meeting was called during the Conference for this purpose. It was decided that the Fourth Conference would be held February 2001 in Central Japan, while the Fifth Conference would be held May 2002 in Philadelphia.

Hai-Yang Cheng
Wei-Shu Hou

Acknowledgements

The Third International Conference on B Physics and CP Violation was held on the campus of National Taiwan University during December 3–7, 1999. We thank the National Science Council, Academia Sinica, National Taiwan University, the Ministry of Education, the Asia Pacific Center for Theoretical Physics (Seoul), and the Physical Society of R.O.C. for support.

Evelyn Lin and Winnie Sam, our hard working secretaries, deserve much of the credit for the organizational work of the Conference, while Michelle Chen helped finalize this Proceedings. Many of our colleagues, especially Xiao–Gang He, provided instrumental help. The young HEP group at National Taiwan University provided much manpower support. We thank especially our students John Y. Chao and Jack K.–F. Chen for their great help on computing and web related matters.

We thank the scientific secretaries, G. Bhattacharyya, A. Bozek, C.–H. Chen, C.–H. Cheng, C.–W. Chiang, R. Chistov, C.–L. Chou, C.–K. Chua, H.–C. Huang, C.–W. Hwang, Y.G. Kim, H.–L. Lai, W.–P. Lai, C.–C. Lih, I. Low, B. Melic, H.–H. Shih, R. Sinha, B. Tseng, C.–Y. Wu, and all the session chairs, for their hard work in making the Conference a success.

International Advisory Committee

A.I. Sanda (Nagoya), **Chair**
R. Aleksan (Saclay)
K. Berkelman (Cornell)
D. Chang (NTHU)
L.L. Chau (Davis)
Y.M. Cho (APCTP)
M. Danilov (ITEP)
J. Dorfan (SLAC)
M. Gronau (Technion)
D. Hitlin (Caltech)
Y.B. Hsiung (Fermilab)
C. Jarlskog (Stockholm)
C.W. Kim (KIAS)
K. Kleinknecht (Mainz)
M. Kobayashi (KEK)
T.D. Lee (Columbia)
K.B. Luk (Berkeley)
L. Maiani (CERN)
T. Nakada (PSI)
H. Quinn (SLAC)
W. Schmidt-Parzefall (DESY)
A. Soni (BNL)
B. Stech (Heidelberg)
S. Stone (Syracuse)
H. Sugawara (KEK)
F. Takasaki (KEK)
E. Thorndike (Rochester)
B. Wicklund (Argonne)
B. Winstein (Chicago)
L. Wolfenstein (Carnegie Mellon)
C.N. Yang (Stony Brook)

Local Organizing Committee

C.–H.V. Chang (NTNU)
Y.–H. Chang (NCU)
A.E. Chen (NCU)
H.–Y. Cheng (AS), **Co–Chair**
C.–Y. Cheung (AS)
C.–Q. Geng (NTHU)
R.–S. Guo (NKNU)
X.–G. He (NTU)
G.W.–S. Hou (NTU), **Co–Chair**
P.W.–Y. Hwang (NTU)
S.–C. Lee (AS)
H.–n. Li (NCKU)
G.–L. Lin (NCTU)
W.T. Lin (NCU)
P.–K. Teng (AS)
C.–H. Wang (NLIT)
M.–Z. Wang (NTU)
H.T. Wong (AS)
S.–N. Yang (NTU)
H. Yao (NTNU)
W.–M. Zhang (NCKU)

CONTENTS

CKM Matrix Elements

Chair: B. K. Chung (APCTP, Seoul)

Contributed Evening Session I

Chairs: C. Kao (University of Wisconsin) & K. Terasaki (YITP, Kyoto University)

Lifetime and Spectroscopy
Chair: S. Suzuki (UC Berkeley)

***B* Decays**
Chair: G. Eilam (Technion)

Charmless *B* Decays/Direct CP Violation
Chair: N. Lockyer (University of Pennsylvania)

Factorization and Final State Interactions
Chair: C. P. Yuan (NCTS & Michigan State University)

ε'/ε

Chair: R. Forty (CERN)

Determination of Unitarity Phases

Chair: T. Browder (University of Hawaii)

Lattice Results/Kaon Decay

Chair: D. Chang (National Tsing Hua University)

Rare B and D Decays

Chair: C. S. Huang (ITP, Beijing)

Contributed Evening Session II
Chair: J. Trampetic (RBI, Zagreb)

Other T and CP Violation
Chair: S. C. Lee (Academia Sinica)

Outlook for Hadronic Machines
Chair: G. Eigen (University of Bergen)

STATUS OF e^+e^- B-FACTORY ACCELERATORS

SHIN-ICHI KUROKAWA
KEK, High Energy Accelerator Research Organization
1-1 Oho, Tsukuba-shi, Ibaraki-ken, 305 Japan

ABSTRACT

Two asymmetric-energy, two-ring B-Factories, KEKB and PEP-II, and a symmetric-energy, single-ring B-Factory, CESR, are outlined. The status of commissioning of KEKB and PEP-II are introduced. The history of increase of luminosity at CESR is reviewed and an upgrade plan of CESR to a two-ring collider, CESR Phase IV, is given.

1. Introduction

At present three B-Factories are running in the world: namely, KEKB at KEK, PEP-II at SLAC, and CESR Phase III at Cornell University. KEKB and PEP-II are asymmetric-energy, two-ring, electron-positron colliders, whereas CESR Phase III is a symmetric-energy, single-ring collider. All of these B-Factories are aimed at reaching a luminosity higher than a few times $10^{33}cm^{-2}s^{-1}$.

The luminosity L of these colliders in the unit of $cm^{-2}s^{-1}$ is given by the following formula:

$$L = 2.2 \times 10^{34} \xi_y (1+r) EI/\beta_y^* \pm ,$$

where ξ_y is the beam-beam tuneshift, r the ratio between the vertical beam size (σ_y^*) to the horizontal beam size (σ_x^*) at the interaction point, IP (usually r<<1 and negligible), I the current in ampere, E the beam energy in GeV, and β_y^* the β-value at the IP in cm. The ± sign means that this formula is applicable to both electrons and positrons. In order to reach a high luminosity, ξ_y and I should be maximized, and β_y^* minimized. If we insert, for an example, typical values of $E = 5$ GeV, $\beta_y^* =$ 1.5 cm and $\xi_y = 0.03$ to the formula, we find that we need to store 1.4 A in a ring to get a luminosity of $3 \times 10^{33}cm^{-2}s^{-1}$. In order to store this large current in a ring, the current should be distributed into many bunches, since there is a limit of single-bunch current at a few mA to a few times 10 mA. Two-ring collider schemes with a single interaction point (KEKB and PEP-II) are straightforward solution to this; while at a single-ring collider such as CESR, special tricks should be adopted to avoid spurious beam collisions other than at IP, since if N_b bunches per beam are stored in a ring, beams collide at 2 x N $_b$ points along the ring. In an asymmetric energy collider, electrons have higher energy to avoid ion trapping; thus the electron ring is called high-energy ring, HER, and the positron ring low-energy ring, LER.

At B-Factories, it is of vital importance to minimize injection time and maximize the efficiency. Beams are injected into B-Factory rings at full energy. Topping on the beam is usual way of injection.

2. KEKB and PEP-II

2.1 KEKB

Table 1 summarizes main design parameters of KEKB and PEP-II. KEKB is an asymmetric-energy, two-ring, electron-positron collider housed in the 3 km TRISTAN tunnel[1]. Eight-GeV electrons and 3.5-GeV positrons are stored in separate rings. They circulate in opposite directions, and collide at one interaction point (IP) at Tsukuba experimental hall. The BELLE detector[2] surrounds the IP. The cross section of the tunnel is large enough to enable a side-by-side installation of HER and LER. The two rings change their inner-outer position at the IP and at a crossover point opposite to the IP, where HER and LER have different heights and electrons and positrons do not collide. An interchange of inner and outer position of the rings is necessary to make the two rings have the same circumference.

Table 1 Main parameters of KEKB and PEP-II

	KEKB	PEP-II
Energy E_-/E_+ (GeV)	8.0/3.5	9.0/3.1
Circumference (m)	3016	2199
Luminosity (10^{33}cm^{-2}s^{-1})	10	3
Beam-beam parameter ξ_x/ξ_y	0.039/0.052	0.03/0.03
β_x^*/β_y^* at IP (cm)	33/1.0	67/2.0 (HER) 50/1.5 (LER)
σ_x^*/σ_y^* at IP (μm)	90/1.9	220/6.8
Stored current I_-/I_+	1.1/2.6	0.99/2.14
Number of bunches	5000	1658
Bunch spacing (cm)	59	124
Crossing	±11 mrad	Head-on

The injector linac of KEKB has been upgraded from 2.5 GeV to 8 GeV in order to facilitate the direct injection of electrons and positrons to the rings and to increase the intensity of positrons by increasing the electron energy impinging on a positron production target[3]. Figure 1 shows a schematic layout of KEKB.

In order to reach the design luminosity of $10^{34}cm^{-2}s^{-1}$, even with challenging values of $\beta_y^* = 1$ cm and $\xi_y = 0.052$, the currents of 1.1 A and 2.6 A should be stored in its HER and LER. These currents are distributed into 5000 bunches with 59 cm bunch spacing, which corresponds to one wave-length of 508 MHz RF frequency of KEKB. Every RF bucket is filled with beam except for about 10% gap necessary to avoid ion trapping in HER.

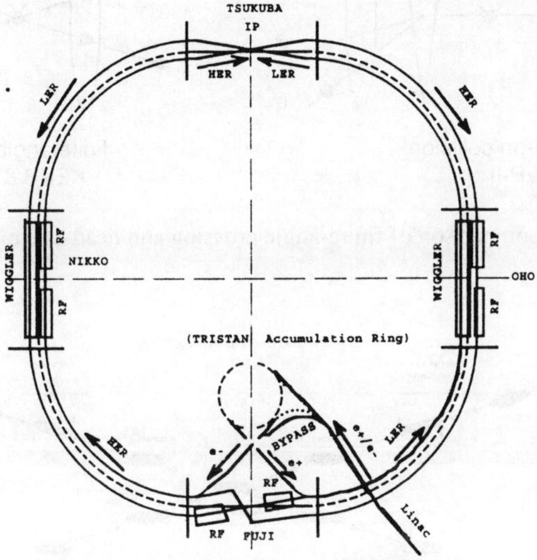

Fig. 1 Schematic layout of KEKB

One of the features of KEKB is its accelerating cavities. To combat high beam loading and prevent coupled-bunch instabilities due to higher-modes (HOM) and also fundamental mode, HOM free cavities with a large stored energy are used at KEKB: normal conducting cavities called ARES[4] and superconducting cavities[5]. ARES consists of three cells: an accelerating cell is connected to an energy-storage cell via a coupling cell. The stored energy of the system is increased by one-order of magnitude by the use of the energy-storage cell. On the other hand, due to its high field gradient, a superconducting cavity has a large stored energy by itself. These large stored energies of cavities mitigate beam loading due to high stored currents and prevents the coupled-bunch instability due to the fundamental mode.

In order to cope with the small bunch spacing and make the interaction region simpler, KEKB adopts a finite-angle crossing of ±11 mr at IP (see Fig. 2 for the difference between finite-angle and head-on collisions). Although simulation shows no degradation of luminosity due to the finite-angle crossing, crab-crossing scheme is a fall-back option at KEKB[6]. In crab crossing, both beams are kicked side-wise by a crab cavity before the interaction point, collide head-on at the IP and kicked-back to its original orientation by another crab cavity (see Fig. 3). Superconducting crab cavities are being developed[7].

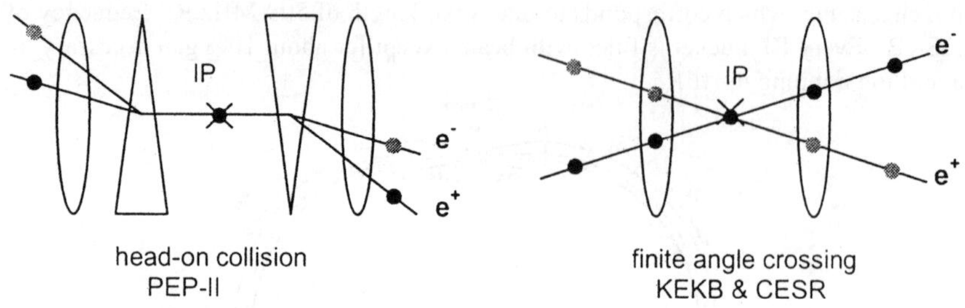

Fig. 2 Comparison of finite-angle crossing and head-on collision at IP

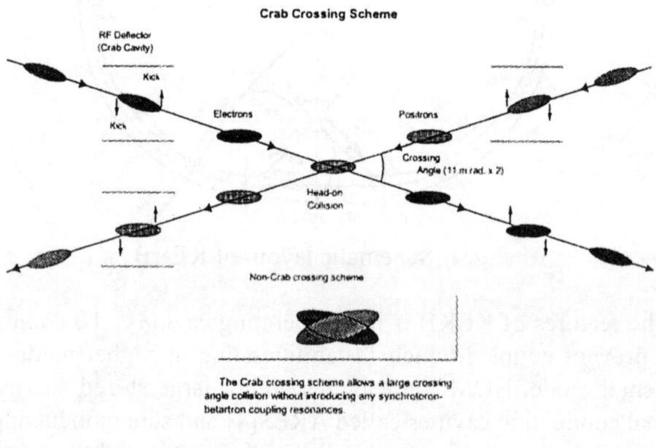

Fig. 3 Principle of crab crossing

Final focus quads should be located under a detector solenoid; the use of iron magnets is precluded. KEKB uses a pair of superconducting final focus quads at IP[8]. In front of a quad, superconducting anti-solenoid magnet is housed in the same cryostat as the quad. In addition to its tunability, it is straightforward for superconducting quads to have a few superconducting correction coils. X and Y steering coils and a skew quad coil are equipped.

Measurement of beam size is a strong tool for tuning of colliders. KEKB uses double-slit synchrotron light interferometers in visible light region for this purpose[9]. By measuring the ratio of peak and valley of the interference pattern made by a double-slit they can monitor the size of the beam; both the vertical and horizontal beam sizes are measured by this method at KEKB.

2.2 PEP-II

PEP-II is also an asymmetric-energy, two-ring, electron-positron collider[10]. Electron rings and positron rings are vertically stacked, the latter on top of the former, and they are housed in the 2.2-km PEP tunnel. 9-GeV electrons and 3.1-GeV positrons collide head-on at a single IP, where BaBar[11] detector surrounds. Figure 4 shows a layout of PEP-II.

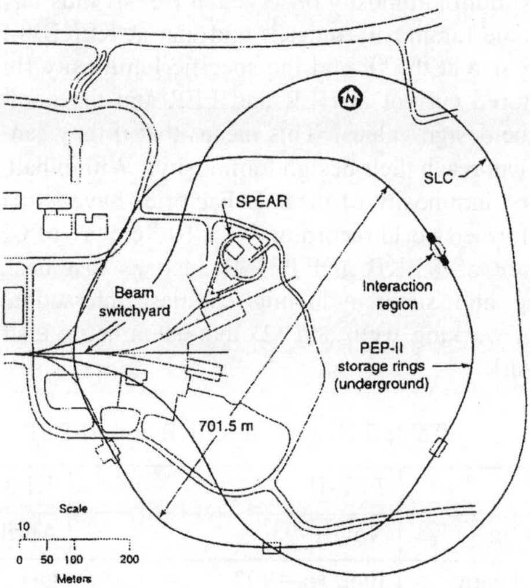

Fig. 4 Layout of PEP-II

PEP-II also uses single-cell, single-mode cavities to combat coupled-bunch instabilities. Although these cavities at PEP-II do not have a large stored energy, coupled-bunch instabilities due to the fundamental mode are suppressed by the use of a RF feedback system [12].

Injector linac to PEP-II is quite strong. Positrons are produced by 30-GeV electrons and its intensity is as strong as that of electrons. Damping of emittances by damping rings makes efficient injection possible. It takes only a few minutes for topping up the current.

PEP-II adopts a head-on collision scheme at the IP. Head-on collision needs to have dipole magnets close to IP to separate beams (see Fig.2). Permanent separation magnets and permanent final focus quads are used at PEP-II. Minimum bunch spacing at PEP-II is 1.24 m, which is twice as long as the 476-MHz RF wavelength at PEP-II. A head-on collision scheme needs longer space for beam separation than a finite-angle crossing scheme; this prevents from filling every bucket with beam.

2.3 Commissioning of KEKB and PEP-II

Table 2 summarizes the history of KEKB and PEP-II from the start of construction up to now (November1999). Although both B-Factories observed the first event in late May to early June 1999, the start of the commissioning of PEP-II is by one and half years earlier at electron rings and by half years at positron rings than that of KEKB. The maximum luminosity observed at PEP-II thus far is 1.43 x 10^{33} $cm^{-2}s^{-1}$ (this is a remarkable luminosity indeed) and that at KEKB 0.52 x $10^{33}$$cm^{-2}s^{-1}$. At PEP-II, the beam size at the IP and the specific luminosity (luminosity divided by the product of stored current in HER and LER and the number of bunches) are almost equal to the design values. This means that if they can increase the number of bunches, they can reach their design luminosity. Within half a year from the start of physics run, the luminosity of these B-Factories have surpassed (PEP-II) and is approaching the former world record of 8.3 x $10^{32}$$cm^{-2}s^{-1}$ of CESR. This shows that (1) basic hardware of KEKB and PEP-II such as fast beam feedback systems, normalconducting and superconducting cavities, interaction region final focus magnets, etc., are working well, and (2) that asymmetric energy electron-positron colliders really work.

Table 2 History of KEKB and PEP-II

	PEP-II	KEKB
Start of construction	Nov. 1993	April 1994
First stored HER beam	June 16, 1997	Dec. 13, 1998
First stored LER beam	July 16, 1998	Jan. 12, 1999
First beam collisions	July 23, 1998	Feb. 5, 1999
Detector installed	May 10, 1999 BaBar	May 24, 1999 BELLE
First detected events	May 26, 1999	June 1, 1999
Max. luminosity observed by November 30,1999	1.43 x $10^{33}$$cm^{-2}s^{-1}$	0.52 x $10^{33}$$cm^{-2}s^{-1}$

2.4 Photoelectron Instability

Photoelectron instability is the most serious instability at KEKB and limits its performance. Figure 5 shows the beam-size at the IP measured by the double-slit interferometer with respect to the stored current in KEKB LER with the total number of bunches of 928 and every fourth bucket filled with beam. At higher current, the beam blows up. This is a single-beam phenomenon and the blow-up is strongly dependent on the bunch spacing. Wider the bunch spacing is, the weaker the instability becomes.

We believe that this is caused by clouds of photoelectrons around the beam; thus called the photoelectron instability. Synchrotron light from the positron beam hits the inner wall of vacuum chambers and produces photoelectrons. These photoelectrons are attracted by the beam and form clouds around the beam. The clouds then excites head-tail type oscillation within a bunch and the beam blows up [13].

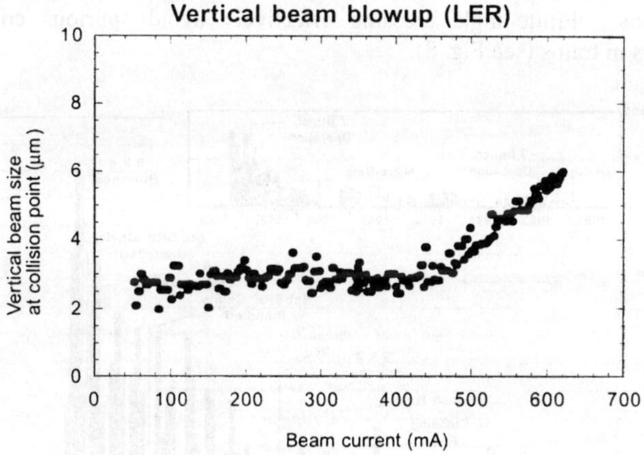

Fig. 5 Beam size blow-up measured at KEKB LER. The vertical axis shows the beam size at the IP measured by double-slit interferometer and the horizontal axis the stored current in LER.

At KEKB, dipole field of 150 gauss was imposed on the vacuum chambers by the use of permanent magnets in order to trap photoelectrons close to its birth place. Unfortunately the effect was not clear. How to fight against the photoelectron instability is the most urgent task at KEKB.

3. CESR

3.1 History of CESR and CESR Phase III

CESR is a single-ring collider and 5.3-GeV electrons and positrons circulate in the same ring and collide at an IP. To increase the number of bunches and at the same time avoid spurious collisions at other than IP is a key to go higher luminosity in a single-ring collider.

Figure 6 shows the history of the increase of the luminosity at CESR. CESR reached its maximum luminosity of 8.3 x $10^{32}cm^{-2}s^{-1}$ in January 1999. At the same time it observed the maximum beam-beam tuneshift of 0.052. The increase of the CESR luminosity after the adoption of micro-beta insertion in 1986 has been mainly due to the increase of the number of bunches from 7 (Phase I), 9 x 3 or 9 x 4 (Phase II), and 9 x 5 in Phase III . This was achieved first by adopting the Pretzel scheme, where spurious collisions between bunches are avoided by separating beams horizontally by the use of horizontal electrostatic separators (see Fig. 7). In addition to the Pretzel scheme, CESR also adopts a final-angle crossing of ±2.3 mr and bunch trains. Finite-angle crossing effectively avoid spurious collisions between bunches in trains (see Fig. 8).

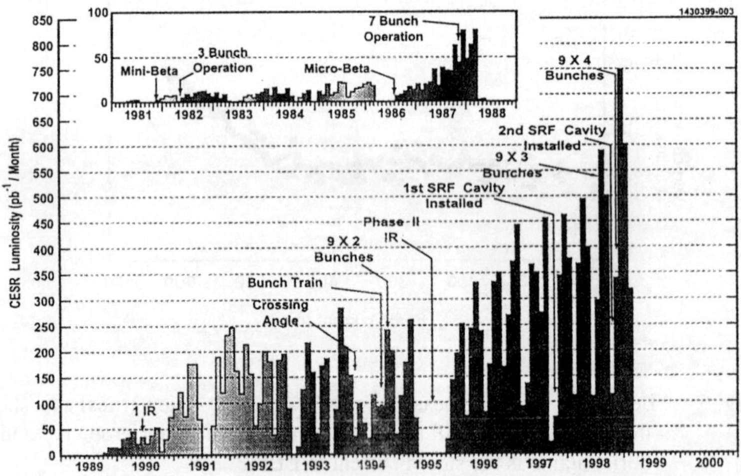

Fig. 6 Increase of CESR integrated luminosity per month from 1981 to 1999. Increase is roughly proportional to the number of bunches stored in the ring.

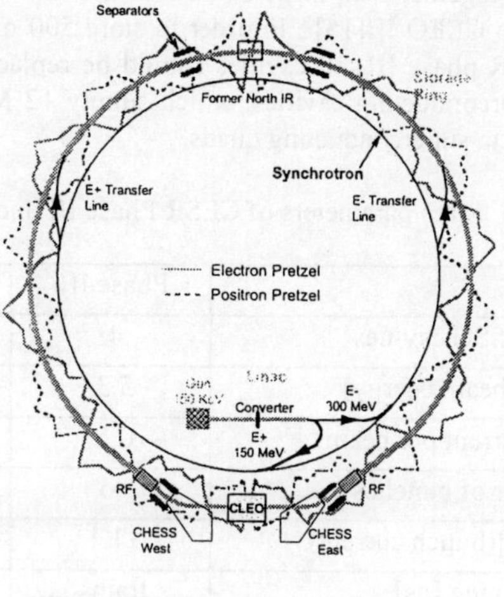

CESR Accelerator Complex with 9 Bunch Pretzels

Fig. 7 CESR layout with 9 bunches of Pretzel scheme

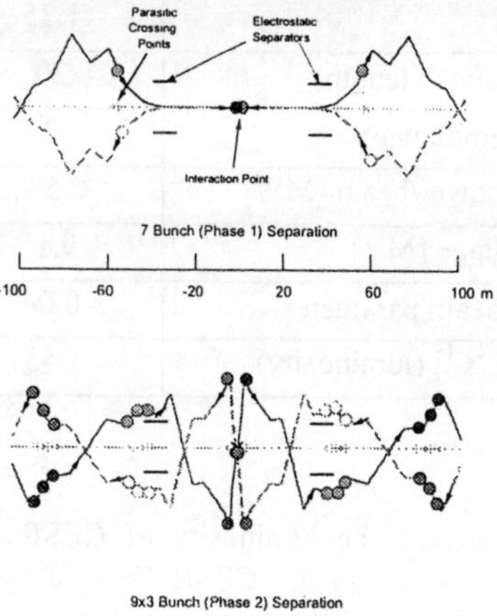

7 Bunch and Phase II Bunch Train Separation. The shaded bunches indicate crossings which occur at a time $t = D_b/2c$ later, where D_b is the bunch spacing.

Fig. 8 Beam separation and avoidance of spurious collision by Pretzel and finite-angle crossing at CESR.

Table 3 summarizes parameters of CESR Phase III with a goal luminosity of 1.55 x $10^{33}cm^{-2}s^{-1}$[14] together with those of CESR Phase IV (see Sec. 3.2). CLEO II is being upgraded to CLEO III[15]. In order to store 500 mA electrons and 500 mA positrons in CESR phase III, all cavities should be replaced with four single-cell, single-mode superconducting cavities, which supply 12 MV[16]. Final-focus quads will be replaced to superconducting quads.

Table 3 Main parameters of CESR Phase III and Phase IV

Parameter	Phase III	Phase IV
Number of SC cavities	4	10
E [GeV] (beam energy)	5.3	5.3
I_{tot} [A] (current per beam)	0.5	3.06
n_b (number of bunches)	45	180
I_{bunch} [mA](bunch current)	11.1	17
Bunch spacing [ns]	trains	14
Crossing half-angle [mr]	2.3	2.3
β^*_y [mm]	19	7
β^*_x [m]	1.25	1
σ_z [mm] (bunch length)	20	7
ε_x [10^{-7}] (emittance)	2	2
Syn. rad. power/beam [MW]	0.58	3.4
Cavity voltage [MV]	9.1	30
ξ_y (beam-beam parameter)	0.04	0.06
L [$10^{33}cm^{-2}s^{-1}$] (luminosity)	1.55	30

3.2 CESR Phase IV

In order to further increase the luminosity of CESR, CESR Phase IV is now considered as a future option [17]. CESR IV is a symmetric-energy, two-ring collider placed on top of the CESR injector synchrotron. The parameters of CESR Phase IV are summarized also in Table 5. The increase of the luminosity comes from (1) the increased total beam current, (2) stronger focusing at IP (β_y^* of 7 mm) and (3) increased beam-beam parameter ξ_y (0.06). Twenty-fold increase of the luminosity from 1.55 x $10^{33}cm^{-2}s^{-1}$ of CESR Phase III to 30 x $10^{33}cm^{-2}s^{-1}$ is envisioned.

4. Conclusions

KEKB and PEP-II have been commissioned quickly. At PEP-II a remarkable luminosity of 1.43 x $10^{33}cm^{-2}s^{-1}$ has been already achieved. At KEKB, how to fight against the photoelectron instability and raise the luminosity of present 0.52 x $10^{33}cm^{-2}s^{-1}$ beyond $10^{33}cm^{-2}s^{-1}$ is the most urgent task. This quick increase of the luminosity in KEKB and PEP-II is an evidence of the fact that basic hardware of these B-Factories is working well at KEKB and PEP-II and the basic designs of these new asymmetric B-Factories are correct.

Fast turn-on of KEKB and PEP-II and their good performance at this early stage make a good basis of confidence that their goals can be met, although it is not easy.

CESR Phase III has started its operation in November 1999. It is promising that it will reach beyond $10^{33}cm^{-2}s^{-1}$ soon on the basis of the accumulated experiences how to increase the luminosity at CESR.

To raise the luminosity higher than $10^{34}cm^{-2}s^{-1}$ may need further innovation in terms of ξ_y and β_y^*, since the increase of stored currents further would be too costly.

5. Addendum

From December 1999 to present (May 2000) both KEKB and PEP-II have shown rapid progresses. Figures 9 and 10 summarize the increase of the luminosity at KEKB and PEP-II from the start of their operation. At KEKB after the luminosity reached 1.0 x $10^{33}cm^{-2}s^{-1}$ on February 18, 2000, the luminosity was increased rapidly, and on May 27, 2000, it observed the peak luminosity of 1.92 x $10^{33}cm^{-2}s^{-1}$. The squeeze of the β values at the IP from 100 cm(β_x^*) and 10 mm(β_y^*) to 70 cm and 7 mm, respectively, in March 2000, has culminated in this improvement. By the end of May 2000, BELLE has logged 4.3/fb. The main obstacles against further improvement of the luminosity at KEK are: (1) heating and arching of some of movable masks and temperature rise at pieces of bellows near IP, and (2) beam blow-up in LER due to the photoelectron instability.

The luminosity of PEP-II has been improved steadily and reached 2.04 x $10^{33}cm^{-2}s^{-1}$. PEP-II enjoys short injection time due mainly to good performance of its injector linac and has a good ratio of average to peak luminosity. The maximum integrated luminosity per week of PEP-II is rather high (700/pb/week) compared to that of KEKB of 470/pb/week. BaBar has already logged more than 8/fb.

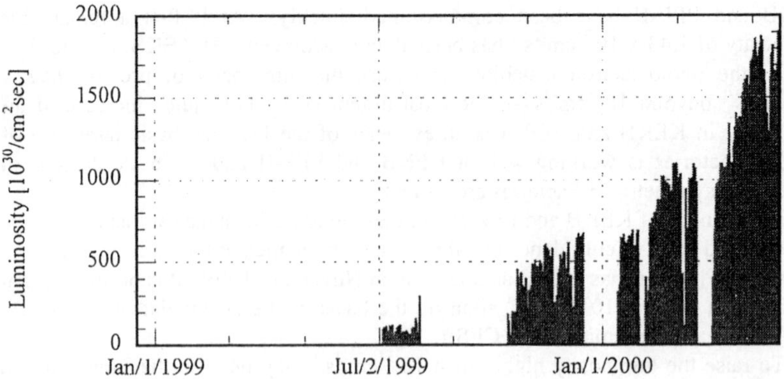

Fig. 9 Luminosity improvement at KEKB

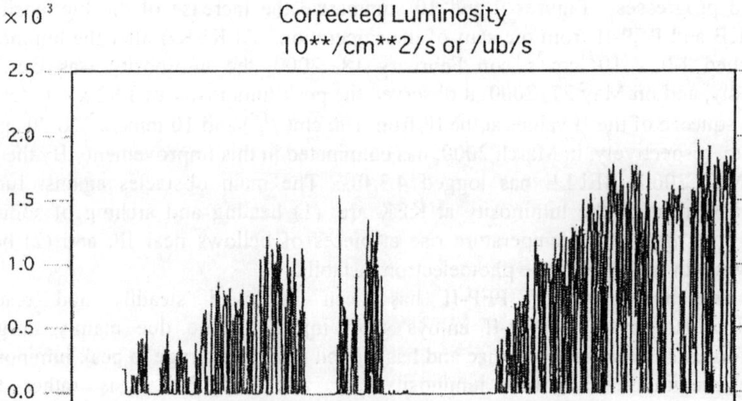

Fig. 10 Luminosity improvement at PEP-II

Acknowledgements

Special thanks to Dr. John Seeman of SLAC and Dr. Dave Rice of Cornell University for supplying me of information of PEP-II and CESR. I also thank to the members of KEKB for their continuous efforts during the construction and commissioning of KEKB.

References

1. KEKB B-Factory Design Report, KEK Report 95-7, August 1995.
2. A Study of CPViolation in B Meson Decays, KEK Report 95-1, April 1995.
3. A. Enomoto, Proceedings of the First Asian Particle Accelerator Conference, pp.292-294.
4. Y. Yamazaki and T. Kageyama, Particle Accel., 44, 107-127 (1994).
5. T. Furuya, et al, Proceedings of 5th European Particle Accelerator Conference (EPAC96), pp.2100-02.
6. K. Oide and K. Yokoya, Phys. Rev. A40 315 (1989).
7. H. Nakai, et al, Proceedings of Cryogenic Engineering Conference, Montreal, Canada, 1999, 45, Plenum Press, pp.853-860.
8. T. Ogitsu, et al., Proceedings of 6th European Particle Accelerator Conference (EPAC98), pp.2038-2040.
9. T. Mitsuhashi, Proceedings of the Joint US-CERN-Japan-Russia School on Particle Accelerators on Beam Mesurement, Montreux 1998, pp.399-427.
10. PEP-II An Asymmetric B Factory, SLAC-PUB-5397, June 1993.
11. BaBar Technical Design Report, SLAC-R-95-457, March 1995.
12. P. Corredoura, Proceedings of 1999 Particle Accelerator Conference (PAC99), pp.435-439.
13. Z.Y. Guo, Proceedings of 1999 Particle Accelerator Conference (PAC99), pp.1566-68..
14. S. Peck, et al., Proceedings of 1999 Particle Accelerator Conference (PAC99), pp.285-287.
15. CLEO III Detector Design and Physics Goals, CLNS 94 (1994).
16. S. Belomestnykh, et al., Proceedings of 1999 Particle Accelerator Conference (PAC99), pp.980-982.
17. D. Dugan, Proceedings of 1997 Particle Accelerator Conference (PAC97), pp.318-320.

STATUS OF THE BABAR DETECTOR

DOMINIQUE BOUTIGNY
On Behalf of the BABAR Collaboration

Laboratoire d'Annecy le Vieux de Physique des Particules – CNRS IN2P3

E-Mail : boutigny@in2p3.fr

The BABAR experiment, now installed on the SLAC PEP II B-Factory has been taking data since May 26, 1999. Luminosity in excess of 10^{33} cm^{-2}s^{-1} is routinely recorded. The status of the various sub-detectors, as well as the performance of the accelerator, is presented.

1 Introduction:

The main goal of the BABAR experiment [1] is the measurement of CP violation parameters. It will also extensively study B, charm, τ and two-photon physics [2]. First analysis studies have been presented at this conference [3].

CP violation measurements, as well as other B physics studies, require:

- Excellent tracking performance and vertex reconstruction.
- The ability to identify and measure γ and π^0.
- An excellent particle identification capability
- The identification and measurement of muons and neutral hadrons.

The BABAR experiment is installed on the SLAC PEP II B-Factory. This facility is a double-ring asymmetric energies collider. A 9 GeV electron beam and a 3.1 GeV positron beam provides collisions on the Y_{4S} resonance with a Lorentz boost $\beta\gamma$=0.56. The first hadronic event has been recorded May 26,1999.

2 PEP II performances

Since the first recorded event in May 1999, PEP II has continuously increased the current stored in the rings and the luminosity. Luminosity in excess of 10^{33} cm^{-2}s^{-1} is now regularly recorded.

Figure 1 shows the evolution of the integrated luminosity delivered by PEP II and recorded by BABAR. Through November 1999, 1.7 pb^{-1} has been recorded on tape for 2 pb^{-1} delivered by PEP II.

The overall efficiency of the BABAR Data Acquisition system is now around 90%.

PEP II is delivering clean beams, and the radiation dose received by the detector is well below the allowed budget. For instance, the radiation dose, recorded on the most exposed PIN diode located on the Silicon Vertex Tracker was of the order of 100 kRad for the period between BABAR startup and December, while the allowed budget was fixed to 170 kRad.

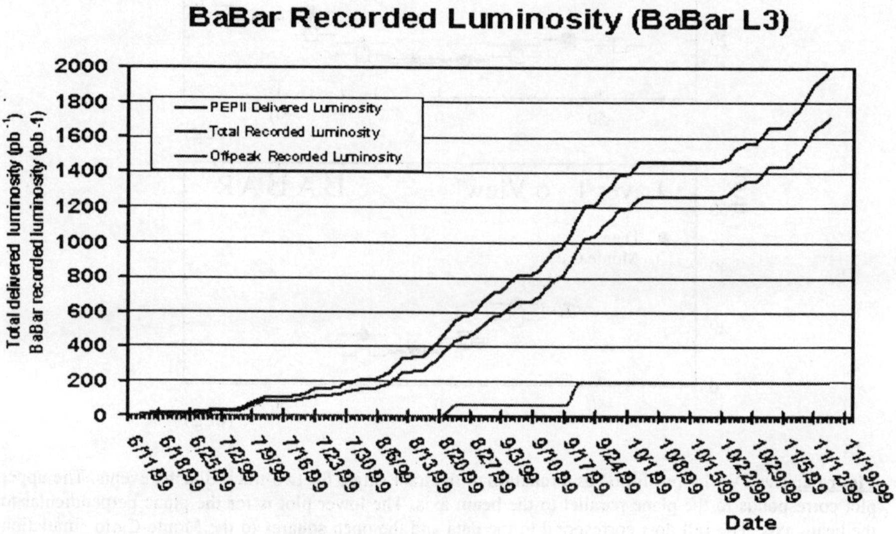

Figure 1: The upper curve is the PEPII delivered luminosity as a function of time. The middle curve is the luminosity recorded by BABAR and the lower curve is the luminosity recorded at center of mass energy off the resonance.

3 The Silicon Vertex Tracker:

The Silicon Vertex Tracker (SVT) is made of 143000 channels organized in five double-sided layers. BABAR has used a radiation hard technology allowing the SVT to sustain a maximum accumulated radiation dose of 2 Mrad. It is intended to provide the excellent vertex resolution necessary to reconstruct the distance between the two B mesons produced in the Y_{4S} decay. This resolution is crucial to accurately measure time dependent CP asymmetries.

Figure 2 shows the r.m.s of the residuals of the hits associated to tracks as a function of the incident track angle for both the plane parallel and perpendicular to the beam axis in the first layer of the Silicon Vertex Tracker. Two-track Bhabha events are used to select a clean sample of charged tracks. The observed resolution is of the order of 10 μm for tracks perpendicular to the silicon plane. The measured resolutions are in good agreement with the Monte-Carlo prediction, which assumes a perfect detector alignment in this case.

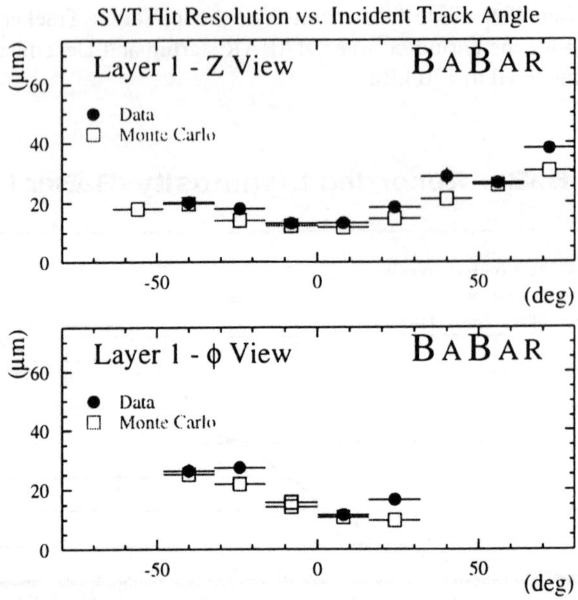

Figure 2: The SVT hit resolution as a function of the track angle for two-track Bhabha events. The upper plot corresponds to the plane parallel to the beam axis. The lower plot is for the plane perpendicular to the beam axis. The full dots correspond to the data and the open squares to the Monte-Carlo simulation assuming a perfect detector alignment.

4 The Drift Chamber:

The drift chamber is located inside a super-conducting magnet, providing a 1.5 Tesla magnetic field. It is made of 7100 hexagonal cells, organized in 40 layers corresponding to 10 axial or stereo super layers. The track trigger uses information from the axial layers while the stereo layers allow a precise determination of the longitudinal coordinate.

The maximum drift distance in the cells is equal to 11 mm.

The gas is a mixture of 80% of Helium and 20% of Isobutane, providing low density and good dE/dx resolution.

Figure 3 (left) shows the single hit resolution as a function of the drift distance to the wire. The result is obtained using an inclusive track selection on a physics sample dominated by Bhabha events. The average resolution is equal to 125 μm, which can be compared to the 140 μm of the design.

Figure 3 (right) shows the dE/dx distribution as a function of the particle momentum together with the Bethe-Bloch parameterization. We clearly identify lines corresponding to deuterons, protons, Kaons, pions or muons and electrons. The deuterons originate from beam-gas events. The clusters around 3 GeV and 9 GeV correspond to electrons and positrons from Bhabha events. When all corrections are applied, the Drift Chamber will be able to provide a separation between pions and Kaons, better than 3 standard deviations up to P=700 MeV/c.

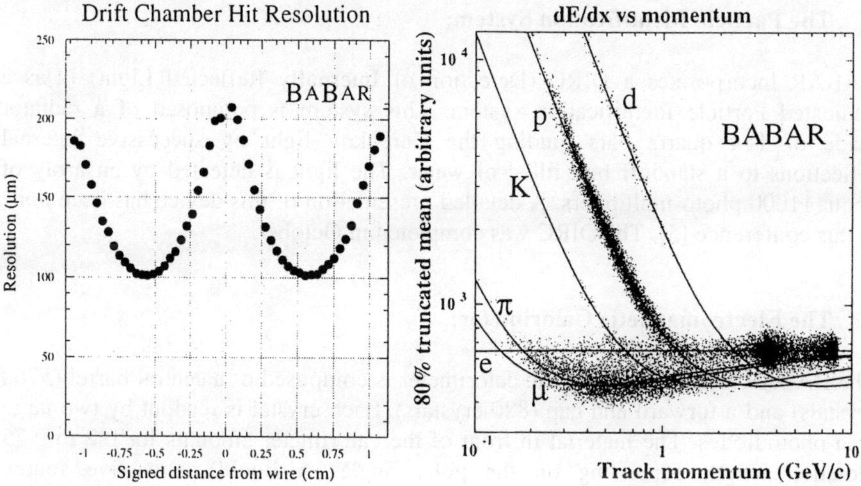

Figure 3: Left: Drift chamber single hit resolution as a function of the drift distance to the wire. Right: dE/dx distribution as a function of the track momentum. The Bethe-Bloch parameterization is also shown.

Using a sample of Bhabha events, we measure a dE/dx resolution of 7.5% at present. Applying all the corrections to the dE/dx, BABAR will ultimately achieve a 7% resolution.

5 Tracking performance:

The resolution on the distance of closest approach has been computed from a sample of di-muon events, using the tracking information provided by both the Silicon Vertex Detector and the Drift Chamber. At high momentum the resolution is equal to 34μm.

The transverse momentum resolution obtained with both the Silicon Vertex Tracker and the Drift Chamber is:

$$\frac{\sigma_{P_t}}{P_t} = 0.30\% P_t \text{ for } P_t \geq 3.5 GeV / c$$

Which increases to 0.45%P_t for the Drift Chamber only.

6 The Particle Identification System:

BABAR incorporates a DIRC (Detection of Internally Reflected Light) [4]as a dedicated Particle Identification system. This system is composed of a radiator made of 144 quartz bars guiding the Cerenkov light by successive internal reflections to a standoff box filled of water. The light is detected by an array of about 11000 photo-multipliers. A detailed presentation of this device has been done at this conference [5]. The DIRC was completed in October.

7 The Electromagnetic Calorimeter:

The BABAR CsI electromagnetic calorimeter is composed of a central barrel (5760 crystals) and a forward end-cap (820 crystals). Each crystal is readout by two large area photodiodes. The material in front of the calorimeter amounts for 0.2 to 0.25 radiation length depending on the polar angle. A liquid radioactive source circulating in front of the crystals is used for calibration purpose.

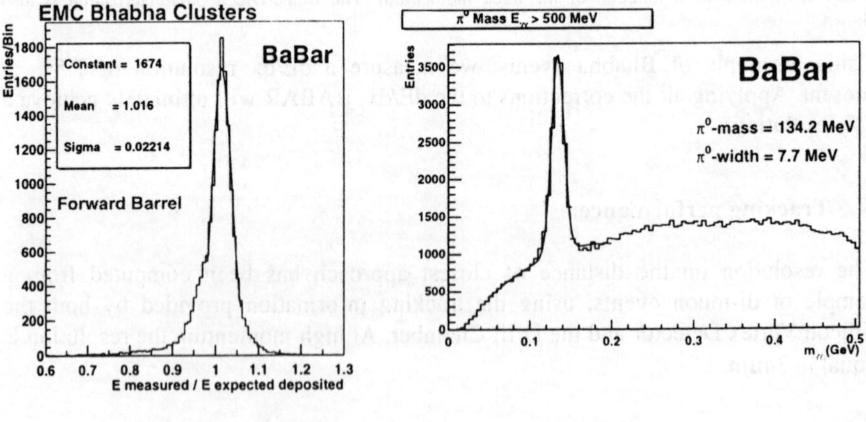

Figure 4: Left: Ratio of the measured to the expected energy for Bhabha events – Right: π^0 mass distribution from an hadronic event sample.

Figure 4 (left) shows the ratio of the measured to the expected energy for a sample of Bhabha events. The expected electron or positron energy is computed from the track angle. The low energy tail is due to the bremhsstrahlung. The resolution on the peak is equal to 2.2%.

Figure 4 (right) is the invariant mass distribution for photon pairs for E_γ>100 MeV and for a 500 MeV cut on the energy of the photon pair. A clear peak around the π^0 mass is visible with a resolution of 7.7 MeV. At this moment, the resolution is compromised both by noise (Incoherent noise: 420 KeV and coherent noise: 380 KeV) and non-linearities (10-15%) in the front-end electronic readout system. The additional noise and non-linearities will be removed by hardware fixes. We anticipate a 200 KeV noise level and <0.25% non-linearity will result in a 5.7% π^0 width.

8 The Instrumented Flux Return:

The Instrumented Flux Return (IFR) is designed to provide muon identification and tracking, as well as neutral hadron detection. The K^0_L identification capability will add significantly to the available event sample from the CP channel B → J/Ψ K_S. The IFR is composed of 19 Resistive Plate Chamber (RPC) layers in the barrel region and of 18 RPC layers in the end-cap regions. Two double layers of cylindrical RPC are located inside the magnet coil and increase the lever-arm for the muon measurements.

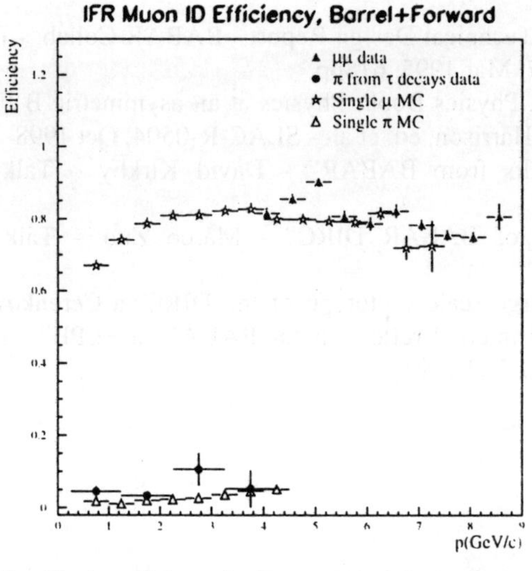

Figure 5: Top: Muon identification efficiency for dimuon data (full triangles) compared to Monte carlo (stars). **Bottom**: Pion contamination from τ decay data (full circles) compared to Monte-Carlo (open triangles)

Figure 5 shows the muon identification efficiency as a function of the momentum for di-muon data events. The observed 80% efficiency is well reproduced by single muon Monte-Carlo simulation. The detector geometry accounts for 10% of the total inefficiency. This figures also shows the pion contamination determined from tau decay data. This contamination comes mainly from pion decays in the drift chamber and in the electromagnetic calorimeter (2.4% / P (GeV)) and from punch-through (1.2%).

9 Conclusions and prospects:

The PEP II collider is working well and is routinely delivering luminosity above 10^{33} $cm^{-2}s^{-1}$. It will probably achieve the design luminosity of 3.10^{33} $cm^{-2}s^{-1}$ in the year 2000.

The BABAR detector is complete. All the sub-detectors already have demonstrated performances at, or close to design values.

After a one-month shutdown in December 1999, necessary to install permanent solutions to fix some leaks in the vacuum chamber, PEP II is now again delivering collisions.

The short-term goal for BABAR and PEP II is to accumulate 10 fb^{-1} by the time the first run ends next summer.

References:

1. The BABAR Technical Design Report - BABAR Collab. - D. Boutigny et al. - SLAC-R-0457, Mar 1995. 618pp.
2. "The BABAR Physics Book: Physics at an asymmetric B Factory" – BABAR Collab - P.F. Harrison, ed. et al. - SLAC-R-0504, Oct 1998
3. "Physics results from BABAR" - David Kirkby – Talk presented at this conference.
4. "Performance of BABAR DIRC" - Marco Zito – Talk presented at this conference
5. "Test of a large scale prototype of the DIRC, a Cerenkov imaging detector based on total internal reflection for BABAR at PEPII" - H. Staengle et al. NIM A397 261-282,1997

STATUS OF BELLE

K. ABE

KEK, Oho 1-1 Tsukuba, Ibaraki 305-0801 Japan

E-mail: kazuo.abe@kek.jp

1 Running status

First physics run of BELLE (Figure 1) took place during June-August of 1999. The peak luminosity recorded by KEKB was $3 \times 10^{32} cm^{-2} sec^{-1}$. The integrated luminosity during this run was $25 pb^{-1}$ at $\Upsilon(4s)$ resonance. We experienced rather harsh beam related background for the silicon vertex detector(SVD) and the central drift chamber(CDC). Innermost SVD-layer developed radiation damage due to an intense 5 keV X-rays which was generated by a few of the high energy ring (HER) steering magnets located upstream of the interaction region. Unfortunately, the radiation dose monitors were placed in such a way that they were not sensitive to 5 keV X-rays. Innermost three CDC-layers had to be turned off most of the time due to an intense 30 keV X-rays. They were generated at the HER focusing magnets just upstream of the interaction region, reflected at the downstream vacuum chamber, and entered the CDC gas volume. The CDC occupancy was high due to a large cross-talk contribution. The rest of BELLE and data acquisition system worked well. A fine tuning of the detector and physics analysis started using the accumulated data.

Based on the findings from this run, we upgraded several hardwares during the summer shutdown. We replaced the HER downstream vacuum chamber with newly designed copper chamber for reducing the reflection of X-rays. We replaced SVD with a new, but identical spare. The beryllium section of the beam pipe was wrapped with 20μ gold foil and the SVD preamp section was wrapped with 300μ gold foil. We reconfigured the location of dose monitors so that they are sensitive down to 5 keV. We reinforced the ground lines of CDC preamp boards, replaced signal cables for inner layers with coaxial type, and shortened the conversion time of CDC data. More shielding was added at several places around the interaction region to prevent spent-particle background.

We are running again since mid-October. Figures 2 show the SVD 1st layer energy spectrum, measured duing normal run before the summer, after the steering magnet strength were increased before the summer, and after the

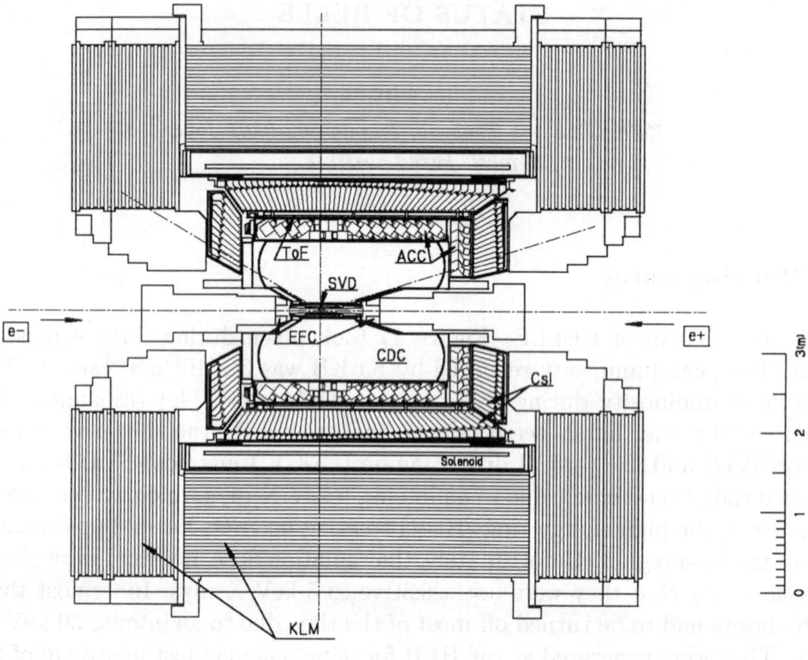

Figure 1.

summer. A factor of 100 reduction of the X-ray flux can be seen. Total CDC dark current vs HER beam current is plotted in figure 3 from the runs before the summer, after the summer at two time periods. A factor of 10 reduction is also apparent and it continues to improve as the ring vacuum improves. A limit which is typically quoted for preventing the wire aging in the drift chambers, 0.1 $C/cm/10year$, corresponds to 1600μA in this plot assuming uniform dark current distribution. Considering a large fraction of the dark current occurs in the inner part, the inner layers might be reaching this limit already. We need a close monitoring.

2 Detector Performance

2.1 Vertexing

An SVD hit pattern of typical hadronic event is shown in figure 4. The S/N ratio of 30(p-side) and 25(n) was achieved for the first layer, and 17~20 for

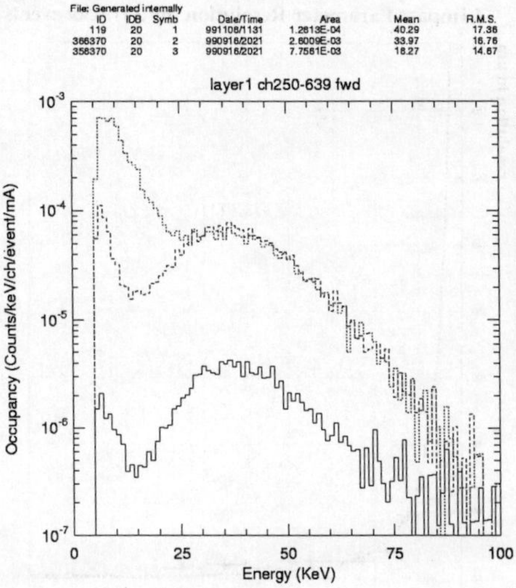

Figure 2.

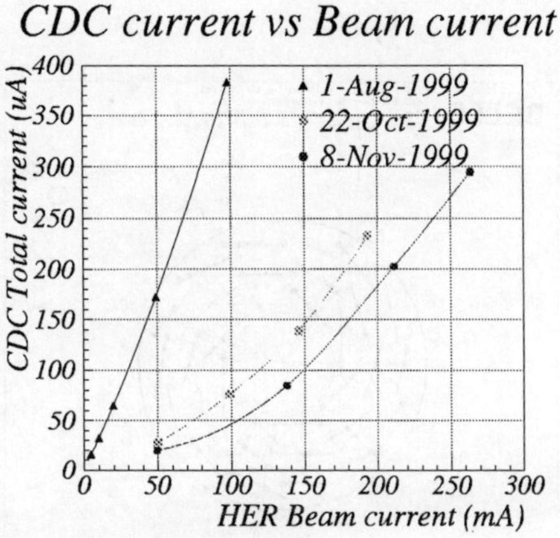

Figure 3.

other layers. A good charge correlation between p-side and n-side has been seen. The track matching efficinecy between CDC and SVD which we obtained is 97% and 96.7% for Bhabhas and for hadrons, respectively. The z impact paramer resolution was determined to be about 40μ at high momentum region (figure 5).

2.2 Tracking

We measured the momentum resolution vs p_t of the tracking system using cosmic tracks taken during physics runs (Figure 6). The track-finding efficiency was determined using a technique of embedding the Monte Carlo muon tracks in the data (Figure 7).

A good performance of the tracking system reflects to the reconstructed mass widths in various modes. The σ width and S/N of $K_S \to \pi^+\pi^-$ are 3.6 MeV and 6.5 below $p_K = 0.5$, 4.3 MeV and 8.6 in $0.5 \leq p_K \leq 1.5$, and 5.4 MeV and 7.6 above $p_K = 1.5$ GeV/c. The σ mass resolutions for $D^0(\bar{D}^0) \to K^\pm\pi^\mp$ and $D^\pm \to K^\pm K^\mp \pi^\pm$ are typically around 7 MeV.

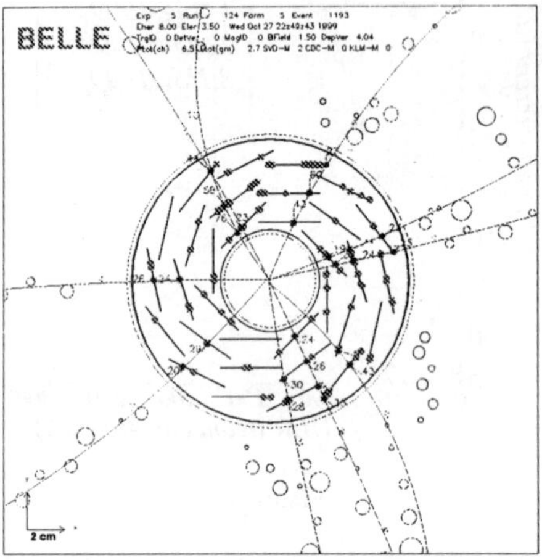

Figure 4.

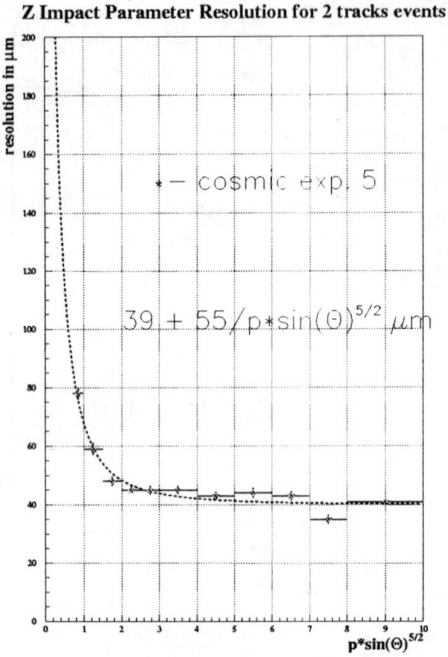

Figure 5.

2.3 Photons and π^0s

At the moment we require a cluster energy in the CsI calorimeter greater than 50 MeV for barrel and 100 MeV for endcap for the photon detection. Eventually we want to lower this down to 20 MeV. We then add the requirements of, i) CDC track veto, ii) shower shape variables being consistent with those of Monte Carlo, and iii) RMS shower width being consistent with that of Monte Carlo. Figure 8 shows the $\gamma\gamma$ mass distributions at each stage of the selection. The σ mass width of π^0 and S/N ratio are 4.6 MeV and 0.3 for $0 \leq p^*_{\gamma\gamma} \leq 0.5$, 4.7 MeV and 1.2 for $0.5 \leq p^*_{\gamma\gamma} \leq 1$, 5.4 MeV and 2.3 for $1 \leq p^*_{\gamma\gamma} \leq 2$, 6.4 MeV and 3.9 for $2 \leq p^*_{\gamma\gamma} \leq 3$, and 8.1 MeV and 5.5 above $p^*_{\gamma\gamma} = 3$ GeV/c.

Figure 9 shows the π^0 detection efficiency which includes geometrical acceptance.

Finely segmented and tower-structured CsI allows a good separation of single photons from π^0s even above 3 GeV. This was confirmed using radiative Bhabha data.

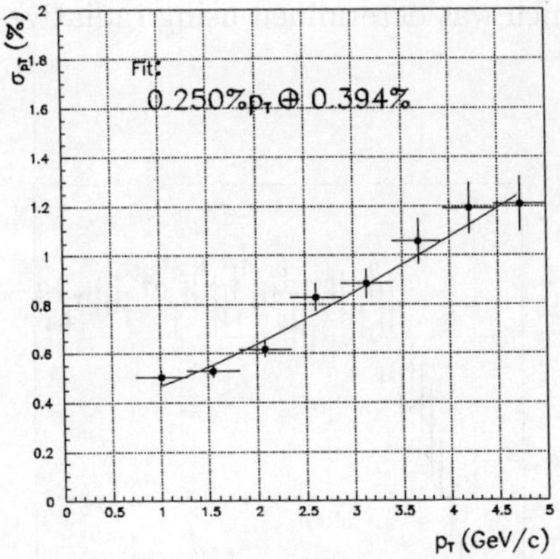

Figure 6.

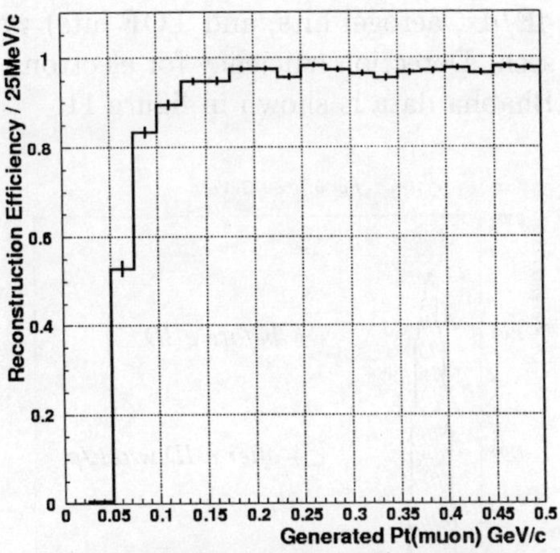

Figure 7.

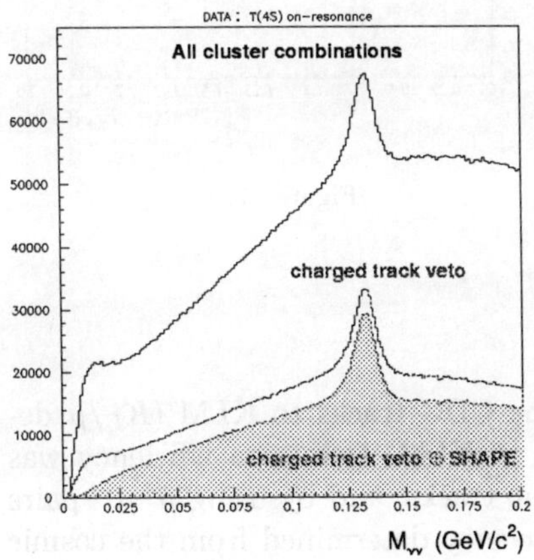

Figure 8.

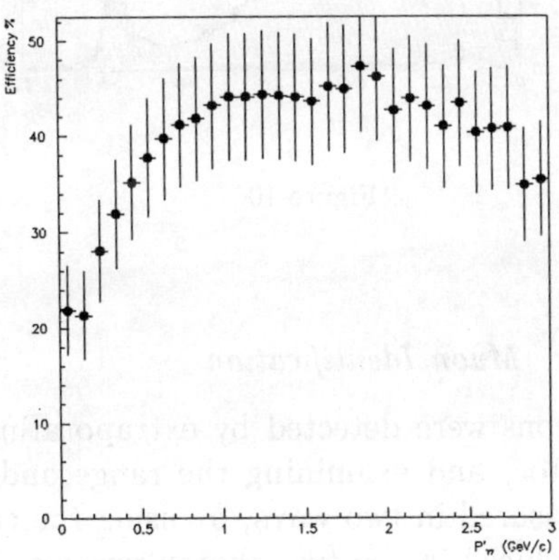

Figure 9.

2.4 Electron Identification

Figure 10 shows the E/p distributions for charged tracks above 1 GeV/c in the hadron event sample before any electron ID was made, and after the electron ID criteria excluding E/p (shower shape variables, CsI-CDC track association, dE/dx, aerogel hits, and TOF hits) was made. A clear electron peak can be seen. Detection efficiency for electrons, which was determined using radiative Bhabha data is shown in figure 11.

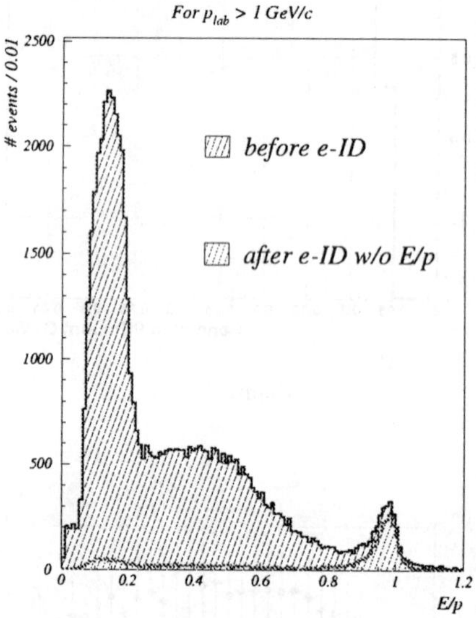

Figure 10.

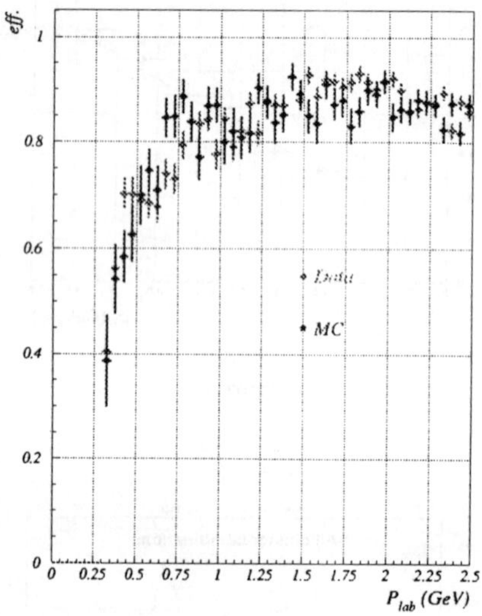

Figure 11.

2.5 Muon Identification

Muons were detected by extraporating the CDC tracks to KLM (K_L/μ detector) and examining the range and χ^2 at KLM. Detection efficiency was measured in two ways, by using the cosmic tracks, and by using $\mu^+\mu^-$ pairs produced in the two-photon process. Efficiency determined from the cosmic tracks shows higher values because this method uses a limited region of the detector. Each was compared with its Monte Carlo and agreement is good (Figure 12). Fake rate was determined using the pion tracks from $K_S \to \pi^+\pi^-$ in the hadron data and is shown in figure 13.

2.6 K_L detection

Figure 14 shows a 2.06 GeV K_L which was produced in a reaction $e^+e^- \to \gamma\phi$ followed by $\phi \to K_L K_S$. We collected only a few of these events so far. With

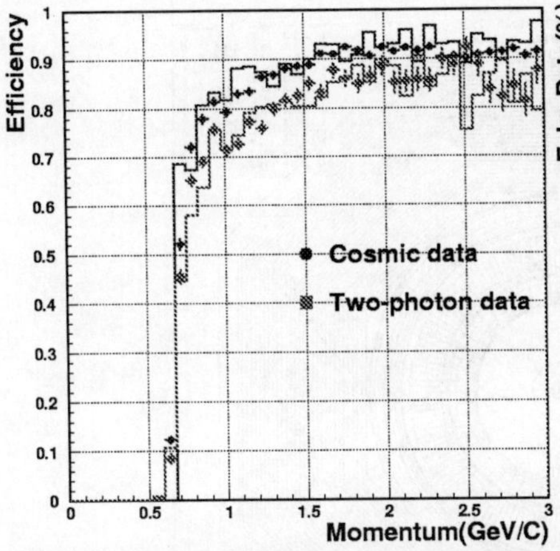

Figure 12.

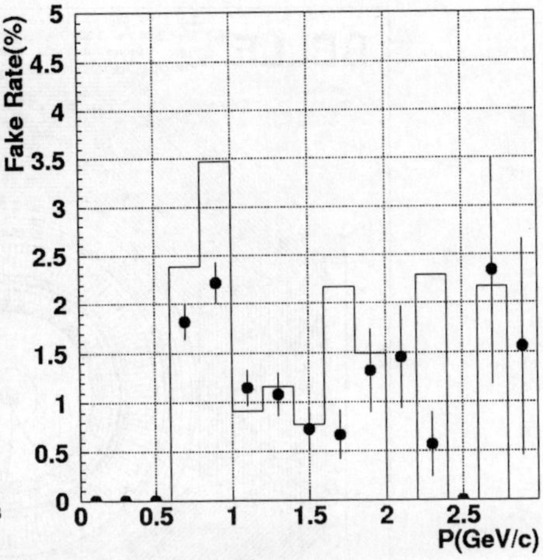

Figure 13.

a few hundreds of these events, we should be able to determine the efficiency and angular resolution of the K_L detection. Our main interest for the K_L detection is to measure $B^0 \to J/\psi K_L$, where most of K_L end up in the momentum region between 1 and 2.5 GeV. According to Monte Carlo, the efficiency is on the order of 70% in this region.

3 Summary and Prospect

Detector performance is approaching the design values. The CDC spatial resolution for Bhabha has reached $\sigma_x(\text{CDC}) = 140\mu\text{m}$. The CsI energy resolution for $e^+e^- \to \gamma\gamma$ is $\sigma_E/E = 2\%$. The TOF resolution has reached $\sigma_t =$ 100 ps for $e^+e^- \to \mu^+\mu^-$. The dE/dx resolution of $\sigma = 9\%$ for pions and 7% for Bhabha is obtained. The Impact parameter resolution $\sigma_z(\text{SVD}) = 40\mu\text{m}$ for cosmic tracks above $p \times \sin\theta^{5/2}$=2 Gev/c was obtained.

Effort for optimizing the reconstruction codes is making a good progress. Momentum resolution for the tracking system is now $\sigma(p_t)/p_t = 0.25\% p_t \oplus 0.39\%$. Track-finding efficiency is 97% above p_t=150 MeV/c and 60% at 50 MeV/c. The SVD-CDC track matching efficiency is 97%. We have good detection efficiencies for electrons above 500 MeV/c, for muons above 800 Mev/c. Fake rejection for both cases is good. We reached the π^0 mass width of $\sim$5 MeV, K_S mass width of $\sim$4.5 MeV, and D mass width of $\sim$7MeV.

We acquired a good understanding of the beam background. We learned

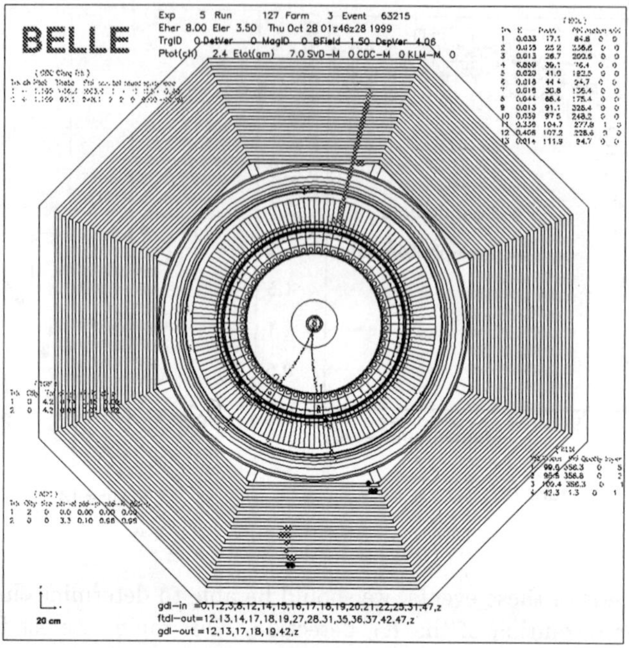

Figure 14.

the hard way. Spent-particles dominate the beam background now, and it continues to improve as the ring vacuum improves. The SVD should last for another 2 years even if the background stays at the present level. We face no significant problem for data acquisition system nor event reconstruction from the present level of background. Since we have now a realistic information of beam background, we can guess what will happen if it happens to increase. If it doubles from the present level, we might begin to see about 10% loss of CP mode reconstruction mainly due to CDC noise hits. Resolution and efficiency of SVD vertexing alone can take several times more background. KEKB reached $L_{\text{peak}} = 6 \times 10^{32}$. Blow up of LER beam size above 300 mA beam current limits from going higher at the moment, and aggresive machine study continues. Detector and analysis codes are in good shape.

KAON IDENTIFICATION IN BELLE

TORU IIJIMA
High Energy Accelerator Research Organization (KEK)
1-1 Oho, Tsukuba, Ibaraki 305-0801, Japan
E-mail: toru.iijima@kek.jp

Identification of the charged kaon plays an important role in the study of CP violation phenomena in B-factory experiments. In the BELLE experiment at the KEK B-factory, the kaon identification is carried out by using dE/dx measurement in a central drift chamber, a time-of-flight counter and a threshold aerogel Cherenkov counter. This paper briefly describes the detector system and the performance obtained in presently available BELLE physics run data.

1 Introduction

In the studies of CP violation phenomena in B-factory experiments, identification of charged kaons, against the large background of charged pions, is of great importance in the following two aspects. One aspect is to determine the flavor of parent B mesons, decaying into specific CP eigen modes such as $J/\psi K_s$, which is useful to extract the angle ϕ_1 in the unitarity triangle. The other aspect is to reconstruct some interesting exclusive decay modes. Typical examples are the two-body decays such as $B \to \pi\pi$ and πK, which are often discussed for the extraction of the angle ϕ_2 and possible finding of decay rate asymmetry (direct CP violation).

An experiment must be furnished with devices which enable the π/K separation in a wide momentum range. At an asymmetric e^+e^- B-factory, the momentum of a final state kaon from B cascade decays, used for the flavor tagging, distributes up to around 1.5 GeV/c, whereas the momentum region of interest for the two-body decays ranges from 1.5 to 3.8 GeV/c. While, the kaon identification in the relatively low momentum region, can be readily achieved by conventional technologies such as dE/dx measurement with a drift chamber or time-of-flight measurement, new development based on Cherenkov technology is necessary to meet the requirement in the high momentum region. In the BELLE experiment at the KEK B-factory, a threshold aerogel Cherenkov counter is used to extend the momentum coverage up to 3.5 GeV/c, while a DIRC counter and a ring imaging Cherenkov detector are used in the BaBar and CLEO-III experiments, respectively.

2 Kaon ID System in BELLE

Figure 1 shows the kaon identification system in the BELLE experiment [1]. The detector system and thier expected performance are described in this section.

The central drift chamber (CDC) [2] uses a gas mixture of 50% He and 50% C_2H_6, and has the dE/dx resolution of about 5% in design. It is expected to provide π/K separation up to 0.8 GeV/c and also in the region of relativistic rise (above 2.5 GeV/c).

The time-of-flight counter (TOF) [3] is a plastic scintillator array, divided into 128 ϕ segments. Each scintillator bar is 4cm thick, 6cm wide and 255cm long, and equipped with 2-inch fine-mesh PMT at both ends. The designed time-of-flight resolution is about 100ps to provide more than 3σ π/K separation up to 1.2 GeV/c.

The aerogel Cherenkov counter (ACC) [4] is divided into a barrel and a forward endcap parts. The barrel ACC is composed of 960 counter modules, segmented into 60 cells in the ϕ direction. The endcap ACC is composed of 228 counter modules, arranged in 5 concentric layers. Each counter module is equipped with one or two fine-mesh PMT, whose diameter is either 2, 2.5 or 3 inch. Silica aerogel, having a colloidal form of SiO_2 and more than 95% porosity, is a unique material to fulfill the index gap between compressed gas and liquid materials, and thus to enable the π/K separation in GeV/c region. We have successfully developed a new method to produce low-index aerogel radiators with high optical quality and also with hydrophobic surface, as reported in detail elsewhere [5]. The refractive index (n) of aerogels ranges from 1.010 to 1.028 for the barrel ACC, depending on the polar angular regions, to cover the momentum range from 1.2 to 3.5 GeV/c. For the endcap ACC, $n = 1.030$ has been chosen to serve for the flavor tagging. The expected light yield for light velocity particles is about 15 for the barrel ACC and about 30 for the endcap ACC.

Performance of the BELLE kaon identification system in some physics channels has been evaluated with a GEANT based Monte Carlo simulation. The simulation accommodates the detailed geometry of each detector component, and is tuned up by taking account of beam test results for prototype detectors. As for the flavor tagging, the b flavor is determined by the sum of kaon charges in the final state, when information from a high momentum lepton ($p_{cm} \geq 1.1$ GeV/c) is not available in the event. The expected kaon tagging efficiency is 29.0%, whereas the wrong tag fraction is 16.6%, leading to the effective tagging efficiency of 13.0%. As for the studies of two-body decays, $B \to \pi\pi$ and πK, we expect that the competing background process is reduced to 8%, while saving about 80% of the signal decays, by applying the kaon identification cut.

3 Performance in Collision Data

The BELLE detector has been commissioned with e^+e^- beams since May 1999. The performance of the kaon identification, obtained from the data collected by December 1999, are briefly described in this section.

Figure 2-a) is a two dimensional plot of the measured dE/dx and the particle momentum. The measured dE/dx resolution is 8.3% for minimum ionizing pions in the momentum region from 0.3 to 0.7 GeV/c, while the expected resolution from a Monte Carlo simulation is 7.4%. More accurate calibration, taking account of the dE/dx dependence on chamber gas pressure and beam induced wire current, is in progress to improve the dE/dx resolution.

The TOF counter has been calibrated with μ-pair events precisely, and the TOF resolution of about 100 ps has been already obtained. Figure 2-b) is a spectrum of hadron mass, calculated from the measured momentum and TOF, in the momentum region below 1.2 GeV/c. The figure shows that charged pions, kaons and protons are clearly identified.

Figure 2-c) displays the distribution of photoelectrons measured by the barrel ACC for $e^{\pm}$ tracks in Bhabha events and also for $K^{\pm}$ candidates, which are selected by TOF and dE/dx measurements. Excellent agreement between data and Monte Carlo simulations has been obtained, as shown in the figure. The average number of photoelectrons for the $e^{\pm}$ tracks is about 17 for the $n = 1.015$ module.

In our physics analyses, the kaon identification is carried out by combining information from the three detectors with a likelihood method. The power of kaon identification can be demonstrated in some of inclusive particle spectra. For example, Figure 3-a) presents the invariant mass of two charged tracks with kaon mass assigned for both tracks. One can clearly see a peak corresponding to the ϕ-meson as soon as the kaon identification is required for both tracks. Efficiency for signals is estimated to be 88%, while the background is suppressed by a factor of about 50.

The kaon identification has been further proven successfully in the reconstruction of the charm decay, $D^{*+} \to D^0\pi^+$, followed by $D^0 \to K^-\pi^+$. This decay can be selected with a good S/N ratio of more than 20 without relying on kaon identification. The kaons and pions can be identified with the relative charge with respect to the slow π^+ from the D^{*+} decay. Figure 3-b) shows a two dimensional plot of kaon probability and measured momenta for the kaon and pion tracks. The figure indicates that the kaons and pions are clearly

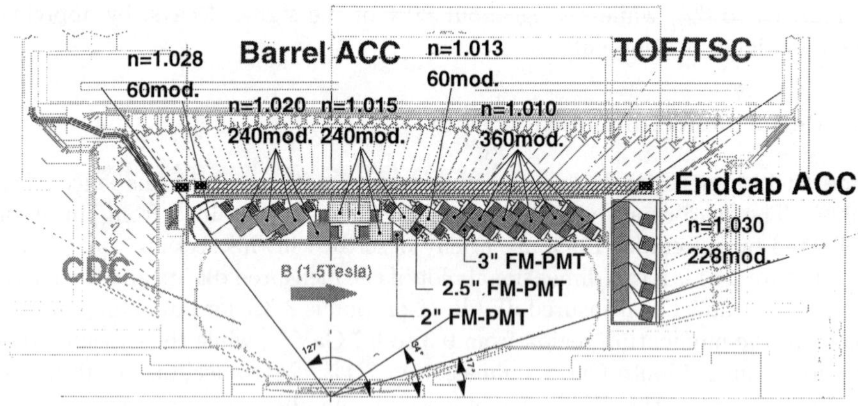

Figure 1. Schematic drawing of the kaon identification system of the BELLE experiment.

separated in their kaon probabilities up to around 4 GeV/c.

4 Summary

In this paper, the kaon identification in the BELLE experiment, based on dE/dx measurement in the central drift chamber, the time-of-flight counter and the threshold aerogel Cherenkov counter, is briefly described. The results of presently available collision data demonstrates that each detector component has performance close to the design already. A clear π/K separation in the momentum region up to 4 GeV/c has been demonstrated using the decay $D^0 \to K^-\pi^+$ tagged by the slow π^+ from the D^{*+}. Further calibration efforts are in progress to improve the performance. We hope to provide meaningful physics results with this excellent kaon identification in near future.

Acknowledgments

The author would like to thank the BELLE collaboration, in particular the members of CDC, TOF and ACC sub-detector groups, for the data analyses

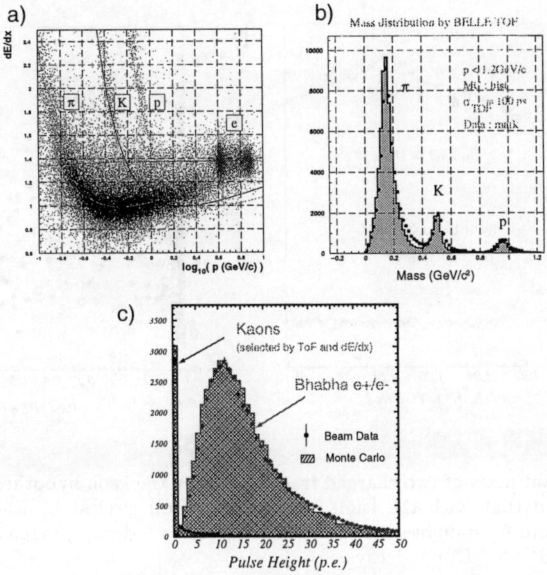

Figure 2. a) dE/dx versus momentum observed in collision data. b) Distribution of hadron mass calculated from the measured time-of-flight and the momentum. c) Pulse-height spectra (in units of photoelectrons) observed by ACC for electrons and kaons. Kaon candidates are obtained by dE/dx and TOF measurements. The Monte Carlo expectations are superimposed.

whose results are presented in this paper.

References

1. BELLE Collaboration, Technical Design Report for A study of *CP* Violation in *B* Meson Decays, March, 1995, KEK report 95-1.
2. S.Uno, *Nucl. Instrum. Methods* A **379**, 421 (1996), K.Emi *et al.*, *Nucl. Instrum. Methods* A **379**, 225 (1996), S.Uno *et al.*, NIMA **330**, 55 (1993).
3. H.Kichimi *et al.*, Proceedings of the 7th International Conference on Instrumentation for Colliding Beam Physics, Hamamatsu, Japan, November 15-19, 1999, submitted to NIMA., H.Kichimi *et al.*, NIMA **325**, 451 (1993).
4. T.Iijima *et al.*, *Nucl. Instrum. Methods* A **379**, 457 (1996), T.Iijima *et al.*, *Nucl. Instrum. Methods* A **387**, 64 (1997), R.Suda *et al.*, *Nucl. Instrum. Methods* A **406**, 213 (1998), M.H.R.Khan *et al.*, *Nucl. Instrum.*

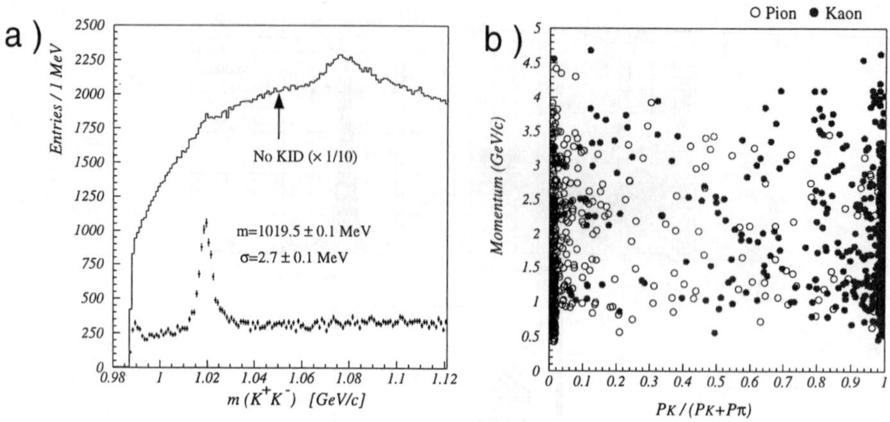

Figure 3. a) Invariant mass of two charged tracks based on the kaon hypothesis (histogram, scaled by 1/10) and that with the kaon identification (full circles) b) Kaon probability versus the momentum for daughter tracks from $D^0 \to K^-\pi^+$ decays, tagged by the charge of the slow π^+ in $D^{*+} \to D^0\pi^+$ decays.

Methods A **413**, 201 (1998), T.Sumiyoshi *et al.*, *Nucl. Instrum. Methods* A **433**, 385 (1999).

5. I.Adachi *et al.*, *Nucl. Instrum. Methods* A **355**, 390 (1995), S.K.Sahu *et al.*, *Nucl. Instrum. Methods* A **382**, 441 (1996), T.Sumiyoshi *et al.*, Journal of Non-Crystalline Solids, **225** 369 (1998).

DIRC : the particle identification detector of BaBar

Marco Zito
DAPNIA/SPP, CE-Saclay,
91191 Gif-sur-Yvette CEDEX, France
zito@hep.saclay.cea.fr

on behalf of the BaBar DIRC collaboration

The DIRC is a new type of Cerenkov detector based on the total internal reflection inside the radiator. Its construction and installation as part of the BaBar detector at SLAC have been completed in October 1999. The first data show that it performs according to the design. The DIRC will provide an excellent particle identification for the BaBar physics program.

1 Introduction

The goal of the BaBar experiment[1] at the PEP-II e^+e^- asymmetric collider at SLAC is to study CP violation in the B meson system. This requires selecting $e^+e^- \to B^0\bar{B}^0$ events where the flavor of one B meson is tagged and the other decays to a CP specific mode.

For the flavor tagging particle identification is required to select the inclusive kaons from $b \to c \to s$, with a momentum spectrum mostly below 2 GeV/c. Moreover it is necessary to separate K and π up to 4.5 GeV/c to select $B^0 \to \pi^+\pi^-$ from the more abundant $B^0 \to K^-\pi^+$.

To accomplish these tasks, BaBar uses the DIRC[2] (Detection of Internally Reflected Cerenkov light), a new type of Cerenkov detector.

2 The DIRC design

The DIRC is a novel ring imaging detector based on the total internal reflection of the Cerenkov light produced in the radiator (fig. 1).

It is a three dimensional device as it measures for each photon two coordinates on the detection surface and the arrival time.

The DIRC (fig. 2) consists of

- a radiator made of quartz bars, which is the only component inside the active detection volume of BaBar
- an expansion region : a tank containing almost 6 m^3 of pure water
- a photon detection surface comprising 10751 photomultipliers.

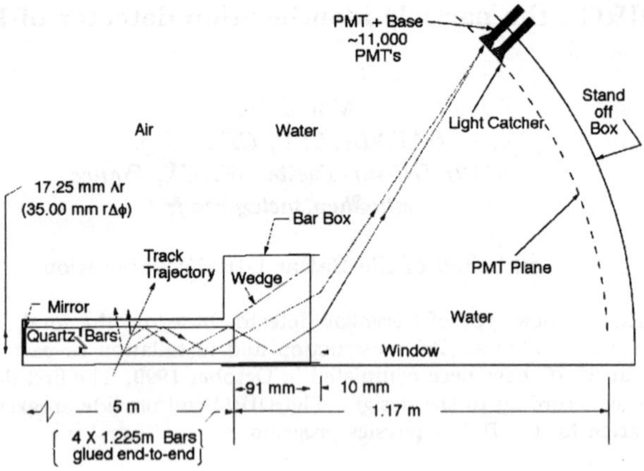

Figure 1: Schematic of the DIRC principle.

The DIRC Cerenkov radiator is made of 144 fused silica (*quartz*) bars (refractive index 1.47), 4.9 meter long, packed in twelve aluminium boxes. The bars are 3.5 *cm* wide and 1.7 *cm* high.

The quartz has been chosen for its optical properties : the bulk transmission is $99.9 \pm 0.1\%/m$ at 442 *nm* and thanks to the high quality polishing of the surfaces the reflectivity is $99.96 \pm 0.01\%$.

A quartz wedge is fixed at each bar exit in order to increase the detection acceptance and a mirror is glued on the opposite end to reflect the photons travelling away from the detection region.

Part of the Cerenkov light produced ($\lambda \approx 380$ *nm*) is trapped by total reflection and eventually reaches the image expansion region. That region, called the *standoff box*, is filled with water for good optical index matching with quartz.

The photon detection is performed by an array of 10751 photomultipliers (ETL 9125 [3]) 1.2 meter away from the end of the bars. The photomultipliers have a high gain (10^7) to provide a good single photon detection, a high quantum efficiency of 30 % at 400 *nm* and a good time resolution of 1.4 *ns*.

The DIRC is a remarkably thin detector: its total radial occupation in the sensitive region being only 8 *cm* (0.18 radiation lengths at normal incidence).

During the October 1999 BaBar/PEP II shutdown the installation of all the quartz bars has been completed.

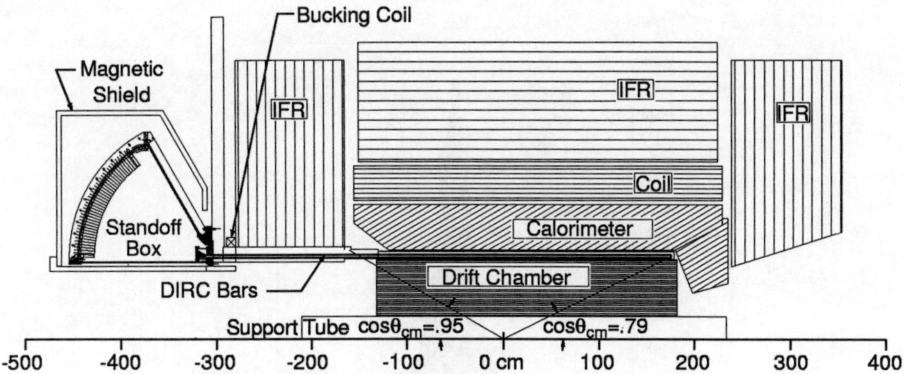

Figure 2: Half transverse view of the BaBar detector.

3 The DIRC performances

The expected single photon angle resolution is 9 *mrad* coming dominantly from the geometry, namely the size of the bar exit window, the lever arm and the photocathode area (7 *mrad*), and from the chromatic dispersions of quartz and water (5.4 *mrad*) (fig. 3).

The average number of photons is of the order of 35 increasing with the incidence angle of the particle with respect to the bar. Due to the angle-momentum correlation given by the boost, the higher momentum tracks from two body decays of B mesons are in the forward region. Therefore, the DIRC geometry is well optimized for a good identification of these tracks.

The overall time resolution is 1.8 *ns*, a powerful tool to reject background both from the machine and the other tracks in the event. In multihadronic events, after strict time cuts the ratio of background to signal photons in the Cerenkov angle distribution is 12 %.

The resolution on the Cerenkov angle has been measured using Bhabha events to be 3.0 *mrad*. It is dominated by residual systematic effects like internal alignments and the precision of the tracking extrapolation to the DIRC. This translates into a $K - \pi$ separation of 3.8 σ at a momentum of 3 GeV/c.

The performance of the DIRC as a particle identification device has been tested using a sample of reconstructed $D^{*+} \to D^0\pi^+$, with $D^0 \to K^-\pi^+$. Figure 4 shows the distribution of the measured Cerenkov angle versus the momentum for these kinematically identified tracks. Most of the tracks outside the expected band are accounted for by the combinatorial background of the sample which is at the 20 % level. The kaon identification efficiency in the

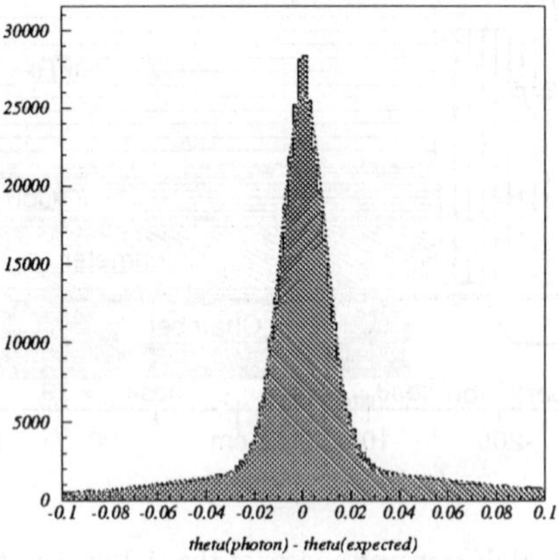

Figure 3: Single photon Cerenkov angle resolution measured in cosmic ray data. The width of the narrow gaussian is 9.3 *mrad*. The broad gaussian is due to background photons, mostly due to the track itself.

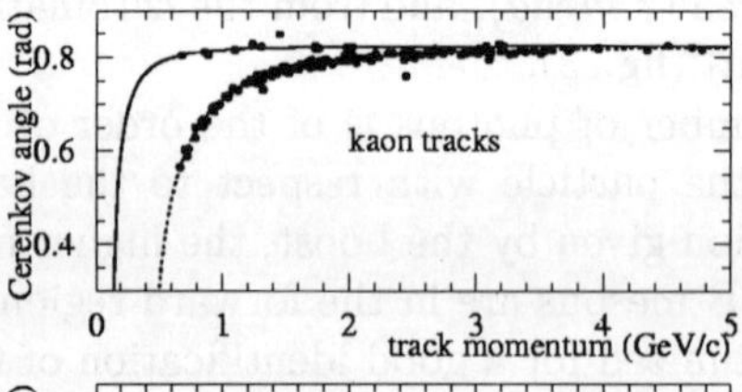

Figure 4: Cerenkov angle vs momentum distribution for tracks kinematically identified as kaons (upper plot) and pions (lower plot) from a sample of reconstructed $D^{*+} \to D^0\pi^+$ with $D^0 \to K^-\pi^+$. The solid (dashed) line shows the expected Cerenkov angle for pions (kaons).

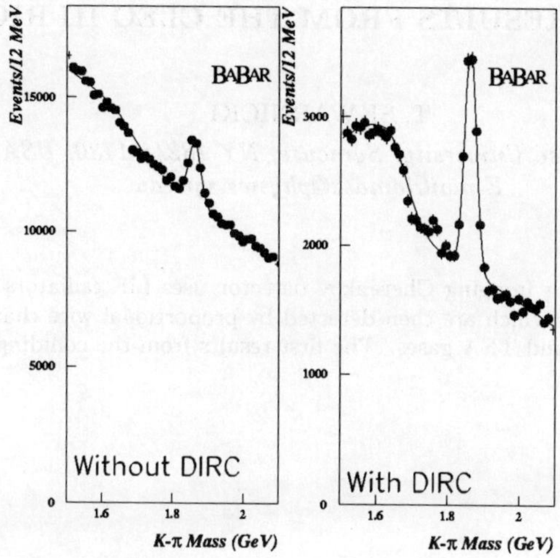

Figure 5: $K - \pi$ mass distribution without DIRC (left) and with DIRC (right) identification.

momentum interval $0.6 - 3.5\ GeV/c$ is 84 % for a cut on the number of detected photons at 10, with a pion misidentification at the 1 % level.

The power of DIRC identification is illustrated in figure 5. The $K - \pi$ mass distribution with the DIRC shows a clean signal from inclusive D^0 production with a reduction of a factor 5 of the combinatorial background.

4 Conclusion.

The DIRC is a new type of Cerenkov detector optimized for hadron identification at a B factory.

The first data have demonstrated that the detector performs according to the design. The Cerenkov angle resolution has been measured to be 3.0 *mrad* for high momentum tracks, which corresponds to 3.8 σ K/π separation at a momentum of 3 GeV/c.

These performances will provide BaBar with a powerful tool for a rich physics program.

1. The BaBar Collaboration, *Technical Design Report*, SLAC-REP-950457 (1995)
2. P. Coyle *et al.*, *Nucl. Instrum. Methods* **A343** (1994) 292
3. Electron Tubes Limited (formerly: Thorn EMI Electron Tubes), Bury Street, Ruislip, Middlesex HA47TA, U.K.

FIRST RESULTS FROM THE CLEO III RICH

T. SKWARNICKI

Syracuse University, Syracuse, NY 13244-1130, USA
E-mail:tomasz@physics.syr.edu

The CLEO III Ring Imaging Cherenkov detector uses LiF radiators to generate Cherenkov photons which are then detected by proportional wire chambers using a mixture of CH_4 and TEA gases. The first results from the colliding beam data are presented.

1 Introduction

The CLEO program at the CESR e^+e^- storage ring has been studying heavy flavor physics (b, c, τ) for the last 20 years. The CLEO II.5 run ended in February 1999 with 9 fb^{-1} of data collected in 4 years of running. Currently, the detector and machine are being upgraded. Upgrade to the next phase of the program is now in progress. The CESR luminosity is being raised from 6×10^{32} cm^{-2}s^{-1} to $> 2 \times 10^{33}$ cm^{-2}s^{-1}. The CLEO detector is being improved by the RICH detector replacing the Time of Flight system. The silicon vertex detector and the drift chamber are also being replaced to allow changes in the interaction point optics and space for the RICH detector. An engineering run with the RICH detector has recently ended.

The RICH detector will greatly improve K/π separation which, in turn, will allow CLEO III to distinguish suppressed processes like $b \to \mathbf{u}$, $c \to \mathbf{s}$, $\tau \to \bar{u}\mathbf{s}$ (tree level decays) and $b \to \mathbf{d}$ (penguin decays) from their dominant backgrounds $b \to c \to \mathbf{s}$, $c \to \mathbf{d}$, $\tau \to \bar{u}\mathbf{d}$ and $b \to \mathbf{s}$, respectively.

The K/π momenta from b decays reach up to $m_b/2 \approx 2.7$ GeV at CESR. The choice of LiF-TEA RICH, initially prototyped by the Séguinot and Ypsilantis[1], was dictated by the available space and desire to minimize amount of material in front of the electromagnetic calorimeter. At normal incidence, the CLEO III RICH detector is about 13% of radiation length thick. The CLEO III RICH is the first full scale implementation of such system. It uses an innovative sawtooth radiator[2] to avoid total internal reflection in the central part of the detector. Also, analog rather than binary readout has been employed.

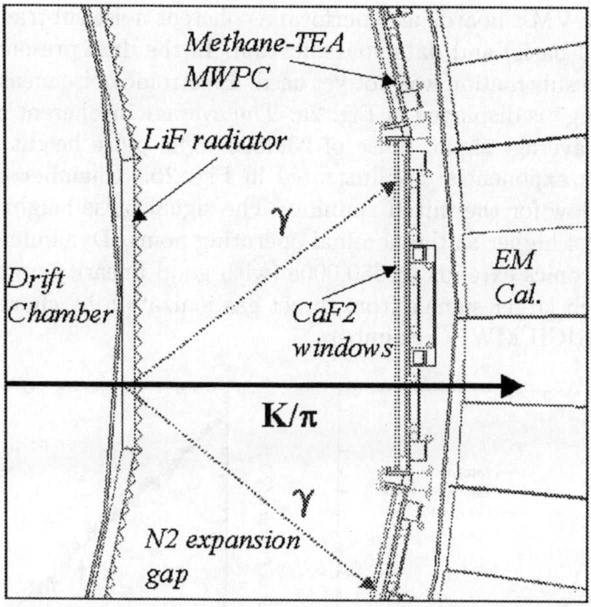

Figure 1. r-ϕ cross-section view of the CLEO III detector.

2 Detector Elements

Detector elements are shown in Figure 1. Charged tracks pass through LiF radiators (on average 1 cm thick) and produce Cherenkov photons. Out of 14 rows of LiF crystals 4 central rows are equipped with the sawtooth radiators. The photons travel through the expansion gap filled with nitrogen and enter multi-wire proportional chamber (MWPC) through CaF_2 windows. The chambers are filled with methane-TEA gas mixture. The Cherenkov photons in a narrow VUV range (135-165 nm) are converted into photo-electrons. The charge is multiplied by the usual proportional wire chamber mechanism. Position of the photon conversion is measured via charge induced on cathode pads (there are 230,400 pads in total).

3 Readout electronics and electronic noise performance

Each pad is input to a VA-RICH chip residing on the detector in which the pad signals are amplified, shaped and multiplexed before being sent to VME based electronics. There the signals are digitized. In the final CLEO III configuration, the VME boards also perform a coherent noise subtraction (on per chip/per ADC basis) and data sparsification. In the data presented here the coherent noise subtraction was not yet used. Electronic noise measured during the data taking[a] is displayed in Fig. 2a. The average incoherent noise is 500e compared to average signal pulse of 23,000e. The pulse height distribution for photons is exponential as illustrated in Fig. 2b. Chamber gain was set intentionally low for the initial running. The signal pulse height is expected to be about 2.4 higher at the nominal operating point. Dynamic range of the readout electronics extends to 650,000e (with good linearity up to 300,000e) to detect much larger signals from direct gas ionization by charged particles crossing the RICH MWPC chambers.

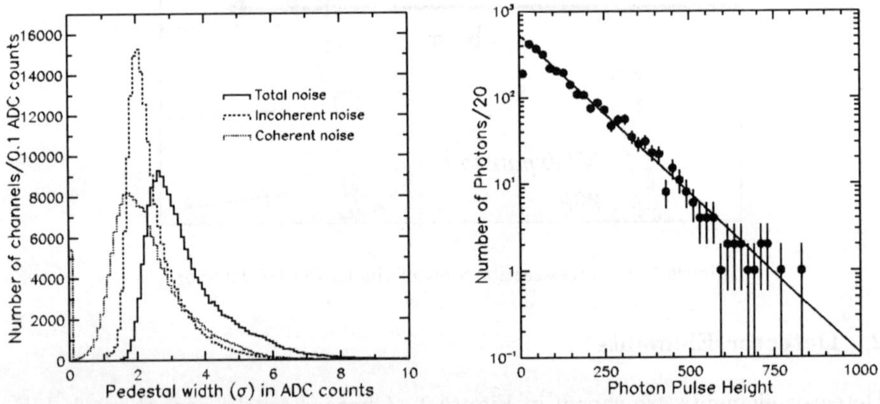

Figure 2. (a) Electronic noise as measured during data taking (left plot). Each entry represents one electronic channel. One ADC count corresponds to about 190e. (b) Pulse height distribution for photons (right plot).

4 Preliminary Analysis of the First Bhabha Events

The first colliding data with fully installed CLEO-III RICH detector were taken in mid November 1999. Cherenkov images for Bhabha electrons crossing sawtooth radiators are shown in Fig. 3. Preliminary analysis of the first Bhabha events $e^+e^- \to e^+e^-$ is presented below. Earlier results based on test beam of the first detector units exposed to high energy muons have been published elsewhere.[3] Since the drift chamber is not calibrated yet, we have developed a special Bhabha "tracker" based on hits from electrons crossing

[a] On-board sparsification was switched off for the noise measurement.

the RICH wire chambers. The results depend on the relative alignment of the RICH wire chambers and LiF radiators, which was fixed to the measurements performed during the detector assembly.

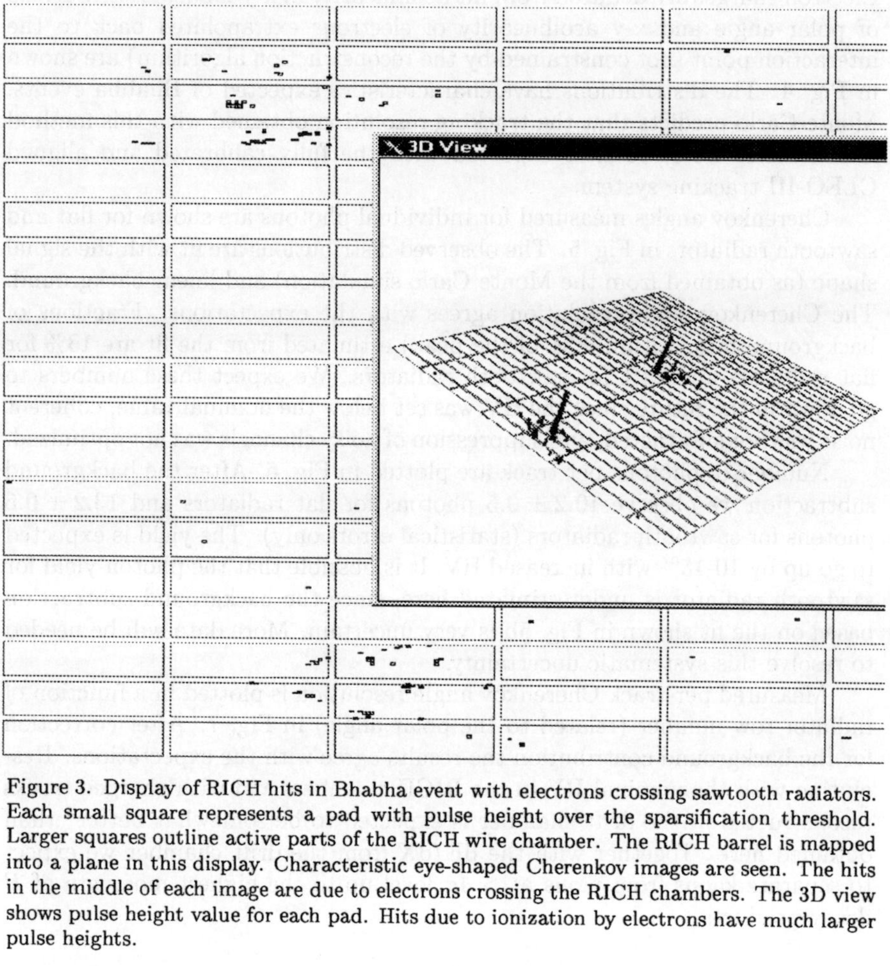

Figure 3. Display of RICH hits in Bhabha event with electrons crossing sawtooth radiators. Each small square represents a pad with pulse height over the sparsification threshold. Larger squares outline active parts of the RICH wire chamber. The RICH barrel is mapped into a plane in this display. Characteristic eye-shaped Cherenkov images are seen. The hits in the middle of each image are due to electrons crossing the RICH chambers. The 3D view shows pulse height value for each pad. Hits due to ionization by electrons have much larger pulse heights.

We look for two large-pulse-height RICH hits, which are characteristic for charged tracks passing the detector, and use them to estimate the electron four-momenta. We assume that electron energies are equal the beam energy and that the electrons originate from the center of the detector in the x-y plane (perpendicular to the beam), as the beam sizes are quite small in these directions. We measure collision point position along the beam by using

RICH hits and assuming that the electrons are back-to-back in the $r-z$ view ($r=\sqrt{x^2+y^2}$). We suppress radiative Bhabhas and background events by requiring that such measured position is within 5cm of the detector origin. Electron charges are deduced from hit acollinearity in x-y plane. Distributions of polar angle and x-y acollinearity of electrons extrapolated back to the interaction point (not constrained by the reconstruction algorithm) are shown in Fig. 4. The distributions have characteristics expected of Bhabha events. Monte Carlo predicts that the tracking resolution obtained with this method is somewhat worse to that expected with the fully calibrated and aligned CLEO-III tracking system.

Cherenkov angles measured for individual photons are shown for flat and sawtooth radiators in Fig. 5. The observed distributions are fit with the signal shape (as obtained from the Monte Carlo simulation) and linear background. The Cherenkov angle resolution agrees with the expectations. Fractions of background hits (within $\pm 3\sigma$ of the peak) estimated from the fit are 13% for flat radiators and 24% for sawtooth radiators. We expect these numbers to go down in the future since the HV was set below the nominal value, coherent noise was not subtracted, and suppression of noisy channels was not optimized.

Number of photons per track are plotted in Fig. 6. After the background subtraction we observe 10.2 ± 0.5 photons for flat radiators and 13.2 ± 0.6 photons for sawtooth radiators (statistical errors only). The yield is expected to go up by 10-15% with increased HV. It is possible that the photon yield for sawtooth radiator is underestimated here, since the background subtraction based on the fit shown in Fig. 5b is very uncertain. More data will be needed to resolve this systematic uncertainty.

Measured per-track Cherenkov angle resolution is plotted in a function of radiator row number (related to the polar angle) in Fig. 7. After correction for the background contribution the results agree with the expectations. Resolution with the nominal HV in the RICH chambers and with charged tracks reconstructed in the drift chamber is expected to be somewhat better[3] than obtained here. Together with the dE/dX from the drift chamber we expect to separate kaons from pions at $> 4\sigma$ level up to the highest momenta of B decay products.

References

1. C T. Ypsilantis and J. Séguinot 1994 *Nucl. Instr. Meth.* **A343** 30
2. A. Efimov and S. Stone 1996 *Nucl. Instr. and Meth.* **A371** 79
3. M. Artuso, R. Ayad, F. Azfar, A. Efimov, S. Kopp, R. Mountain, G. Majumder, S. Schuh, T. Skwarnicki, S. Stone, G. Viehhauser, J. C. Wang,

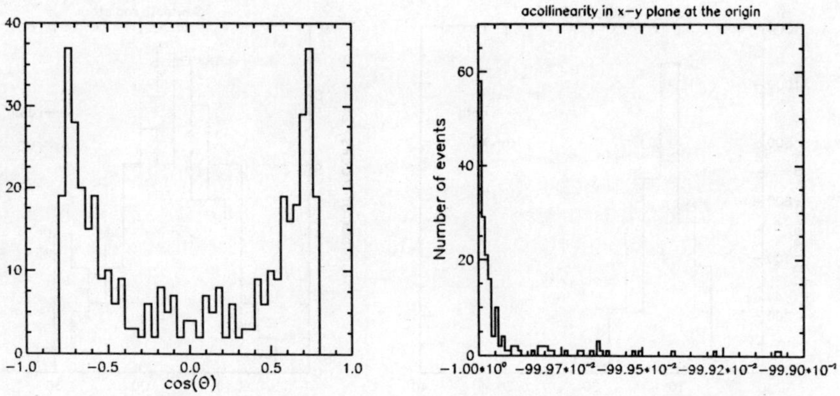

Figure 4. Distributions of cosine of the polar angle (left plot) and of the cosine of the angle in x-y plane between electrons extrapolated into the interaction point (right plot). The data peak at larger polar angles and for back-to-back configuration as expected for Bhabha electrons.

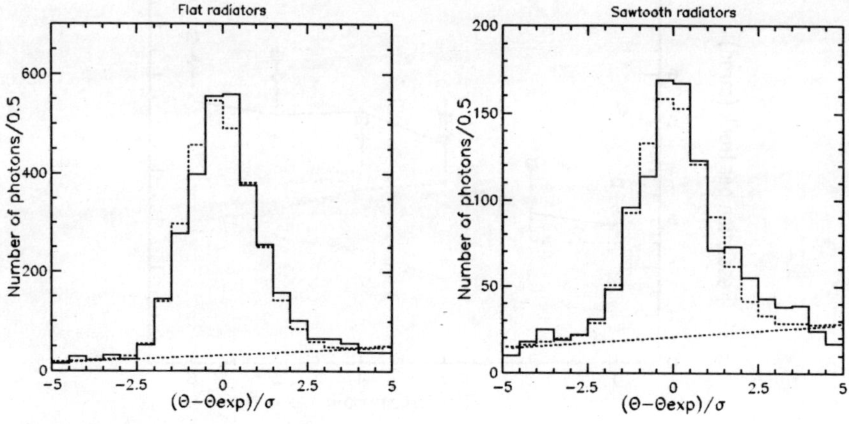

Figure 5. Cherenkov angles measured for individual photons for flat (left plot) and sawtooth (right plot) radiators. Difference between measured and expected angles divided by the expected resolution is plotted. The signal is expected to be approximately a standard Gaussian in this representation. The observed distributions (solid lines) are fit with the signal shape as obtained from the Monte Carlo simulation (dashed lines) and linear background (lower dashed lines).

T. Coan, V. Fadeyev, I. Volobouev, J. Ye, S. Anderson, Y. Kubota, A. Smith, E. Lipeles 1999 *Syracuse University Preprint HEPSU 99-5*; accepted by *Nucl. Instr. and Meth.*; hep-ex/9910054.

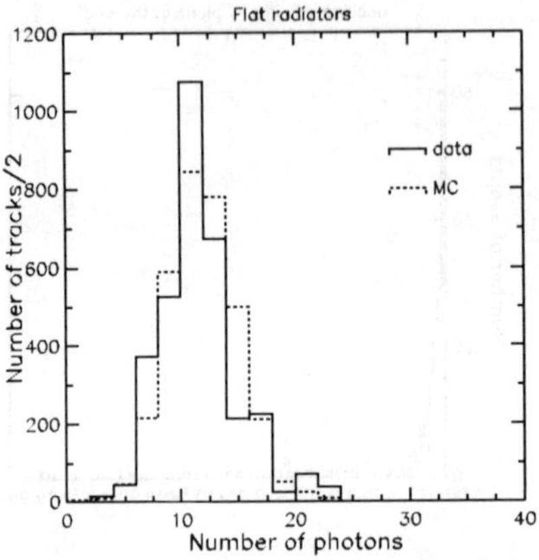

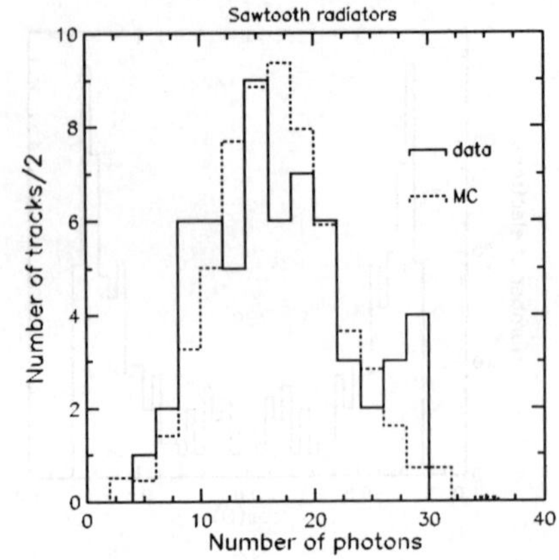

Figure 6. Number of Cherenkov photons per electron with the Cherenkov angle within $\pm 3\sigma$ of the expected value for flat (left plot) and sawtooth radiators (right plot) including background hits. The data are shown by solid histogram whereas the dashed histogram represents Monte Carlo distribution.

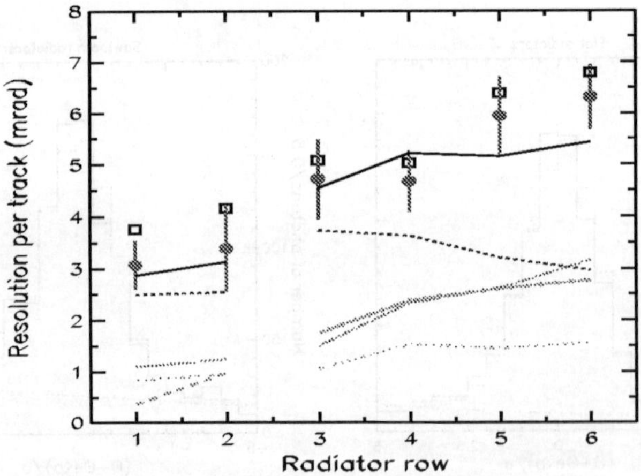

Figure 7. Measured per-track Cherenkov angle resolution plotted in function of radiator row number (1=center, 6=detector end). The first two rows are sawtooth radiators. The detector has one more row of flat radiators which is not populated by our data since for these crystals the electron misses the RICH wire chamber and cannot by reconstructed by our algorithm. We explain the graphic symbols as ordered from top to bottom at the left side of the plot. The squares represent the measured resolution. The points with error bars represent the measured resolution corrected for the background contribution to be compared to the solid line, which represents the expected values. The data agree with the expectations within the statistical errors. The following lines show composition of the expected resolution: chromatic error, emission point error, photon position error and tracking error.

STATUS AND RESULTS FROM HERA-B

OLAF STEINKAMP
Physik-Institut der Universität Zürich, Winterthurerstrasse 190, 8057 Zürich
E-mail: olaf.steinkamp@desy.de

HERA-B is a hadronic B factory at DESY. $b\overline{b}$ quark pairs are produced in proton-Nucleon interactions on thin wire targets in the halo of the HERA proton beam. Charged decay products are reconstructed in a forward magnetic spectrometer, which is complemented by detectors for precise vertex reconstruction and particle identification. The experiment has been optimized for a measurement of the unitarity angle $\sin 2\beta$ in the "golden" decay channel $B_d^0 \to J/\psi K_S^0$. However, the copious production of all flavors of beauty and charm hadrons opens the field of investigation to a wide range of heavy-quark physics. The experimental challenge is due to the overwhelming inelastic cross section that puts high demands on detector technologies and trigger design. The detector is essentially completely installed, trigger components are in hand and commissioning is in progress. The experiment should be ready for physics data taking early in 2000.

1 Introduction

The HERA-B experiment is a hadronic B factory at the HERA electron-proton collider at DESY. Thin wire targets, introduced into the halo of the proton beam, allow to produce competitive rates of all flavours of B mesons, in proton-Nucleon interactions. Target stations with 8 wires of $50 \times 500\ \mu$m diameter have been routinely operated since 1994, at nominal interaction rates and in parallel to the other HERA experiments.

The HERA proton beam momentum is 920 GeV/c, corresponding to a center-of-mass energy of 40 GeV. A major uncertainty in the estimation of the HERA-B physics reach lies in the poorly known $b\overline{b}$ production cross section at this energy. Performance studies assume a cross section of 12 nb, which is in accordance with theoretical predictions and measurements, but suffers from an uncertainty of at least a factor of two.

At an interaction rate of 40 MHz, several 10^8 B_d and B_s mesons are produced per year. The primary aim of HERA-B is a measurement of $\sin 2\beta$ in the golden decay $B_d^0 \to J/\psi K_S^0$, where J/ψ and K_S^0 will be detected in their charged decays to two leptons and pions, respectively. Taking into account acceptances and efficiencies, about 1500 fully reconstructed golden decays will be recorded in one year of data taking (10^7 s). With a tagging power of 0.31, this allows the determination of $\sin 2\beta$ with an error of 0.13.

The main experimental challenge is the total inelastic pN cross section of

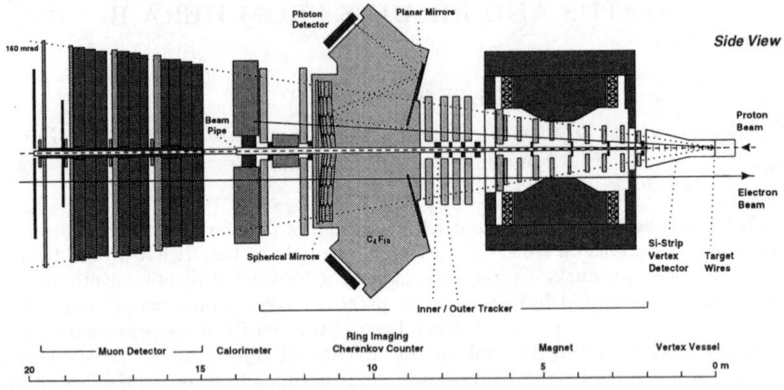

Figure 1. The HERA-B detector

10 mb, six orders of magnitude above the $b\bar{b}$ cross section. A sophisticated multi-level trigger system must provide a reduction factor of more than 10^5 to reduce this background to a rate suitable for mass storage, while retaining a good efficiency, of the order of 50%, for the channels of interest. Furthermore, the high rates of high-multiplicity hadronic events, with typically 200 charged tracks through the last tracking stations, put strict demands on pattern-recognition capabilities and the radiation tolerance of the detectors.

In the past years, the experiment suffered significant delays due to ageing problems in the tracking detectors. By 1999, renewed R&D efforts demonstrated workable solutions to all observed problems. Installation of the trackers is now nearing completion, while commissioning of the remaining detector components is well advanced. Commissioning of the first-level trigger, which is based on the reconstruction of high-p_T charged tracks, has commenced and defines the critical path for the startup of physics data taking.

2 The Detector

The experimental apparatus is depicted in figure 1. The circulating proton beam traverses the detector through a 500 μm thin aluminium beam pipe. Downstream of the target stations, a seven super-layer silicon vertex detector is located in retractable Roman pots inside the HERA vacuum vessel. For data taking, the detectors approach the proton beam to 1 cm, they are retracted during beam insertion. An eighth super-layer of silicon is situated just downstream of a thin exit window.

The main tracking system consists of 13 stations. Six of these are located in the field-free region downstream of the 2.1 Tm normal-conducting dipole magnet. They serve for triggering and pattern recognition. Additional stations inside the magnet help to follow tracks through the magnetic field and link them to the vertex detector. Per station, the Inner Tracker, in MSGC-GEM technology, covers the innermost 20 cm around the beam pipe, where particle densities are highest. The Outer Tracker, built from Honeycomb Wire Chambers, covers the remaining part of the acceptance, reaching out to 160 mrad vertically and 250 mrad in the bending plane of the magnet.

A gaseous RICH detector is situated in front of the last two tracking stations. Its main purpose is the identification of Kaons for tagging purposes. A sampling electromagnetic calorimeter follows downstream of the tracking system. It allows to trigger on high-p_T electrons and provides offline reconstruction for decays involving photons. Electron identification in the very forward direction is enhanced by a transition radiation detector.

The muon detector consists of four super-layers, which are separated by iron/concrete absorbers. In each super-layer, the innermost part around the beam pipe is covered by gas pixel detectors, the outer part by gas tube chambers. The last two super-layers are equipped with a pad readout for triggering on high-p_T muons. Finally, a system of dedicated pixel and pad gas detectors inside the magnet will allow to trigger on charged high-p_T hadrons.

3 Detector Status

The vertex detector employs single- and double-sided silicon strip detectors, with a readout pitch of 50 μm . The system is completely installed and routinely operated. It performs well, with a measured signal-to-noise ratio of about 20 or better. Using a preliminary alignment, primary vertex resolutions of 60 μm perpendicular to the beam direction and 600 μm along the beam axis have been demonstrated. These numbers approach design values.

The Inner Tracker was originally designed as a Micro-Strip Gas Chamber (MSGC). The detectors showed good aging behaviour in x-ray irradiation tests but they were destroyed, due to sparking, when exposed to irradiation with hadrons. The problem was solved in a revised design that employs the use of Gas Electron Multiplier (GEM) foils. Positioned in between the drift volume and the MSGC plane, the moderate gas amplification obtained in the GEM allows to operate the MSGC at a reduced gain at which no sparking occurs. Several of these detectors have been installed and are successfully operated in HERA-B. Production of the remaining detectors is underway, installation is scheduled to be completed in March.

The first Outer Tracker detectors, up to 4.5 m long Honeycomb Drift Chambers with hexagonal cells of 5 mm and 10 mm wire pitch, suffered from various ageing problems when operated in HERA-B. A considerable R&D effort was necessary to show that these problems could be overcome by different choices of materials and gas mixture, and improved production techniques. No major redesign of the detector concept was necessary. Since June 1999, the full detector has been installed in a series of monthly 3-day "mini-shutdowns". The system is now complete, commissioning is in progress.

The RICH detector employs multi-anode photo tubes to detect Cherenkov light produced in a 2.5 m long C_4F_{10} radiator. The detector is fully installed and commissioned. The average photon yield, 35 photons for $\beta = 1$ rings, as well as the angular resolution per photon, match design values. An early analysis, comparing measured RICH ring radii and track momenta, shows a clear p/K/π separation.

The electromagnetic calorimeter uses a shashlik design with scintillating fibres and two different absorber materials: tungsten in the innermost section, lead in the outer sections. Three sections of readout granularity match the radially decreasing occupancy. The detector is fully installed. A preliminary calibration, using the $\pi^0 \rightarrow 2\gamma$ invariant mass, has yielded a resolution of 7%. Further improvement is expected from the use of electron momenta provided by the tracking detectors. Clear $\eta \rightarrow 2\gamma$ and $\omega \rightarrow 3\gamma$ signals are seen.

The muon detector is fully installed and operational. Observed occupancies and coincidence rates between the trigger layers agree with expectations.

4 Trigger and Data Acquisition System

The HERA-B multi-level trigger system is based on the reconstruction of high-p_T tracks, two-particle invariant masses and detached vertices.

High-p_T pretriggers are generated by three sources: clusters in the calorimeter, coincidences in the two trigger layers of the muon system, or three-fold coincidences from the pad detectors inside the magnet. Any such pretrigger defines a geometrical "Region of Interest" (RoI) which is sent as a message to the first-level trigger. At nominal interaction rates, several such RoIs will be found for each HERA bunch crossing.

The first-level trigger (FLT) must provide a reduction factor of about 200, within a latency of less than 12 μs. Starting from the RoIs defined by the pretrigger systems, it attempts to follow tracks through four of the six tracking stations downstream of the magnet. A track candidate is passed as a message between custom-made processors, each of which covers a part of the acceptance of one super-layer. On reception of a message, the processor searches the corresponding RoI for matching hits and, if found, generates an updated message which it sends to the appropriate processors in the next super-layer. Tracks that could be followed through all super-layers, are collected in a programmable Trigger Decision Unit. The trigger decision is based on single high-p_T tracks and on the invariant mass of track pairs (J/ψ trigger).

The second- and third-level triggers are integrated with the data-acquisition system. They are implemented on a 240-processor farm of Pentium processors running the Linux operating system, a high-bandwidth switch, and a system of so-called "second-level" buffers. Switch and buffers are built from the same DSP-based board.

Together, the two trigger levels have to provide a reduction factor of about 1000. The second-level trigger re-processes FLT tracks. Additional tracking layers are used and a track fit is performed. Successfully fit candidates are extrapolated and tracked through the vertex detector, and a vertex fit is performed. The trigger decision is based on single-track impact parameters or on secondary vertices. The third-level trigger is foreseen for such trigger types where cuts based on the FLT tracks alone do not provide sufficient reduction, for example if no secondary vertex can be required.

The 4th-level trigger is intended for full online event reconstruction. It runs on a 200-node farm of Pentium processors, which is connected via standard Ethernet technology. It operates at a data throughput rate of 50 Hz.

5 Trigger and DAQ Status

Almost the full acceptance of the calorimeter is equipped with pretrigger electronics and the system is in routine operation. Muon pretrigger electronics are installed on a significant part of the detector and perform according to expectations. Both systems should be completed end February. The high-p_T system is under construction, installation is expected to begin in March.

Most of the FLT hardware has been produced, installation and commissioning are under way. The system is planned to reach full power end March. The Trigger Decision Unit has been in routine operation since last summer. Pretrigger messages from the calorimeter were routed directly to the unit and a trigger was issued when two such messages were received for an event.

The second-level farm has been operating routinely for more than a year. All 240 processor nodes are installed, most of them are cabled to the switch. Portions of the second-level algorithm have been exercised routinely. Figure 2 shows a $J/\psi \rightarrow e^+e^-$ signal as it was reconstructed by the second-level trigger. Two calorimeter clusters with p_T above 1 GeV and with matching tracks in the vertex detector were demanded. A loose vertex cut was made and an

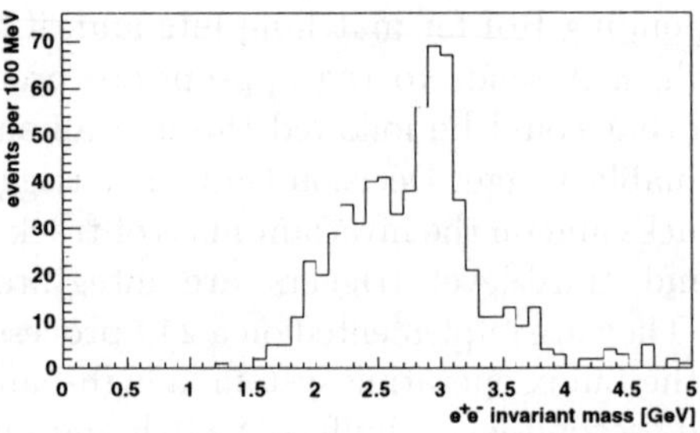

Figure 2. $J/\psi \rightarrow e^+e^-$ signal, reconstructed by the second-level trigger

associated bremsstrahlung cluster was required for at least one of the tracks. No information from the main tracking detectors was used yet.

The 4th-level trigger farm is complete. The event stream has gone routinely through the farm on its way to mass storage. Several nodes have been in use for monitoring purposes and partial reconstruction. The full reconstruction code exists and is being tuned for online running.

6 Summary

HERA-B has suffered significant delays compared to its ambitious original schedule. These delays were mainly due to operational problems of the tracking detectors in a harsh high-radiation environment. Solutions to these problems have been found and construction of the spectrometer is nearing completion. The remaining detector has been operated routinely over the last year and performs up to expectations. The critical path is now defined by the installation and commissioning of the first-level trigger. In view of the impressive progress that HERA-B has shown, DESY management has decided to extend this year's HERA running by four months, until September. We are looking forward to using this time to accumulate a solid sample of B decays.

References

1. *HERA-B Proposal*, T. Lohse et al., DESY-PRC 94/02 (1994).
2. *HERA-B Design Report*, E. Hartouni et al., DESY-PRC 95/01 (1995).
3. *HERA-B Status and Commissioning Results*, C. Padilla, Proceedings Beauty99, Bled, Slovenia (in press).
4. *HERA-B Status*, M. Medinnis, Proceedings Lepton-Photon99, Standford, USA (in press).

BABAR PHYSICS: STATUS REPORT

D. KIRKBY
FOR THE BABAR COLLABORATION
Varian 156, Dept of Physics, Stanford University, Stanford CA 94305, USA
E-mail: kirkby@hep.stanford.edu

The BaBar experiment at SLAC recorded 1.7 fb^{-1} by the end of November 1999, and is currently in the process of tuning up and understanding the detector's performance. This talk reviews some examples of this tuning process, emphasizing some key ingredients of a measurement of the CP violating parameter $\sin(2\beta)$ using $B^0 \to J/\psi K^0_S$ decays.

1 Introduction

The BaBar experiment at SLAC is designed primarily to perform a systematic study of CP violating asymmetries in B^0 decays. The BaBar detector[1] recorded its first collisions on 26 May 1999, and had logged 1.7 fb^{-1} by the end of November. At this early stage in the experiment, the collaboration is in the process of tuning up and understanding the detector's performance. Therefore, this paper will not present "physics results", but instead will review some examples of this tuning process. Reports on the BaBar detector are presented elsewhere[2] in these proceedings.

The first major result in our program to systematically study CP violation in B^0 decays will be a measurement of the unitarity triangle angle, β, through the interference of B_d mixing and decays into the CP eigenstate B^0, $\overline{B^0} \to J/\psi K^0_S$. The main ingredients of this measurement are to: fully reconstruct the decay of one B into the CP eigenstate $J/\psi\ K^0_S$, tag the flavor of the other B at its decay time using charge correlations with the remaining tracks (those not associated with the reconstructed $J/\psi\ K_S$ decay), and measure the difference in B decay times, Δt, from the difference in B decay z vertices, Δz, and using $\Delta z \simeq \langle \beta_z \gamma \rangle\, c\Delta t$. I will focus on our progress towards the first of these tasks here.

1.1 Data Samples

BaBar recorded 1.7 fb^{-1} during its initial run, from June to November 1999, including 0.2 fb^{-1} below the $\Upsilon(4S)$ peak. This data sample includes approximately 1.6 million $\Upsilon(4S)$ decays and 5.5 million continuum hadronic events. Most of this data was recorded with partial coverage from the instrumented flux return (for muon and K^0_L identification) and from the Cerenkov detector (for $\pi^\pm/K^\pm$ discrimination at high momentum).

1.2 Resonance Energy Scan

During June 15–17 1999, PEP-II varied the energy of the electron beam in 13 steps to scan the shape of the $\Upsilon(4S)$ resonance. The data from each center-of-mass energy were analyzed to calculate the ratio, r, of multihadron candidates to Bhabha candidates. We expect an essentially flat contribution from Bhabha and continuum hadronic events, with a resonant contribution from $\Upsilon(4S)$ decays superimposed. We use a multihadron selection which includes some event shape cuts to enhance the $\Upsilon(4S)$ contribution over the continuum contribution.

The values of r calculated from each sample are shown in Figure 1. We fit these values to a model including the $\Upsilon(4S)$ lineshape and the known spread in the PEP-II center-of-mass energy. The results of the fit are:

$$m_{\Upsilon(4S)} = 10.5841 \pm 0.0007\,\mathrm{GeV} \ , \ \Gamma_{\Upsilon(4S)} = 11.1 \pm 3.4\,\mathrm{MeV} \ ,$$

where the fitted mass is measured on the PEP-II energy scale.

2 Inclusive Particle Reconstruction

2.1 D* *Reconstruction*

D* mesons are reconstructed using the decay chain $D^* \to D^0\pi^+$, $D^0 \to K^-\pi^+$, which has an overall branching fraction product of 2.7%. For this analysis, we do not use any particle identification criteria and consider any pair of oppositely charged tracks from a common vertex as a D^0 candidate. D* candidates are formed by combining D^0 candidates with any other charged track (slow pion candidate) and requiring that the D^0 candidate mass and D*-D^0 mass difference be consistent with expected values. Adding a constraint that the D* originate from the beam spot region (measured run by run using $\mu^+\mu^-$ and e^+e^- events) improves the resolution on the D*-D^0 mass difference from 354 keV (narrow Gaussian component, containing 27% of signal peak) to 280 keV (47% of signal peak). The corresponding improvement in the width of the wide Gaussian component is from 908 keV to 609 keV.

2.2 K^0_S *Reconstruction*

We identify $K^0_S \to \pi^+\pi^-$ candidates as pairs of oppositely charged tracks from a common vertex and with the expected mass. For a cleaner selection, we also

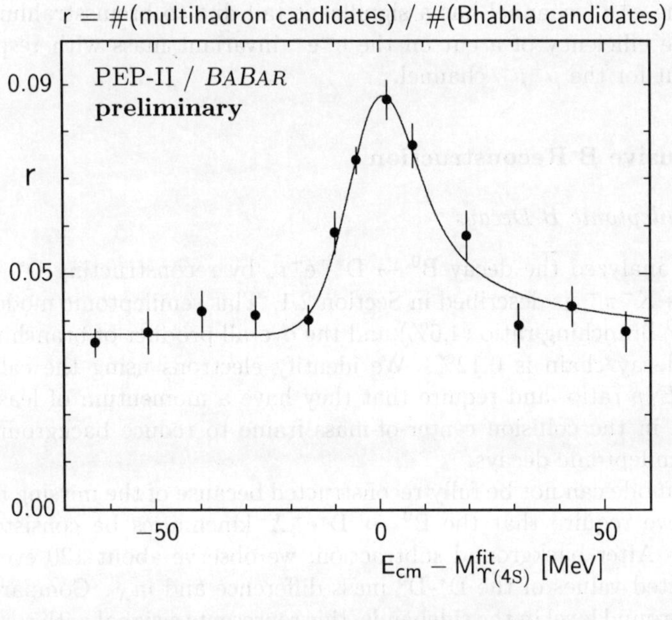

Figure 1. The ratio of multihadron to Bhabha candidates selected from samples recorded at different center-of-mass energies. The curve shows the best fit to these data. The multihadron selection enhances the contribution of $\Upsilon(4S)$ decays over continuum hadronic events.

require that the K_s^0 candidate point back to the interaction point (the average K_s^0 flight length in BaBar is about 4 cm) and require a minimum transverse momentum in the decay (to suppress combinatoric and Λ backgrounds). With this tight selection, we obtain a resolution on the reconstructed K_s^0 mass of 2.8 MeV (11 MeV) for the narrow (wide) component of the peak, which contains 69% (31%) of the signal events.

2.3 J/ψ *Reconstruction*

We reconstruct J/ψ candidates in the e^+e^- and $\mu^+\mu^-$ decay modes which each have a 6% branching ratio. We identify electrons using the pattern of energy deposits in the calorimeter and by requiring a ratio, E/p, of the energy measured in the calorimeter to the measured track momentum close to one. We identify muons from their minimum-ionizing signature in the calorimeter. J/ψ candidates are required to have a common vertex and have the expected mass. The e^+e^- channel has a significant tail due to bremsstrahlung which lowers the efficiency of a cut on the e^+e^- invariant mass with respect to a similar cut for the $\mu^+\mu^-$ channel.

3 Exclusive B Reconstruction

3.1 Semileptonic B Decays

We have analyzed the decay $B^0 \to D^{*-}e^+\nu_e$ by reconstructing $D^* \to D^0\pi^+$ and $D^0 \to K^-\pi^+$ as described in Section 2.1. This semileptonic mode has the largest B^0 branching ratio (4.6%) and the overall product of branching ratios for this decay chain is 0.12%. We identify electrons using the calorimeter and the E/p ratio, and require that they have a momentum of least 1 GeV measured in the collision center-of-mass frame to reduce backgrounds from charm semileptonic decays.

This mode can not be fully reconstructed because of the missing neutrino. Instead, we require that the $B^0 \to D^*e^+X$ kinematics be consistent with $m_X = 0$. After background subtraction, we observe about 120 events with the expected values of the D^*-D^0 mass difference and m_X. Comparing with the background level in the sidebands, this represents a signal with a statistical significance of 6σ.

3.2 Charged B Decays into Charmonium

We have studied $B^\pm \to J/\psi K^\pm$ decays by inclusively selecting events with a J/ψ decay (see Section 2.3) and then requiring that it combine with any other track from the same vertex (Kaon candidate) to form a $B^\pm$ candidate with the expected mass. This mode is interesting because it is the charged B equivalent of the golden $B^0 \to J/\psi K^0_s$ mode and so is an important control sample for a measurement of $\sin(2\beta)$. The branching fraction product for the charged decay is 1.2×10^{-4}, which is about three times larger than the neutral mode's product.

We perform a fit to improve the resolution on the $B^\pm$ candidate's energy and momentum using the constraints of the known J/ψ and $K^\pm$ masses. We then require that the difference, ΔE, between the fitted candidate energy (in the collision center of mass) and the measured beam energy be less than 30 MeV. Figure 2 shows the distribution of candidates in ΔE and the beam-substituted mass

$$m'_B \equiv \sqrt{E(\text{beam})^2 - p_B^2} \; .$$

Figure 3 shows the m'_B distribution of candidates with $\Delta E < 30$ MeV. We require that m'_B be compatible with the known $B^{\pm}$ mass within our experimental resolution using a 3σ cut. We find 19 candidates which pass these cuts on ΔE and m'_B from an analysis of 500 pb^{-1}. By assuming a uniform density of background candidates in Figure 2, we estimate that these 19 candidates include a background of 1.8 events.

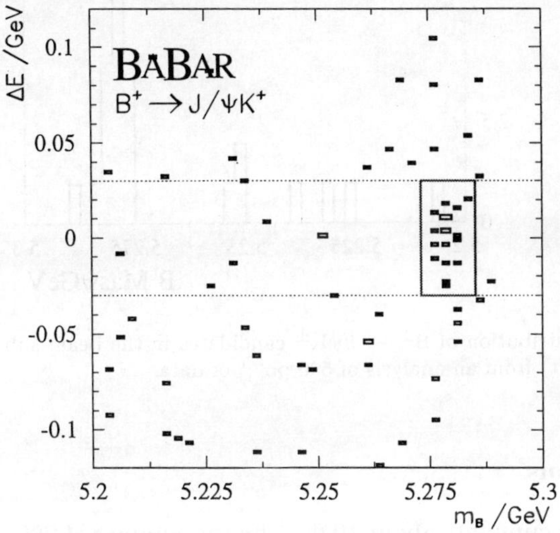

Figure 2. The distribution of $B^{\pm} \to J/\psi K^{\pm}$ candidates in the variables ΔE and m'_B (described in the text) from an analysis of 500 pb^{-1} of data. The signal box contains 19 events of which an estimated 1.8 are due to background.

3.3 Neutral B Decays into Charmonium

The decay $B^0 \to J/\psi K^0_s$ is the golden mode for measuring the CP violation parameter $\sin(2\beta)$. The study of this mode is very similar to the study of $B^{\pm}$ decays into charmonium described in the previous section. The main difference is the K^0_s reconstruction which was already described in Section 2.2 (we use the loose selection for this analysis).

Figures 4 and 5 show the distributions of ΔE and m'_B for our B^0 candidates. We find 8 events from 450 pb^{-1} in the signal box defined by $|\Delta E| < 30$ MeV and $|m'_B - m_B| < 3\sigma$, of which we estimate that 0.9 are due to background.

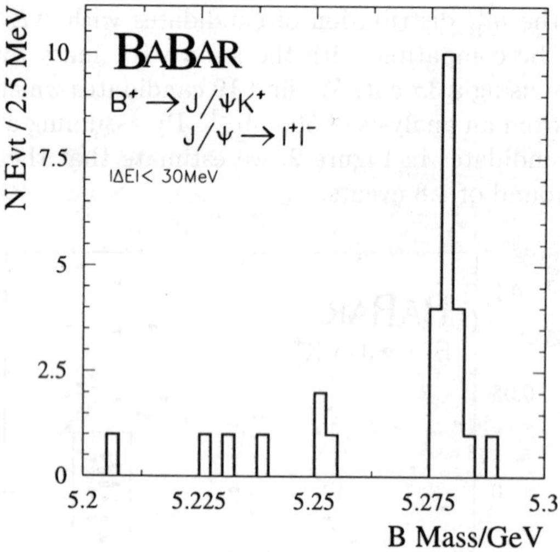

Figure 3. The distribution of $B^{\pm} \to J/\psi K^{\pm}$ candidates in the beam-substituted mass (described in the text) from an analysis of 500 pb^{-1} of data.

4 Conclusions

We expect to accumulate about 10 fb^{-1} by the summer of 2000. This data will be the basis for our first measurement of the CP violation parameter $\sin(2\beta)$. We are now in the process of tuning up and understanding our detector's performance.

Acknowledgments

It is a pleasure to acknowledge the support of the National Science Foundation through a grant PHY–9876867.

References

1. BaBar Collaboration, BaBar Technical Design Report, SLAC-R-95-457, March 1995.
2. see the contributions of D. Boutigny and M. Zito in these proceedings.

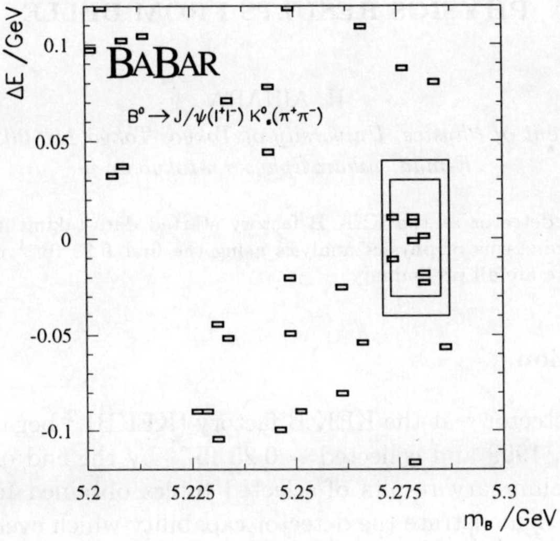

Figure 4. The distribution of $B^0 \to J/\psi K^0_S$ candidates in the variables ΔE and m'_B (described in the text) from an analysis of 450 pb^{-1} of data. The signal box contains 8 events of which an estimated 0.9 are due to background.

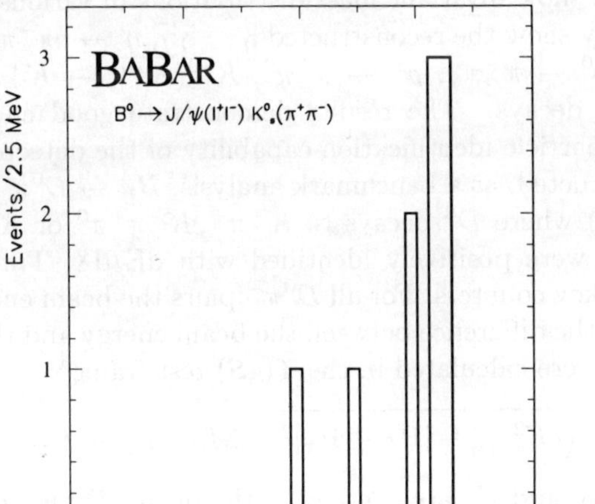

Figure 5. The distribution of $B^0 \to J/\psi K^0_S$ candidates in the beam-substituted mass (described in the text) from an analysis of 450 pb^{-1} of data.

PHYSICS RESULTS FROM BELLE

H. AIHARA

Department of Physics, University of Tokyo, Tokyo 113-0033, Japan
E-mail: aihara@phys.s.u-tokyo.ac.jp

The BELLE detector at the KEK B factory started data-taking in June, 1999. We report the status of physics analysis using the first 0.25 fb^{-1} data. Results presented here are all preliminary.

1 Introduction

The BELLE detector [1] at the KEK B factory (KEKB) [2] began physics data-taking in June, 1999 and collected $\sim 0.25\ fb^{-1}$ by the end of the year. We report very preliminary results of selected topics obtained from this initial data sample to demonstrate the detector capability which eventually leads to observation of CP violation in the B meson.

2 Spectroscopy

Figures 1 and 2 show invariant mass distributions of various particle decays. The plots clearly show the reconstructed $\eta \to \gamma\gamma$, $\eta' \to \eta\pi^+\pi^-$, $\omega \to \pi^+\pi^-\pi^0$, $\phi \to K^+K^-$, $\rho^0 \to \pi^+\pi^-$, $\rho^\pm \to \pi^\pm\pi^0$, $K^{*0}(\overline{K^{*0}}) \to K^+\pi^-(K^-\pi^+)$, and $K^{*\pm} \to K^\pm\pi^0$ decays. The results demonstrate good energy/momentum resolution and particle identification capability of the detector.

We reconstructed, as a benchmark analysis, $B^- \to D^0\pi^-$ (and its charge conjugate state) where D^0 decays to $K^-\pi^+, K^-\pi^+\pi^0$ or $K^-\pi^+\pi^-\pi^+$. The $K^\pm$ candidates were positively identified with dE/dX, Time of Flight and Aerogel Cherenkov counters. For all $D^0\pi^-$ pairs the beam energy constrained mass, M_B, and the difference between the beam energy and the reconstructed B energy, ΔE, were calculated in the $\Upsilon(4S)$ rest frame[a]:

$$M_B \equiv \sqrt{E_{\text{beam}}^2 - |\mathbf{P}_D + \mathbf{P}_\pi|^2}; \quad \Delta E \equiv E_D + E_\pi - E_{\text{beam}} \tag{1}$$

Here, E_{beam}, E_D, and E_π are energies of the beam, D meson candidate, and π, respectively, while $\mathbf{P}_D$ and $\mathbf{P}_\pi$ are momenta of D and π. The resolution of ΔE for the above decay channels was found to be ~ 36 MeV/c^2. Figure 3 shows the M_B distribution, after $|\Delta E| < 40$ MeV/c^2 was imposed, obtained

[a]Because the KEK B factory is an asymmetric collider, the $\Upsilon(4S)$ is boosted along the beam axis by $\beta\gamma = 0.425$.

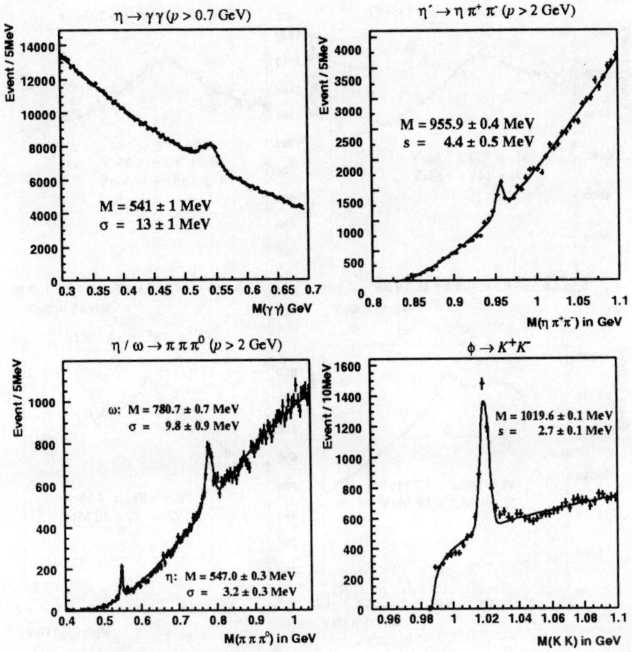

Figure 1. Invariant mass distributions (I).

from $\sim$ 0.125 fb^{-1} data at $\Upsilon(4S)$. From this distribution we estimate the yield of the signal to be 28 ± 7 events. The resolution of M_B was about $\sigma \sim 3$ MeV/c^2.

3 Measurement of Lifetime of D^0 Meson

In this section we report a measurement of the lifetime of the D^0 meson. The measurement is interesting [3] in its own right and, at the same time, serves as an important performance check of the central drift chamber (CDC) and the silicon vertex detector (SVD). We reconstruct D^0 mesons in the decay mode $D^0 \to K^-\pi^+$. Charged tracks forming a D^0 candidate were required to be well-reconstructed in the tracking system and to originate from a common vertex. Kaons were positively identified. In order to avoid bias due to B meson lifetime, D^0 mesons originating from B mesons were removed by requiring D^0 mesons have a momentum of at least 2.5 GeV/c in the $\Upsilon(4S)$ rest frame.

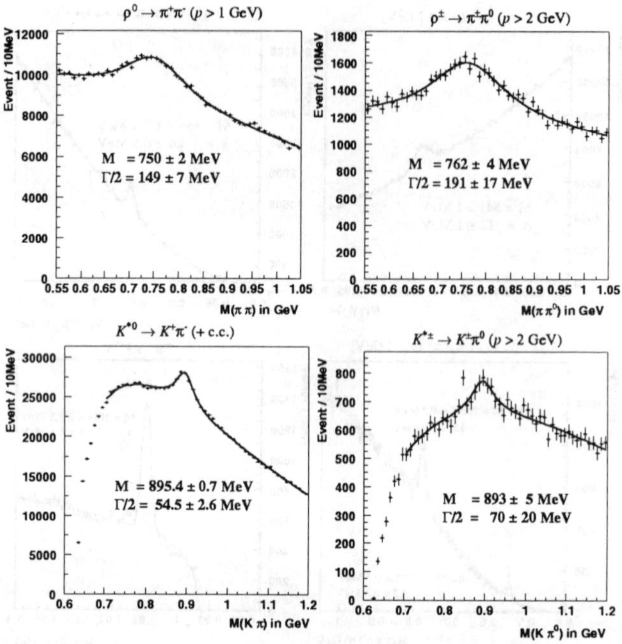

Figure 2. Invariant mass distributions.

Figure 4 (b) shows the invariant mass distribution of $D^0 \to K^-\pi^+$ decays obtained from ~ 0.25 fb^{-1} data. By fitting the distribution with a Gaussian for the signal plus a linear function for the background, we found the width of the signal distribution to be 7.6 ± 0.2 MeV/c and the number of signal events to be 1730 ± 50. The background fraction was calculated from the same fit and was about 10% under a signal peak (the region within $\pm 3\sigma$).

As illustrated in Fig. 4(b) the lifetime of the charmed meson was computed from the flight distance between the reconstructed D^0 vertex and the interaction point(IP) determined for each KEKB beam fill. The profile of the IP is known to have, from the machine optics and the beam size monitoring system, $\sigma_z \sim 6$ mm along the beam direction (z), and $\sigma_x \sim 100$ μm, and $\sigma_y \sim 5-10$ μm in the direction perpendicular to the beam direction. The D^0 flight length was measured in the $x-y$ plane to take advantage of the small IP size in the plane. The projected decay length ℓ^{xy}_{decay} of the D^0 meson was calculated from the D^0 momentum $\mathbf{p}^{xy}_{D^0}$, the D^0 decay vertex $\mathbf{r}^{xy}_{D^0}$, and the D^0 production point $\mathbf{r}^{xy}_{IP}$ reconstructed by extrapolating the D^0 momentum

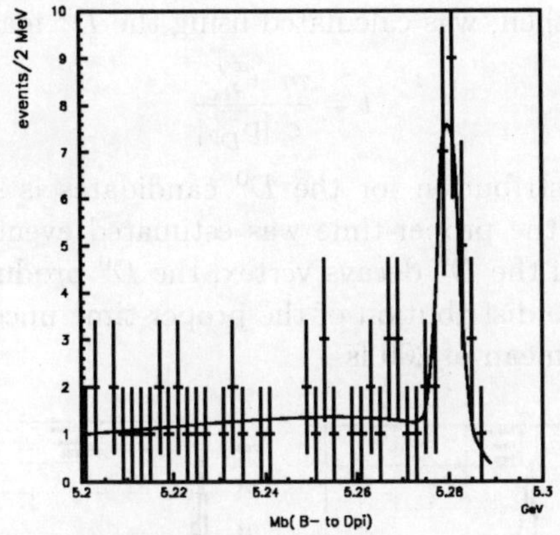

Figure 3. Reconstructed $B^{\pm}$ mass distribution in $B^- \to D^0\pi^-$ (and its charge conjugate state) decays. Beam energy was used as B meson energy and beam-energy constrained $|\Delta E| < 40\text{MeV}/c^2$ was imposed.

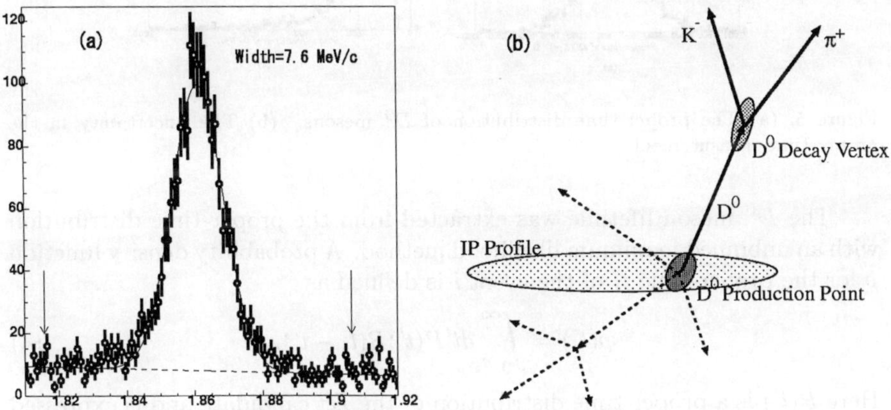

Figure 4. (a)Invariant mass distributions of $D^0 \to K^-\pi^+$ decays. (b)Illustration to show how the D^0 meson flight length is reconstructed in the $(x-y)$ plane perpendicular to the beam (z)axis.

back to the IP profile:

$$\ell^{xy}_{decay} = (\mathbf{r}^{xy}_{D^0} - \mathbf{r}^{xy}_{IP}) \cdot \frac{\mathbf{p}^{xy}_{D^0}}{|\mathbf{p}^{xy}_{D^0}|} \tag{2}$$

The proper time t, then, was calculated using the D^0 mass m as

$$t = \frac{m}{c} \frac{\ell^{xy}_{dec}}{|\mathrm{p}^{xy}_{D^0}|} \tag{3}$$

The proper-time distribution for the D^0 candidates is shown in Fig. 5(a). The uncertainty of the proper-time was estimated event-by-event based on the error matrices of the D^0 decays vertex, the D^0 production point and the D^0 momentum. The distribution of the proper-time uncertainty is shown in Fig. 5(b). with the mean of 200 fs.

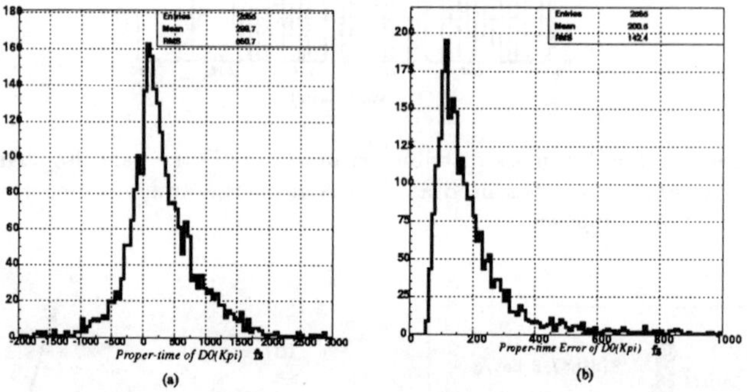

Figure 5. (a) The proper-time distribution of D^0 mesons. (b) The uncertainty in the proper-time measurement.

The D^0 meson lifetime was extracted from the proper-time distribution with an unbinned maximum likelihood method. A probability density function ρ for the proper-time t_i of the event i is defined as

$$\rho(t_i) = \int_0^\infty dt' P(t') R(t_i - t'). \tag{4}$$

Here $P(t')$ is a proper-time distribution of the D^0 candidate and is expressed as a sum of the signal and the background contributions:

$$P(t') = p_{sig} E(t'|\tau_{sig}) + (1 - p_{sig})(f_{\tau_{BG}}) E(t'|\tau_{BG}) + (1 - f_{\tau_{BG}})\delta(t')), \tag{5}$$

where p_{sig} is the probability of the candidate event being a signal, $f_{\tau_{BG}}$ is the fraction of the background events with a finite lifetime of τ_{BG} and

$$E(t'|\tau) \equiv \frac{1}{\tau} \exp(-\frac{t'}{\tau}). \tag{6}$$

The detector resolution is mostly described by a single Gaussian with a small correction for events which might have had hard multiple Coulomb scattering and/or wrong CDC-hit and SVD-hit association. The measurement resolution $R(t_i - t')$ was, therefore, assumed to be represented by the formula

$$R(t) = (1 - f_{mis})G(t|S\sigma) + f_{mis}G(t|\sigma_{mis}), \tag{7}$$

where $G(t|S\sigma)$ is a Gaussian with the mean 0 and the standard deviation given by $S\sigma$. The sale factor S is introduced as a correction factor for the event-by-event predicted measurement error σ. Figure 6 shows the fit to proper lifetime data. The preliminary reulst for the lifetime of D^0 meson is

$$405.3 \pm 13.3 \text{ fs},$$

where the error includes only the statistical error. This is in good agreement with the PDG value of 415 ± 4 fs.

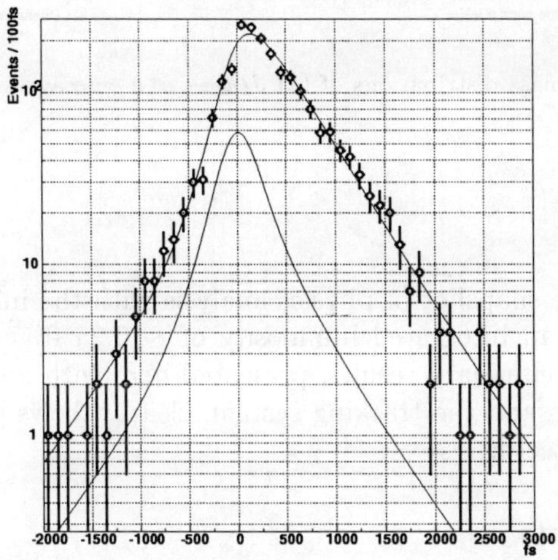

Figure 6. The proper-time distribution of the D^0 mesons superimposed with the fit result. The solid line with data points is the overall fit result and the solid line without data points indicate the contribution from the background.

4 $B \to J/\psi$ inclusive decays

Figure 7 shows the invariant mass distributions of $J/\psi \to \mu^+\mu^-$ and $J/\psi \to e^+e^-$ pairs reconstructed at $\Upsilon(4S)$. We obtained 83 ± 11 events and 133 ± 13 events in dielectron and dimuon channels, from 0.215fb^{-1} data, respectively. This yield corresponds to ~ 1 $J/\psi \to$ dilepton event per 0.001 fb^{-1} and consistent with the prediction based on the Monte Carlo simulation.

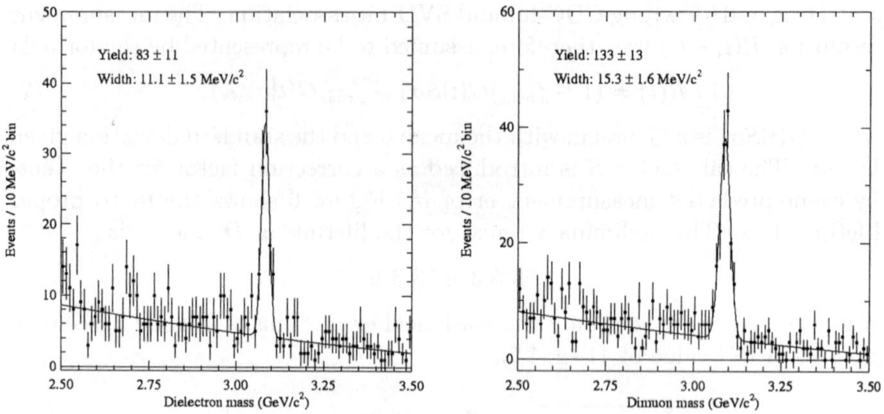

Figure 7. Invariant mass distributions of (a) $J/\psi \to e^+e^-$ decays and (b) $J\psi \to \mu^+\mu^-$ decays.

5 Conclusion

We have sown the snapshot of physics analysis with the initial data sample, corresponding to an integrated luminosity of ~ 0.25 fb^{-1}. Analysis is well under way and preliminary results, presented here with an emphasis on the particle ID system and the tracking system, clearly shows physics capability of the BELLE detector.

Acknowledgments

This work was supported by Grant-in-Aid for Scientific Research on Priority Areas (Physics of CP violation) from Ministry of Education, Science and Culture of Japan, and by Japan-Taiwan Cooperative Program of the Interchange Association.

References

1. Performance of the BELLE detector, K. Abe, In this proceedings.
2. Status of e^+e^- B Factory Accelerators, S. Kurokawa, In this proceedings.
3. G.Bonvicini et al. (CLEO Collaboration), Phys.Re.Lett.**82**,4586,1999.

HIGHLIGHTS OF 10 YEARS OF LEP B PHYSICS

SAU LAN WU
University of Wisconsin–Madison, Madison, WI 53706, USA
E-mail: wu@wisconsin.cern.ch

Highlights from the b physics results produced by the four LEP collaborations in the last decade are presented.

1 Introduction

The Large Electron-Positron collider was commissioned in 1989 and the first Z decay was recorded on 14 August. Since that time, LEP has provided over 200 pb^{-1} of integrated luminosity to each of the four LEP experiments ALEPH, DELPHI, L3 and OPAL, resulting in data samples of more than 4×10^6 decays of $Z \rightarrow$ hadrons for each experiment. With these bountiful data sets, the four collaborations have observed, measured and probed many previously inaccessible areas of high energy physics. Heavy flavor physics, and in particular studies of the b quark, have produced many interesting and sometimes unanticipated results. In this area, the following topics are briefly discussed: the Z hadronic decay width ratio R_b, observations of the B_s^0 and Λ_b^0 hadrons, the lifetimes of the b hadrons, B_d^0 and B_s^0 oscillations, measurements of the CKM matrix elements V_{ub} and V_{cb}, observation of the electromagnetic penguin decay $b \rightarrow s\gamma$, and initial hints of CP non-conservation in B_d^0 decays. This list is by no means complete and only serves as an indication of the wealth of b physics that has come from LEP in the past decade.

2 R_b

The quantity R_b is defined as the ratio of the partial width of Z decays to $b\bar{b}$ to the partial width of Z decays to hadrons,

$$R_b \equiv \frac{\Gamma(Z \rightarrow b\bar{b})}{\Gamma(Z \rightarrow q\bar{q})} . \tag{1}$$

The measured ratio $\sigma(e^+e^- \rightarrow b\bar{b})/\sigma(e^+e^- \rightarrow \text{hadrons})$ is corrected by a small amount to account for photon exchange. There are large radiative corrections due to coupling to the top quark, making R_b sensitive to the existence of new physics such as supersymmetric top quarks or charginos, Z/Z' mixing, b/b' or t/t' mixing, and exotic fermions or scalars.

As of March 1996 at the 31st Rencontres de Moriond conference, the measured value of R_b was three standard deviations from the Standard Model prediction. Initial measurements had been performed using samples of leptons with high momentum and high momentum transverse to the nearest jet to identify or 'tag' $Z \to b\bar{b}$ decays, achieving precisions of 5–10%. With the growing deviation of measurement from the Standard Model, the hemisphere method was developed, isolating the two hemispheres of each $Z \to b\bar{b}$ event, thereby reducing the reliance on simulation for correlations, the largest systematic uncertainties. In this method, a b quark tag is applied to the data, and the number of singly- and doubly-tagged hemispheres are counted: N_t and N_{tt}, respectively. Thus the efficiency of the tag may be measured in data simultaneously with R_b

$$\epsilon_b \simeq \frac{2N_{tt}}{N_t} \tag{2}$$

$$R_b \simeq \frac{N_t^2}{4N_{tt}N_{hadrons}} \,. \tag{3}$$

This enabled measurements on the order of 3% precision. A b tag exploiting the lifetime information in b decays was then developed, enabling measurements of approximately 1% precision. In the present measurements of R_b, much effort has been directed to understanding the dominant systematic effect, correlations between the two b quark hemishperes, and in improving the lifetime-based tags. In addition, new multivariate and event-shape tags for $Z \to b\bar{b}$ decays as well as uds events and $Z \to c\bar{c}$ decays have been developed. These advances have allowed measurements of 0.5% precision or better to be achieved. [1,2,3,4,5]

The results of the current combined LEP and SLD electroweak fit for R_b and R_c, the ratio of the Z partial widths to $c\bar{c}$ and to hadrons, are shown in Figure 1. The fitted value is [6] $R_b = 0.21642 \pm 0.00073$ and the Standard Model prediction is [2] $R_b^{SM} = 0.21584 \pm 0.00018$, assuming $m_t = 173.8 \pm 5.2$ GeV$/c^2$.

3 Discovery of Λ_b^0 and B_s^0

Leptons with high momentum and high momentum transverse to the nearest jet form a signature for b quark decay. In addition, semileptonic b decays have large branching fractions: $Br(b \to \ell X) = (10.62 \pm 0.17)\%$ in Z decays. [6] The semileptonic decay of the Λ_b^0 produces such a lepton as well as a final baryon. Examining yields of "right sign" $\Lambda\ell^-$ and "wrong sign" $\Lambda\ell^+$ combinations, the ALEPH Collaboration found an excess of $54.5 \pm 13.0 \pm 8.2$ events in 1991, [7] subsequently confirmed by the OPAL and DELPHI Collaborations. [8,9] The

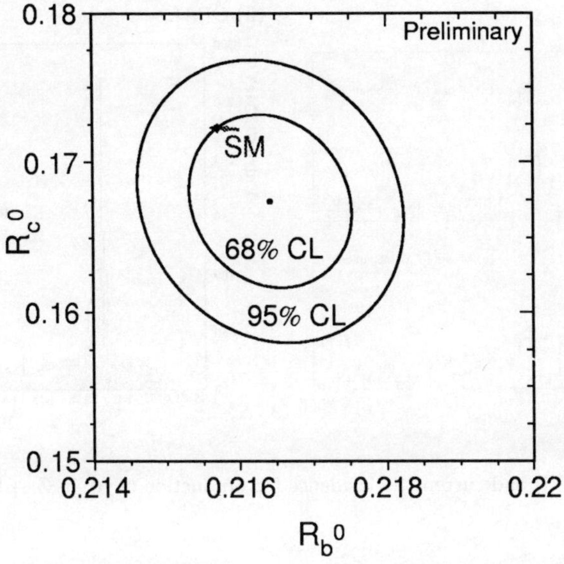

Figure 1. Results for R_b and R_c from the LEP Electroweak Working Group combined 14 parameter fit to the LEP and SLD data. [6]

ALEPH data are shown in Figure 2(i). The current OPAL mass peak based on the full LEP 1 data set is shown in Figure 2(ii). [10] The LEP average product branching fraction is

$$f(\mathrm{b} \to \Lambda_\mathrm{b}^0) \times \mathrm{Br}(\Lambda_\mathrm{b}^0 \to \Lambda \ell^- \overline{\nu} \mathrm{X}) = (3.04 \pm 0.25) \times 10^{-3}\,, \tag{4}$$

which is the mean of the electron and muon decay modes. [11]

More recently, full reconstruction of a few Λ_b^0 decays has permitted measurement of the Λ_b^0 mass by the ALEPH, DELPHI, and CDF Collaborations. [12,13,14] The world average is $m(\Lambda_\mathrm{b}^0) = (5624 \pm 9)\ \mathrm{MeV}/c^2$. [15]

Evidence for the strange b meson $\mathrm{B_s^0}$ was established at LEP via the decay $\mathrm{B_s^0} \to \mathrm{D_s^+} \ell^- \overline{\nu}(\mathrm{X})$. The $\mathrm{D_s^+}$ was reconstructed in the two decay modes $\mathrm{D_s^+} \to \phi\pi^+$ and $\mathrm{D_s^+} \to \overline{\mathrm{K}}^*(892)\mathrm{K}^+$. [16,17,18] The ALEPH data are shown in Figure 3(i). The current OPAL mass peak based on the full LEP 1 data set is shown in Figure 3(ii). [19] The LEP average product branching fraction is

$$f(\mathrm{b} \to \overline{\mathrm{B}}_\mathrm{s}^0) \times \mathrm{Br}(\overline{\mathrm{B}}_\mathrm{s}^0 \to \mathrm{D_s^+} \ell^- \overline{\nu} \mathrm{X}) = (8.3^{+2.3}_{-2.4}) \times 10^{-3}\,, \tag{5}$$

which is the mean of the electron and muon decay modes. [11]

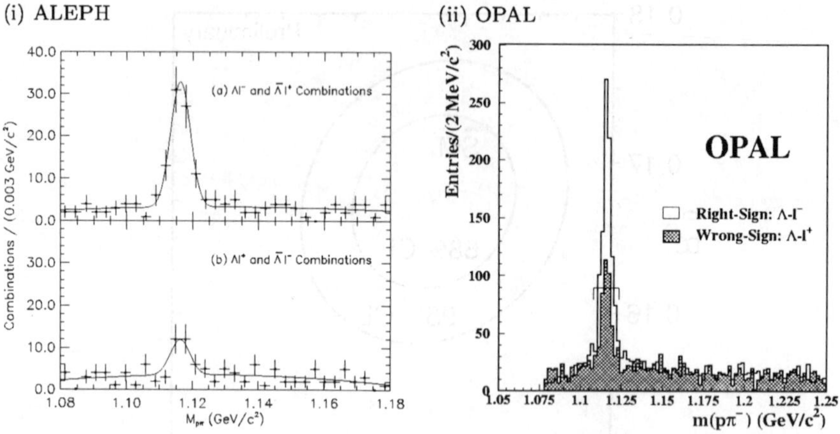

Figure 2. First (i) and current (ii) evidence for production of Λ_b^0 in $Z \to b\bar{b}$ decays. [7,10]

In 1993, the ALEPH Collaboration fully reconstructed two B_s^0 decays, permitting measurement of the mass. [20] Subsequently, the DELPHI, OPAL, and CDF Collaborations have also measured the B_s^0 mass. [21,22,23] The world average is $m(\Lambda_b^0) = (5369.3 \pm 2.0)\ \text{MeV}/c^2$. [15]

4 B Hadron Lifetimes

In the Spectator Model, all b hadrons have equal lifetimes. Non-spectator model effects are expected to contribute at the rate of f_B^2/m_b^2, where f_B is the decay constant, giving rise to lifetime differences on the order of 10% among species. [24,25] Specifically, I. Bigi *et al.* expect

$$\frac{\tau(B^+)}{\tau(B_d^0)} \simeq 1.0 \pm 0.05 \left(\frac{f_B}{200\text{MeV}}\right)^2 \tag{6}$$

$$\frac{\tau(B_s^0)}{\tau(B_d^0)} \simeq 1.0 \pm \mathcal{O}(0.01) \tag{7}$$

$$\frac{\tau(\Lambda_b^0)}{\tau(B_d^0)} \simeq 0.9\text{–}1.0 \tag{8}$$

See also I. Bigi's talk, these proceedings.

Measurements of the average b hadron lifetime were pioneered by experiments on the PEP storage ring at SLAC in 1983, based on lepton im-

(i) ALEPH (ii) OPAL

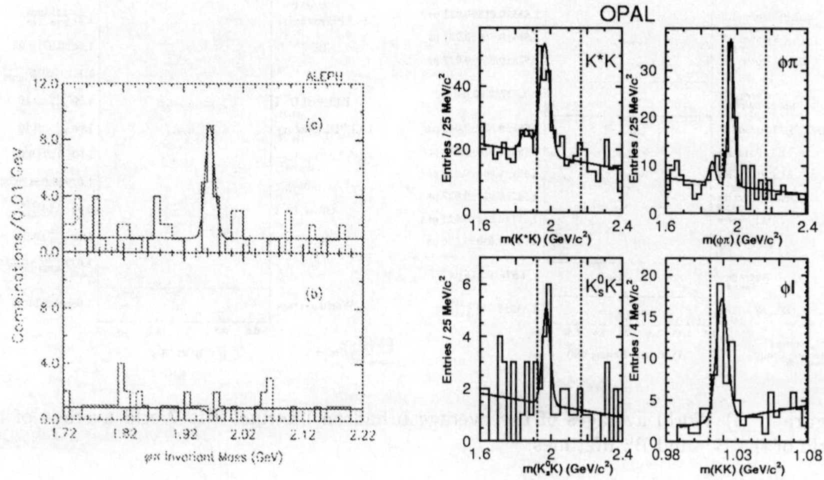

Figure 3. (i) Example of the first evidence for the B^0_s, ALEPH data: (a) right sign $D^+_s\ell^-$ combinations (b) wrong sign $D^-_s\ell^+$ combinations. [17] (ii) Example of a current event sample of B^0_s decays, OPAL data: four decay modes are shown. [19]

pact parameters and also on secondary, detached decay vertices. [26,27] The current measurements and world averages are shown in Figure 4(i). The average of measurements based on lepton impact parameters (and therefore a sample composed slightly differently than the production fractions) is $\langle\tau_b\rangle^{lept} = (1.537 \pm 0.020)$ ps. [28] The average of measurements based on inclusive secondary vertices is $\langle\tau_b\rangle^{vtx} = (1.577 \pm 0.016)$ ps. [28]

The B^0_d and B^+ lifetimes have been measured with several methods. Samples have been isolated based on the semileptonic decays $B \to D^{(*)}\ell\nu X$ with either full or partial reconstruction of the D or D*. Additionally, samples of fully reconstructed B^0_d and B^+ mesons have been used. And finally, inclusive secondary vertex samples have been used in which the charge of the vertex has been measured. The current world average of the ratio of the lifetimes of the charged and neutral B mesons is shown in Figure 4(ii). It is $\tau(B^+)/\tau(B^0_d) = 1.066 \pm 0.024$. [28] The individual world average lifetimes are $\tau(B^0_d) = (1.553 \pm 0.029)$ ps and $\tau(B^+) = (1.639 \pm 0.025)$ ps. [28]

The B^0_s lifetime has been measured at LEP with the decays $B^0_s \to D^-_s\ell^+\nu X$ and $B^0_s \to D^-_s h^+ X$, where h^+ is an unidentified charged hadron. [29,30,31,32] Sev-

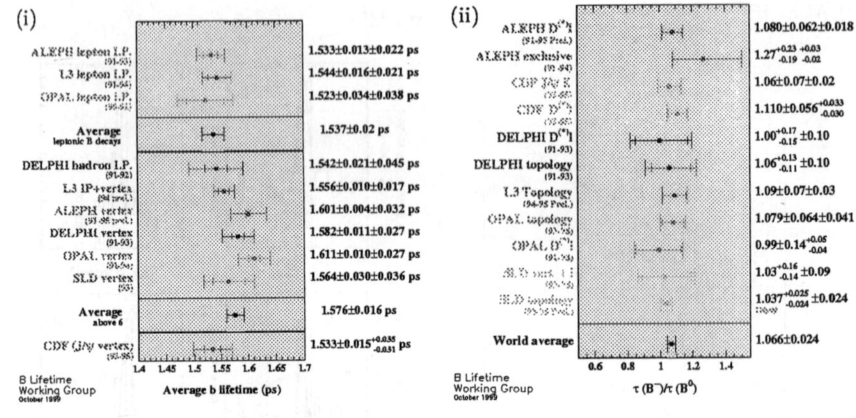

Figure 4. (i) World averages of the average b hadron lifetime. (ii) World average of the ratio of the B^+ and B_d^0 lifetimes. [28]

en decay modes of the D_s^- have been reconstructed; samples are now several hundred events. The current world measurements and their average are shown in Figure 5(i). The CDF Collaboration have also measured $\tau(B_s^0)$ using exclusively reconstructed $J/\psi\phi$ decays, a CP eigenstate. The world average of all of these measurements is $\tau(B_s^0) = (1.460 \pm 0.056)$ ps. [28]

The lifetime of the Λ_b^0 baryon has been measured using the semileptonic decay $\Lambda_b^0 \to \Lambda_c^+ \ell^- \overline{\nu} X$. Up to five decay modes of the Λ_c^+ have been reconstructed, and event samples exceed two hundred events. [31,33,34,35] The average b baryon lifetime has been measured using the semileptonic decay $\Lambda_b^0 \to \Lambda \ell^- \overline{\nu} X$. [10,33,34,36] These samples are an unknown mixture of b baryons, but are expected to be dominated by Λ_b^0. The current measurements are shown in Figure 5(ii). The world average of all the b baryon lifetime measurements is is $\tau(\text{b baryon}) = (1.208 \pm 0.051)$ ps. [28]

The comparison of the measured b hadron lifetimes with theory is shown in Figure 6. The charged B meson lifetime has been shown to be significantly longer than $\tau(B_d^0)$, in accordance with expectation. The B_s^0 lifetime is at one standard deviation from equality with $\tau(B_d^0)$, also in accordance with expectation. However, the Λ_b^0 lifetime and more precisely, the average b baryon lifetime, are significantly lower than $\tau(B_d^0)$; it is three standard deviations from the lower end of theoretical expectation. This constitutes a puzzle unexplained by current techniques such as Heavy Quark Effective Theory and the operator product expansion applied to b decays, [24,25,37] though at least

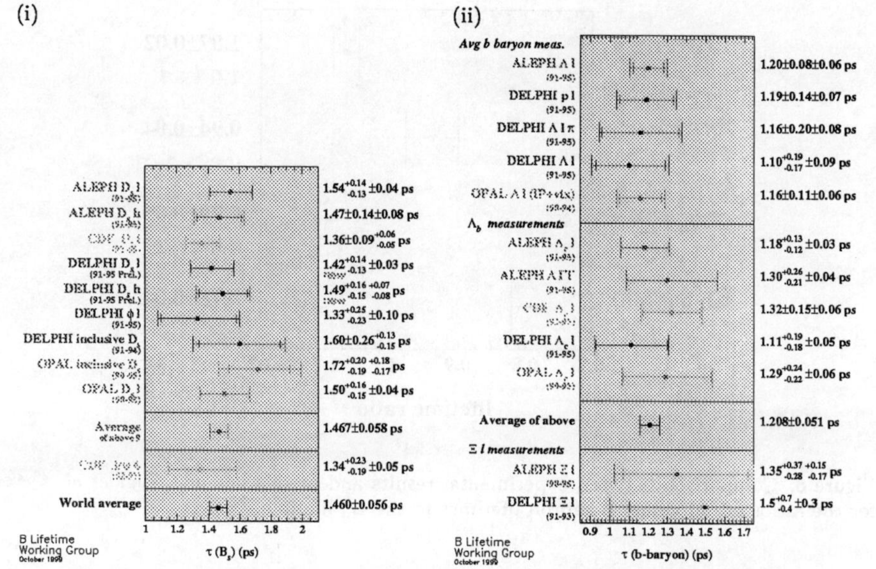

Figure 5. (i) World average of the B_s^0 lifetime. (ii) World average of b baryon lifetime measurements. [28]

one author has accommodated current measurements at the very edge of an allowed range. [38]

5 B_d^0 and B_s^0 Oscillations

As in the neutral kaon system, the neutral B mesons are allowed to oscillate between particle and anti-particle states via second-order weak interactions, as shown in Figure 7. If CP violation is neglected, the CP eigenstates are also the mass eigenstates

$$B_{S,L} = \frac{B^0 \pm \overline{B}^0}{\sqrt{2}} \quad \text{and} \quad \Delta m = |m_L - m_S| \,. \tag{9}$$

Assuming the differences in the decay widths are small, $\Gamma \approx \Gamma_L \approx \Gamma_S$, the decay probability at time t for an initially pure B^0 or $\overline{B}^0$ state is

$$P(t) = \frac{\Gamma}{2} e^{-\Gamma t} (1 \pm \cos \Delta m t) \,, \tag{10}$$

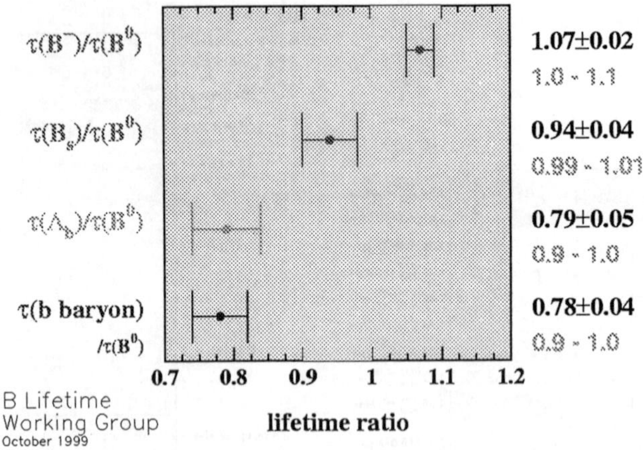

Figure 6. Comparison between experimental results and predictions of I. Bigi *et al.* [24,25] for the ratios of individual b hadron lifetimes to the B_d^0 lifetime. [28]

in which (+) signifies a decay in the same state ("unmixed") and (−) a decay in the opposite state ("mixed"). The oscillation rates are proportional to the CKM matrix elements of the coupling to the quarks in the internal loops. By taking the ratio of the B_d^0 and B_s^0 oscillation frequencies, the proportionality may be estimated more precisely

$$\frac{\Delta m_s}{\Delta m_d} = \frac{m(B_s^0)}{m(B_d^0)}\xi^2 \left|\frac{V_{ts}}{V_{td}}\right|^2 \frac{\eta(B_s^0)}{\eta(B_d^0)} \tag{11}$$

The quantities η_i are the short-distance QCD correction factors and ξ contains the ratio of the decay constants and non-perturbative QCD corrections.

Many methods have been developed both to measure Δm_d and to search for Δm_s. As an example of the former, a result from the OPAL Collaboration for Δm_d is shown in Figure 8 which is based on jet charge and a high momentum, high-p_t lepton to estimate the initial and final states of the B_d^0 mesons in the sample. They find $\Delta m_d = (0.444 \pm 0.029^{+0.020}_{-0.017})$ ps^{-1}, the most precise determination at this time. [39]

All of the current results from LEP for Δm_d are shown in Figure 9. The LEP average is $\Delta m_d = (0.470 \pm 0.018)$ ps^{-1}. [40] Including measurements from the CDF and SLD experiments, the world average is $\Delta m_d = (0.482 \pm 0.016)$ ps^{-1}, or $\Delta m_d = (3.17 \pm 0.11) \times 10^{-4}$ eV. [40]

The ALEPH Collaboration set the first lower limit on the B_s^0 oscillation frequency, in 1993. They found $\Delta m_s > 1.8$ ps^{-1} at 95% C.L. from a sample

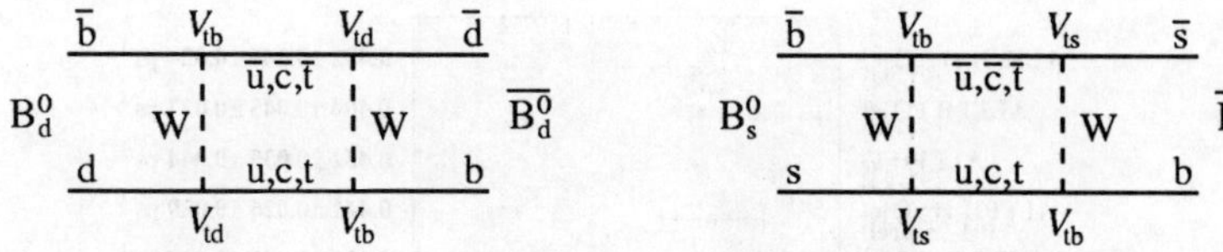

Figure 7. Two of the box diagrams responsible for B^0_d-$\overline{B}^0_d$ and B^0_s-$\overline{B}^0_s$ oscillations. There are similar contributions with the internal W and internal quark lines exchanged. The CKM matrix elements shown correspond only to the case of top quarks in the internal loop.

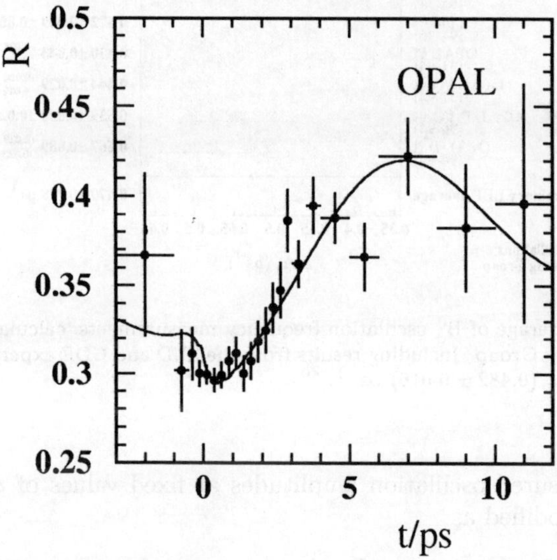

Figure 8. The OPAL B^0_d oscillation frequency analysis based on leptons and jet-charge. The ratio R is the fraction of mixed events over total number of events. The oscillation frequency is measured to be $\Delta m_d = (0.444 \pm 0.029^{+0.020}_{-0.017})$ ps^{-1}. [39]

of dilepton events. [41] The following year, based on an analysis of single lepton events using jet-charge, ALEPH set the limit at $\Delta m_s > 6.1$ ps^{-1} at 95% C.L., reported at the 1994 ICHEP Glasgow conference. [42]

Initially, the result of each analysis was a fit to or scan of the likelihood as a function of Δm_s, and a limit was set from this curve. However, it was found that combining the likelihoods from different analyses in order to produce a single world limit on Δm_s was difficult, and a new method was developed,

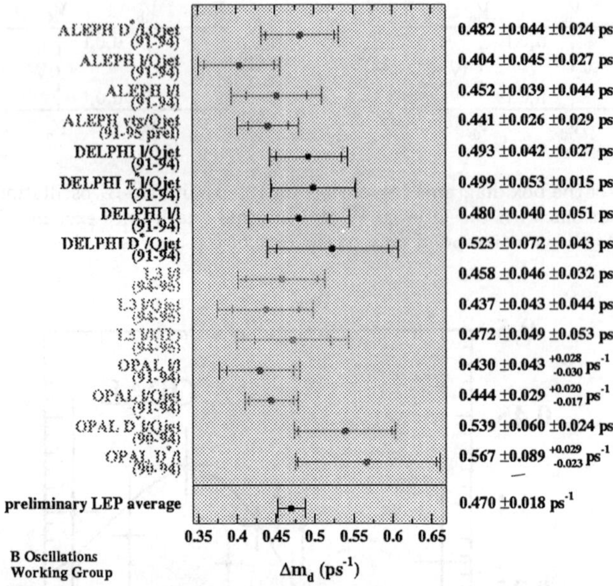

Figure 9. LEP average of B_d^0 oscillation frequency measurements, calculated by the B Oscillations Working Group. Including results from the SLD and CDF experiments, the world average is $\Delta m_d = (0.482 \pm 0.016)$ ps^{-1}. [40]

based on measured oscillation amplitudes at fixed values of Δm_s. [43] Equation (10) is modified as

$$P(t) = \frac{\Gamma}{2} e^{-\Gamma t}(1 \pm \mathcal{A} \cos \Delta m t) , \tag{12}$$

in which $\mathcal{A}$ is the oscillation amplitude. At the true oscillation frequency $\mathcal{A} = 1$ is expected and elsewhere $\mathcal{A} = 0$ is expected. Figure 10 shows the measured amplitude at $\Delta m_s = 10$ ps^{-1} of each B_s^0 oscillation analysis that is part of the current world combination. Figure 10 also shows the sensitivity of each analysis, defined as the largest frequency which would be excluded if all measured amplitudes were zero.

As of this conference, the combined world limit on B_s^0 oscillations is $\Delta m_s > 14.3$ ps^{-1}, or 9.4×10^{-3} eV, at 95% C.L. [40] This is shown in Figure 11. At 16.3 ps^{-1} the combined amplitude is 1.96 standard deviations from zero; however, this is beyond the sensitivity of 14.5 ps^{-1}. The world average of Δm_d and limit on Δm_s together imply a limit on the ratio of CKM matrix

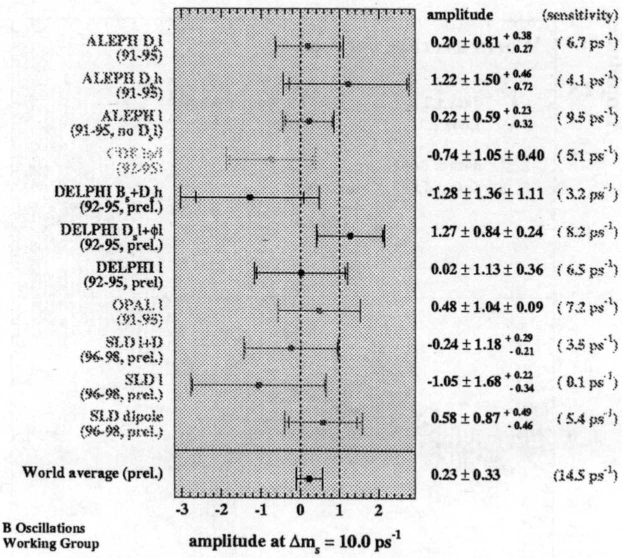

Figure 10. LEP average of B_s^0 oscillation amplitude measurements at $\Delta m_s = 10\ \mathrm{ps}^{-1}$, calculated by the B Oscillations Working Group. [40] The sensitivity of each analysis is also shown.

elements $|V_{ts}/V_{td}| > 4.5$ at 95% C.L.

6 V_{cb}

There are two methods of measuring $|V_{cb}|$ at LEP. The first method is based on the measured decay rate of the exclusive semileptonic decay $\overline{B}_d^0 \to D^{*+}\ell^-\overline{\nu}$. The second is based on the measured decay rate of the inclusive semileptonic decay $b \to \ell^-\overline{\nu}X$.

In the exclusive method, the decay rate is measured as a function of the product of the four-velocities of the hadrons, ω, which is related simply to the four-momentum transfer of the decay, q^2. From Heavy Quark Effective Theory, the differential decay rate is

$$\frac{d\Gamma}{d\omega} = \mathcal{K}(\omega)\mathcal{F}^2_{D^*}(\omega)|V_{cb}|^2 \tag{13}$$

in which $\mathcal{K}(\omega)$ is a known phase space term and $\mathcal{F}^2_{D^*}(\omega)$ is the hadronic form factor. [44,45] At the point of zero recoil, the D^* is produced at rest in the B_d^0

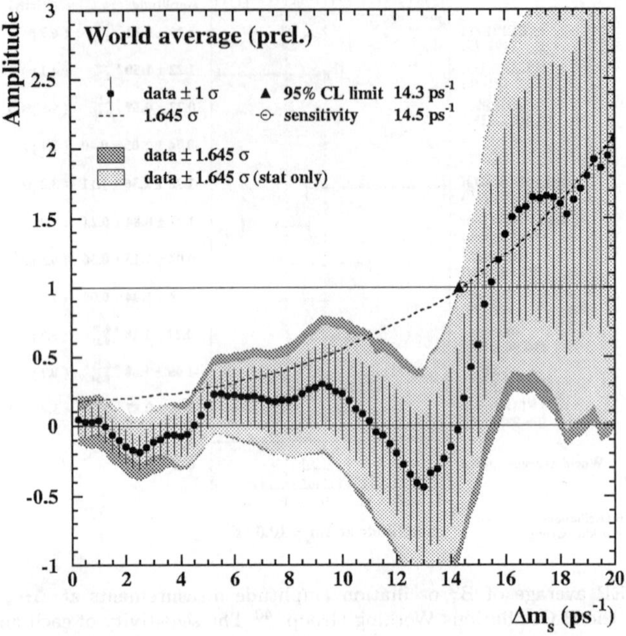

Figure 11. Combined B_s^0 oscillation amplitude spectrum and limit on Δm_s, calculated by the B Oscillations Working Group for this conference. [40]

rest frame thus $\omega = 1$. In the limit $m_b \to \infty$, $\mathcal{F}^2_{D^*}(\omega)$ becomes the Isgur-Wise function which is normalized to unity at this point of zero recoil ($\omega = 1$). [46,47] Therefore, $|V_{cb}|$ may be measured by extrapolating the differential decay rate to $\omega = 1$. As m_b is finite, this method is accurate up to the knowledge of $\mathcal{F}^2_{D^*}(1)$.

The ARGUS Collaboration first performed such a measurement in 1993, followed by the CLEO Collaboration. [48,49] At the $\Upsilon(4S)$ resonance, the two B mesons are produced almost at rest in the laboratory frame and thus reconstruction of the B_d^0 rest frame is straightforward. At LEP, it was only the advent of the precise determination of the interaction point that allowed a constraint to reconstruct the missing neutrino's momentum and therefore reconstruct the B_d^0 rest frame. ALEPH first published a measurement using this technique in 1995, which has since been updated. [50,51] The OPAL and DELPHI Collaborations have also measured $|V_{cb}|$ using this technique. [52,53]

The LEP V_{cb} Working Group have performed an average of these exclusive measurements taking into account the correlated systematic uncertainties. In addition, they have imposed a unified set of theoretical estimates and assume $\mathcal{F}^2_{\mathrm{D}^*}(1) = 0.88 \pm 0.05$. [54] The LEP average is $|V_{\mathrm{cb}}|^{excl} = (38.4 \pm 1.1 \pm 2.2 \pm 2.2) \times 10^{-3}$, in which the final uncertainty is due to the assumed value of $\mathcal{F}^2_{\mathrm{D}^*}(1)$. [54] For comparison, assuming the same value of $\mathcal{F}^2_{\mathrm{D}^*}(1)$ the CLEO Collaboration have measured $|V_{\mathrm{cb}}|^{excl} = (39.9 \pm 2.2 \pm 2.0 \pm 2.3) \times 10^{-3}$. [55]

The second method of measuring the CKM matrix element V_{cb} at LEP is based on inclusive semileptonic decays of b hadrons. The total semileptonic decay rate is proportional to $|V_{\mathrm{cb}}|^2$, thus measurements of the semileptonic branching fractions and the lifetimes may be used to infer $|V_{\mathrm{cb}}|$. From the current LEP averages $\mathrm{Br}(\mathrm{b} \to \ell^- \overline{\nu} \mathrm{X}) = (10.58 \pm 0.18)\%$, $\mathrm{Br}(\mathrm{b} \to \ell^- \overline{\nu} \mathrm{X_u}) = (1.67 \pm 0.55) \times 10^{-3}$, the average b hadron lifetime $\langle \tau_{\mathrm{b}} \rangle = (1.554 \pm 0.015)$ ps and some assumptions about the theoretical uncertainties inherent in the proportionality, the LEP V_{cb} Working Group have calculated $|V_{\mathrm{cb}}|^{incl} = (40.76 \pm 0.41 \pm 2.04) \times 10^{-3}$, in which the first uncertainty is from experiment and the second is from theory. [54]

7 V_{ub}

At LEP, the CKM matrix element $|V_{\mathrm{ub}}|$ is measured from the inclusive yield of $\mathrm{b} \to \mathrm{u}$ transitions in $\mathrm{b} \to \ell^- \overline{\nu} \mathrm{X}$ decays,

$$|V_{\mathrm{ub}}|^2 \propto \mathrm{Br}(\mathrm{b} \to \mathrm{X_u} \ell^- \overline{\nu}) \,. \tag{14}$$

The three collaborations L3, DELPHI, and ALEPH have each performed this measurement. [56,57,58] The LEP V_{ub} Working Group has adjusted these measurements to a common set of inputs and theoretical assumptions. The LEP average for the branching fraction is $\mathrm{Br}(\mathrm{b} \to \mathrm{X_u} \ell^- \overline{\nu}) = (1.67 \pm 0.35 \pm 0.38 \pm 0.20) \times 10^{-3}$, in which the first uncertainty is experimental, the second is derived from the measured rate for $\mathrm{b} \to \mathrm{c} \ell^- \overline{\nu}$ transitions, and the third is due to theoretical uncertainties in the modelling of $\mathrm{b} \to \mathrm{u} \ell^- \overline{\nu}$ transitions. With the common set of inputs, the LEP V_{ub} Working Group find for the LEP average $|V_{\mathrm{ub}}| = (4.04^{+0.62}_{-0.73}) \times 10^{-3}$. [59]

The CLEO and ARGUS Collaborations measured $|V_{\mathrm{ub}}|$ from this inclusive method first. [60,61] The average of these measurements is $|V_{\mathrm{ub}}| = (3.16 \pm 0.81) \times 10^{-3}$. [62] More recently, the CLEO Collaboration have observed the exclusive decays $\mathrm{B} \to \pi \ell \nu$ and $\mathrm{B} \to \rho \ell \nu$, and derive their most precise measurement from the sample of $\mathrm{B} \to \rho \ell \nu$ decays, measuring $|V_{\mathrm{ub}}| = (3.25^{+0.61}_{-0.64}) \times 10^{-3}$. [63,64]

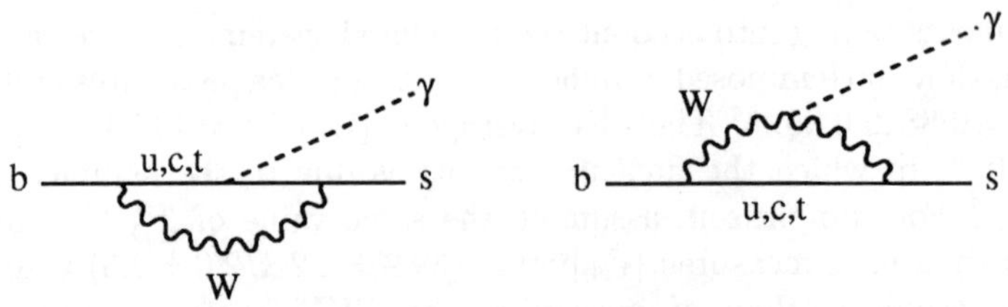

Figure 12. The two diagrams for the penguin decay $b \to s\gamma$.

8 $b \to s\gamma$

The flavor-changing neutral current decay $b \to s\gamma$ proceeds via an electromagnetic penguin diagram, shown in Figure 12. This decay is sensitive to new physics as the loop may contain a charged Higgs boson or a supersymmetric particle, which could enhance or suppress the decay with respect to the Standard Model expectation.

The CLEO Collaboration were the first to observe evidence of a penguin decay, $B \to K^*(892)\gamma$, in 1993. [65] At LEP, the ALEPH Collaboration have observed inclusive evidence for the $b \to s\gamma$ transition via an excess in the photon energy spectrum between 2.2 and 2.8 GeV. [66] The ALEPH data are shown in Figure 13. For the branching fraction, ALEPH measure $\mathrm{Br}(b \to s\gamma) = (3.11 \pm 0.80 \pm 0.72) \times 10^{-4}$. For comparison, the most recent measurement of this quantity by the CLEO Collaboration is $\mathrm{Br}(b \to s\gamma) = (3.15 \pm 0.35 \pm 0.32 \pm 0.26) \times 10^{-4}$, in which the final uncertainty is due to model dependence. [67] The Standard Model prediction for this rate is $\mathrm{Br}(b \to s\gamma) = (3.28 \pm 0.33) \times 10^{-4}$. [68]

9 CP Violation

Both the B_d^0 and the $\overline{B}_d^0$ mesons decay to the CP eigenstate $J/\psi K_S^0$. The asymmetry in these decay rates is proportional to $\sin 2\beta$

$$A(t) = \frac{\Gamma(B_d^0 \to J/\psi K_S^0) - \Gamma(\overline{B}_d^0 \to J/\psi K_S^0)}{\Gamma(B_d^0 \to J/\psi K_S^0) + \Gamma(\overline{B}_d^0 \to J/\psi K_S^0)} \tag{15}$$

$$= -\sin 2\beta \sin \Delta m_d t \,, \tag{16}$$

in which β is one of the angles of the CKM unitarity triangle

$$\beta = \arg\left(\frac{-V_{cd}V_{cb}^*}{V_{td}V_{tb}^*}\right) . \tag{17}$$

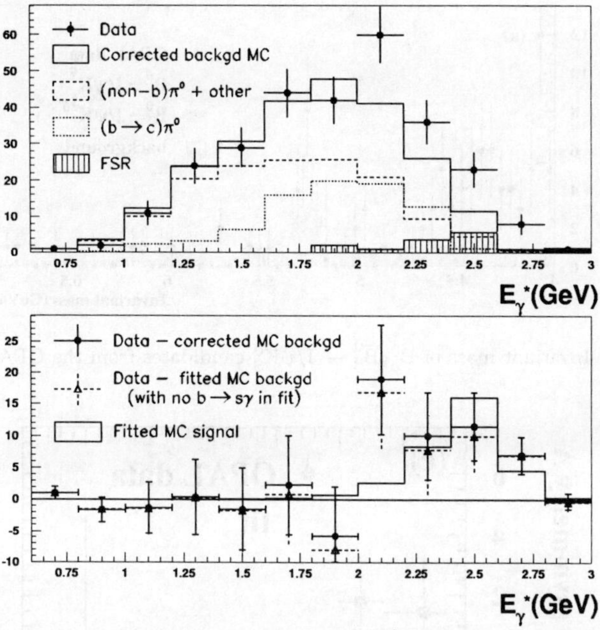

Figure 13. Evidence for the penguin decay $b \to s\gamma$ in the ALEPH data. [66] The top plot shows the measured photon energy spectrum for the data and for the simulation with no penguin transitions included. The lower plot shows the background-subtracted spectrum, with the fitted signal simulation.

From measurements of Δm_{d}, limits on Δm_{s}, measurements of $|V_{\mathrm{cb}}|$ and $|V_{\mathrm{ub}}|$, and measurements of CP violation in the neutral kaon system, this angle is expected to lie in the range $\sin 2\beta = (0.75 \pm 0.09)$. [69]

The OPAL Collaboration were the first to study CP violation in the $\mathrm{B_d^0}$ system at LEP, in 1998. [70] The $\mathrm{B_d^0}$ mass peak from the OPAL data is shown in Figure 14. There are 24 events in the signal region with an estimated signal purity of $(60 \pm 8)\%$. The fit is shown in Figure 15; the fitted angle is $\sin 2\beta = (3.2^{+1.8}_{-2.0} \pm 0.5)$.

The CDF Collaboration have studied CP violation in $\mathrm{B_d^0}$ decays this past year, finding $\sin 2\beta = (0.79^{+0.41}_{-0.44})$. [71]

The ALEPH Collaboration have recently studied CP violation in the $\mathrm{B_d^0}$ system — the results were presented for the first time at this conference. [72] The $\mathrm{B_d^0}, \overline{\mathrm{B}}{}_{\mathrm{d}}^0 \to \mathrm{J}/\psi \mathrm{K_S^0}$ mass peak is shown in Figure 16. There are 23 events in

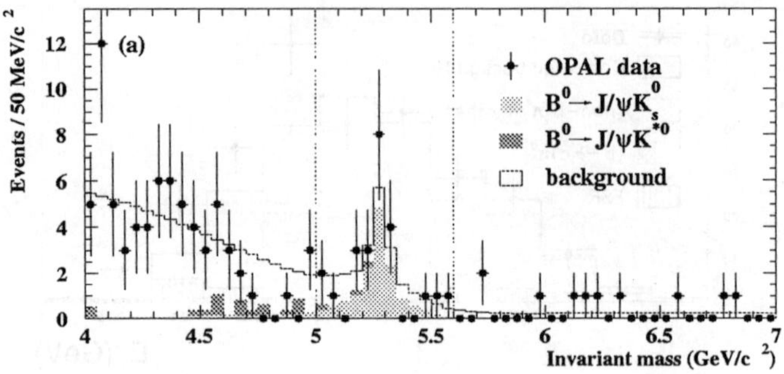

Figure 14. Invariant mass of $B_d^0, \overline{B}_d^0 \to J/\psi K_S^0$ candidates from the OPAL data. [70]

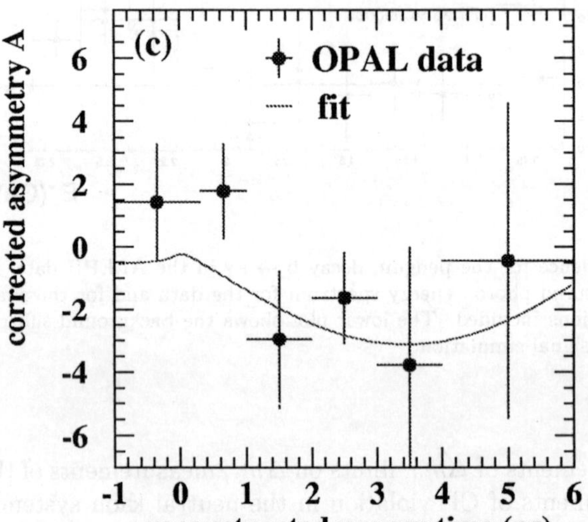

Figure 15. CP asymmetry fit to candidate $B_d^0, \overline{B}_d^0 \to J/\psi K_S^0$ decays in the OPAL data. [70]

the signal region with an estimated purity of $(71 \pm 12)\%$. Using a multivariate discriminant, 15 events are ascribed to $\overline{B}_d^0$ and 8 to B_d^0. The fitted angle is $\sin 2\beta = (0.93^{+0.64+0.36}_{-0.88-0.24})$; the fit with the data superimposed is shown in Figure 17.

The three measurements may be combined, with the result $\sin 2\beta = (0.91 \pm 0.35)$. [72] From the CDF data alone, the probability that $\sin 2\beta < 0$

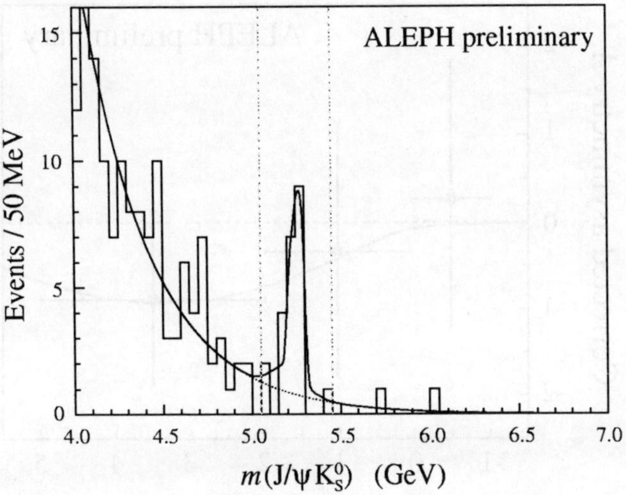

Figure 16. Invariant mass of $B_d^0, \overline{B}_d^0 \to J/\psi K_S^0$ candidates in the ALEPH data. [72]

is 7%. With the ALEPH and OPAL data, the probability that $\sin 2\beta < 0$ decreases to 1.5%. This may be interpreted as a confidence level of almost 99% that CP violation in the B_d^0 system has been observed. [72]

10 Conclusion

In the past decade, the four LEP experiments ALEPH, DELPHI, L3, and OPAL have each collected approximately 0.9 million $Z \to b\bar{b}$ events. With these data, the collaborations have produced an impressive array of results, advancing the field of b physics greatly. Among the highlights have been: the discovery of b baryons and the strange b meson, measurement of the masses and lifetimes of a number of the b hadrons (and the low Λ_b^0 lifetime is still a theoretical puzzle), the first observation of time-dependent B_d^0 oscillations and the precise determination of this frequency, the first and currently best lower limit on the B_s^0 oscillation frequency Δm_s, contributions to the measurement of the CKM matrix elements $|V_{cb}|$ and $|V_{ub}|$, and contributions to the measurement of the unitarity triangle angle β.

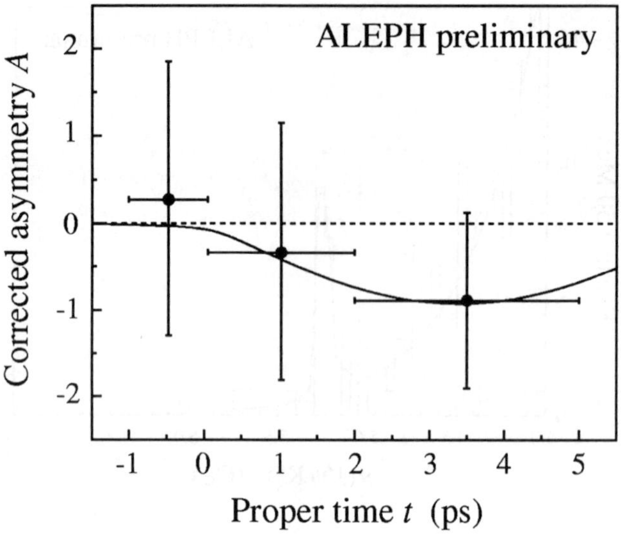

Figure 17. CP asymmetry fit to candidate $B_d^0, \overline{B}_d^0 \rightarrow J/\psi K_S^0$ decays in the ALEPH data. [72]

Acknowledgments

I sincerely thank Professors Hai-Yang Cheng and Wei-Shu Hou for organizing this outstanding and most productive conference on B physics and for their wonderful hospitality. I would like to thank the CERN SL Division for the stellar performance of the LEP collider throughout the last decade. I would also like to thank my colleagues in the ALEPH Collaboration, as well as the DELPHI, OPAL, and L3 Collaborations for the large number of high quality b physics results that they have produced in the last decade. I must acknowledge the invaluable assistance of Dr. Owen Hayes during the preparation of this talk and these proceedings. I would also like to thank Dr. Roger Forty for his assistance. This work was supported by the US Department of Energy, grant DE-FG0295-ER40896.

References

1. ALEPH Collaboration, *Phys. Lett.* B **401**, 150 (1997); *Phys. Lett.* B **401**, 163 (1997).
2. DELPHI Collaboration, *Eur. Phys. J.* C **10**, 41 (1999).

3. L3 Collaboration, contribution **5_282** to the International Europhysics Conference on High Energy Physics, EPS-HEP 99, 15–21 July 1999, Tampere, Finland.
4. OPAL Collaboration, *Eur. Phys. J.* C **8**, 217 (1999).
5. SLD Collaboration, contribution **EPS-118** to the International Europhysics Conference on High-Energy Physics, EPS-HEP 97, 19–26 August 1997, Jerusalem, Israel; V. Serbo for the SLD Collaboration, "Electroweak measurements with heavy quarks at SLD", III International Conference on Hyperons, Charm and Beauty Hadrons, 30 June–3 July 1998, Genova, Italy.
6. ALEPH, DELPHI, L3, OPAL, and SLD Collaborations, "A Combination of Preliminary Electroweak Measurements and Constraints on the Standard Model", preprint CERN-EP/2000-016 (2000).
7. ALEPH Collaboration, *Phys. Lett.* B **278**, 209 (1992).
8. OPAL Collaboration, *Phys. Lett.* B **281**, 394 (1992).
9. DELPHI Collaboration, *Phys. Lett.* B **311**, 379 (1993).
10. OPAL Collaboration, *Z. Phys.* C **69**, 195 (1996).
11. Averages performed by Owen Hayes.
12. ALEPH Collaboration, *Phys. Lett.* B **380**, 442 (1996).
13. DELPHI Collaboration, *Phys. Lett.* B **374**, 351 (1996).
14. CDF Collaboration, *Phys. Rev.* D **55**, 1142 (1997).
15. Particle Data Group, R.M. Barnett *et al.*, *Eur. Phys. J.* C **3**, 197 (1998); 1999 update at `http://pdg.lbl.gov/`.
16. DELPHI Collaboration, *Phys. Lett.* B **289**, 199 (1992).
17. ALEPH Collaboration, *Phys. Lett.* B **294**, 145 (1992).
18. OPAL Collaboration, *Phys. Lett.* B **295**, 357 (1992).
19. OPAL Collaboration, *Phys. Lett.* B **426**, 161 (1998).
20. ALEPH Collaboration, *Phys. Lett.* B **311**, 425 (1993).
21. DELPHI Collaboration, *Phys. Lett.* B **324**, 500 (1994).
22. OPAL Collaboration, *Phys. Lett.* B **337**, 196 (1994).
23. CDF Collaboration, *Phys. Rev.* D **53**, 3496 (1996).
24. I. Bigi, "Heavy Quark Expansions for Inclusive Heavy-Flavour Decays and the Lifetimes of Charm and Beauty Hadrons", preprint hep-ph/9612293 (1996).
25. G. Bellini, I. Bigi, P. Dornan, *Phys. Rept.* **289**, 1 (1997) and references therein.
26. MAC Collaboration, *Phys. Rev. Lett.* **51**, 1022 (1983).
27. MARK II Collaboration, *Phys. Rev. Lett.* **51**, 1316 (1983).
28. LEP B Lifetimes Working Group, world averages performed for this conference, at `http://home.cern.ch/claires/lepblife.html`.

29. ALEPH Collaboration, *Phys. Lett.* B **377**, 205 (1996); *Eur. Phys. J.* C **4**, 367 (1998).
30. DELPHI Collaboration, contribution **4_520** to the International Europhysics Conference on High Energy Physics, EPS-HEP 99, 15–21 July 1999, Tampere, Finland.
31. OPAL Collaboration, *Phys. Lett.* B **426**, 161 (1998).
32. OPAL Collaboration, *Eur. Phys. J.* C **2**, 407 (1998).
33. ALEPH Collaboration, *Eur. Phys. J.* C **2**, 197 (1998).
34. DELPHI Collaboration, *Eur. Phys. J.* C **10**, 185 (1999).
35. CDF Collaboration, *Phys. Rev. Lett.* **77**, 449 (1996).
36. DELPHI Collaboration, *Z. Phys.* C **71**, 199 (1996).
37. J. Rosner, *Phys. Lett.* B **379**, 267 (1996);
    G. Altarelli *et al.*, *Phys. Lett.* B **382**, 409 (1996);
    P. Colangelo and F. De Fazio, *Phys. Lett.* B **387**, 371 (1996);
    T. Ito *et al.*, *Prog. Theor. Phys.* **99**, 271 (1998);
    M. Di Pierro and C.T. Sachrajda, *Nucl. Phys. Proc. Suppl.* **73**, 384 (1999).
38. M. Neubert, "Theory of Beauty Lifetimes", preprint hep-ph/9707217 (1997); "Heavy-Quark Effective Theory and Weak Matrix Elements", preprint hep-ph/9801269 (1998).
39. OPAL Collaboration, *Z. Phys.* C **76**, 401 (1997).
40. LEP B Oscillations Working Group, averages and combinations performed for this conference, at `http://www.cern.ch/LEPBOSC/`.
41. ALEPH Collaboration, *Phys. Lett.* B **322**, 441 (1994).
42. Y. Pan for the ALEPH Collaboration, invited talk at 1994 International Conference on High Energy Physics, August 1994, Glasgow, UK.
43. H.-G. Moser and A. Roussarie, *Nucl. Instrum. Meth.* A **384**, 491 (1997).
44. M.A. Shifman and M.B. Voloshin, *Sov. J. Nucl. Phys.* **47**, 511 (1988).
45. M. Neubert, *Phys. Lett.* B **264**, 455 (1991); *Phys. Lett.* B **338**, 84 (1994).
46. N. Isgur and M. Wise, *Phys. Lett.* B **232**, 113 (1989); *Phys. Lett.* B **237**, 527 (1990).
47. A.F. Falk, H. Georgi, B. Grinstein, and M.B. Wise, *Nucl. Phys.* B **343**, 1 (1990).
48. ARGUS Collaboration, *Z. Phys.* C **57**, 533 (1993).
49. CLEO Collaboration, *Phys. Rev.* D **51**, 1014 (1995); *Phys. Rev. Lett.* **79**, 2208 (1997); *Phys. Rev. Lett.* **82**, 3746 (1999).
50. ALEPH Collaboration, *Phys. Lett.* B **359**, 236 (1995).
51. ALEPH Collaboration, *Phys. Lett.* B **395**, 373 (1997).
52. DELPHI Collaboration, *Z. Phys.* C **71**, 539 (1996).

53. OPAL Collaboration, *Phys. Lett.* B **395**, 128 (1997); "Measurement of $|V_{\rm cb}|$ using $\overline{\rm B}^0_{\rm d} \to {\rm D}^{*+}\ell^-\overline{\nu}$ decays", preprint CERN-EP/2000-032 (2000).
54. LEP $V_{\rm cb}$ Working Group, results prepared from contributions to the 1999 summer conferences.
55. CLEO Collaboration, *Phys. Rev.* D **51**, 1014 (1995).
56. L3 Collaboration, *Phys. Lett.* B **436**, 174 (1998).
57. DELPHI Collaboration, contribution #241 to the International Conference on High-Energy Physics, ICHEP 98, July 1998, Vancouver, Canada.
58. ALEPH Collaboration, *Eur. Phys. J.* C **6**, 555 (1999).
59. LEP $V_{\rm ub}$ Working Group, results prepared from contributions to the 1999 summer conferences.
60. CLEO Collaboration, *Phys. Rev. Lett.* **64**, 16 (1990); *Phys. Rev. Lett.* **71**, 4111 (1993).
61. ARGUS Collaboration, *Phys. Lett.* B **234**, 409 (1990); *Phys. Lett.* B **241**, 278 (1990); *Phys. Lett.* B **255**, 297 (1991).
62. K. Österberg, "Measurement of $|V_{\rm cb}|$ and $|V_{\rm ub}|$ at LEP", talk at the International Europhysics Conference on High Energy Physics, EPS-HEP 99, 15–21 July 1999, Tampere, Finland.
63. CLEO Collaboration, *Phys. Rev. Lett.* **77**, 5000 (1996).
64. CLEO Collaboration, "Measurement of $B \to \rho\ell\nu$ decay and $|V_{ub}|$", preprint CLNS-99/1611 (1999), also as preprint hep-ex/9905056 (1999).
65. CLEO Collaboration, *Phys. Rev. Lett.* **71**, 674 (1993).
66. ALEPH Collaboration, *Phys. Lett.* B **429**, 169 (1998).
67. CLEO Collaboration, "$\rm b \to s\gamma$ Branching Fraction and CP Asymmetry", preprint CLEO-CONF-99-10 (1999), also as preprint hep-ex/9908022v2 (1999).
68. A.L. Kagan and M. Neubert, *Eur. Phys. J.* C **7**, 5 (1999).
69. S. Mele, *Phys. Rev.* D **59**, 113011 (1999).
70. OPAL Collaboration, *Eur. Phys. J.* C **5**, 379 (1998).
71. CDF Collaboration, *Phys. Rev.* D **61**, 072005 (2000).
72. R. Forty for the ALEPH Collaboration, these proceedings.

A REVIEW OF B_s^0 MIXING: PAST, PRESENT AND FUTURE

STÉPHANE WILLOCQ
Physics Department, University of Massachusetts, MA 01003, USA
E-mail: willocq@physics.umass.edu

We review the experimental status of B_s^0–$\overline{B_s^0}$ mixing. After a brief historical overview, current studies of the time dependence of B_s^0 oscillations are described, with an emphasis on the different experimental techniques used by the ALEPH, CDF, DELPHI, OPAL, and SLD Collaborations. To conclude, the outlook for future experiments is presented.

1 Introduction

In analogy to the K^0–$\overline{K^0}$ system, the B^0–$\overline{B^0}$ system consists of B^0 and $\overline{B^0}$ flavor eigenstates, which are superpositions of heavy and light mass eigenstates B_H and B_L. Due to the difference in mass and width, the mass eigenstates evolve differently as a function of time, resulting in time-dependent B^0–$\overline{B^0}$ flavor oscillations with a frequency equal the mass difference $\Delta m \equiv m_H - m_L$. As a consequence, an initially pure $|B^0\rangle$ state may be found to decay as $|B^0\rangle$ or $|\overline{B^0}\rangle$ at a later time t with a probability density equal to $P(B^0 \to B^0) = \frac{\Gamma}{2}e^{-\Gamma t}(1 + \cos \Delta m\, t)$ or $P(B^0 \to \overline{B^0}) = \frac{\Gamma}{2}e^{-\Gamma t}(1 - \cos \Delta m\, t)$. (Here we have taken $\Gamma_L \simeq \Gamma_H \simeq \Gamma$ since $\Delta\Gamma << \Delta m$ in the Standard Model.)

The oscillation frequency Δm_q ($q = d$ and s for B_d^0 and B_s^0) can be computed via the second order box diagrams that induce $B^0 \leftrightarrow \overline{B^0}$ transitions, see Fig. 1. Calculations yield[1]

$$\Delta m_q = \frac{G_F^2}{6\pi^2} m_{B_q} m_t^2 F(m_t^2/m_W^2) f_{B_q}^2 B_{B_q} \eta_{QCD} \left|V_{tb}^* V_{tq}\right|^2, \qquad (1)$$

where G_F is the Fermi constant, m_{B_q} is the B_q^0 hadron mass, m_t is the top quark mass, m_W is the W boson mass, F is the Inami-Lim function,[2] and η_{QCD} is a perturbative QCD parameter. The "bag" parameter B_{B_q} and the decay constant f_{B_q} parameterize hadronic matrix elements. Therefore, a measurement of the B_d^0 (B_s^0) oscillation frequency allows the CKM matrix element V_{td} (V_{ts}) to be determined. However, Lattice QCD calculations[3] of the product $f_{B_q}\sqrt{B_{B_q}}$ are plagued by an uncertainty of 20-25%. This uncertainty limits the precision of the extraction of V_{td} from the fairly precise measured value[4] of the B_d^0 oscillation frequency $\Delta m_d = 0.476 \pm 0.016$ ps^{-1}. Theoretical uncertainties are significantly reduced in the ratio between B_s^0

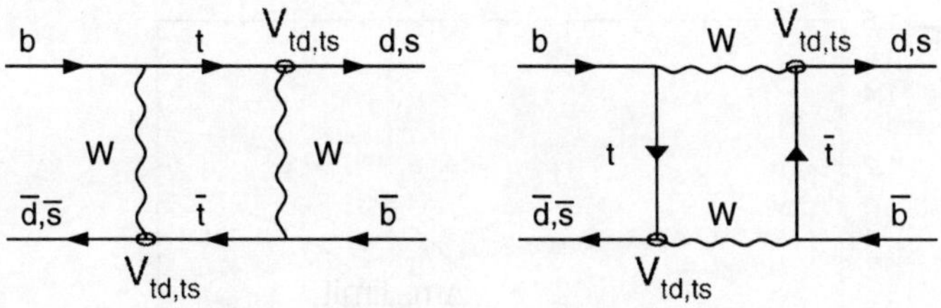

Figure 1. *Box diagrams leading to B^0–$\overline{B^0}$ mixing. Only the dominant top quark contribution is shown.*

and B_d^0 oscillation frequencies:[3]

$$\frac{\Delta m_s}{\Delta m_d} = \frac{m_{B_s} f_{B_s}^2 B_{B_s}}{m_{B_d} f_{B_d}^2 B_{B_d}} \left| \frac{V_{ts}}{V_{td}} \right|^2 = (1.14 \pm 0.06)^2 \left| \frac{V_{ts}}{V_{td}} \right|^2 . \tag{2}$$

Invoking CKM unitarity to obtain $|V_{ts}|^2$ from the measured value for $|V_{cb}|^2$, one can then extract V_{td} with good precision.

Determination of the CKM element V_{td} is of great importance since it is sensitive to the CP violating phase in the Standard Model. In the Wolfenstein parameterization of the CKM matrix, $|V_{td}|^2 = A^2\lambda^6[(1-\rho)^2+\eta^2]$ and $|V_{ts}|^2 = A^2\lambda^4$. The parameters $\lambda \equiv \sin\theta_c$ and A are well-known but ρ and η are not. A non-vanishing value for η implies the existence of CP violation in weak decays. The impact of Δm_d and Δm_s measurements on the knowledge of the fundamental parameters ρ and η is presented in Fig. 2, along with the constraints from the measurement of CP violation in the K^0–$\overline{K^0}$ system (ϵ_K) and the measurement of $b \to u$ transitions ($|V_{ub}/V_{cb}|$). From the above parameterization, it is clear that B_s^0 oscillations are very fast: $\Delta m_s/\Delta m_d \simeq 1/\lambda^2$, which is of order 20. Resolving these rapid oscillations thus poses a serious experimental challenge.

2 Past

The first evidence for B^0–$\overline{B^0}$ mixing was reported by UA1 in 1987 with a study of like-sign muon pairs produced in $\overline{p}p$ collisions.[5] The rate for like-sign pairs was found to exceed the expected background and was thus interpreted as evidence (2.9 σ) for B^0–$\overline{B^0}$ mixing. Later in the same year, conclusive evidence was presented by ARGUS in a study of like-sign dileptons produced by e^+e^- annihilation at the Υ_{4S} resonance.[6] ARGUS determined the time-integrated B_d^0–$\overline{B_d^0}$ mixing probability to be $\chi_d = 0.17 \pm 0.05$. Such a large

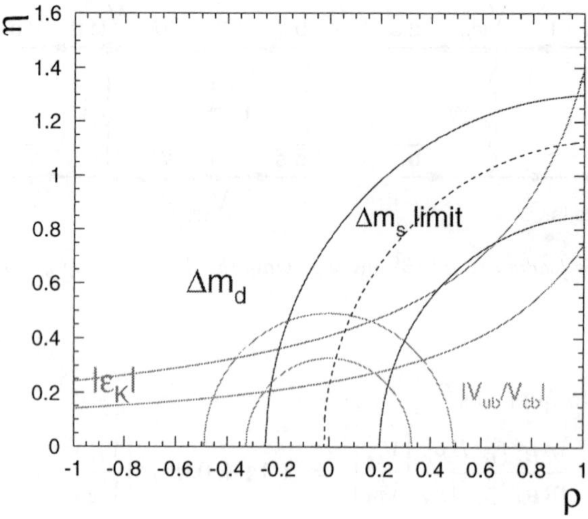

Figure 2. *Constraints on the apex (ρ, η) of the unitarity triangle. The area to the left of the dashed line is excluded by the current lower limit on Δm_s.*

mixing probability came as a surprise and indicated that the top quark mass had to be very large.

Further studies of like-sign dilepton events produced in e^+e^- annihilation at the Υ_{4S} and Z^0 resonances, as well as in $\bar{p}p$ collisions, confirmed the large rate of mixing. In 1993, the world averages were[7] $\chi_d = 0.162 \pm 0.021$ (Υ_{4S}) and $\langle\chi\rangle = 0.117 \pm 0.010$ (Z^0), where $\langle\chi\rangle$ represents the average mixing probability over the different types of b hadrons produced in Z^0 decays, $\langle\chi\rangle = f(B_d^0)\,\chi_d + f(B_s^0)\,\chi_s$, where $f(B_d^0)$ and $f(B_s^0)$ are the fractions of B_d^0 and B_s^0 in the selected sample, respectively. Combining these two results, a lower limit on the B_s^0 oscillation frequency was obtained:[7] $\Delta m_s > 0.5$ ps^{-1} at 90% C.L.

In 1994, ALEPH extended the like-sign dilepton technique by incorporating the B^0 proper decay time to investigate the time dependence of B_s^0–$\overline{B_s^0}$ mixing for the first time. Using a sample of 1 million hadronic Z^0 decays, a direct limit of $\Delta m_s > 1.8$ ps^{-1} (95% C.L.) was obtained.[8] Shortly after, OPAL improved the limit to $\Delta m_s > 2.2$ ps^{-1} in a similar study based on 1.5 million Z^0 decays.[9] These initial studies were limited mostly by the low efficiency of the dilepton event selection. Analyses were later improved by incorporating new vertex selection and tagging algorithms, as described in the next section.

3 Present

Studies of time-dependent oscillations require three ingredients: (i) reconstruction of the B_s^0 decay proper time, (ii) determination of the B_s^0 or $\overline{B_s^0}$ flavor at production, and (iii) determination of the flavor at decay. Decays for which the production and decay flavors are different are tagged as "mixed", otherwise they are tagged as "unmixed". The significance for a B_s^0 oscillation signal can be approximated by[10]

$$S = \sqrt{\frac{N}{2}}\, f(B_s^0)\, [1 - 2\,w]\, e^{-\frac{1}{2}(\Delta m_s \sigma_t)^2}, \quad (3)$$

where N is the total number of decays selected, w is the probability to incorrectly tag a decay as mixed or unmixed (i.e. the mistag rate) and σ_t is the proper time resolution. The proper time resolution depends on both the decay length resolution σ_L and the momentum resolution σ_p according to $\sigma_t^2 = (\sigma_L/\gamma\beta c)^2 + (t\,\sigma_p/p)^2$. The ability to resolve rapid B_s^0 oscillations thus requires excellent decay length and momentum resolution, and benefits from having a low mistag rate and a high B_s^0 purity.

Tagging of the production flavor is performed by combining several techniques. The most powerful technique exploits the large polarized forward-backward asymmetry of $Z^0 \to b\,\bar{b}$ decays (available at SLD only). In this case, a left- (right-) handed incident electron tags the forward hemisphere quark as a b ($\bar{b}$) quark. Other tags used by most experiments rely on charge information from the hemisphere opposite that of the B_s^0 decay candidate (i.e. the hemisphere expected to contain the other b hadron in the event): (i) charge of lepton from the direct transition $b \to l^-$, (ii) momentum-weighted jet charge, (iii) secondary vertex charge, and (iv) charge of kaon from the dominant decay transition $b \to c \to s$. Information from the same hemisphere is also used: (i) unweighted (or weighted) jet charge, and (ii) charge of fragmentation kaon. The different tags are combined to provide effective mistag rates of $\sim 25\%$ for LEP experiments and up to $\sim 15\%$ for SLD.

The various analyses differ mostly in the way B_s^0 decay candidates are reconstructed, which in turn affects the quality of the decay flavor tag and the B_s^0 purity. Analyses can be grouped in three main categories: inclusive, semi-exclusive, and fully exclusive. Inclusive methods have the advantage of large statistics but suffer from low purity, whereas more exclusive methods yield small sample sizes but benefit from a much increased sensitivity per event.

3.1 Inclusive Methods

Inclusive reconstruction of semileptonic decays has been investigated by ALEPH, DELPHI, OPAL, and SLD. The method typically relies on the selection of identified leptons (e or μ) with sufficiently large momentum transverse to the b jet (the minimum p_T is usually 1 GeV/c) in order to reduce the contribution from cascade decays ($b \to c \to l^+$). Direct leptons from $b \to l^-$ transitions contribute $\sim$ 90% of all selected leptons. As a result, the decay flavor tag is very clean. The charm decay vertex is reconstructed topologically and the resultant "D" track is intersected with the lepton trajectory to define the B decay point.

This method benefits from high statistics and a low mistag rate for the decay flavor tag but suffers from a low B_s^0 purity (typically 10-15%). The sensitivity of the method is enhanced by estimating the mistag rates, the B_s^0 purity and the proper time resolution event by event. For example, the most sensitive analysis by ALEPH selects 33023 events, with an estimated B_s^0 purity of 10.4% (close to the B_s^0 production fraction). The sample is divided into 11 subsamples with B_s^0 purity varying between 5% and 24%, depending upon the charm vertex track multiplicity, the charge and momentum of tracks in the vertex, as well as the presence of identified kaons in the vertex.

SLD has devised novel inclusive methods relying on the excellent tracking resolution provided by its CCD pixel vertex detector. In particular, the lepton+D vertex analysis achieves a decay length resolution $\sigma_L = 67\,\mu$m (60% fraction) and $\sigma_L = 273\,\mu$m (40%). The Charge Dipole analysis attempts to reconstruct the charged track topology of $B_s^0 \to D_s^- X$ decays by reconstructing both secondary ("B") and tertiary ("D") vertices. The charge difference between the B and D vertices $\delta q = Q_D - Q_B$ tags the decay flavor ($\delta q < 0$ for B_s^0 and $\delta q > 0$ for $\overline{B_s^0}$) with a mistag rate of 21%. This rate is considerably larger than that achieved with semileptonic analyses but it is compensated by the increase in statistics due to the fully inclusive selection.

3.2 Semi-Exclusive Methods

Semi-exclusive methods enhance the sensitivity to B_s^0 oscillations mostly by improving the B_s^0 purity and, to a lesser extent, the proper time resolution. This, however, comes at the cost of much lower efficiency. ALEPH, CDF, and DELPHI perform partial B_s^0 reconstruction in the modes $B_s^0 \to D_s^- l^+ \nu_l X$ and $B_s^0 \to D_s^- h^+ X$, where h represents any charged hadron and the D_s decay is either fully or partially reconstructed in the modes $D_s^- \to \phi\pi^-, K^{*0}K^-, K^0K^-, \phi\pi^-\pi^+\pi^-, \phi l^-\overline{\nu_l}$, etc.

The most sensitive single analysis performed by DELPHI selects 436 $D_s^- l^+$

events. Despite the low statistics the analysis is competitive due to its high $B_s^0 \to D_s^- l^+ \nu_l$ purity, estimated to be $\sim 53\%$, and its good decay length and momentum resolution, $\sigma_L = 200\,\mu$m (82% fraction) and 670 μm (16%), $\sigma_p/p = 0.07$ (82% fraction) and 0.16 (16%).

Analyses reconstructing $D_s^- h^+$ final states benefit from increased statistics but their overall sensitivity is somewhat reduced due to lower B_s^0 purity and worse resolution.

3.3 *Exclusive Methods*

DELPHI has performed an exploratory analysis in which the B_s^0 decays are fully reconstructed in the modes $B_s^0 \to D_s^- \pi^+, D_s^- a_1^+, \overline{D^0} K^- \pi^+$, and $\overline{D^0} K^- a_1^+$, where the D_s^- and $\overline{D^0}$ decays are fully reconstructed. The analysis selects 44 candidates with an estimated B_s^0 purity of approximately 50% and an excellent decay length resolution of $\sigma_L = 117\,\mu$m (58% fraction) and 216 μm (42%). The uncertainty in momentum is essentially negligible and thus the oscillation amplitude is not damped at large proper time. Despite a high sensitivity per event, the analysis is limited by the available statistics. Nevertheless, it is clearly the method of choice for future studies of B_s^0–$\overline{B_s^0}$ mixing at hadron colliders (see Sec. 4).

3.4 *World Average*

The fit for the B_s^0 oscillation frequency is performed using the amplitude method.[10] In this method, the unmixed (mixed) probability density is expressed as $P(B^0 \to B^0(\overline{B^0})) = \frac{\Gamma}{2} e^{-\Gamma t}(1 \pm A \cos \Delta m\, t)$. A fit is then performed to determine the oscillation amplitude "A" at a series of fixed frequencies. Amplitude values of $A = 0$ are expected for frequencies sufficiently different from the true oscillation frequency and a value of $A = 1$ is expected at the true frequency. The amplitude method is thus similar to a normalized Fourier transform.

The measured amplitude at $\Delta m_s = 15\,\mathrm{ps}^{-1}$ for the various analyses is shown in Fig. 3. Also shown is the sensitivity of each analysis to set a 95% C.L. lower limit on Δm_s. These analyses have been combined,[4] taking correlated systematic uncertainties into account and the resulting world average amplitude spectrum is shown in Fig. 4. Mixing (A=1) is excluded for $\Delta m_s < 14.3\,\mathrm{ps}^{-1}$ at 95% C.L., a limit close to the expected sensitivity of 14.5 ps^{-1} (obtained by setting the measured amplitude to zero). Beyond that limit, the uncertainties become too large (due to the limited proper time resolution) to discriminate between mixing and no mixing.

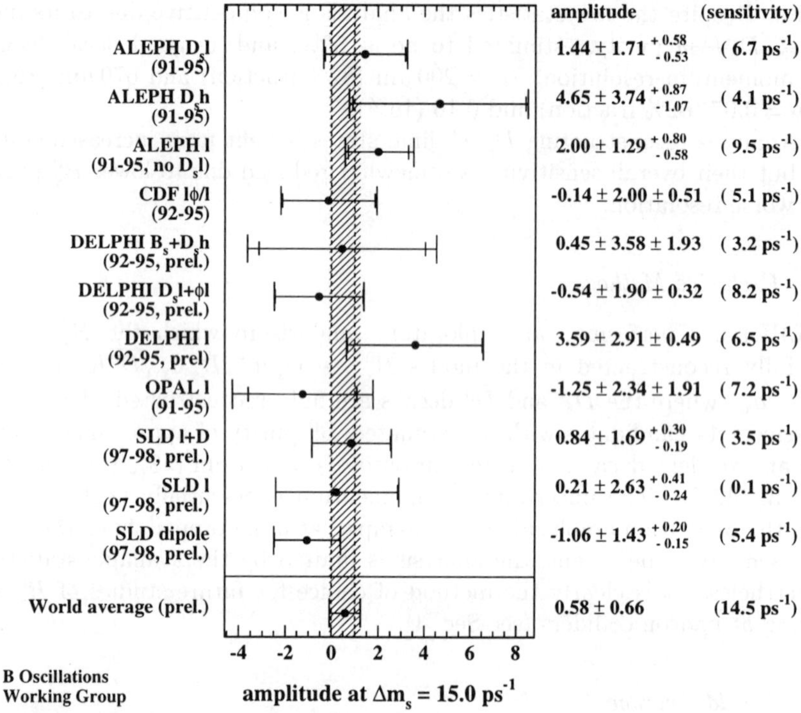

Figure 3. *Measurements of the B_s^0 oscillation amplitude at $\Delta m_s = 15\,ps^{-1}$.*

4 Future

In the near future, ALEPH, DELPHI, OPAL, and SLD expect to further improve their sensitivity by adding new analysis techniques and refining existing analyses. Beyond LEP and SLD, future experiments at hadron machines are expected to bring the study of B_s^0 oscillations to a new level. By exploiting the tremendous cross section for b hadrons at those machines and designing new trigger schemes aimed at identifying secondary vertices, HERA-B expects to observe a $3\,\sigma$ signal for Δm_s up to ~ 27 ps^{-1}, whereas CDF, BTeV, and LHC-b expect to observe a $5\,\sigma$ signal for Δm_s up to about 40 ps^{-1}, 40 ps^{-1}, and 48 ps^{-1}, respectively. If a signal is found, the statistical precision on Δm_s is predicted to be excellent (e.g., LHC-b expects to achieve a precision better than 0.1%).

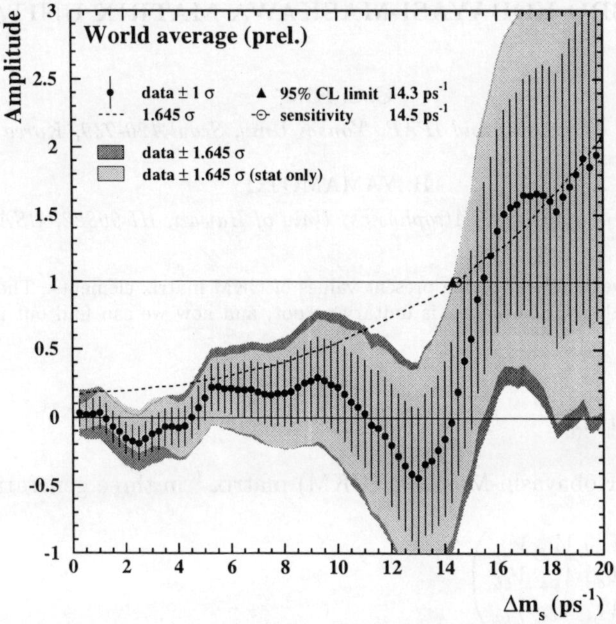

Figure 4. *Measured oscillation amplitude as a function of Δm_s.*

Acknowledgements

I wish to thank the following people for their help: G. Blaylock, R. Forty, R. Hawkings, M. Paulini, P. Roudeau, O. Schneider, and A .Stocchi.

References

1. A.J.Buras, W.Slominski, and H.Steger, *Nucl. Phys.* B **345**, 369 (1984).
2. T.Inami & C.S.Lim, *Prog. Theor. Phys.* **65**, 297 (1981).
3. T. Draper, *Nucl. Phys. Proc. Suppl.* **73**, 43 (1999).
4. B Oscillations Working Group, CERN-EP preprint, in preparation; see also `http://www.cern.ch/LEPBOSC/`.
5. C. Albajar *et al.*, *Phys. Lett.* B **186**, 247 (1987).
6. H. Albrecht *et al.*, *Phys. Lett.* B **192**, 245 (1987).
7. H.-G. Moser, CERN-PPE-93-164, September 1993.
8. D. Buskulic *et al.*, *Phys. Lett.* B **322**, 441 (1994).
9. R. Akers *et al.*, *Z. Phys.* C **66**, 555 (1995).
10. H.-G. Moser & A. Roussarie, *Nucl. Instrum. Methods* A **384**, 491 (1997).

IS CABIBBO-KOBAYASI-MASKAWA MATRIX UNITARY?

C. S. KIM*

Dept. of Physics and IPAP, Yonsei Univ, Seoul 120-749, Korea

H. YAMAMOTO

Dept. of Physics & Atrophysics, Univ of Hawaii, HI 96822, USA

First, we give summary of the present values of CKM matrix elements. Then, we discuss whether CKM matrix is unitary or not, and how we can find out if it is not unitary.

1 Introduction

The Cabibbo-Kobayashi-Maskawa (CKM) matrix [1] in three generation is

$$V_{\rm CKM} = \begin{pmatrix} V_{ud} & V_{us} & V_{ub} \\ V_{cd} & V_{cs} & V_{cb} \\ V_{td} & V_{ts} & V_{tb} \end{pmatrix} \tag{1}$$

$$= \begin{pmatrix} c_{12}c_{13} & s_{12}c_{13} & s_{13}e^{-i\delta_{13}} \\ -s_{12}c_{23} - c_{12}s_{23}s_{13}e^{i\delta_{13}} & c_{12}c_{23} - s_{12}s_{23}s_{13}e^{i\delta_{13}} & s_{23}c_{13} \\ s_{12}s_{23} - c_{12}c_{23}s_{13}e^{i\delta_{13}} & -c_{12}s_{23} - s_{12}c_{23}s_{13}e^{i\delta_{13}} & c_{23}c_{13} \end{pmatrix}.$$

Directly measured values [2] of the elements of 2×2 Cabibbo matrix, $V_{\rm C}$, are

$$V_{\rm C} = \begin{pmatrix} V_{ud} & V_{us} \\ V_{cd} & V_{cs} \end{pmatrix} = \begin{pmatrix} 0.9740 \pm 0.0010 & 0.2196 \pm 0.0023 \\ 0.224 \pm 0.016 & 1.04 \pm 0.16 \end{pmatrix}. \tag{2}$$

Only after assuming 3 generation unitarity, $V_{\rm C}$ becomes well known, and can be parametrized with one parameter, $\lambda = \sin\theta_c \approx 0.22$, within 90% CL

$$V_{\rm C} = \begin{pmatrix} 0.9745 \sim 0.0010 & 0.217 \sim 0.224 \\ 0.217 \sim 0.224 & 0.9737 \sim 0.9753 \end{pmatrix} \simeq \begin{pmatrix} 1 - \frac{\lambda^2}{2} & \lambda \\ -\lambda & 1 - \frac{\lambda^2}{2} \end{pmatrix}. \tag{3}$$

We note that the directly measured values (2) still have relatively large uncertainties of $\sim 5\%$. Extension to (unitarized) three generation by Kobayashi and

*KIM@KIMCS.YONSEIAC.KR, HTTP://PHYA.YONSEI.AC.KR/~CSKIM/, TALK GIVEN BY C. S. KIM.

Maskawa leads one non-trivial phase angle $\delta_{\rm KM}$. In Wolfenstein parametrization [3], it can be approximated as

$$V_{\rm KM} = \begin{pmatrix} & & V_{ub} \\ & & V_{cb} \\ V_{td} & V_{ts} & V_{tb} \end{pmatrix} \approx \begin{pmatrix} & & A\lambda^3(\rho - i\eta) \\ & & A\lambda^2 \\ A\lambda^3(1-\rho-i\eta) & -A\lambda^2 & 1 \end{pmatrix}, \tag{4}$$

where $\rho + i\eta \approx \sqrt{\rho^2+\eta^2}\exp(i\delta_{13})$, and we asumed $V_{tb} = 1$. In coming discussions, we first assume $V_{\rm CKM}$ being unitary, but later we will investigate possible non-unitarity of $V_{\rm CKM}$.

2 Theoretical Determination of Elements of $|V_{\rm CKM}|$

2.1 $|V_{cb}| = A\lambda^2$

From the exclusive $B \to D^* l\nu$ decay,

$$\frac{d\mathcal{BR}}{dw}(B \to D^* l\nu) = \tau_B \frac{d\Gamma}{dw} = \tau_B \times [....] \times \sqrt{w^2-1}|V_{cb}|^2\mathcal{F}^2(w), \tag{5}$$

where [4]

$$\mathcal{F}(w) = \mathcal{F}(1)(1 - \hat{\rho}^2(w-1) + ...), \qquad \mathcal{F}(1) = 0.924 \pm 0.027, \tag{6}$$

with $w \equiv v_B \cdot v_{D^*}$, LEP and CLEO [5] measured $\frac{d\mathcal{BR}}{dw}|_{w\to 1}, \hat{\rho}^2, \tau_B$ to obtain the value

$$|V_{cb}|(B \to D^* l\nu) = 0.0387 \pm 0.0031, \quad \text{(i.e. } A \sim 0.8).$$

Inclusive measurement of semileptonic total decay rate, $\Gamma_{s.l.}(B \to X_c l\nu)$, can also give [6] the value $|V_{qb}|$ from

$$\begin{aligned} \Gamma_{s.l.}(B \to X_q l\nu) &= \gamma_q^{\rm th} \times |V_{qb}|^2 \qquad (7) \\ &\equiv \left(\frac{G_F^2 m_b^5}{192\pi^3}\right)\left[z_0\left(1 - \frac{2\alpha_s}{3\pi}g(\epsilon_q)\right) + \frac{1}{2}z_0(G_b - K_b) - 2z_1 G_b + .\right]|V_{qb}|^2, \end{aligned}$$

where

$$\begin{aligned} g(\epsilon) &= (\pi^2 - \frac{31}{4})(1-\epsilon)^2 + \frac{3}{2}, \\ z_0(\epsilon) &= 1 - 8\epsilon^2 + 8\epsilon^6 - \epsilon^8 - 24\epsilon^4 \ln\epsilon, \quad z_1(\epsilon) = (1-\epsilon^2)^4, \\ K_b &= -\lambda_1/m_b^2, \quad G_b = 3\lambda_2/m_b^2, \quad \alpha_s = \alpha_s(m_b^2), \end{aligned}$$

with $\epsilon_q = \frac{m_q}{m_b}$, $-\lambda_1 = \mu_\pi^2 = (0.1 \sim 0.6)$ GeV^2, $\lambda_2 = \frac{1}{4}(m_{B^*}^2 - m_B^2) = 0.12$ GeV^2. Now from the total semileptonic decay width, we can measure

$$|V_{cb}| = \left[\frac{\Gamma_{s.l.}(B \to X_c l\nu)_{\rm exp}}{\Gamma_{tot}(B)_{\rm exp}\tau_B^{\rm exp}\gamma_c^{\rm th}}\right]^{1/2} = 0.0419\sqrt{\frac{\mathcal{BR}(B \to X_c l\nu)}{0.105}}\sqrt{\frac{1.55}{\tau_B}}(1 \pm 0.04)\,.$$

We remark that the uncertainty in determination of $\lambda_1 = -\mu_\pi^2$, the average kinetic energy of b-quark in B meson, is still large.

2.2 $|V_{ub}| = A\lambda^3\sqrt{\rho^2+\eta^2}$

This is probably the most important element, and at the same time one of the most difficult to be measured. Exclusively, we use the semileptonic decays, $B \to \rho l\nu, \to \pi l\nu, \ldots$, which invoke large theoretical uncertainties from hadronic form factors, and their model dependences. Recently by using data of large q^2 $(14 < q^2 < 21\ \mathrm{GeV}^2)$ and large $E_l > 2.3$ GeV in $B \to \rho l\nu$, CLEO derived [7]

$$|V_{ub}| = (3.23 \pm 0.24 \pm 0.25 \pm 0.58) \times 10^{-3},$$

where the last error is from model dependence.

As we can easily see from Eq. (7), if we also measure the total decay width of $\Gamma_{s.l.}(B \to X_u l\nu)$, then we can extract [8] $|V_{ub}|$ from

$$\left|\frac{V_{ub}}{V_{cb}}\right| \simeq (0.81 \pm 0.06) \times \left[\frac{\mathcal{BR}(B \to X_u l\nu)}{\mathcal{BR}(B \to X_c l\nu)}\right]^{1/2},$$

where the error is from m_b, α_s, μ_π^2. However, the separation of $B \to X_u$ from the dominant $B \to X_c$ is experimentally difficult. The promising method is to use M_X, hadronic invariant recoiled mass. This is because [9]

$$m_u \ll m_c \quad\Longrightarrow\quad m_{\pi,\rho} \ll m_{D,D^*} \quad\Longrightarrow\quad m_{X_u} \ll m_{X_c}.$$

In Fig. 1 we show the double differential distribution [10], $\frac{d\Gamma}{dm_X dE_l}$.

2.3 $|V_{ts}|$ *and* $|V_{td}|$

$B_d - \overline{B}_d$ and $B_s - \overline{B}_s$ mixings can give values of $|V_{td}|$ and $|V_{ts}|$ from

$$\begin{aligned}\Delta m_q &\equiv m(B_q^H) - m(B_q^L)\\ &= \left(\frac{G_F^2}{6\pi^2}m_W^2 m_{B_q}\right)[f_{B_q}^2 B_{B_q}][\eta_{\mathrm{QCD}}\mathcal{F}_{\mathrm{sd}}^q]|V_{tq}^* V_{tb}|^2, \end{aligned} \tag{8}$$

where $f_{B_d} \approx (170 \sim 180) \pm 30$ MeV, $f_{B_s} \approx (195 \sim 205) \pm 30$ MeV, $\frac{f_{B_s}}{f_{B_d}} = 1.15 \pm 0.05$, $B_{B_d} = 0.75 \pm 0.15 \approx B_{B_s}$ and $f_{B_s}\sqrt{B_{B_s}}/f_{B_d}\sqrt{B_{B_d}} = 1.14 \pm 0.06$. The recent experimental values [11] are

$$\Delta m(B_d) = 0.471 \pm 0.016\ \mathrm{ps}^{-1}, \quad \Delta m(B_s) > 14.3\ \mathrm{ps}^{-1}\ (95\%\ \mathrm{CL}). \tag{9}$$

If we assume the Standard Model short distance (sd) interaction, $\mathcal{F}_{\mathrm{sd}}^d = \mathcal{F}_{\mathrm{sd}}^s$, then:

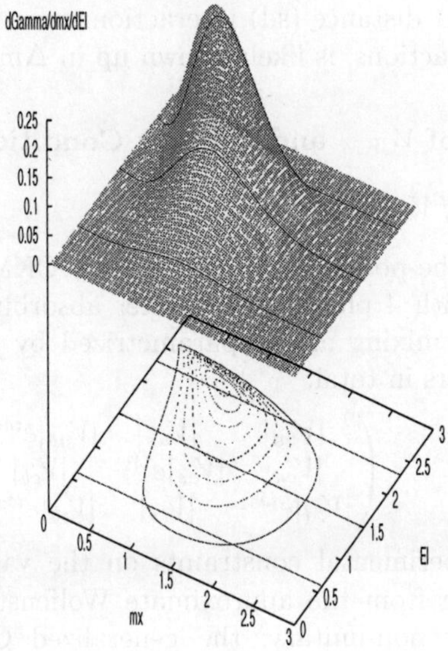

Figure 1. 3-dimension plot of $\frac{d\Gamma}{dm_X dE_l}$ in $B \to X_u l\nu$.

- $\Delta m(B_d)_{\text{exp}}$ gives $|V_{td}| \approx (8.1 \pm 1.8) \times 10^{-3}$.
- Asumming $|V_{ts}| = |V_{cb}|$ and $|V_{td}| = (0.004 \sim 0.012)$, we get
$$\left(\frac{\Delta m(B_d)}{\Delta m(B_s)}\right)_{\text{SM}} = 0.056 \pm 0.08,$$
which is quite larger than the present experimental ratio, < 0.0341 (95% CL).
- Assuming the Standard Model is correct, then the present data (9) gives $|V_{td}/V_{ts}| < 0.217$ (95% CL).

As is well known, the mass difference, Δm_q, can be easily poluted by new physics, and so we can approach differently to find new physics, instead of determining the Standard Model parameters, like V_{ts}, V_{td}. Any new physics, that has different short distance (sd) interaction structure, or has spectator-quark depending interactions, is likely shown up in Δm_q.

3 Generalization of $V_{\rm CKM}$ and Unitary Conditions

3.1 Generalization of $V_{\rm CKM}$

If we are considering the possibility of non-unitary CKM matrix, Eq. (1) can be generalized now with 4 phase angles (after absorbing 5 phases to quark fields), and nine real mixing angles parametrized by $|V_{ij}|$, resulting to 13 independent parameters in total:

$$V_{\rm CKM}^{\rm general} = \begin{pmatrix} |V_{ud}| & |V_{us}| & |V_{ub}|e^{i\delta_{13}} \\ |V_{cd}| & |V_{cs}|e^{i\delta_{22}} & |V_{cb}| \\ |V_{td}|e^{i\delta_{31}} & |V_{ts}| & |V_{tb}|e^{i\delta_{33}} \end{pmatrix}. \tag{10}$$

Based on present experimental constraints on the values of CKM matrix elements, and starting from the approximate Wolfenstein parameterization (4), which is already non-unitary, the generalized CKM matrix can be parametrized as (V_{tb} is assumed as unity)

$$\begin{aligned} V_{\rm CKM}^{\rm general} &= \begin{pmatrix} 1-\frac{\lambda_1^2}{2} & \lambda_2 & A_{13}\lambda_3^3(\rho_1 - i\eta_1) \\ -\lambda_1 & 1-\frac{\lambda_2^2}{2}e^{ia} & A_{23}\lambda_3^2 \\ A_{31}\lambda_1^3(1-\rho_2-i\eta_2) & A_{32}\lambda_2^2 & A_{33}e^{ib} \end{pmatrix} \\ &\approx \begin{pmatrix} 1-\frac{\lambda^2}{2} & \lambda & A_1\lambda^3(\rho_1 - i\eta_1) \\ -\lambda & 1-\frac{\lambda^2}{2} & A_1\lambda^2 \\ A_1\lambda^3(1-\rho_2-i\eta_2) & A_2\lambda^2 & 1 \end{pmatrix}. \end{aligned} \tag{11}$$

We note that

- By assuming that $V_{tb} = 1$ and V_{cs} is real as in the case of Wolfenstein parametrization, in Eq. (11) we now have only 5 real parameters, $\lambda, A_1, A_2, \rho_1, \rho_2$ and 2 phase parameters, η_1, η_2, instead of full 9 real parameters. (Present experimental results (2) show even the possibilty of $\lambda_1 \neq \lambda_2$.)
- $|V_{cb}|$ and $|V_{ts}|$ can be different because of $A_1 \neq A_2$. This will produce consequently many new results in analysis of CKM constraints, such as $B_s - \overline{B}_s$ mixing, $B \to X_s + \gamma$ decay, $b \to s$ penguin, etc.
- To check the unitarity of CKM matrix, we have to see whether the unitary relations of the generalized CKM matrix (10,11), instead of already-unitarized parametrization of CKM matrix (1,4), are satisfied. This is

because if we use (1,4) almost all the experimental observables are adjusted to the (assumed) unitary conditions.

- If there exists a massive singlet down quark b', then the columns of $V_{\rm CKM}^{\rm general}$ remain to be unitary, but the rows do not.

Usual unitary relation for the matrix (11) becomes

$$V_{k1}V_{k3}^* = V_{11}V_{13}^* + V_{21}V_{23}^* + V_{31}V_{33}^* = 0\ ? \qquad (12)$$

$$0 \simeq V_{13}^* - A_1\lambda^3 + V_{31},$$

$$0 \approx (A_1\lambda^3) \times \{(\rho_1 + i\eta_1) - 1 + (1 - \rho_2 - i\eta_2)\}, \qquad (13)$$

and now we have only 5 unknowns,

$$A_1\lambda^3,\ \ \rho_1,\ \ \eta_1,\ \ \rho_2\ \ \text{and}\ \ \eta_2,$$

or equivalently

$$\lambda|V_{cb}|,\ \ |V_{ub}|,\ \ \sin\gamma,\ \ |V_{td}|\ \ \text{and}\ \ \sin\beta.$$

Those 5 unknowns are exactly the same as the 5 sufficient conditions for the drawing a triangle uniquely – three sides and two angles. As is well known, the usual 3 conditions, two sides and one angle, which are $A_1\lambda^3$, ρ, η (or $\lambda|V_{cb}|$, $|V_{ub}|$, $\sin\gamma$ or equivalently $A_1\lambda^3$, $|V_{td}|$, $\sin\beta$) in unitarized CKM matrix, are only the necessary conditions for drawing a triangle. In order to see that Eq. (13) equals zero, we have to measure those 5 independent observables.

3.2 (Minimum and Complete) Unitary Conditions

As is explained, we have to measure precisely 5 observables to check if just (most popular) one of unitary triangles is really triangle. Eq. (13) can be written in two equations by using the usual sine and cosine rules,

$$\text{(A)}\quad \frac{|V_{ub}|}{\sin\beta} = \frac{|V_{td}|}{\sin\gamma} \quad \text{or} \quad \frac{|V_{td}|}{|V_{ub}|} = \frac{\sin\gamma}{\sin\beta}, \qquad (14)$$

$$\text{(B)}\quad \frac{\lambda|V_{cb}|}{|V_{ub}|} = \cos\gamma + \frac{|V_{td}|}{|V_{ub}|}\cos\beta.$$

Here we have, instead of one sine (or cosine) rule, two equations because the third angle α is *not* independent due to the relation $\alpha+\beta+\gamma=\pi$. We remark here a few comments on Eq. (14):

- The angles β, γ need to be measured independently without assuming unitarity of CKM matrix. Recent direct measurement of β at CDF still relies on the presumed unitary assumption of $|V_{cb}| = |V_{ts}|$ in the analysis of $B \to J/\psi K_s$ decay. If we ignore the small unitary violation in $|V_{cb}| \neq |V_{ts}|$, CDF's direct result [12] can be one solid ingredient for Eq. (14) as

$$\beta_{\rm CDF} = 0.79 \pm 0.43.$$

- Future measurements on $\frac{|V_{ub}|}{|V_{cb}|}$ at Babar and Belle, as explained in section 2, would be one of the most important ingredient in the test of Eq. (14B).

- For the ratio $\frac{|V_{td}|}{|V_{ub}|}$, we may use the relation [13],

$$\frac{\mathcal{BR}(B^\pm \to \rho^\pm \nu\bar{\nu})}{\mathcal{BR}(B^\pm \to \rho^0 e^\pm \nu)} = 6\left(\frac{\alpha^2}{4\pi^2}\right)|C_{10}^\nu|^2 \times \frac{|V_{td}|^2}{|V_{ub}|^2}, \tag{15}$$

with

$$C_{10}^\nu \equiv \frac{X(m_t^2/m_W^2)}{\pi \sin^2\theta_W}, \quad \text{where} \;\; X(x_t) \text{ is Inami Lim function.}$$

In Eq. (15), the constant 6 comes from 3 neutrino species and from isospin relation in form factors of $B \to \rho^\pm$, $B \to \rho^0$. Because of complete cancelation of the hadronic form factors in the ratio of branching fractions of those two decays, there is not any theoretical uncertainties in Eq. (15), though it would be an experimental challenge to measure the small branching fraction, $\mathcal{BR}(B \to \rho\nu\nu) \sim 4 \times 10^{-7}$.

Measuring angle γ would be very difficult, if we donot assume any unitarity in the analyses. However, if we assume the unitarity priorly, then we can calculate it from the relation (14A), and then compare the value with the independently measured values from, as an example, Neubert-Rosner bound [14] to check if the unitarity holds.

3.3 *Comments on the Discrepancy in recently extracted Values of* γ

As is well known, the discrepancy in extracted values of γ from CKM-fitting [15] at $\rho - \eta$ plan and from the χ^2 analysis [16] of non-leptonic B decays has aroused hot debates over what is going on underneath of the unitary triangle. The value of γ has been obtained as

$$\begin{aligned}\gamma &= 60^0 \sim 80^0 \quad \text{(from unitary triangle fitting)},\\ &= 90^0 \sim 140^0 \quad \text{(from nonleptonic analysis)}.\end{aligned}$$

In both analyses, the unitary conditions have been assumed and extensively. If we believe both analyses are correctly performed with correct theoretical assumptions, one of the most plausible answer would be "non-unitarity of CKM matrix". An easy answer from our previous argument, explained in the second item after Eq. (11), is that $|V_{cb}| \neq |V_{ts}|$. If those two elements are not equal, then:

- We can not simply add the constraint from $B_s - \overline{B}_s$ mixing result on $\rho - \eta$ plane to get the allowed smaller circular region of $|V_{td}|$.
- Adding of two circular regions (from measurements $|V_{ub}|$, $|V_{td}|$) does not reduce to the overlapped small region. Instead, we will have the summed large region in the unitary plane. Therefore, without the direct measurement on γ we cannot decide which value is correct.

Very soon, we will have a flood of experimental results on CKM elements from many experiments, asymmetric and symmetric e^+e^- colliders, and hadronic machines. Without having correct (theoretical) stratage of coping the data, we will be easily fooled. We argue the present discrepancy on γ is just an early example.

Acknowledgments

CSK was supported in part by Grant No. 1999-2-111-002-5 from the Interdisciplinary research program of the KOSEF, in part by the BSRI Program of MOE, Project No. 99-015-DI0032, and in part by the KRF Sughak-research program, Project No. 1997-011-D00015.

References

1. N. Cabibbo, *Phys. Rev. Lett.* **10**, 531 (1963); M. Kobayashi, T. Maskawa, *Prog. Theo. Phys.* **49**, 652 (1973).
2. Particle Date Group, *Eur. Phys. J.* C **3**, 1 (1998).
3. L. Wofenstein, *Phys. Rev. Lett.* **51**, 1945 (1983).
4. M. Neubert, hep-ph/9801269.
5. P. Drell, hep-ex/9711020.
6. M. Luke *et al., Phys. Lett.* B **321**, 88 (1994); M. Shifman *et al., Phys. Rev.* D **51**, 2217 (1995); P. Ball *et al., Phys. Rev.* D **52**, 3929 (1995).
7. CLEO Collab., *Phys. Rev.* D **61**, 052001 (2000).
8. N. Uraltzev, *Int. J. Mod. Phys.* A **11**, 515 (1996); C.S. Kim, *Phys. Lett.* B **414**, 347 (1997).

9. V. Barger, C.S. Kim, R.J.N. Phillips, *Phys. Lett.* B **251**, 629 (1990).
10. K.K. Jeong, C.S. Kim, Y.G. Kim, work in progress.
11. $|V_{ts}|, |V_{td}|$ from Lepton-Photon Symposium (1999).
12. CDF Collab., hep-ex/9909003.
13. T. Aliev, C.S. Kim, *Phys. Rev.* D **58**, 013003 (1998).
14. M. Neubert, J.R. Rosner, *Phys. Rev. Lett.* **81**, 5076 (1998).
15. A. Stocchi, talk given in this Conference.
16. N.G. Deshpande, X. He, W. Hou, S. Pakvasa, *Phys. Rev. Lett.* **82**, 2240 (1999).

Determination of the CKM unitarity triangle parameters by end 1999

F. Caravaglios
Laboratoire de Physique Théorique, LPT-Orsay, Université de Paris-Sud

F. Parodi
Dipartimento di Fisica, Universitá di Genova and INFN

P. Roudeau and A. Stocchi
Laboratoire de l'Accélérateur Linéaire, IN2P3-CNRS et Université de Paris-Sud

Within the Standard Model, a review of the current determination of the CKM unitarity triangle parameters is presented, using experimental constraints from the measurements of $|\epsilon_K|$, $\left|\frac{V_{ub}}{V_{cb}}\right|$, Δm_d and from the limit on Δm_s, available by end 1999.

1 Introduction.

This paper contains an update of the determination of the CKM parameters, mainly $\overline{\rho}$ and $\overline{\eta}$, of the Wolfenstein parametrization and of the angles of the unitarity triangle, using experimental results and theoretical estimates available at the end of 1999. Previous analyses using the same approach can be found in [1 2]. Other determinations of these parameters can be found in [3]. As compared to end 1998, new experimental results concern mainly first averages of LEP measurements of $|V_{cb}|$ and $|V_{ub}|$ [4] and an improved limit on Δm_s [5].

2 Determination of the CKM unitarity triangle parameters in the Standard Model framework.

Results have been obtained using the constraints from the measurements of $|\epsilon_K|$, $\left|\frac{V_{ub}}{V_{cb}}\right|$, Δm_d and Δm_s. The central values and the uncertainties for the relevant parameters used in this analysis are given in Table 1. As compared with the similar analysis done in [1], the following changes have been made:

i) the uncertainty on $|V_{cb}|$[4] has been increased from 1.5×10^{-3} to 1.9×10^{-3}. This uncertainty is entirely dominated by theoretical errors and the current belief is that these errors amount to 5%[6]. The quoted central value for $|V_{cb}|$, in Table 1, corresponds to the present LEP average of $|V_{cb}|$ measurements from inclusive and exclusive B semileptonic decays. Correlations between theoretical

Table 1: Values of the most relevant quantities entering into the expressions of $|\epsilon_K|$, $\left|\frac{V_{ub}}{V_{cb}}\right|$, Δm_d and Δm_s. In the third column the Gaussian and the flat part of the errors are given explicitly.

Parameter	Values	Gaussian − Flat error	Ref.
λ	0.2205 ± 0.0018	$\pm 0.0018 - \pm 0.000$	1
$\|V_{cb}\|$	$(40.5 \pm 1.9) \times 10^{-3}$	$(\pm 1.9 - \pm 0.0) \times 10^{-3}$	4
$\frac{\|V_{ub}\|}{\|V_{cb}\|}$ (CLEO)	0.085 ± 0.018	$\pm 0.009 - \pm 0.016$	7
$\frac{\|V_{ub}\|}{\|V_{cb}\|}$ (LEP)	0.104 ± 0.019	$\pm 0.011 - \pm 0.015$	4
Δm_d	$(0.476 \pm 0.016)\ ps^{-1}$	$\pm 0.016 - \pm 0.000$	5
Δm_s	> 14.3 ps^{-1} at 95% $C.L.$	sensit. at 14.5 ps^{-1}	5
$\overline{m_t}(m_t)$	$(167 \pm 5)\ GeV/c^2$	$\pm 5 - \pm 0$	8
B_K	0.86 ± 0.10	$\pm 0.06 - \pm 0.08$	1
$f_{B_d}\sqrt{B_{B_d}}$	$(210 \pm 42)\ MeV$	$\pm 29 - \pm 31$	1
$\xi = \frac{f_{B_s}\sqrt{B_{B_s}}}{f_{B_d}\sqrt{B_{B_d}}}$	$1.11^{+0.06}_{-0.04}$	$\pm 0.02 - ^{+0.06}_{-0.04}$	1

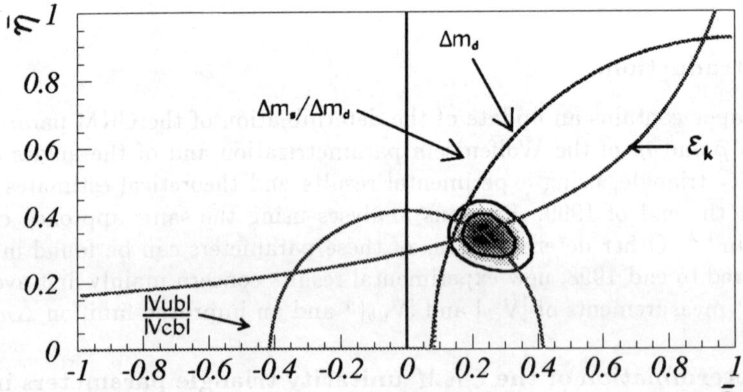

Figure 1: *The allowed region for $\bar{\rho}$ and $\bar{\eta}$ using the parameters listed in Table 1. The contours at 68% and 95% are shown. The full lines correspond to the central values of the constraints given by the measurements of $\frac{|V_{ub}|}{|V_{cb}|}$, $|\epsilon_K|$ and Δm_d. The dotted curve (which is hidden behind the curve of Δm_d) corresponds to the 95% C.L. upper limit obtained from the experimental limit on Δm_s.*

errors explain why the quoted uncertainty happens to be slightly below 5%.
ii) the values quoted for the ratio $\left|\frac{V_{ub}}{V_{cb}}\right|$ have been obtained using the value of $|V_{cb}|$ mentioned before and the values of $|V_{ub}|$ measured at CLEO with exclusive decays and at LEP from inclusive analyses.
iii) The limit on Δm_s, $\Delta m_s > 14.3\ ps^{-1}$ at 95% C.L.[5], is higher than the one quoted at the EPS-HP99 Conference after the addition of recent measurements from the SLD Collaboration. The sensitivity[5] of present measurements is equal to 14.5 ps^{-1}.

2.1 Values of $\overline{\rho}$, $\overline{\eta}$ and of the angles of the unitarity triangle

The region in the ($\overline{\rho}$, $\overline{\eta}$) plane selected by the measurements of $|\epsilon_K|$, $\left|\frac{V_{ub}}{V_{cb}}\right|$, Δm_d and from the limit on Δm_s, is given in Figure 1. The measured values of the two parameters are:

$$\overline{\rho} = 0.240^{+0.057}_{-0.047},\ \overline{\eta} = 0.335 \pm 0.042 \tag{1}$$

Fitted values for the angles of the unitarity triangle have been also obtained:

$$\sin(2\beta) = 0.750^{+0.058}_{-0.064},\ \sin(2\alpha) = -0.38^{+0.24}_{-0.28} \text{ and } \gamma = (55.5^{+6.0}_{-8.5})^{\circ}. \tag{2}$$

The accuracy on $\sin 2\beta$ obtained with the present analysis, is better than the one expected from B factories but it is valid only in the Standard Model framework. First measurements of $sin(2\beta)$ are now available. The world average is: $sin(2\beta) = 0.91 \pm 0.35$[9]. The 68% C.L. region is shown in Figure 2 and the constraints on $\overline{\rho}$ and $\overline{\eta}$ from this measurement are shown in Figure 3.
The angle γ is known with a 15% relative error. Values of $\gamma > 90^{\circ}$ (or $\overline{\rho} < 0$) are excluded at 99.6% C.L.. The origin of asymmetric errors on $\overline{\rho}$, γ and $\sin 2\beta$ is that the information brought by the limit on Δm_s is quite efficient in constraining values of Δm_s up to 15 ps^{-1} but no information can be obtained for values above 20 ps^{-1}.
The values of the non-perturbative QCD parameters can be also obtained:

$$f_B\sqrt{B_B} = (232 \pm 13) MeV \quad ; \quad B_K = 0.80^{+0.15}_{-0.17} \tag{3}$$

after having removed, in turn, these constraints from the analysis. It can be noticed that $f_B\sqrt{B_B}$ is better determined than the present evaluation of this parameter from lattice QCD calculations (Table 1). As a consequence, it is important to observe, contrarily to common belief, that a large uncertainty attached to $f_B\sqrt{B_B}$ has no real impact on the present analysis. Evaluation of this parameter with 5-10% relative error is needed to bring additional information. On the contrary, the parameter B_K is determined with a 20% relative

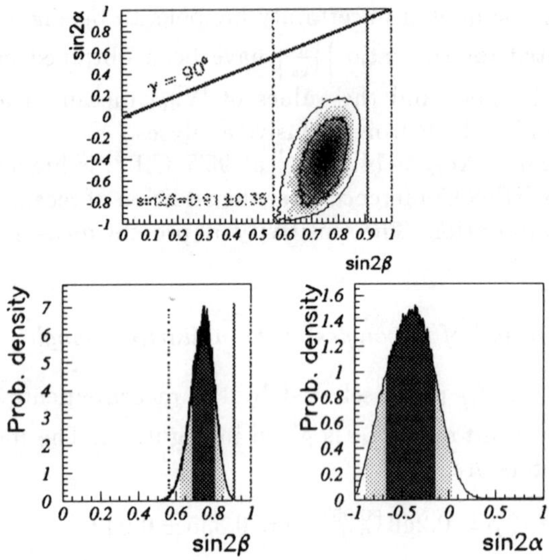

Figure 2: *The sin2α and sin2β distributions have been obtained using the constraints corresponding to the values of the parameters listed in Table 1. The dark-shaded and the clear-shaded intervals correspond, respectively, to 68% and 95% confidence level regions. The lines corresponding to the* $\pm 1\sigma$ *errors from the measurement of sin 2β using* $J/\Psi, K^0_s$ *events are also shown. The line obtained for* $\gamma = 90°$ *is also drawn (sin2β = sin2α).*

error using the other constraints. The present estimate of this parameter from lattice QCD calculations, given in Table 1, has thus an impact on the present analysis.
A region of the $(\overline{\rho}, \overline{\eta})$ plane can be selected without using the $|\epsilon_K|$ constraint [10]. The result is shown in Figure 3. This test shows that the $(\overline{\rho}, \overline{\eta})$ region selected by the measurements in the B sector (of the two sides of the unitarity triangle) is very well compatible with the region selected from the measurement of CP violation in the kaon sector. Using B decay and oscillations properties only, the values of $\overline{\eta}$=0.325± 0.054 and $sin2\beta = 0.747^{+0.067}_{-0.084}$ are determined.

3 Often asked questions

We would like to dedicate this section to two often asked questions.
Question 1: The results from this analysis are precise since, both from the constraints and for the parameters, Gaussian distributions have been taken for the experimental and the theoretical errors.
The answer is that we do not use only Gaussian distributions ! Values for the different parameters entering into the equations of constraints are extracted us-

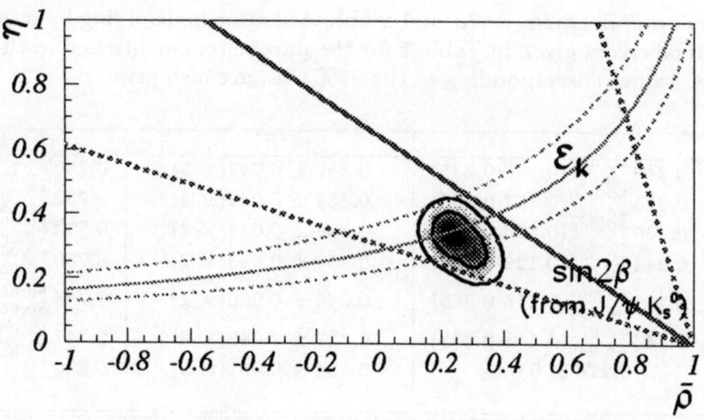

Figure 3: *The allowed region for $\overline{\rho}$ and $\overline{\eta}$ using the parameters listed in Table 1. The contours at 68% and 95% are shown. The constraint from $|\epsilon_K|$ has not been used and the dotted lines correspond to 68% C.L.. The lines corresponding to the $\pm 1\sigma$ errors from the measurement of sin 2β using $J/\Psi, K^0_s$ events are also shown ($\sin 2\beta = 0.91 \pm 0.35$)[9].*

ing random generations from Gaussian/non Gaussian distributions depending on the source of the error (Table 1). This has been shown at several seminars and conferences and explained in detail in [1]. For any parameter or constraint, entering into the fitting procedure, a detailed study has been done [1 2].
Question 2: How the results are affected if theoretical errors are multiplied by a factor 2 ?

We believe that the aim of this work is to try to use at best the available measurements and theoretical estimates. Nevertheless it is a simple basic exercise to quantify the dependence of the uncertainties on the final results from the variation of the errors attached to the different quantities used in the analysis. This was already done in [2 1]. The exercise presented here consists in multiplying by a factor two all the ranges (flat distributions) used for the theoretical estimates given in Table 1 and taking for V_{cb} an error of $\pm 3.0\ 10^{-3}$. Multiplying theoretical errors by a factor two seems to be rather extreme. In some cases (as for V_{cb} or some lattice QCD parameters) this is equivalent to ignore more than 10 years of theoretical efforts. Just two examples. For the extraction of V_{cb} using exclusive decays $\overline{\mathrm{B}} \rightarrow \mathrm{D}^* \ell^- \overline{\nu_\ell}$, the conservative value of $F_{D^*}(w=1) = 0.90 \pm 0.05$ has been already taken (in [11] the suggested value is 0.913 ± 0.042). In the following exercise the error of $\pm$ 0.10 is used ! In this exercise the theoretical error on $f_B\sqrt{B_B}$, coming from the evaluation of the quenched approximation, is taken to be $\pm$ 62 MeV. This type of errors circulated within the Lattice QCD community in the late '80. Table 2 summarizes

Table 2: Errors on $\overline{\rho}$, $\overline{\eta}$, sin2β, sin2α and γ obtained after multiplying by a factor 2 the flat part of theoretical errors given in Table 1 for the parameters mentioned in the first column. Inside brackets, values corresponding to the 95% C.L. are also given.

Parameter	$\overline{\rho}$	$\overline{\eta}$	sin2β
Standard	$0.240^{+0.057}_{-0.047}$(0.150-0.335)	0.335 ± 0.042(× 2)	$0.750^{+0.058}_{-0.064}$(0.618-0.858)
$f_B\sqrt{B_B}$	$0.245^{+0.060}_{-0.045}$(0.157-0.364)	0.334 ± 0.042(× 2)	$0.752^{+0.058}_{-0.064}$(0.621-0.860)
B_K	$0.240^{+0.058}_{-0.047}$(0.153-0.356)	0.334 ± 0.044(× 2)	$0.753^{+0.058}_{-0.064}$(0.620-0.860)
ξ	$0.244^{+0.060}_{-0.059}$(0.129-0.356)	0.338 ± 0.043(× 2)	$0.750^{+0.056}_{-0.067}$(0.622-0.857)
$\frac{\|V_{ub}\|}{\|V_{cb}\|}$	$0.251^{+0.067}_{-0.056}$(0.115-0.386)	0.345 ± 0.050(× 2)	$0.784^{+0.077}_{-0.089}$(0.591-0.916)
$\|V_{cb}\|$	$0.243^{+0.066}_{-0.049}$(0.158-0.373)	0.322 ± 0.049(× 2)	$0.747^{+0.059}_{-0.078}$(0.582-0.860)
All	0.278 ± 0.079	0.334 ± 0.061(× 2)	$0.800^{+0.073}_{-0.120}$(0.575-0.919)

Parameter	sin2α	γ	γ <90°
Standard	$-0.38^{+0.24}_{-0.28}$(-0.89-0.06)	$(55.5^{+6.0}_{-8.5})^\circ(38.4\text{-}67.3)^\circ$	99.6% C.L.
$f_B\sqrt{B_B}$	$-0.41^{+0.24}_{-0.30}$(-0.91-0.04)	$(55.1^{+5.7}_{-9.5})^\circ(37.3\text{-}66.4)^\circ$	99.6% C.L.
B_K	$-0.37^{+0.25}_{-0.29}$(-0.91-0.07)	$(55.8^{+6.1}_{-9.3})^\circ(37.8\text{-}67.1)^\circ$	99.6% C.L.
ξ	-0.41±0.30 (× 2)	(54.5± 8.3)°(×2)	99.6% C.L.
$\frac{\|V_{ub}\|}{\|V_{cb}\|}$	$-0.37^{+0.25}_{-0.29}$(-0.91-0.09)	$(55.0^{+6.0}_{-8.8})^\circ(38.2\text{-}66.7)^\circ$	99.5% C.L.
$\|V_{cb}\|$	$-0.45^{+0.27}_{-0.35}$(-1.0-0.0)	$(54.3^{+6.9}_{-10.7})^\circ(33.6\text{-}67.1)^\circ$	99.5% C.L.
All	$-0.54^{+0.35}_{-0.33}$(-1.0-0.14)	(50.7± 10.3)°(× 2)	99.4% C.L.

the results of this exercise. As a general conclusion, if theoretical errors on all the parameters given in Table 1 are simultaneously multiplied by a factor 2, the errors on the values of $\overline{\rho}$, $\overline{\eta}$, sin2β, sin2α and γ increase by only a factor 1.5.

4 Conclusions

Impressive improvements have been accomplished during the last ten years in B physics. Figure 4 illustrates the progress on the measurements of the two sides of the Unitarity Triangle. Some conclusions can be drawn. The selected region in the ($\overline{\rho}, \overline{\eta}$) plane, using B physics only, is very well compatible with the measurement of CP violation in the Kaon system. sin2β is measured with an accuracy better than 10% within the SM framework. The angle γ is smaller than 90° at 99.6% C.L.. This result is very slightly affected by multypling theoretical errors by a factor two and is essentially due to the impressive improvement on the limit on Δm_s obtained during the last 4 years. The situation will still improve by summer 2000. Thanks to the achieved accuracy, future measurements of CP violation in the B sector would further test the consistency of the Standard Model by determining directly the angles of the Unitarity Triangle.

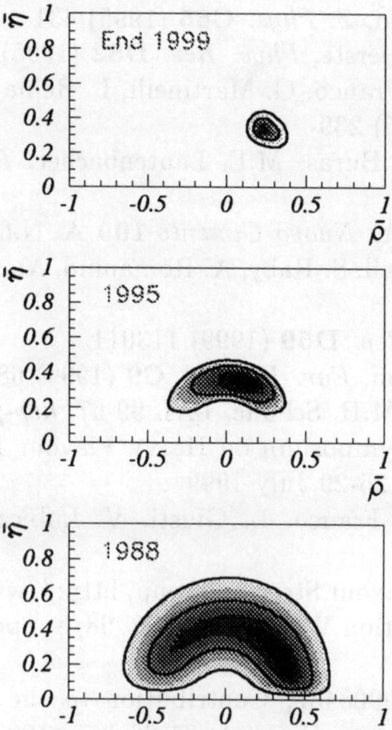

Figure 4: ***The allowed region for $\bar{\rho}$ and $\bar{\eta}$ from 1988 to the end of 1999***

Acknowledgements

Many thanks to the organizers for arranging such a nice Conference in a very stimulating atmosphere. The speaker has benefited from useful discussions with R. Forty, T. Nakada, R. Peccei, S. Willocq and M. Zito; they are warmly thanked.

References

1. F. Parodi, P. Roudeau and A. Stocchi, *Il Nuovo Cimento* **112 A**, N.8. (1999) 833
2. P. Paganini, F. Parodi, P. Roudeau and A. Stocchi, *Phys. Scripta* **58** (1998) 556

3. a) A. Ali, D. London, *Z. Phys.* **C65** (1995) 431
b) S. Herrlich, U. Nierste, *Phys. Rev.* **D52** (1995) 6505
c) M. Ciuchini, E. Franco, G. Martinelli, L. Reina, L. Silvestrini, *Z. Phys.* **C68** (1995) 239
d) G. Buchalla, A. Buras, M.E. Lautenbacher, *Rev. Mod. Phys.* **68** (1996) 1125
e) A. Ali, D. London, *Nuovo Cimento* **109 A**, N.6-7. (1996) 957
f) R. Barbieri, L. Hall, S. Raby, A. Romanino, *Nucl. Phys.* **B493** (1997) 3
g) S. Mele, *Phys. Rev.* **D59** (1999) 113011
h) A. Ali, D. London, *Eur. Phys. J.* **C9** (1999) 687
i) S. Plaszczynski, M.H. Schune, LAL 99-67, *hep-ph/9911280*, in: Proc. 8th International Symposium on Heavy Flavour Physics, University of Southampton, UK, 25-29 July 1999
j) M. Ciuchini, E. Franco, L. Giusti, V. Lubicz, G. Martinelli, *hep-ph/9910236*
4. The LEP Heavy Flavour Steering group, http://www.cern.ch/LEPHFS/
5. The LEP B Oscillation WG, LEPBOSC 98/3, updated for the Summer Conferences 1999.
6. Z. Ligeti, hep-ph/9908432, Contribution to the Conference on Kaon Physics (K99), Chicago, IL, USA, 21-26 June 1999.
7. CLEO Collaboration, hep-ex/9910417
8. a) F. Abe et al. (CDF Collaboration), *Phys. Rev. Lett.* **82** (1999) 271
b) B. Abbott et al. (D0 Collaboration), *Phys. Rev.* **D60** (1999) 052001
9. R. Forty, *these proceedings*
10. a) R. Barbieri, L. Hall, A. Stocchi, N. Weiner, *Phys Lett.* **B425** (1998) 119
b) P. Checchia, F. Simonetto, E. Piotto, *Europhys. Lett.* **47** (1999) 656
11. "The BaBar Physics Book", SLAC-R-504 October,1998

MEASUREMENTS OF CKM PARAMETERS AT LEP

GERALD EIGEN (DELPHI Collaboration)
Department of Physics, University of Bergen, Allegaten 55, 5007 Bergen, Norway
E-mail: eigen@asfys2.fi.uib.no

A novel technique for measuring $| V_{ub}/V_{cb} |$ by DELPHI is discussed and LEP results for $| V_{ub} |, | V_{cb} |$ and $| V_{cs} |$ are summarized.

The origin of *CP* violation is a key issue in the forthcoming years. In the Standard Model (SM) *CP* violation originates from the phase of the unitary CKM matrix. The unitarity conditions yield six triangular relations, of which $V_{ub}^* V_{ud} + V_{cb}^* V_{cd} + V_{tb}^* V_{td} = 0$ is experimentally the most useful. Since $V_{ud} \simeq 1$, $V_{tb} \simeq 1$ and $V_{cd} = \lambda = \cos\theta_c$, the knowledge of V_{ub}, V_{cb} and V_{td} completely specifies this triangle via its sides. This is obtained from measurements of $b \to u$ and $b \to c$ semileptonic decays, $B^0\bar{B}^0$ oscillations and *CP* violation in $K^0 \to \pi^+\pi^-$ (ϵ). The extraction of all three CKM parameters suffers from large theoretical uncertainties, *e.g.* $| V_{ub}/V_{cb} |$ measured in both the lepton-endpoint spectrum and exclusive modes has a model dependence of $\leq 20\%$.[1] A precise knowledge of the triangle sides is important, since an additional angle measurement, such as β obtained from *CP* violation measurements in $B \to J/\psi K_S^0$, overconstrains the triangle and thus provides a crucial SM test.

DELPHI has performed a novel inclusive B semileptonic analysis by exploring the mass of the hadronic system recoiling against $l\nu$. A b-enriched sample is selected from 2.8×10^6 hadronic Z^0 decays, consisting of 2-jet or 3-jet events plus one identified e or μ ($p > 3.5$ GeV). In the hemisphere of the identified lepton a secondary hadronic vertex is reconstructed iteratively, considering all charged tracks with a low probability of originating from the primary vertex, as well as K_S^0's and up to two π^0's. Events are retained if the hadronic system having a total charge of $| Q_X | \leq 1$ contains at least one charged particle and if the hadronic invariant mass computed for particles with a vertex fit contribution of $\Delta\chi^2 < 4$ satisfies $M_X \leq 3.0$ GeV. After reconstructing ν from the missing energy and momentum in the signal hemisphere, the mass of the $Xl\nu$ system is calculated. For $B \to X_u l\nu$ modes and fully reconstructed $B \to X_c l\nu$ decays $M_{Xl\nu}$ peaks at 5 GeV, whereas for candidates with partially reconstructed D-decays $M_{Xl\nu}$ peaks at 4.5 GeV. After removing events with $M_{Xl\nu} < 3.0$ GeV, the hadronic mass of candidates with $3.0 \leq M_{Xl\nu} \leq 4.5$ and $M_X < 1.6$ GeV is rescaled by $M_B/M_{Xl\nu}$. To further separate $B \to X_c l\nu$ from $B \to X_u l\nu$ an inclusive D^* search is performed by evaluating the mass difference $\Delta M = M_{X\pi} - M_X$ in events with soft pions. For events consistent

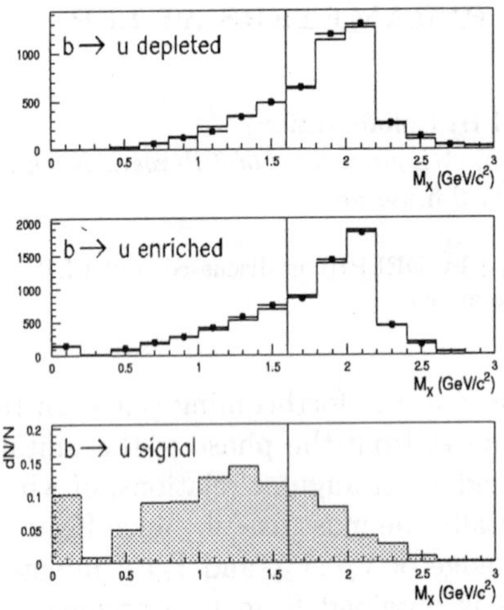

Figure 1. The M_X distributions for the $b \to u$ depleted and $b \to u$ samples in comparison with a Monte Carlo model.

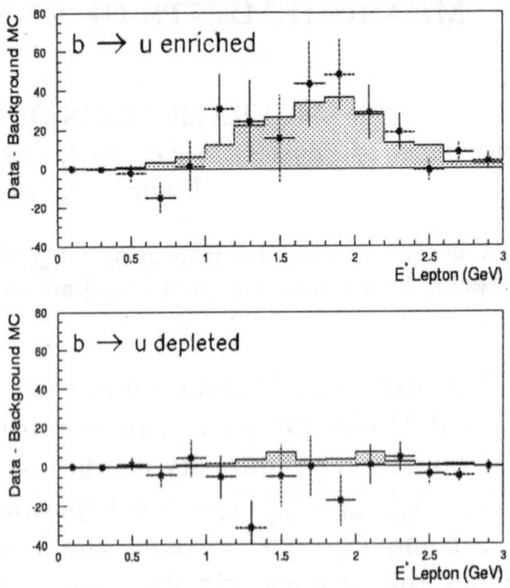

Figure 2. The E_l^* distributions for $b \to u$ depleted and $b \to u$ samples for $M_X < 1.6$ GeV.

with the D^* hypothesis (0.14GeV $< \Delta M <$ 0.16) M_X is constrained to 2.01 GeV. After determining the lepton energy in the B rest frame (E_l^*), all candidates are classified into four subsamples according to their hadronic mass ($M_X <$ or $>$ 1.6 GeV) and their $b \to u$ enrichment or depletion, where $b \to u$ depleted events are those that have an identified kaon in the signal hemisphere together with a lepton impact parameter wrt the hadronic vertex of < -0.015 cm. Further details of the DELPHI analysis are given in reference [2].

Figure 1 shows the M_X distributions for the $b \to u$ depleted and $b \to u$ enriched samples in comparison to the $b \to u$ signal shape expected in a specific model.[2] In the low (high) M_X and $b \to u$ enriched sample 2292 ± 48 (5017 ± 71) events are observed compared to estimated backgrounds of 2078 ± 30 (5019 ± 60) events and expected signal yields of 68% (17%). In the corresponding $b \to u$ depleted classes 1081 ± 33 (3744 ± 61) events are found compared to background estimates of 1118 ± 19 (3618 ± 46) events and expected signal yields of 9% (6%), respectively. In the low M_X and $b \to u$ enriched sample a significant excess of 214 ± 56 events is seen, while the yields in the other subsamples are consistent with background. Figure 2 shows the background-subtracted E_l^* distributions for the $b \to u$ enriched and $b \to u$ depleted samples for $M_X < 1.6$ GeV.

A binned maximum-likelihood fit is performed to extract $\mid V_{ub}/V_{cb} \mid$ from the event yields in each subsample and their E_l^* distributions. The result is $\mid V_{ub}/V_{cb} \mid = 0.103^{+0.011}_{-0.012\ stat} \pm 0.016_{sys} \pm 0.010_{th}$. The overall normalization of the data to the Monte-Carlo estimated background, which is also kept as a free parameter in the fit, is $N = 1.013 \pm 0.011$. The largest systematic uncertainties result from the lepton impact parameter determination (0.007), hadron multiplicity (0.0065), amount of non-resonant $D^{(*)}\pi$ to resonant D^{**} or $D^{(*)\prime}$ (0.0065), and $D \to K^0X$ branching fraction (0.0062). The theoretical error includes contributions from the hadronization model (0.0060), QCD corrections (0.0050), b-quark mass (0.0047), b-quark kinetic energy (0.0028) and b kinematic model (0.0025). Using the world average[3] of $| V_{cb} |= 0.0402 \pm 0.002$ yields $| V_{ub} |= (4.1 \pm 0.8_{stat+sys} \pm 0.4_{b\to u} \pm 0.2_{V_{cb}}) \times 10^{-3}$.

ALEPH[4] and L3[5] measured the charmless semileptonic branching fraction from which $| V_{ub} |$ was extracted. ALEPH via a neural net that enhances $B \to X_u l\nu$ measured $| V_{ub} |= (4.1^{+0.8}_{-1.1}{}_{stat+sys}{}^{+0.2}_{-0.3}{}_{b\to u} \pm 0.2_{HQE}) \times 10^{-3}$, while L3 with a standard selection obtained $| V_{ub} |= (5.7^{+1.4}_{-2.4}{}_{stat+sys}{}^{+0.4}_{-0.5}{}_{b\to u} \pm 0.2_{HQE}) \times 10^{-3}$. The latter two errors represent uncertainties in the $b \to u$ model and from HQE.[7] Combining the three LEP measurements yields a LEP average[6] of $| V_{ub} |= (4.13^{+0.59}_{-0.70}{}_{stat+sys}{}^{+0.23}_{-0.27}{}_{b\to u} \pm 0.16_{HQE}) \times 10^{-3}$, where the $b \to c$ systematics alone contributes $\delta =^{+0.44}_{-0.52}$. This is consistent with the CLEO $B \to \pi(\rho)l\nu$ exclusive result[1] of $| V_{ub} |= (3.25 \pm 0.14_{stat}{}^{+0.21}_{-0.29}{}_{sys} \pm 0.55_{th}) \times 10^{-3}$. The LEP results are dominated by systematic errors ($b \to c$ background). The reduced model dependence wrt CLEO is partially due to a questionable treatment of theoretical errors.[2]

At LEP $| V_{cb} |$ is extracted either from the inclusive branching fraction $B(B \to X_c l\nu)$ or by extrapolating the differential decay rate $d\Gamma(B \to D^* l\nu)/d\omega$ to the point of zero-recoil ($\omega = 1$). All LEP experiments measured $B(B \to X_c l\nu)$ yielding an average value of $(10.58 \pm 0.07_{stat} \pm 0.17)_{sys}\%$. Using heavy quark expansion[7] and an average b life time of $\tau_b = (1.554 \pm 0.014)$ ps, yields a LEP average[8] of $| V_{cb} |= (40.76 \pm 0.41_{stat+sys} \pm 2.04_{th}) \times 10^{-3}$. ALEPH, DELPHI and OPAL measured $d\Gamma(B \to D^* l\nu)/d\omega$, which in HQET is just given by phase space, $| V_{cb} |^2$ and the Isgur-Wise function $F^2(\omega)$.[9] For infinite b-quark mass $F(1) = 1$ while for finite b-quark mass $F(1)$ is reduced to 0.88 ± 0.05.[10] Figure 3 shows the ALEPH, DELPHI and OPAL results for $F(1)\times | V_{cb} |$ at zero recoil. The LEP average is $F(1)\times | V_{cb} |= (33.8 \pm 0.9_{stat} \pm 1.9_{sys}) \times 10^{-3}$, yielding $| V_{cb} |= (38.4 \pm 1.1_{stat} \pm 2.2_{sys} \pm 2.2_{th}) \times 10^{-3}$. Including the inclusive result from above gives a combined LEP average[8] of $| V_{cb} |= (40.2 \pm 1.9) \times 10^{-3}$, making $| V_{cb} |$ the fourth best-known CKM parameter. The error is almost entirely due theoretical uncertainties. Further details of $| V_{ub} |$ and $| V_{cb} |$ measurements are given in the reviews by Thaler and Wu.[11]

Since the production of $e^+e^- \to W^+W^-$ at LEP2, the determination of $| V_{cs} |$ has improved considerably. All LEP experiments have determined $| V_{cs} |$ both from W branching fraction measurements and from flavor tagging of c-quarks (ALEPH, L3, OPAL) or $c\bar{s}$ quark combinations (DELPHI). Figure 4

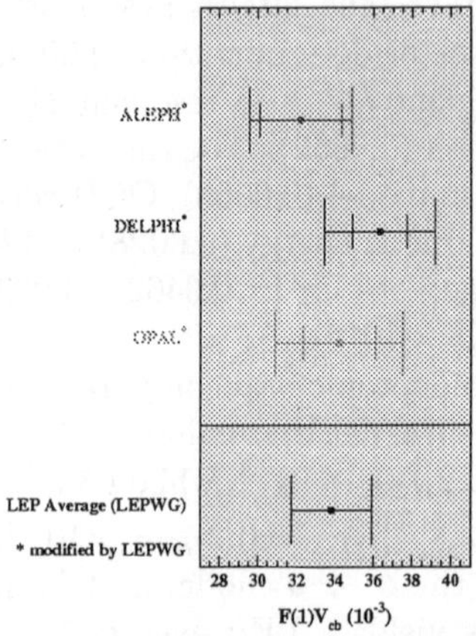

Figure 3. Measurements of $F(1)\times \mid V_{cb} \mid$ in $B \to D^* l\nu$ at LEP.

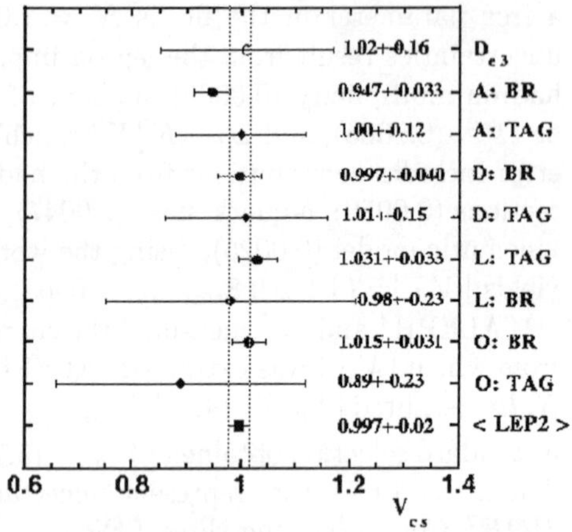

Figure 4. LEP $\mid V_{cs} \mid$ measurements.

shows the $\mid V_{cs} \mid$ measurements from the branching fraction for 172-189 GeV data and from flavor tagging for 172-183 GeV data.[12] The LEP average is $\mid V_{cs} \mid = 0.997 \pm 0.02$, making V_{cs} the third best-measured CKM parameter.

Acknowledgments

This work has been supported by the Research Council of Norway. I thank G. Hou for organizing a fruitful and excellent conference. I acknowledge M. Battaglia, E. Barbiero and J. Timmermanns for fruitful discussions.

References

1. B.H. Behrens *et al.* (CLEO), *Phys. Rev.* D **61**, 052001 (2000); J. Bartelt *et al.* (CLEO), *Phys. Rev. Lett.* **71**, 4111 (1993).
2. P.Abreu *et al.* (DELPHI), submitted to *Phys. Lett.* B.
3. C. Caso *et al.* Particle Data Group (1999).
4. R.Barate *et al.* (ALEPH), *Eur. Phys. J.* C **6**, 555 (1999).
5. M. Acciarri *et al.* (L3), *Phys. Lett.* B **436**, 174 (1998).
6. V_{ub} LEP working group (http://home.cern.ch/ battagl/vub/vub.html).
7. N. Uraltsev *et al.*, *Eur. Phys. J.* C **4**, 453 (1998) and hep-ph/9907270.
8. LEP V_{cb} working group (http://www.cern.ch/LEPVCB/).
9. A.F. Falk *et al.*, *Phys. Lett.* B **343**, 1 (1990).
10. M. Luke *Phys. Lett.* B **252**, 447 (1990).
11. See reviews by J. Thaler and S.L. Wu, these proceedings.
12. LEP Electroweak working group (http://www.cern.ch/LEPEWWG/).

MODEL-INDEPENDENT DETERMINATION OF $|V_{ub}|$

CHANGHAO JIN

School of Physics, University of Melbourne, Victoria 3010, Australia
E-mail: jin@physics.unimelb.edu.au

The decay distribution of the kinematic variable ξ_u in inclusive charmless semileptonic decays of B mesons is unique. The novel method for a model-independent determination of $|V_{ub}|$ is described.

1 Problems and Solutions

The fundamental Cabibbo-Kobayashi-Maskawa (CKM) matrix element $|V_{ub}|$ has been determined from inclusive charmless semileptonic decays of B mesons, $B \to X_u \ell \nu$. However, there are experimental and theoretical problems that obstruct a precise determination of $|V_{ub}|$. Experimentally, it is very difficult to separate signals from the rare $B \to X_u \ell \nu$ decay from large $B \to X_c \ell \nu$ background. Theoretically, QCD uncertainties arise in calculations that relate the measured quantity to $|V_{ub}|$. The potential theoretical uncertainties from perturbative and nonperturbative QCD can be comparable.

The solutions for the problems are provided by a novel method.[1,2] It has been proposed to use the kinematic cut on the variable $\xi_u = (q^0 + |\mathbf{q}|)/M_B$ (q is the momentum transfer to the lepton pair) above the kinematic limit for $B \to X_c \ell \nu$, $\xi_u > 1 - M_D/M_B$, to separate $B \to X_u \ell \nu$ signal from $B \to X_c \ell \nu$ background. Most of $B \to X_u \ell \nu$ events pass the above cut. This kinematic requirement provides a very efficient way for background suppression. $|V_{ub}|$ can then be extracted from the weighted integral of the measured ξ_u spectrum via the sum rule for inclusive charmless semileptonic decays of B mesons with little theoretical uncertainty. The sum rule is derived from the light-cone expansion and beauty quantum number conservation. Thus a model-independent determination of $|V_{ub}|$ can be achieved, minimizing the overall (experimental and theoretical) error.

2 Sum Rule

Because of the large B meson mass, the light-cone expansion is applicable to inclusive B decays that are dominated by light-cone singularities. For inclusive charmless semileptonic decays of B mesons, the light-cone expansion and beauty quantum number conservation lead to the sum rule [1]

$$S \equiv \int_0^1 d\xi_u \, \frac{1}{\xi_u^5} \frac{d\Gamma}{d\xi_u}(B \to X_u \ell \nu) = |V_{ub}|^2 \frac{G_F^2 M_B^5}{192\pi^3} \, . \tag{1}$$

This sum rule has the following advantages:

- Independent of phenomenological models
- No perturbative QCD uncertainty
- Dominant hadronic uncertainty avoided

The sum rule (1) establishes a relationship between $|V_{ub}|$ and the observable quantity S in the leading twist approximation of QCD. The only remaining theoretical uncertainty in the relation comes from higer-twist corrections to the sum rule, which are suppressed by a power of $\Lambda_{\rm QCD}^2/M_B^2$.

3 The ξ_u Spectrum

Now let me explain why the decay distribution of the kinematic variable ξ_u is unique and why the kinematic cut on ξ_u is very efficient in the discrimination between $B \to X_u \ell \nu$ signal and $B \to X_c \ell \nu$ background.

Without QCD corrections, the tree-level ξ_u spectrum in the free quark decay $b \to u \ell \nu$ in the b-quark rest frame is a discrete line at $\xi_u = m_b/M_B$. This is simply a consequence of kinematics that fixes ξ_u to the single value m_b/M_B, no other values of ξ_u are kinematically allowed in $b \to u \ell \nu$ decays. This discrete line at $\xi_u = m_b/M_B \approx 0.9$ lies well above the charm threshold, $\xi_u > 1 - M_D/M_B = 0.65$.

The $O(\alpha_s)$ perturbative QCD correction to the ξ_u spectrum has been calculated.[2] The ξ_u spectrum remains a discrete line at $\xi_u = m_b/M_B$, even if virtual gluon emission occurs. Gluon bremsstrahlung generates a small tail below the parton-level endpoint $\xi_u = m_b/M_B$.

To calculate the real physical decay distribution in $B \to X_u \ell \nu$, we must also account for hadronic bound-state effects. In the framework of the light-cone expansion, the leading nonperturbative QCD effect is incorporated in the b-quark distribution function [3]

$$f(\xi) = \frac{1}{4\pi} \int \frac{d(y \cdot P)}{y \cdot P} \, e^{i\xi y \cdot P} \langle B | \bar{b}(0) \not{y} \mathcal{P} exp[ig_s \int_y^0 dz^\mu A_\mu(z)] b(y) | B \rangle |_{y^2=0} \, , \tag{2}$$

where $\mathcal{P}$ denotes path ordering. Although several important properties of it are known[3] in QCD, the form of the distribution function has not been completely determined. The distribution function $f(\xi)$ has a simple physical interpretation: It is the probability of finding a b-quark with momentum ξP inside the B meson with momentum P. The real physical spectrum is then obtained from a convolution of the hard perturbative spectrum with the soft nonperturbative distribution function:

$$\frac{d\Gamma}{d\xi_u}(B \to X_u \ell\nu) = \int_{\xi_u}^{1} d\xi\, f(\xi) \frac{d\Gamma}{d\xi_u}(b \to u\ell\nu, p_b = \xi P), \tag{3}$$

where the b-quark momentum p_b in the perturbative spectrum is replaced by ξP. The interplay between nonperturbative and perturbative QCD effects has been accounted for.

Bound-state effects lead to the extension of phase space from the parton level to the hadron level, also stretch the spectrum downward below m_b/M_B, and are solely responsible for populating the spectrum upward in the gap between the parton-level endpoint $\xi_u = m_b/M_B$ and the hadron-level endpoint $\xi_u = 1$. The interplay between nonperturbative and perturbative QCD effects eliminates the singularity at the endpoint of the perturbative spectrum, so that the physical spectrum shows a smooth behaviour over the entire range of ξ_u, $0 \leq \xi_u \leq 1$.

Although the monochromatic ξ_u spectrum at tree level is smeared by gluon bresstrahlung and bound-state effects around $\xi_u = m_b/M_B$, about 80% of $B \to X_u \ell\nu$ events remain above the charm threshold. The uniqueness of the decay distribution of the kinematic variable ξ_u implies that the kinematic cut on ξ_u is very efficient in disentangling $B \to X_u \ell\nu$ signal from $B \to X_c \ell\nu$ background.

4 How Do You Measure S?

To measure the observable S defined in Eq. (1), one needs to measure the weighted ξ_u spectrum $\xi_u^{-5} d\Gamma(B \to X_u \ell\nu)/d\xi_u$, using the kinematic cut $\xi_u > 1 - M_D/M_B$ against $B \to X_c \ell\nu$ background. S can then be obtained from an extrapolation of the weighted spectrum measured above the charm threshold to low ξ_u.

While the normalization of the weighted spectrum given by the sum rule (1) does not depend on the b-quark distribution function $f(\xi)$, thus being model-independent, the shape of the weighted spectrum does. The detailed analysis is presented in Ref. [2]. Gluon bremsstrahlung and hadronic bound-state effects strongly affect the shape of the weighted ξ_u spectrum. However, the shape of the weighted ξ_u spectrum is insensitive to the value of the strong coupling α_s, varied in a reasonable range. The overall picture appears to be that the weighted ξ_u spectrum is peaked towards larger values of ξ_u with a narrow width. The contribution below $\xi_u = 0.65$ is small and relatively insensitive to forms of the distribution function. This suggests that extrapolating the weighted ξ_u spectrum down to low ξ_u would not introduce a considerable uncertainty in the value of S.

5 Summary

The kinematic cut on ξ_u, $\xi_u > 1 - M_D/M_B$, and the semileptonic B decay sum rule, Eq. (1), make an outstanding opportunity for the precise determination of $|V_{ub}|$ from the observable S. This method is both exceptionally clean theoretically and very efficient experimentally in background suppression.

There remain two kinds of theoretical error in the model-independent determination of $|V_{ub}|$. First, higher-twist (or power suppressed) corrections to the sum rule cause an error of the order $O(\Lambda^2_{\rm QCD}/M_B^2) \sim 1\%$ in $|V_{ub}|$. Second, the extrapolation of the weighted ξ_u spectrum to low ξ_u gives rise to a systematic error in the measurement of S. The size of this error depends on how well the weighted spectrum can be measured, since the measured spectrum would directly determine the form of the distribution function. In addition, the form of the universal distribution function can also be determined directly by the measurement of the $B \to X_s\gamma$ photon energy spectrum.[3] The experimental determination of the distribution function would provide a model-independent way to make the extrapolation, allowing an error reduction.

Eventually, the error in $|V_{ub}|$ determined by this method would mainly depend on how well the observable S can be measured. To measure S experimentally one needs to be able to reconstruct the neutrino. This poses a challenge to experiment. The unique potential of determining $|V_{ub}|$ warrants a feasibility study for the experiment.

Acknowledgments

I would like to thank Hai-Yang Cheng and Wei-Shu Hou for the very stimulating and enjoyable conference. This work was supported by the Australian Research Council.

References

1. C.H. Jin, *Mod. Phys. Lett.* A **14**, 1163 (1999).
2. C.H. Jin, hep-ph/9911396.
3. C.H. Jin, *Eur. Phys. J.* C **11**, 335 (1999).

STUDY OF THE CP ASYMMETRY OF $B^0 \to J/\psi K^0_S$ DECAYS IN ALEPH

ROGER FORTY
CERN, Geneva
E-mail: Roger.Forty@cern.ch
Representing the ALEPH Collaboration

From the full ALEPH dataset at LEP1 of about 4 million hadronic Z decays, 23 $B^0 \to J/\psi K^0_S$ candidates are selected with an estimated purity of 71%. They are used to measure the CP asymmetry of this decay, with the result $\sin 2\beta = 0.93\,^{+0.64}_{-0.88}\,^{+0.36}_{-0.24}$. This is combined with other existing measurements, and increases the confidence level that CP violation has been observed to almost 99%.

1 Introduction

The channel $B^0 \to J/\psi K^0_S$ has long been recognised as the "gold-plated" mode for the observation of CP violation in the B system. The final state is a CP eigenstate, to which both B^0 and $\overline{B}^0$ can decay. The interference between their direct and indirect decays via B^0–$\overline{B}^0$ mixing leads to a time-dependent CP asymmetry given by:

$$A(t) \equiv \frac{\Gamma(B^0 \to J/\psi K^0_S) - \Gamma(\overline{B}^0 \to J/\psi K^0_S)}{\Gamma(B^0 \to J/\psi K^0_S) + \Gamma(\overline{B}^0 \to J/\psi K^0_S)} = -\sin 2\beta \sin \Delta m_d t \ . \quad (1)$$

Here $\Gamma(B^0 \to J/\psi K^0_S)$ represents the rate of particles that were *produced* as B^0 decaying to $J/\psi K^0_S$ at proper time t, Δm_d is the oscillation frequency of the B^0, and β is an angle of the "unitarity triangle" of the quark mixing matrix, given by the following combination of matrix elements: $\arg(-V_{cd}V^*_{cb}/V_{td}V^*_{tb})$. Information can be obtained indirectly about the value of $\sin 2\beta$ within the context of the Standard Model, from the combination of other measurements that constrain the matrix elements, such as Δm_d, charmless B decays and CP violation in the K sector. Many such fits have been made, typically preferring large positive values for $\sin 2\beta$ in the range 0.4–0.8. A result of $\sin 2\beta = 0.750\,^{+0.055}_{-0.065}$ was presented at this conference,[1] although arguments continue to rage over the correct handling of theoretical uncertainties.

The analysis presented here is only the third direct measurement of the CP asymmetry of this decay, following OPAL[2] and CDF,[3] and was new at this conference. Full details can be found in the contributed paper.[4]

2 Event selection

Candidates are selected using loose cuts to keep high efficiency, as the background is not too severe in the clean environment of Z decays at LEP. The decay products are reconstructed in the modes $J/\psi \to e^+e^-$ or $\mu^+\mu^-$, and $K^0_S \to \pi^+\pi^-$, with overall reconstruction efficiency of 28%.

To tune the selection cuts for background suppression, 6.5 million hadronic Monte Carlo events were used, corresponding to about 1.5 times the data statistics. The resulting $J/\psi K^0_S$ mass distribution is shown in Fig. 1 (a). There are 20 events in the signal region, of which 11 are from background. The input branching ratio $\mathcal{B}(B^0 \to J/\psi K^0) = 8.0 \times 10^{-4}$ is 10% lower than the latest world average, so the expected number of signal events is 12, compatible with the number observed. The superimposed fit is the sum of an exponential shape to describe the background and the signal shape taken from Monte Carlo simulation. The four fitted parameters are the background slope and normalization, and the signal mass and normalization.

When the selection is applied to the full ALEPH dataset of 4.2 million hadronic Z decays, the resulting reconstructed $J/\psi K^0_S$ mass distribution is shown in Fig. 1 (b). A clear signal is seen for the B^0, with 23 events in the signal region. A similar fit is made to that of the hadronic Monte Carlo, giving a fitted B^0 mass of 5.26 ± 0.01 GeV, slightly lower than, but consistent with, the world average value of 5.279 GeV. Using the fit to assign an event-by-event background probability for each event in the signal region, their sum corresponds to 6.6 background events, whilst the integration of the background curve gives 6.4 events in that region. The fitted shape and normalization of the background are consistent with those seen in the Monte Carlo. The Poisson probability of 6.6 fluctuating to 23 or more is less than 10^{-6}. To account for possible fluctuation of the background, and after studying alternative fitting functions, the background fraction in the signal region is taken to be 0.29 ± 0.12.

After subtracting the background, about 16 signal events remain. This is rather more than the predicted number of 9 ± 2, calculated from the expected production rate and the reconstruction efficiency. The Poisson probability of 9 fluctuating up to 16 or more is only a few percent, so the observed signal favours a production rate towards the upper end of its predicted range. A particularly clean signal candidate is shown in Fig. 2, where there are no other charged tracks in the signal hemisphere, and clear vertex activity in the opposite jet.

The proper time of each candidate decay is measured from the combination of the decay length, determined by vertexing the decay products, and

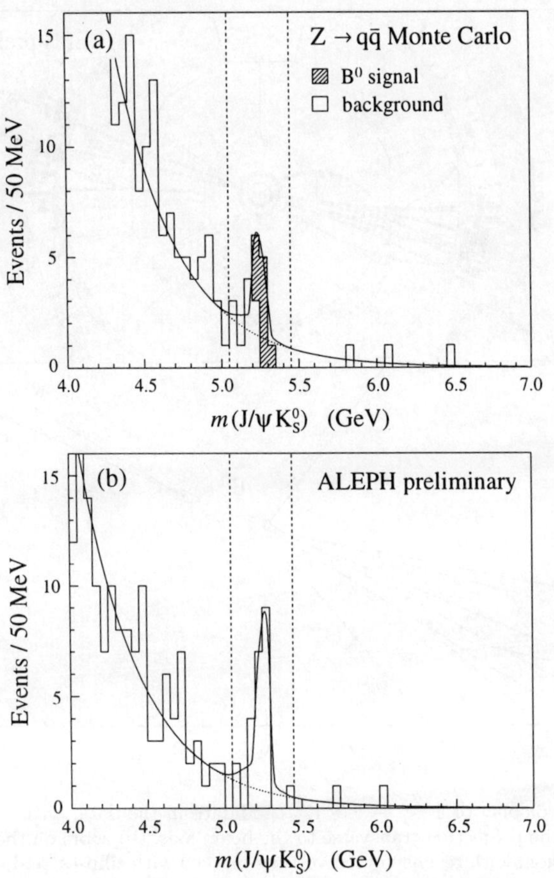

Figure 1. Reconstructed $J/\psi\, K^0_S$ mass (a) for hadronic Monte Carlo, (b) for the data. The superimposed fits are described in the text, the dashed lines indicate the signal region.

the reconstructed momentum, with a typical resolution of 0.1 ps. The production state of the B^0 (or $\overline{B}^0$) is determined by combining the information of 9 discriminating variables, following the techniques developed for B^0_s oscillation studies; they include jet charges, vertex charge, selected lepton and kaon charges etc.,[4] and are combined into a single discriminant variable x_B that peaks at 1 for events produced as B^0 and 0 for events produced as $\overline{B}^0$. Excellent separation is achieved, with a tagging purity of 73% at full efficiency.

The tagging discriminant x_B is plotted against the proper time t in Fig. 3 for the 23 events in the signal region of the data. There are 8 "B^0-like" events

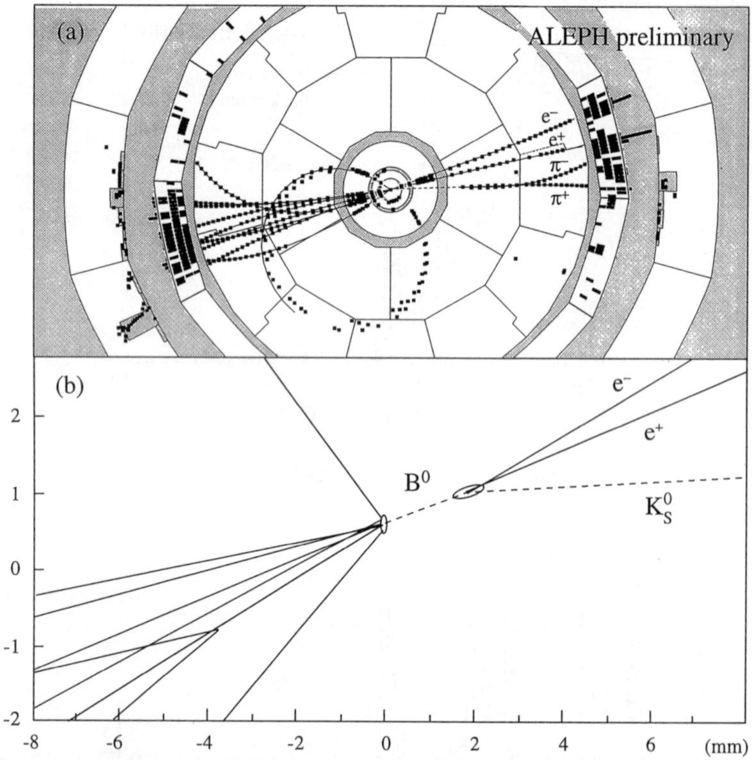

Figure 2. Event display of a $B^0 \to J/\psi\, K^0_S$ candidate in the data, with $J/\psi \to e^+e^-$: (a) fish-eye view of the projection transverse to the beam axis, (b) zoom on the vertex region in the same projection, with reconstructed vertices marked with ellipses, and the reconstructed neutrals marked with dashed lines.

($x_B > 0.5$) and 15 "$\overline{B}^0$-like". Furthermore, the excess of events with low x_B is more noticeable at long proper time, as would be expected for a negative CP asymmetry.

3 Asymmetry measurement

To extract a measurement of the asymmetry, an unbinned likelihood is calculated. The probability density function expected for the signal distribution is

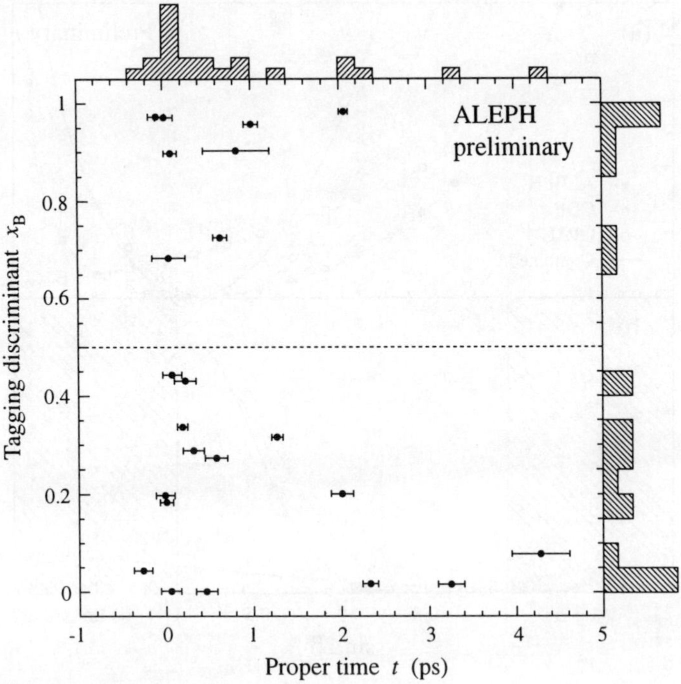

Figure 3. Tagging discriminant x_B versus proper time t for the 23 events in the signal region of the data. The proper time uncertainty is indicated for each point, and histograms of the two variables are shaded.

given by:

$$\mathcal{P}(t,q) = e^{-t/\tau_d}(1 - q\,\sin 2\beta\,\sin \Delta m_d t)/2\,\tau_d\ . \tag{2}$$

Here the tagging variable q is calculated from the discriminant x_B using: $q = 2\,F(x_B)/(F(x_B) + G(x_B)) - 1$, where $F(x_B)$ is the distribution of x_B for events produced as B^0 and $G(x_B)$ is the distribution for $\overline{B}^0$; $q = +1$ for events produced as B^0 and -1 for $\overline{B}^0$ in the case of perfect tagging. The signal distribution is convoluted with a Gaussian resolution function, and a similar term is added to the likelihood for the background component.

Finally the log-likelihood is calculated, summing over the 23 events. This is repeated for different input values of $\sin 2\beta$, and the resulting log-likelihood scan is shown by the line with solid points in Fig. 4 (a). Its minimum is at 0.93, and the statistical error is determined as the point at which the log-likelihood has changed by 0.5 units. Various contributions to the systematic error have been investigated,[4] the dominant ones being the uncertainties in the tagging purity (checked with data using the control channel $B^+ \to J/\psi K^+$), and the

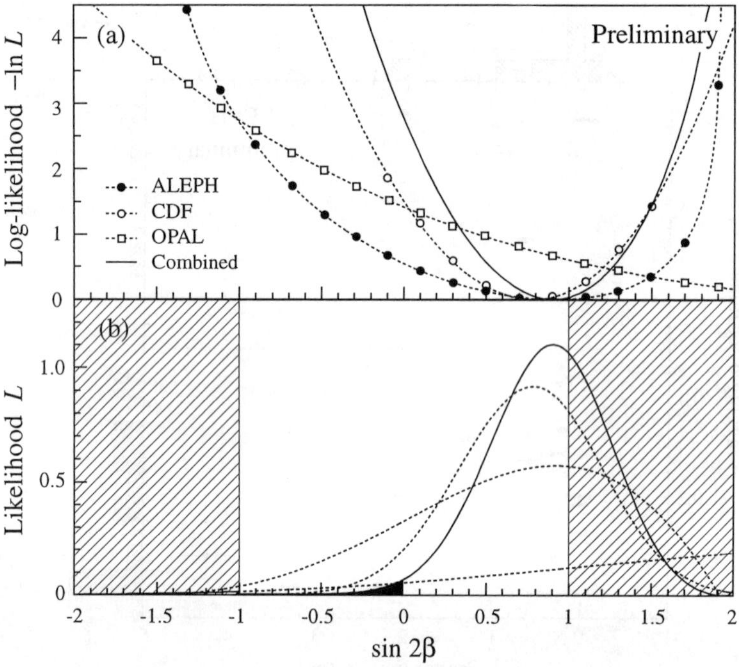

Figure 4. Likelihood distributions versus $\sin 2\beta$ for the current measurement (solid points), other measurements (open points), and their combination (solid line), shown as (a) log-likelihood with respect to the minimum, (b) likelihood. The non-physical region is hatched, and the fraction of the combined likelihood with $\sin 2\beta < 0$ is shaded.

background level. Adding the systematic error contributions in quadrature, the final measurement is $\sin 2\beta = 0.93\ ^{+0.64}_{-0.88}\,(\mathrm{stat})\ ^{+0.36}_{-0.24}\,(\mathrm{syst})$.

The result of the measurement is displayed as a function of proper time in Fig. 5. For the purpose of display the data are divided into three proper-time bins, and the average asymmetry is corrected for the average dilution (from mistagging and background) in each bin. Negative asymmetry, and thus positive $\sin 2\beta$, is favoured.

4 Conclusion

The log-likelihood of this measurement is compared with those of the other two existing measurements in Fig. 4 (a). The log-likelihoods of the three results can be summed to give a combined result, shown by the solid line

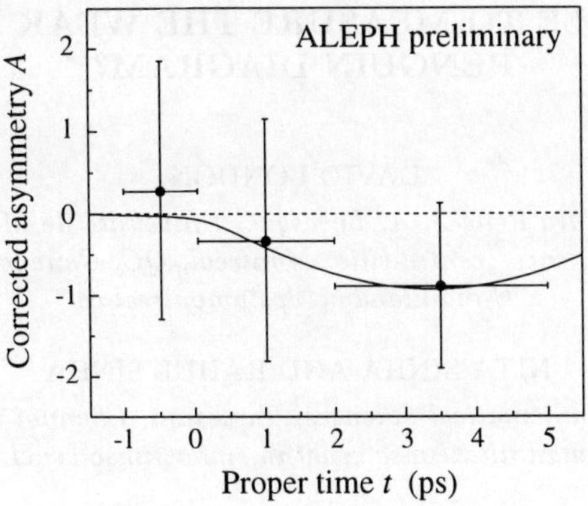

Figure 5. Reconstructed asymmetry versus proper time, after correction for the average dilution in each time bin. The result of the likelihood scan for the asymmetry is shown superimposed as a solid line; note that this is not fit directly to the binned data shown.

in the figure. This corresponds to $\sin 2\beta = 0.91 \pm 0.35$. From the way that CDF fit for the asymmetry, their likelihood includes the systematic error, whilst those of ALEPH and OPAL do not; however, as they are statistically limited the detailed treatment of systematic errors should not substantially effect this result. The log-likelihood curves can be converted back to likelihood distributions, as shown in Fig. 4 (b). The integral of the likelihood for $\sin 2\beta < 0$ is 5% for the CDF result alone, or 7% if the total integral is limited to the physical region $|\sin 2\beta| < 1$. The corresponding fractions are 1.0% and 1.5% for the combination of the three analyses, indicating that the confidence level that CP violation has been seen is increased from 93% to almost 99%.

References

1. A. Stocchi, these proceedings.
2. OPAL Collaboration, K. Ackerstaff *et al.*, Eur. Phys. J. **C5** (1998) 379.
3. CDF Collaboration, "A measurement of $\sin 2\beta$ from $B^0 \to J/\psi K^0_S$ with the CDF detector", FERMILAB–PUB–99/225–E (September 1999); W. Trischuk, these proceedings.
4. ALEPH Collaboration, ALEPH 99–099/CONF 99–054 (November 1999) `http://alephwww.cern.ch/ALPUB/oldconf/oldconf_99.html`.

IS IT POSSIBLE TO MEASURE THE WEAK PHASE OF A PENGUIN DIAGRAM? *

DAVID LONDON

Laboratoire René J.-A. Lévesque, Université de Montréal, C.P. 6128, succ. centre-ville, Montréal, QC, Canada H3C 3J7
E-mail:london@lps.umontreal.ca

NITA SINHA AND RAHUL SINHA

Institute of Mathematical Sciences, Taramani, Chennai 600113, India
E-mail:nita@imsc.ernet.in,sinha@imsc.ernet.in

The $b \to d$ penguin amplitude receives contributions from internal u, c and t-quarks. We show that it is impossible to measure the weak phase of any of these penguin contributions without theoretical input. However, a single assumption involving the hadronic parameters makes it possible to obtain the weak phase and test for the presence of new physics in the $b \to d$ flavour-changing neutral current.

1 Introduction

One of the most compelling features of CP violation in the B system is that all three interior angles of the unitarity triangle, $\phi_1(\equiv \beta)$, $\phi_2(\equiv \alpha)$ and $\phi_3(\equiv \gamma)$[1], can be measured cleanly, i.e. without theoretical hadronic uncertainties. The B system is thereby expected to provide a test of CP violation in the standard model (SM). Any inconsistency with the predictions of the SM will reveal the much sought after signal of new physics (NP).

NP can affect CP violation in one of two possible ways: through contributions to B decays or to B_d^0-$\overline{B_d^0}$ mixing. Most decay modes of the B-meson are dominated by W-mediated tree-level diagrams and will not be much affected by NP, since in most models of NP there are no contributions that can compete with the SM. Thus, with the exception of penguin-dominated decay modes, NP cannot significantly affect the decays. However, new contributions to B_d^0-$\overline{B_d^0}$ mixing can affect the CP asymmetries[2]. Such NP contributions will affect the extraction of V_{td} and V_{ts}, as well as possible measurements of ϕ_1, ϕ_2 and ϕ_3. Thus, NP enters principally through contributions to B_d^0-$\overline{B_d^0}$ mixing[3].

The angles ϕ_1,ϕ_2 and ϕ_3 are to be measured principally through the modes $B_d^0(t) \to \Psi K_s$, $B_d^0(t) \to \pi\pi$ (or $\rho\pi$)[4], and $B^\pm \to DK^\pm$ (or $D^*K^{*\pm}$)[5], respectively. NP in B_d^0-$\overline{B_d^0}$ mixing will then affect the measurements of ϕ_1 and ϕ_2, but in opposite directions[6]. That is, in the presence of a new-physics phase

*Talk presented by Rahul Sinha

ϕ_{NP}, the CP angles are changed as follows: $\phi_1 \to \phi_1 - \phi_{NP}$ and $\phi_2 \to \phi_2 + \phi_{NP}$. Hence the sum $\phi_1 + \phi_2 + \phi_3$ is *insensitive* to the NP. However, if ϕ_3 is measured in the decay $B_s^0(t) \to D_s^{\pm} K^{\mp}$ [7], then $\phi_1 + \phi_2 + \phi_3 \neq \pi$ can be found if there is NP in B_s^0-$\overline{B_s^0}$ mixing.

The most well known method for detecting NP is to compare the unitary triangle as constructed from measurements of the angles with that constructed from independent measurements of the sides. Any inconsistency will be evidence for new physics. However, since at present the allowed region of the unitarity triangle is rather large, the triangle as constructed from the angles could still lie within the allowed region even if NP is present. Furthermore, even if the ϕ_1-ϕ_2-ϕ_3 triangle lies outside the allowed region, one might still be skeptical about the presence of NP: perhaps the theoretical uncertainties which go into the constraints on the unitarity triangle have been underestimated.

Clearly we would like cleaner, more direct tests of the SM in order to probe for the presence of NP. More promising tests for NP are possible by comparing two distinct decay modes which, in the SM, probe the same CP angle. One can compare the rate asymmetries in $B^{\pm} \to DK^{\pm}$ and $B_s^0(t) \to D_s^{\pm} K^{\mp}$, both of which measure ϕ_3. A discrepancy between the extracted values would point to NP in B_s^0-$\overline{B_s^0}$ mixing. Similarly, a discrepancy in ϕ_1, as measured via $B_d^0(t) \to \Psi K_S$ and $B_d^0(t) \to \phi K_S$, implies new physics in the $b \to s$ penguin[8]. One can also measure the CP asymmetry in the decay $B_s^0(t) \to \Psi\phi$, which vanishes to a good approximation in the SM. Such an asymmetry would indicate the presence of new physics in B_s^0-$\overline{B_s^0}$ mixing. Note that all such tests probe NP in the $b \to s$ flavour-changing neutral current (FCNC).

One may then ask the question: are there are any direct tests of NP in the $b \to d$ FCNC? For example, consider pure $b \to d$ penguin decays such as $B_d^0 \to K^0\overline{K^0}$ or $B_s^0 \to \phi K_S$, with the asumption that t-quark contribution dominates among up-type quarks in the loop. In such a case the SM would predict that (i) the CP asymmetry in $B_d^0(t) \to K^0\overline{K^0}$ vanishes, and (ii) the CP asymmetry in $B_s^0(t) \to \phi K_S$ measures $\sin 2\phi_1$ [9]. Any discrepancy between measurements of these CP asymmetries and their predictions would thus imply that there is NP in either B_d^0-$\overline{B_d^0}$ mixing or the $b \to d$ penguin, i.e. in the $b \to d$ FCNC. However, it is well known that $b \to d$ penguins are *not* dominated by the internal t-quark. The contributions of the u- and c-quarks can be as large as 20–50% of that of the t-quark[10]. As a consequence, one cannot probe NP in $b \to d$ FCNC using such modes, and, unfortunately, the answer to the question asked is *no*[11].

2 The CKM Ambiguity

The full $b \to d$ penguin amplitude is a sum of contributions from the three internal up-type quarks in the loop:

$$P = P_u \, V_{ub}^* V_{ud} + P_c \, V_{cb}^* V_{cd} + P_t \, V_{tb}^* V_{td} \ , \tag{1}$$

with $V_{ub} \sim e^{-i\phi_3}$ and $V_{td} \sim e^{-i\phi_1}$. Using the unitarity relation, $V_{ud}V_{ub}^* + V_{cd}V_{cb}^* + V_{td}V_{tb}^* = 0$, the u-quark piece can be eliminated in Eq. (1), allowing us to write

$$P = \mathcal{P}_{cu} \, e^{i\delta_{cu}} + \mathcal{P}_{tu} \, e^{i\delta_{tu}} e^{-i\phi_1} \ , \tag{2}$$

where δ_{cu} and δ_{tu} are strong phases. Now imagine that there were a method in which a series of measurements allowed us to cleanly extract ϕ_1 using the above expression. In this case, we would be able to express $-\phi_1$ as a function of the observables.

On the other hand, we can instead use the unitarity relation to eliminate the t-quark contribution in Eq. (1), yielding

$$P = \mathcal{P}_{ct} \, e^{i\delta_{ct}} + \mathcal{P}_{ut} \, e^{i\delta_{ut}} e^{i\phi_3} \ . \tag{3}$$

Comparing Eqs. (2) and (3), we see that they have the same form. Thus, the same method used to extract $-\phi_1$ from Eq. (2) can be used on Eq. (3) to obtain ϕ_3. That is, we would be able to write ϕ_3 as *the same function* of the observables as was used for $-\phi_1$ above! But this implies that $-\phi_1 = \phi_3$, which clearly does not hold in general.

Due to the ambiguity in the parametrization of the $b \to d$ penguin — which we refer to as the *CKM ambiguity* — we conclude that one cannot cleanly extract the weak phase of any penguin contribution. Indeed, it is *impossible* to cleanly test for the presence of new physics in the $b \to d$ FCNC. Nevertheless, it is instructive to examine in detail a few candidate methods, to see exactly how they fail.

The measurement of the time-dependent rate for the decay $B_d^0(t) \to K^0\overline{K^0}$ can at best allow one to extract the magnitudes and relative phase of $e^{i\phi_1}A$ and $e^{-i\phi_1}\bar{A}$, where A is the amplitude for $B_d^0 \to K^0\overline{K^0}$. With an independent measurement of ϕ_1, there are a total of 4 measurements. Using the form of the $b \to d$ penguin given in Eq. 2, we have $e^{i\phi_1}A = e^{i\phi_1}(\mathcal{P}_{cu} \, e^{i\delta_{cu}} + \mathcal{P}_{tu} \, e^{i\delta_{tu}} e^{-i\phi_1'})$, where we have written the phase ϕ_1' to allow for the possibility of new physics. There are thus 5 theoretical (hadronic) parameters: $\mathcal{P}_{cu}$, $\mathcal{P}_{tu}$, $\delta_{cu} - \delta_{tu}$, ϕ_1, and $\theta_{NP} \equiv \phi_1' - \phi_1$. We see that there are not enough measurements to determine all the theoretical parameters. In fact, there is just one more theoretical unknown than there are measurements.

A similar examination[11] of the $B \to \pi\pi$ isospin analysis, Dalitz-plot analysis of $B \to 3\pi$, angular analysis of $B^0 \to VV$ (where V is a vector meson), and a combined isospin + angular analysis of $B \to \rho\rho$ leads to the same conclusion that there is one more unknown than there are measurements.

We thus conclude that, due to the CKM ambiguity, if one wishes to test for the presence of NP in the $b \to d$ FCNC by comparing the weak phase of the t-quark contribution to the $b \to d$ penguin with that of B_d^0-$\overline{B_d^0}$ mixing, it is necessary to make a single assumption[12] about the hadronic parameters.

Acknowledgments

R.S. would like to thank the organizers of this conference, Prof. H.Y. Cheng and Prof. A.I. Sanda, for financial assistance to attend the conference. The work of D.L. was financially supported by NSERC of Canada.

References

1. C. Caso *et. al.* *Eur. Phys. J.* **C3**, 1 (1998) and 1999 off-year partial update for the 2000 edition (**URL:**http://pdg.lbl.gov/). J. Rosner, A.I. Sanda and M. Schmidt in *Proc. of the Fermilab Workshop on High Sensitivity Beauty Physics at Fermilab*, A.J. Slaughter, N. Lockyer and M. Schmidt, eds., 1987.
2. C.O. Dib, D. London and Y. Nir, *Int. J. Mod. Phys.* **A6**, 1253 (1991).
3. M. Gronau and D. London, *Phys. Rev.* **D55**, 2845 (1997).
4. M. Gronau and D. London, *Phys. Rev. Lett.* **65**, 3381 (1990); A.E. Snyder and H.R. Quinn, *Phys. Rev.* **D48**, 2139 (1993).
5. M. Gronau and D. Wyler, *Phys. Lett.* **265B**, 172 (1991); M. Gronau and D. London, *Phys. Lett.* **253B**, 483 (1991); I. Dunietz, *Phys. Lett.* **270B**, 75 (1991); D. Atwood, I. Dunietz and A. Soni, *Phys. Rev. Lett.* **78**, 3257 (1997); N. Sinha and R. Sinha, *Phys. Rev. Lett.* **80**, 3706 (1998).
6. Y. Nir and D. Silverman, *Nucl. Phys.* **B345**, 301 (1990).
7. R. Aleksan, I. Dunietz, B. Kayser and F. Le Diberder, *Nucl. Phys.* **B361**, 141 (1991); R. Aleksan, I. Dunietz and B. Kayser, *Zeit. Phys.* **C54**, 653 (1992).
8. Y. Grossman and M.P. Worah, *Phys. Lett.* **395B**, 241 (1997); D. London and A. Soni, *Phys. Lett.* **407B**, 61 (1997).
9. D. London and R. Peccei, *Phys. Lett.* **223B**, 257 (1989).
10. A.J. Buras and R. Fleischer, *Phys. Lett.* **341B**, 379 (1995).
11. D. London, N. Sinha and R. Sinha, *Phys. Rev.* **D60**, 074020 (1999).
12. C.S. Kim, D. London and T. Yoshikawa, *Phys. Lett.* **361B**, 458 (1999).

CONSTRAINTS ON SUPERGRAVITY UNIFIED MODELS FROM THE $b \to s\gamma$ DECAY

Chung Kao

Department of Physics, University of Wisconsin
Madison, WI 53706, USA

The branching fraction $B(b \to s\gamma)$ is evaluated in the minimal supergravity (mSUGRA) model and a supersymmetric model with an inverted-scalar-mass hierarchy (ISMH). The CLEO measurement of $B(b \to s\gamma)$ is employed to place tight constraints on the parameter space of both the mSUGRA and the ISMH models.

1 Introduction

In the minimal supergravity (mSUGRA) unified model [1], it is assumed that supersymmetry (SUSY) is broken in a hidden sector with SUSY breaking communicated to the observable sector through gravitational interactions, leading to a common scalar mass (m_0), a common gaugino mass ($m_{1/2}$), a common trilinear coupling (A_0), and a bilinear coupling (B_0), at the grand unified scale ($M_{\rm GUT} \sim 2 \times 10^{16}$ GeV). The mSUGRA model has great predictive power, however, most flavor-changing-neutral current (FCNC) and CP violating observables are assumed to be small without much justification. Recently, several models have been proposed to solve the FCNC and the CP problems [2,3,4,5].

In this article, we consider a supergravity unified model with an inverted-scalar-mass hierarchy (ISMH) [6] providing a satisfactory solution to the FCNC and the CP problems with the following boundary conditions at the $M_{\rm GUT}$:

$$\begin{aligned} &m_{1/2},\ \tilde{m}_1 = \tilde{m}_2 = 10 \text{ TeV},\ \tilde{m}_3 = m_0, \\ &A_1 = A_2 = 10 \text{ TeV},\ A_3 = A_0, \end{aligned} \tag{1}$$

where $m_{1/2}$ = a universal gaugino mass, $\tilde{m}_i$ = the ith-generation sfermion masses, A_i = the ith trilinear couplings, and $m_0 \lesssim 1$ TeV[a].

At the weak scale, we choose $\tan\beta$ and sign of μ as free parameters and evaluate the value of $|\mu|$ and the B parameter with electroweak symmetry breaking (EWSB) conditions. The weak-scale SUSY particle masses and couplings are evaluated by the renormalization group evolution from the $M_{\rm GUT}$.

[a] We take $A_0 = 0$ in our analysis since the value of A_0 does not significantly affect the results.

2 Constraints from $b \to s\gamma$

In the minimal supersymmetric model (MSSM), there are dominant contributions to the $b \to s\gamma$ decay [7] from loop diagrams involving the W boson and the quarks (W–loop), the charged-Higgs boson ($H^{\pm}$) and the quarks ($H^{\pm}$–loop), and the charginos ($\chi_i^{\pm}$) and the squarks ($\chi^{\pm}$–loop). As $\tan\beta$ becomes large, the form factors from the $H^{\pm}$–loop slightly increase and interfere constructively with the W-loop; while the form factors of the $\chi^{\pm}$–loop are greatly enhanced, having constructive interference with the W–loop when $\mu < 0$, but destructive interference with the W–loop when $\mu > 0$. The predicted value of $B(b \to s\gamma)$ in the MSSM is larger than that in the Standard Model when $\mu < 0$ and the experimental limits of $B(b \to s\gamma)$ disfavor most of mSUGRA parameter space when $\tan\beta \gtrsim 10$ and $\mu < 0$ [8].

Figure 1 shows the branching fraction of $b \to s\gamma$ versus $m_{1/2}$ in the ISMH and the mSUGRA models with $\mu > 0$, for $\tan\beta = 3$ and 35. Also shown are the CLEO [9] experimental limits at 95% confidence level (C.L.) ($2.0 \times 10^{-4} < B(b \to s\gamma) < 4.5 \times 10^{-4}$).

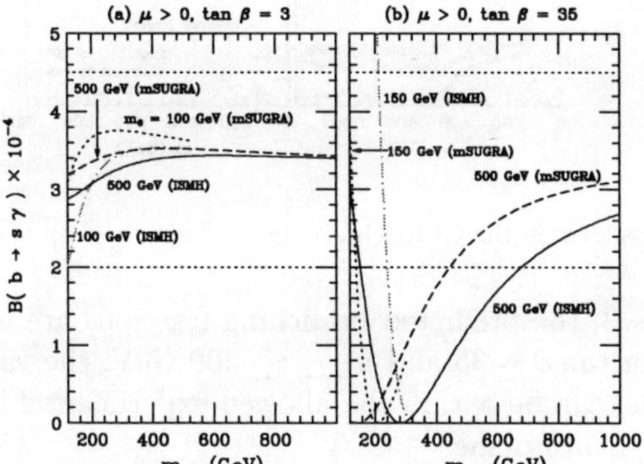

Figure 1: $B(b \to s\gamma)$ versus $m_{1/2}$ for $\mu > 0$, (a) $\tan\beta = 3$ with $m_0 = 100$ GeV and 500 GeV, and (b) $\tan\beta = 35$ with $m_0 = 150$ GeV and 500 GeV. The dotted lines are the CLEO experimental limits at 95% C.L.

When $\tan\beta \sim 3$, the predicted branching fractions are within the experimental limits in both the mSUGRA and the ISMH models. For $\tan\beta \sim 35$ and $m_{1/2} \lesssim 400$ GeV, a destructive interference occurs in both the ISMH and mSUGRA models between loops involving the first two generation squarks and loops involving the third generation squarks. This destructive interference severely reduces the mSUGRA chargino contribution, and makes $B(b \to s\gamma)$ in mSUGRA much smaller than the experimental lower limit.

3 Constraints from $b \to s\gamma$

To place constraints on the parameter space of both the mSUGRA and the ISMH models, we have employed the recent CLEO experimental limits [9]. In Fig. 2, we present the favored region in the $(m_0, m_{1/2})$ plane with $\tan\beta = 35$. Also shown are the regions excluded by the $m_{\chi_1^+} \lesssim 95$ GeV limit from the chargino search [10] at LEP 2, and the regions that do not satisfy the following theoretical requirements: radiative EWSB, the correct vacuum for EWSB obtained (tachyon free), and the lightest neutralino as the lightest SUSY particle.

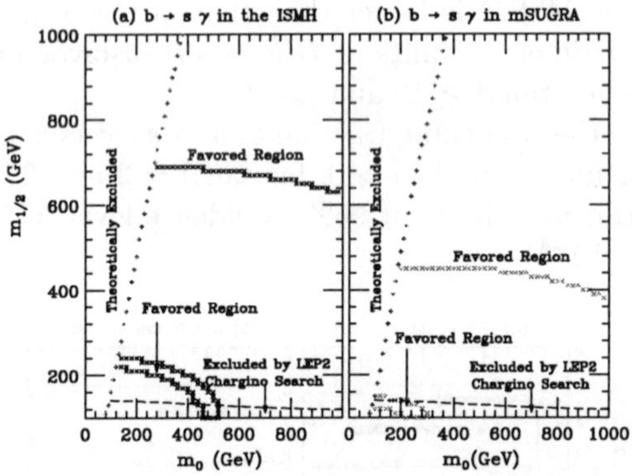

Figure 2: Favored regions by the CLEO limits in the $(m_0, m_{1/2})$ plane with $\tan\beta = 35$.

When $\tan\beta \sim 3$, the predicted branching fractions are within the experimental ranges. For $\tan\beta \sim 35$ and $m_{1/2} \lesssim 400$ GeV, the value of $B(b \to s\gamma)$ in the ISMH model can be within the allowed experimental limits in a sizable region of the $(m_{1/2}, m_0)$ plane.

4 Conclusions

The branching fraction of $b \to s\gamma$, the trilepton signature at the Tevatron, and the relic density of the neutralino dark matter, can be complementary in distinguishing the ISMH and the mSUGRA models [6]. If the nature favors small $\tan\beta \sim 3$, searches for the trileptons from $\chi_1^\pm \chi_2^0$ production at the upgraded Tevatron can be made up to a larger chargino mass in the ISMH model for $m_0 \gtrsim 200$ GeV. The mSUGRA model has a larger parameter space for a cosmologically interesting relic density of the neutralino dark matter. At high $\tan\beta \gtrsim 35$ with $m_{1/2} \lesssim 400$ GeV, the experimental $b \to s\gamma$ constraint allows more parameter space in the ISMH model.

Acknowledgments

I am very grateful to Vernon Barger, Mike Berger, Gi-Chol Cho, Kaoru Hagiwara and Ren-Jie Zhang, for collaborations and discussions. This research was supported in part by the U.S. Department of Energy under Grant No. DE-FG02-95ER40896, and in part by the Wisconsin Alumni Research Foundation.

References

1. A. Chamseddine, R. Arnowitt and P. Nath, Phys. Rev. Lett. **49** (1982) 970; L. Ibañez and G. Ross, Phys. Lett. **B110** (1982) 215; R. Barbieri, S. Ferrara and C. Savoy, Phys. Lett. **B119** (1982) 343; L.J. Hall, J. Lykken and S. Weinberg, Phys. Rev. D **27** (1983) 2359; L. Alvarez-Gaumé, J. Polchinski and M. Wise, Nucl. Phys. **B121** (1983) 495.
2. Y. Nir and N. Seiberg, Phys. Lett. **B309**, 337 (1993); M. Dine, A.E. Nelson and Y. Shirman, Phys. Rev. D **51**, 1362 (1995); L. Randall and R. Sundrum, hep-th/9810155.
3. A.G. Cohen, D.B. Kaplan and A.E. Nelson, Phys. Lett. **B388**, 588 (1996); S. Dimopoulos and G.F. Giudice, Phys. Lett. **B357**, 573 (1995); A. Pomarol and D. Tommasini, Nucl. Phys. **B466**, 3 (1996).
4. G. Dvali and A. Pomarol, Phys. Rev. Lett. **77**, 3728 (1996); P. Binetruy and E. Dudas, Phys. Lett. **B389**, 503 (1996).
5. J. Bagger, J.L. Feng and N. Polonsky, hep-ph/9905292; J.A. Bagger, J.L. Feng, N. Polonsky and R.-J. Zhang, hep-ph/9911255.
6. V. Barger, C. Kao and R.-J. Zhang, University of Wisconsin Report MADPH-99-1138, hep-ph/9911510.
7. S. Bertolini, F. Borzumati and A. Masiero, Nucl. Phys. **B294**, 321 (1987); S. Bertolini, F. Borzumati, A. Masiero and G. Ridolfi, Nucl. Phys. **B353**, 591 (1991).
8. J. Wu, R. Arnowitt and P. Nath, Phys. Rev. D **51**, 1371 (1995); V. Barger, M.S. Berger, P. Ohmann and R.J. Phillips, Phys. Rev. D **51**, 2438 (1995); H. Baer and M. Brhlik, Phys. Rev. D **55** (1997) 3201; H. Baer, M. Brhlik, D. Castano and X. Tata, Phys. Rev. D **58**, 015007 (1998).
9. M.S. Alam *et al.*, the CLEO Collaboration, Phys. Rev. Lett. **74** (1995) 2885; S. Ahmed *et al.*, [CLEO Collaboration], hep-ex/9908022.
10. The LEP2 SUSY working group, http://www.cern.ch/lepsusy/, LEPSUSYWG/99-03.1 (1999).

$B \to X_s\gamma$ and Vector Quark Model

Chia-Hung V. Chang
Physics Department, National Taiwan Normal University, Taipei, Taiwan

Darwin Chang
Physics Department, National Tsing-Hua University, Hsinchu, Taiwan

Wai-Yee Keung
Physics Department, University of Illinois at Chicago, Chicago IL 60607-7059, USA

We study the B meson radiative decay $B \to X_s\gamma$ in the vector quark model. Deviation from the Standard Model arises from the non-unitarity of the charged current KM matrix and related new FCNC interactions. The complete calculation including leading logarithmic QCD correction is carefully carried out. Using the most updated data, the branching fraction of the observed B meson radiative decay places a limit on the mixing angles as stringent as that of the process $B \to X\mu\bar{\mu}$. Direct CP violation is also discuessed.

1 Introduction

A simple extension of Standard Model (SM) is to enlarge the particle content by adding vector quarks, whose right-handed and left-handed components transform in the same way under the weak $SU(2)\times U(1)$ gauge group. In this article, we discuss the B meson radiatve decay in the context of vector quark model and show that the experimental data can be used to constrain the mixing angles.

2 Vector Quark Model

We consider the model in which the gauge structure of SM remains while one charge $-\frac{1}{3}$ is introduced. Large Dirac masses of vector quarks, invariant under $SU(2)_L$, naturally arise. Gauge invariant Yukawa couplings between vector quarks and ordinary quarks are possible, which give rise to mixing between quarks of the same charge. In general, the vector quark will mix with down-type quarks.

3 B Meson Radiative Decay

The $B \to X_s\gamma$ decay, which already exists via one-loop W-exchange diagram in SM, is known to be extremely sensitive to the structure of fundamental interactions at the electroweak scale and serve as a good probe of new physics beyond SM because new interaction contribution can well be at the one-loop level also.

In SM, $b \to s\gamma$ arises at the one loop level from the various W meson mediated penguin diagrams. It is convenient to discuss weak decays using the Effective Hamiltonian formalism [1]. To avoid the uncertainty in m_b, it is customary to calculate the ratio R between the radiative decay and the dominant semileptonic decay. The ratio R is given, to LLA, by

$$R \equiv \frac{\Gamma(b \to s\gamma)}{\Gamma(b \to ce\bar{\nu}_e)} = \frac{1}{|V_{cb}|^2} \frac{6\alpha}{\pi g(z)} |V_{ts}^* V_{tb} C_7(m_b)|^2 .$$

In the vector quark model, deviations from SM result come from various sources: (1) charged current KM matrix non-unitarity, (2) Flavor Changing Neutral Current (FCNC) effects in neutral meson mediated penguin diagrams. We concentrate on these two contributions, which have been discussed in Refs.[2,3,4], but none of them presented the complete result without errors.

KM matrix is not unitary in the presence of a isosinglet vector quark. The mass-independent contributions from the various up-type quarks no longer cancel and could give rise to a significant deviation from SM prediction. The extra contribution to the Wilson coefficient C_7 is given by

$$\frac{(V^\dagger V)_{23}}{V_{ts}^* V_{tb}} \frac{23}{36} = \frac{\delta}{V_{ts}^* V_{tb}} \frac{23}{36} . \tag{1}$$

with $\delta = (V^\dagger V)_{23} = z_{sb}$.

The $b \to s\gamma$ transitions also arise from FCNC Z meson and Higgs boson mediated penguin diagrams. The contribution from internal $i = b, s$ quark, $f_{s,b}^Z$, is given by ($y_i \equiv m_i^2/M_Z^2$):

$$f_b^Z + f_s^Z \approx -\frac{1}{9} - \frac{1}{27} \sin^2\theta_W \simeq -0.12 \tag{2}$$

The Z-mediated penguin diagram with internal D quark can also be calculated.

$$f_D^Z = -\frac{5}{72 y_D} + O(\frac{1}{y_D^2}) . \tag{3}$$

It approaches zero when $y_D \to \infty$ and thus f_D^Z can be ignored, as the leading order approximation. For a gauge invariant result, the unphysical neutral Higgs χ mediated penguin need to be considered together with the Z meson penguin.

$$f_i^\chi = -\frac{Q_d}{8} y_i \left[(2 - y_i) \ln y_i - \frac{5}{6} y_i^3 + 4 y_i^2 - \frac{13}{2} y_i + \frac{10}{3} \right] \frac{1}{(1 - y_i)^4} \tag{4}$$

It is obvious that the light quark contributions are suppressed by the light quak masses and thus negligible. The situation is quite different for the heavy D quark. For $y_D \to \infty$, $f_D^\chi \to -\frac{5}{144} \sim -0.035$. We expand f_D^χ in powers of $1/y_D$ and keep also the next leading term.

$$f_D^\chi \approx -\frac{5}{144} + \frac{1}{36}\frac{1}{y_D} + O(\frac{1}{y_D^2}) \,. \tag{5}$$

The Higgs boson H mediated penguin is similar to that of unphysical Higgs χ.

Put together, the Wilson coefficient $C_7(M_W)$ in the vector quark model is given by

$$C_7^{\rm SM}(M_W) + \frac{z_{bs}}{V_{tb}V_{ts}^*}\left(\frac{23}{36} - \frac{1}{9} - \frac{1}{27}\sin^2\theta_W + \frac{5}{72y_D} + \frac{5}{144} - \frac{1}{36}\frac{1}{y_D} - \frac{7}{144} + \frac{1}{18}\frac{1}{w_D}\right)$$
$$\to -0.193 + \frac{z_{bs}}{V_{tb}V_{ts}^*} \times 0.506 \quad . \tag{6}$$

Here we have used the unitariy relations $z_{4b}z_{4s}^* = -|U_{44}|^2 z_{sb} \approx -z_{sb}$ to leading order in FCNC due to the unitariry of U_L^d and $\delta = z_{sb}$ from Eq. (3).

Similarly the Wilson coefficient of the gluonic magnetic-penguin operator O_8 is modified by the vector quark.

$$C_8^{\rm SM}(M_W) \to -0.096 + \frac{z_{bs}}{V_{tb}V_{ts}^*} \times 0.942 \quad . \tag{7}$$

The above deviation from SM does not include QCD evolution. Actually it is trivial to incorporate LL QCD corrections to these deviations in the framework of effective Hamiltonian. The correction to the ratio R in the vector quark model including its LL QCD contributions is given by

$$\Delta R = \frac{6\alpha}{\pi g(z)} \times 0.163 \times {\rm Re}\left[\frac{V_{cs}^*}{V_{cb}} z_{sb}\right]$$

to leading order in δ.

4 Constraints

The inclusive $B \to X_s\gamma$ branching ratio has been measured by CLEO with the branching ratio [5]

$$\mathcal{B}(B \to X_s\gamma)_{\rm EXP} = (3.15 \pm 0.54) \times 10^{-4} \tag{8}$$

This branching ratio could be used to constrain the mixing in the vector quark model. We calculate the vector quark model deviation to leading logarithmic order. It gives a range of possible vector quark model deviation and hence on z_{sb} (with the input $\mathrm{B}(B \to X_c e\bar{\nu}) = 0.105$):

$$-0.0027 < z_{sb} < 0.0045 \tag{9}$$

Our bound is as strong as that from FCNC. It shows that even though the quark model deviation contributes to the radiative decay rate through one loop, as SM, the data could still put strong bound.

5 Direct CP violation

The CP violating partial rate asymmetry A_{CP} in $b \to s\gamma$ as defined in [6] is related to $\mathrm{Im}\, z_{sb}$.

$$A_{CP}(\delta) = \frac{1}{|C_7|^2} \left\{ a_{27}\, \mathrm{Im}\,[C_2 C_7^*] + a_{87}\, \mathrm{Im}\,[C_8 C_7^*] + a_{28}\, \mathrm{Im}\,[C_2 C_8^*] \right\} \tag{10}$$

In the vector quark model:

$$A_{CP}(\delta) = 5.6\, \mathrm{Im}\, z_{sb} \tag{11}$$

Potentially, a large asymmetry could arise. However $|z_{sb}|$ is constrainted by FCNC to be smaller than 0.0016.

$$A_{CP}(\delta) \approx 1\% \tag{12}$$

It is a bit larger than the SM result of 0.5%.

1. G. Buchalla, A. J. Buras and M. Lautenbacher, Rev. Mod. Phys. **68** 1125 (1996).
2. G. Bhattacharyya, G. C. Branco and D. Choudhury, Phys. Lett. B **336** 487 (1994).
3. L. T. Handoko and T. Morozumi, Mod. Phys. Lett. A **10** 309 (1995).
4. V. Barger, M.S. Berger and R. J. N. Phillips, Phys. Rev. D **52** 1663 (1995).
5. T. Skwarnicki (representing the CLEO Collaboration), talk presented at the XXIVth International Conference on High Energy Physics, Vancouver, B.C. , Canada, 23-29 July 1998.
6. A.L. Kagan and M. Neubert, Phys. Rev. D **58** 094012 (1998).

NEW PHYSICS SEARCH IN B MESON DECAYS

Y.-Y. KEUM

Institute of Physics, Academia Sinica, Nankang, Taipei, Republic of China
E-mail: keum@phys.sinica.edu.tw

We discuss new physics search within the minimal supergravity model by using a possible large direct CP asymmetry in $B^{\pm} \to K^{\pm}\phi, K^0\pi^{\pm}$ decays and B meson rare decays.

1 Introduction

B fatories at KEK and SLAC are taking data to probe the origin of CP violation which is one of main issue in current high energy physics. In the standard model(SM) the CP violation is originated by a physical phase of the Cabbibo-Kobayashi-Maskawa(CKM) matrix [1]. A new source of CP violation can appear in models beyond the SM. If there is new physics beyond SM, we expect to see its effects in CP violating B meson decays.

The most important task for new physics search is to identify decay modes where one can find a large deviations from standard model expectations, and also experimentally accessible in near future. It is really good chance to probe the new physics effects in B-meson decays through indirect ($B^0 - \bar{B}^0$ mixing) and direct CP-violating processes at B factories.

In this talk we discuss a possible new physics impact on B-meson rare decays and the direct CP violation phenomena through the magetic gluon penguin contributions in $B^{\pm} \to K^{\pm}\phi, K^0\pi^{\pm}$ decays.

2 New physics Effects in mSUGRA Model

We investigate the new physics effects in rare B decays: $B \to X_s\gamma$ and $B \to X_s\ell^+\ell^-$ and in direct CP violating modes: $B^{\pm} \to \phi K^{\pm}, K^0\pi^{\pm}$ within the minimal Supergravity model(mSUGRA). In the mSUGRA, there are four new CP violating phases, i.e. phases of the gaugino mass, the higgsino mass parameter, the SUSY breaking Higgs beson mass, and the trilinear scalar coupling constant, of which two combinations are physically independent. When we impose the universal condition at GUT scale, two physical phases at GUT scale comes from, if we take a phase convention, the trilinear coupling constant and higgsino mass parameter, ϕ_A and ϕ_μ, respectively. These phases induce the neutron and electron electric dipole moments (EDMs). When we require the unversality of SUSY breaking term at GUT scale and explicitly solve the renormalization group equations (RGEs) to determine the masses and the mixings of SUSY particles and also require the condition for the radiative electroweak breaking, the phase ϕ_μ is strongly constrained by EDMs and the phase ϕ_A is not constrainted at GUT scale, however, in low energy scale, the phase of A-term for top squarks is strongly supressed, becuase the phase of A-term for top squarks is reduced due to the large top Yukawa coupling constant and aligned to that of the gaugino mass [3].

When we investigate the effect of the SUSY CP violating parameter on rare B decays, $B \to X_s\gamma$ and $B \to X_s\ell^+\ell^-$, in the mSUGRA, we find some interesting results [3] with following SUSY parameters: $0 < m_0 < 1$ TeV, $120 < M_X < 500$ GeV, $|A_x| < 5m_0$, and the bound of EDMs[4] : $|d_n| \leq 0.97 \times 10^{-25} e \cdot cm$, $|d_e| \leq 4.0 \times 10^{-27} e \cdot cm$, in addition, the branching ratio of $B \to X_s\gamma$ [5] : $2.0 \times 10^{-4} < \mathcal{B}(B \to X_s\gamma) < 4.5 \times 10^{-4}$; (i) $\phi_\mu \leq 10^{-2}$ and $0 \leq \phi_A \leq 2\pi$, (ii) As in the case of no SUSY CP violating phase, C_7 and C_8 have large SUSY contributions, however, those to C_9 and C_{10} are small. (iii) CP asymmetry of $B \to X_s\gamma$: $A_{CP}(B \to X_s\gamma) \leq 2\%$ with EDM constraints, however when we consider EDM cancellation in one loop level [6], it can be reached upto 7%. (iv) In $B \to X_s\ell^+\ell^-$ decay, for small $tan\beta$ value, $Im(C_7/C_7^{sm}) \simeq 0$ since $Im(A_t)$ becomes small, however, for large $tan\beta$ value, since chargino and stop loop effect becomes dominated in C_7, $C_7 \simeq \pm C_7^{sm}$. So branching ratio of $B \to X_s\ell^+\ell^-$ can be enchanced when $C_7 \simeq -C_7^{sm}$. (v) The allowed domain of C_8/C_8^{sm} can be extracted from the $B \to X_s g\gamma$ contribution into $B \to X_s\gamma$, shown in Figure 1.

When we investigate the new physics effect through magnetic gluon penguin contributions in the exclusive B meson decays, we find that a possible large direct CP violation can be observed in pure penguin modes, specially $B^\pm \to \phi K^\pm$ and $K^0\pi^\pm$.

The CP asymmetry is defined as :

$$A_{CP} = \frac{\Gamma(B^- \to f) - \Gamma(B^+ \to \bar{f})}{\Gamma(B^- \to f) + \Gamma(B^+ \to \bar{f})}$$

For instance, the amplitude of $B^+ \to \phi K^+$ decay is :

$$A(B^+ \to \phi K^+) = -\frac{G_F}{\sqrt{2}} V_{tb}V_{ts}^* \left[a_3 + a_4 + a_5 - \frac{1}{2}(a_7 + a_9 + a_{10}) + a_{8G}\right] M^{(BK,\phi)}$$

$$M^{(BK,\phi)} = <\phi|(\bar{s}s)_{V-A}|0><K|(\bar{b}s)_{V-A}|B> = 2f_\phi\, m_\phi\, [\epsilon \cdot P_B] F_1^{BK}(m_\phi^2)$$

where

$$a_{8G} = \frac{\alpha_s}{4\pi}\frac{m_b^2}{q^2} C_{8G} \frac{N_c^2 - 1}{N_c^2}\, S_{\phi K} \cdot e^{i\sigma}$$

$$C_{8G} = C_{8G}^{new} + C_{8G}^{sm} = C_{8G}^{sm} \cdot R \cdot e^{i\theta}$$

where $S_{\phi K} = -0.49$, σ is the strong phase difference between O_{8G} and O_{1-10}, $\theta = \delta_{new} - \delta_{sm}$ is the electroweak phase difference between new physics and SM, and $R = |C_{8G}/C_{8G}^{sm}|$. In our analysis we use the factroization approach including non-factorizable contributions into N_c^{eff} and strong phases via Bander-Silverman-Soni mechanism [7] into a_i and C_i^{eff}. We use $(N_c^{eff})_{LL} = 2.0$ for $O_{1,2,3,4,9,10}$ and $(N_c^{eff})_{LR} = 6.0$ for $O_{5,6,7,8}$ as like as H.Y. Cheng et al.[8]. As shown in Figure 2, we have large direct CP asymmetries which is induced by new physics contributions : in $B^\pm \to \phi K^\pm$ decay, we have $0.1 \times 10^{-5} \leq \mathcal{B}(B^\pm \to \phi K^\pm) \leq 0.75 \times 10^{-5}$ and $|A_{CP}| \leq 15\%$ with EDM constrained condition, however, without EDM constraints, $|A_{CP}| \leq 30\%$. For $B^\pm \to K^0\pi^\pm$ decay, we get the branching ratio upto 22.5×10^{-6} which is well agreed with present experimental data [9]: $\mathcal{B}(B^\pm \to K^o\pi^\pm) = (18.2^{+4.6}_{-4.0} \pm 1.6) \times 10^{-6}$ and $|A_{CP}| \leq 20\%$.

In conclusion we have given example of decay modes which can allow an early detection of new physics effects in the minimal supergravity model.

Acknowledgments

The author wishes to thank T. Goto, T.Nihei, Y. Okada and Y.Shimizu for collaboration on some of the work presented here. Y.Y.K. would like to thank M. Kobayashi and H.Y Cheng for their hospitality.

References

1. N. Cabibbo, Phys. Rev. Lett. **10** (1963)531; M. Kobayashi and T. Maskawa, Prog. Th. Phys. **49** (1973) 652.
2. T. Goto et al., Pjys. Rev D **53** (1995) 6662. R.N. Cahn and M. Worah, Phys.Rev. D **60** 076006 (1999).
3. T. Goto, Y.Y.Keum, T.Nihei, Y. Okada and Y. Shimizu, Phys. Lett. B **460** (1999) 333.
4. E.D. Commins et al. Phys. Rev. A **50** (1994) 2960; K. Abdullah et al., Phys. Rev. Lett. **65** (1990) 2347.
5. CELO Collaboration, CONF 98-17, ICHEP98 1011.
6. M. Brhlik, G.J. Good and G.L. Kane, Phys. Rev. **D59**, 115004 (1999).
7. M. Bander, D. Silverman and A. Soni, Phys. Rev. Lett. **43** (1979) 242.
8. H.Y.Cheng and K.C. Yang, hepph/9910291; H.Y.Cheng, hepph/9912372, in these proceedings.
9. CLEO Collaboration, Y. Kwon et al. hep-ex/9908039

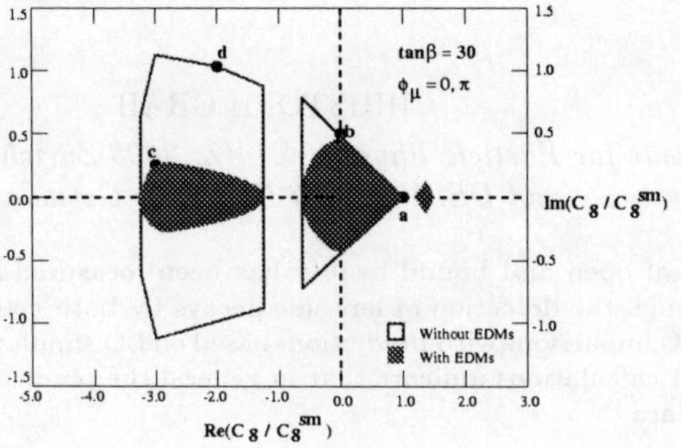

Figure 1. The allowded region of C_8/C_8^{sm} at the bottom mass scale with $tan\beta = 30$.

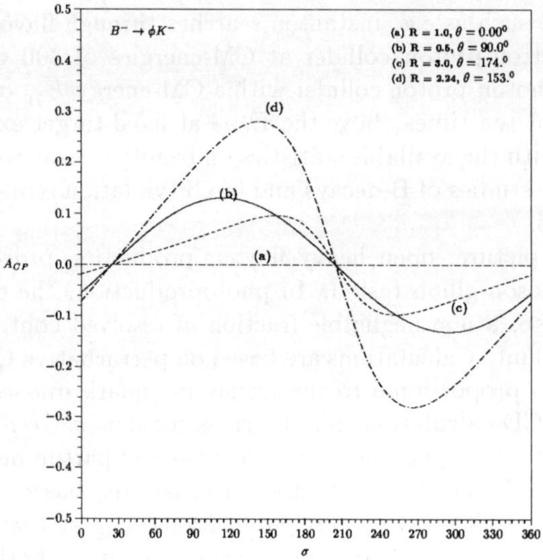

Figure 2. CP asymmetry at four different points: (a) prediction of the Standard Model, (b) the point with pure imaginary of C_8, (c) the point with maximum distance from origin, (d) an example point without EDM constaint.

B-PHYSICS IN H1 AND ZEUS AT HERA

CHRISTOPH GRAB

Institute for Particle Physics, ETHZ, 8093 Zürich, Switzerland, and DESY, 22607 Hamburg, Germany

Production of open and bound beauty has been measured in $e-p$-collisions at HERA through the detection of leptonic decays by both collider experiments H1 and ZEUS. Comparisons with predictions based on LO simulation and NLO perturbative QCD calculations indicate that in general the predictions fall significantly below the data.

Introduction

The main interest in measuring beauty quarks at the HERA $e-p$ collider is twofold: a) investigating the production processes allows detailed comparisons with pQCD calculations and extends the studies to non-perturbative phenomena; b) as a probing tool, it gives access to the parton density functions in the proton and in the photon. Beyond this, it serves as a probe for non-standard physics processes, such as e.g. instanton searches through flavour democracy. HERA is an electron-proton collider at CM-energies of 300 GeV; however, HERA is also a photon-proton collider with a CM-energy $E_{\gamma p}$ of typically 100 to 300 GeV, about ten times above the E_{CM} at fixed target experiments.

Comparing with the available statistics on beauty meson events, it is clear that spectroscopy studies of B-decays and of CP-violation remain the domain of e^+e^- machines.

In the QCD picture, open heavy flavour production proceeds predominantly through boson-gluon fusion. In photoproduction, the domain at low Q^2 considered here, a non-negligible fraction of resolved contributions must be taken into account. Calculations are based on perturbative QCD[1] and typically apply scales proportional to the transverse quark masses. Predictions by such NLO pQCD calculations for the cross sections $\sigma^{tot}(ep \to eb\bar{b})$ are at the level of 5 to 10 nb, depending on quark mass and parton density function assumptions. The Monte Carlo simulation programs, based on LO matrix elements, predict 2 to 4 nb. Previous publications suggest that pQCD calculations are pretty accurate in describing charm production, both in shape and normalisation. Because the beauty quark mass, and hence the scale involved, is considerably larger, pQCD is anticipated to be more accurate in describing the beauty case.

The results presented here are based on data collected in 1995 through 1997, and correspond to luminosities between 8 and 37 pb^{-1}.

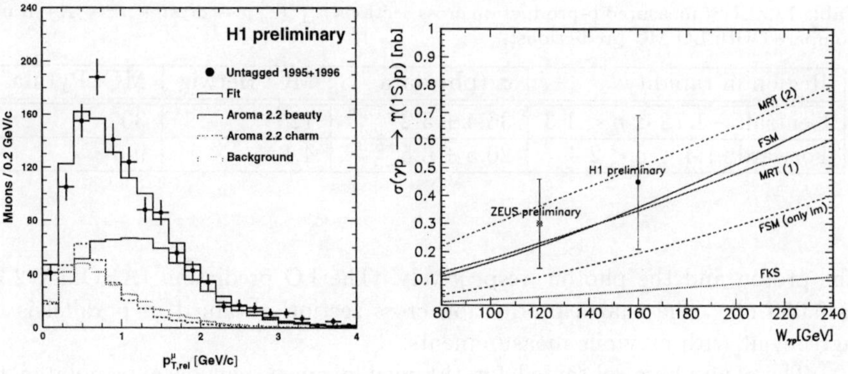

Figure 1. a) H1: fit to the $p^{\mu}_{T,rel}$ distribution using three components, shown separately: beauty (solid), charm (dashed), background (dotted). b) $\Upsilon(1S)$ photoproduction cross section as a function of the γp CMS energy, compared with predictions, based on pQCD approaches.

$b\bar{b}$ Production Cross Sections

The main strategy for finding beauty production is to look for semileptonic decays of the b-quark, $b \rightarrow l\nu_l X$, with its particular decay kinematics of a hard lepton (μ or e). In the H1 analysis[2], events with at least two jets with $E_T^{jet} > 6\,\mathrm{GeV}$ and one high momentum muon with $p_T > 2$ GeV are selected. Jets are reconstructed by means of the cone algorithmus ($R = \sqrt{\Delta\Phi^2 + \Delta\eta^2} < 1$). The discriminating variable to distinguish hard muons of b-decays from other muons of charm- and light quark decays, is the so called $p^{\mu}_{T,rel}$. It describes the relative transverse momentum of the muon with respect to the jet axis (taken as the thrust of the jet). The $p^{\mu}_{T,rel}$ distribution is fitted to a superposition of three components: beauty, charm and background (BG). The shapes for the b and c distributions are modeled by MC. The BG shape is determined from data, and its normalisation is kept fixed for the fit. The fit, shown for the H1 analysis[2] in fig.1a, yields a considerable b-fraction of $(50.8 \pm 4.9)\%$.

The resulting electroproduction cross section in the visible range, defined by $35° < \theta^{\mu} < 130°$, $p^{\mu}_{T,lab} > 2.0$ GeV, $Q^2 < 1\,\mathrm{GeV}^2$, $95 \leq W_{\gamma p} \leq 270\,\mathrm{GeV}$, is found to be $\sigma^{vis}(ep \rightarrow b\bar{b}X \rightarrow \mu X') = (0.176 \pm 0.016^{+0.026}_{-0.017})$ nb. The corresponding predictions by NLO QCD[1] are (0.104 ± 0.017) nb, assuming $m_b = 4.75\,\mathrm{GeV}$ and the MRSG and GRV-HO parton density functions for

Table 1. ZEUS: measured b-production cross sections $\sigma^{vis}_{b\bar{b}}(e^+p \to dijet + \mu^{\pm} + X)$ in nb, compared with LO MC predictions.

Region in rapidity	σ (pb) data	MC: Herwig	MC: Pythia
Central: $-1.75 < \eta < 1.3$	$36.4 \pm 5.2^{+10.4}_{-9.1}$	12	30
Forward: $1.4 < \eta < 2.4$	$20.5 \pm 6.5^{+6.3}_{-5.7}$	4.1	9.6

the proton and the photon, respectively. The LO prediction (AROMA 2.2) is 0.038 nb. The charm production cross section obtained in parallel is in agreement with previous measurements.

These numbers, obtained for the visible range, can be extrapolated to the full phase space by using the NLO calculations and the branching ratio $BR(B \to \mu\nu_\mu X)$. Converting to photoproduction yields finally a total cross section of $\sigma^{tot}(\gamma p \to eb\bar{b}X) = (111 \pm 10^{+16}_{-11} \pm 17)$ nb for the range $95 \leq W_{\gamma p} \leq 270\,\text{GeV}$. A separate independent study of dimuon events confirms this observation.

The ZEUS collaboration[3] has measured b-production in a similar way via the semi-muonic decays in different rapidity regions. In the forward direction a large resolved contribution of order 64% is observed. The measured b-production cross section for the visible range and the corresponding LO MC predictions are given in table 1.

A separate analysis of the ZEUS collaboration[3] is based on the electron semileptonic decay. The final data sample of dijet events with at least one electron is obtained by statistical subtraction of a hadron enriched and an electron enriched sample. The enrichment criteria rely on the energy loss dE/dx and the ratio E_{elmagn}/E_{tot} of electromagnetic to total deposited energy.

The shapes of various kinematical variables are well described by the MC simulations. The absolute normalization is however below the data by a factor of 3.7 for the HERWIG model and about a factor of 2 for the Pythia model. From the fit of the p_T^{rel} distribution, the production cross section for the visible range is calculated, and then extrapolated to the parton level to allow a comparison with the NLO prediction. For the region $p_T[b(\bar{b})] \geq 5\,\text{GeV}$, $|\eta[b(\bar{b})]| < 2$ and $Q^2 < 1\,\text{GeV}^2$, $0.2 < y_{\text{JB}} < 0.8$, the data yield $\sigma^{\text{ext}}_{ep\to b(\bar{b})X} = (4.3 \pm 1.2^{+2.6+0.3}_{-1.8-0.5})$ nb, whereas NLO[1] predicts 1.2 nb (with MRSG, GRV-G HO, $m_b = 4.75\,\text{GeV}$).

It is apparent from these several results, that the observed cross section lies about a factor of 2 above the NLO and roughly 3-5 times above the LO predictions.

Υ Production Cross Sections

Another method of accessing the beauty production is through the detection of elastic Υ production and their dilepton decays. The analyses of both H1 and ZEUS are performed in the regime of photoproduction and of diffractive interactions. The available statistical samples are tiny: H1[4] sees some (8.3 ± 3.9) $\Upsilon \to \mu^+\mu^-$ events in a total of 34.5 pb^{-1} luminosity, and ZEUS[5] has similar statistics in both muon and electron decay channels. Assuming 70% of Υ are produced in the 1S state, and using a MC simulation for acceptance corrections, the total photoproduction cross section for elastic $\Upsilon(1S)$ production is calculated and compared with predictions (see fig.1b). These calculations of the diffractive Υ production describe the exchange of a colourless object between the photon and the proton, based on leading log approximations in pQCD inspired models. The curve labeled FKS[6] in fig.1b corresponds to the hitherto standard model, that describes the J/ψ production very well. One might argue that there is a rise visible towards higher energies, which appears to be steeper than in the J/ψ case, and that the data lie above the predictions. Further details can be found in ref.[7]. A comparison with the J/ψ yields gives $\sigma(\gamma p \longrightarrow \Upsilon(1S)\ p)\ /\ \sigma(\gamma p \longrightarrow J/\psi\ p) \ = \ (4.8 \pm 2.2^{+0.7}_{-0.6}) \cdot 10^{-3}$.

Summary

Beauty production has been measured and the cross section extraction is well understood by both collaborations. Shapes and resolved contributions are found to be consistent with expectations, however the absolute normalisations of the predictions fall below the data, by about a factor of 2 in the case of NLO and about 3-5 for LO calculations. Upsilon production has been measured, and seem to lie above the predictions. These results from HERA confirm previous observations at the Tevatron stating that the NLO pQCD predictions consistently lay below the data.

References

1. S.Frixione *et al*, Phys. Lett. B **348**, 633 (1995).
2. H1 colla., C.Adloff *et al*, DESY 99-126, submitted to Phys. Lett. B.
3. ZEUS collab., contribution 498 to EPS-HEP conference, Tampere 1999.
4. H1 collaboration, contribution 574 to ICHEP98, Vancouver 1998.
5. ZEUS collaboration, J.Breitweg *et al*, Phys. Lett. B **437**, 432 (1998); P.Merkel, talk at EPS-HEP conference, Tampere 1999.
6. L.Frankfurt *et al*, Phys. Rev. **D57**, 512 (1998).
7. P.Merkel, PhD Thesis, Univ. Hamburg 1999, DESY Thesis 99-010.

PHYSICS PROGRAM AT HERA-B WITH HIGH-P$_T$ TRIGGER

V.BALAGURA, R.CHISTOV*, M.DANILOV, V.EIGES, R.MIZUK

Institute for Theoretical and Experimental Physics,
B.Cheremushkinskaya 25, 117259, Moscow, Russia

HERA-B was originally proposed as an experiment for CP violation study in $B^0 \to J/\psi K_s^0$ decay channel. High-p$_t$ trigger considerably extends HERA-B physics program. It selects events with two high-p$_t$ hadrons or high-p$_t$ hadron and lepton. Physics program with high-p$_t$ trigger at HERA-B includes the measurements of CP violation in $B^0 \to K^+\pi^-$, $B^0 \to D^{(*)+}D^{(*)-}$, $B^0 \to a_1^\pm \pi^\mp$ and $B^+ \to \rho^0\pi^+$ decays and measurement of $B_s - \overline{B}_s$ mixing.

1 Introduction

HERA-B is a dedicated experiment to search for CP violation in B meson decays. The main goal of the experiment is to determine sin2β by measuring CP asymmetry in $B^0 \to J/\psi K_s^0$ decay. For that purpose dilepton trigger was developed. The detector description and the present status of the HERA-B experiment is presented in the dedicated plenary talk on this conference[1].

High-p$_t$ trigger was proposed to enrich HERA-B physics program [2]. It selects $b\bar{b}$ events with high-p$_t$ hadron-hadron or hadron-lepton event topologies. The difference between the p$_t$ spectra in min. bias and $b\bar{b}$ events is used to reach necessary background suppression. Hadron-hadron trigger was optimized for the selection of two-prong B decays like $B \to \pi^+\pi^-, K^+\pi^-$. Hadron-lepton trigger, for example, selects the events where one B decays semileptonically and the second B gives high-pt charged track. The hardware implementation of the high-p$_t$ trigger and first run results can be found elsewhere[3].

2 Measurement of CP violation in $B^0 \to K^+\pi^-$ decay

The measurement of CP violating asymmetry in $B^0(\overline{B}^0) \to K^+\pi^-(K^-\pi^+)$ decays which provides information about the CKM angle γ is the most important target for hadron-hadron trigger. At the first level trigger two oppositely charged tracks are required with $p_t^{1,2} > 1.5$ GeV/c, invariant mass $M_{1,2} > 4.5$ GeV/c^2 and energy asymmetry $\frac{|E_1 - E_2|}{E_1 + E_2} < 0.5$. Further suppression of min. bias background will be attained by the cuts on the sec-

*TALK PRESENTED BY R.CHISTOV, E-MAIL: CHISTOV@MAIL.DESY.DE

ondary vertex, presence of a second B in the event and by the cut on the angle between the reconstructed B momentum and its flight direction. Also the cuts on kaon and pion identification should be imposed to suppress kinematic reflections from $B^0 \to \pi^+\pi^-, B_s \to K^+K^-$ decays[4]. The background rejection factor in the mass window of $|M_{1,2} - M_{B^0}| < 50$ MeV/c^2 was roughly estimated to be 0.84×10^{12}. Taking the central value of the CLEO measurement[5] of $\frac{1}{2} \cdot (\mathcal{B}(B^0 \to K^+\pi^-) + \mathcal{B}(\overline{B}^0 \to K^-\pi^+)) = 17.2 \times 10^{-6}$, the sample of $B^0 \to K^+\pi^-$ and $\overline{B}^0 \to K^-\pi^+$ decays was evaluated to 460 events which could be used in the analysis[a]. This gives the relative statistical error $\delta\mathcal{B}(B^0 \to K^+\pi^-)/\mathcal{B}(B^0 \to K^+\pi^-) = 0.15$ and the absolute error in the asymmetry $\delta\mathcal{A}_{CP}(K^\pm\pi^\mp) = 0.10$.

3 $B_s^0 - \overline{B}_s^0$ mixing

$B_s^0 - \overline{B}_s^0$ mixing can be searched for using $B_s^0 \to D_s^-\pi^+(3\pi^\pm)$ and $B_s^0 \to D_s^{*-}\pi^+(3\pi^\pm)$ decays where D_s^+ is reconstructed in 6 hadronic modes. The D_s^{*-} is reconstructed only partially without γ reconstruction from $D_s^{*-} \to D_s^-\gamma$ decay. Such B_s decays will be selected by hadron-lepton trigger[b]. The first level hadron-lepton trigger requirements are $p_t^{electron} > 1.4$ or $p_t^{muon} > 1.0$ and $p_t^{hadron} > 1.8$ GeV/c. In order to further suppress the min. bias background at the trigger level the cuts on hadron and lepton impact parameters of > 100 μm are applied. The expected numbers of reconstructed $B_s \to D_s\pi(3\pi)$ and $B_s \to D_s^*\pi(3\pi)$ tagged decays are 214 and 187 respectively. The mean proper time resolution for B_s was estimated to be 0.050 ps for $B_s \to D_s\pi(3\pi)$ and 0.066 ps for $B_s \to D_s^*\pi(3\pi)$ mode. After the very rough background modeling the background to signal ratio was conservatively estimated to be $B/S = 0.1$. It allows to measure $B_s - \overline{B}_s$ mixing at HERA-B with at least 3σ accuracy for the x_s values up to $x_s = 35$ after one year of operation. $x_s = 25$ is reachable within 7 months of running.

4 sin2β measurement using $B^0 \to D^{(*)+}D^{(*)-}$ decays

The events with $B^0 \to D^{(*)+}D^{(*)-}$ decays along with the tagging lepton will be selected by hadron-lepton trigger. We intend to use both fully reconstructed events and events with one of the $D^{*+}(\to D^0\pi^+)$ reconstructed partially

[a]Everywhere in this article we assume one year(10^7 s) of running, full HERA-B setup, nominal interaction rate 40 MHz and $\sigma_{inel}/\sigma_{b\bar{b}} = 10^{-6}$.

[b]The requirement of high-p_t lepton in the hadron-lepton trigger will tag the B flavour at production.

when only π^+ and D^0 vertex are reconstructed. In order to suppress the background the vertex cuts have to be applied. To form B^0 vertex in case of partial reconstruction the slow pion from D^* smeared by multiple scattering is not used. Instead of it we use the line parallel to π momentum which passed through the well determined D^0 vertex. Neglecting the P-wave contribution for $D^{*+}D^{*-}$ and complications due to the penguin pollution and also probably non zero strong phase between $D^{*+}D^-$ and D^+D^{*-} the estimation yields the error $\delta\sin2\beta = 0.18$ where the ratio of the background to signal in case of partial reconstruction was roughly estimated to be $B/S = 1$. Thus the accuracy of this measurement can be comparable with that of using $B \to J/\psi K_s^0$ decay mode.

5 sin2α measurement using $B^0(\overline{B}^0) \to a_1^\pm \pi^\mp$ decays

Hadron-lepton trigger gives also the possibility to collect 140 $B^0(\bar{B}^0) \to a_1^\pm \pi^\mp$ tagged decays[c]. In order to estimate the reachable level of accuracy in sin2α measurement, we assumed the equality $\mathcal{B}(B^0 \to a_1^- \pi^+) = \mathcal{B}(B^0 \to a_1^+ \pi^-)$ and neglected the penguin contribution to these decays. We also neglected the strong phase between the amplitudes for the $B^0 \to a_1^- \pi^+$ and $B^0 \to a_1^+ \pi^-$ decays. In case of negligible background this leads to the error of $\delta\mathcal{A}_{CP} = \delta\sin(2\alpha) = 0.19$. Recent fit of the Unitarity Triangle angles gives[7] $\sin2\alpha = -0.26^{+0.29}_{-0.28}$. Therefore if α is around 90° the error $\delta\sin(2\alpha) = 0.19$ implies $\delta\alpha \approx 6^o$.

6 Measurement of CP violation using $B^\pm \to \rho^0 \pi^\pm$ decays

Based on the CLEO measurement[6] of $\mathcal{B}(B^+ \to \rho^0 \pi^+) = (15 \pm 5 \pm 4) \times 10^{-6}$ the number of $B^\pm \to \rho^0 \pi^\pm$ decays collected by both hadron-lepton and hadron-hadron triggers is estimated to be 150. This sample provides the possibility to measure CP violation asymmetry in $B^\pm \to \rho^0 \pi^\pm$ decays with the accuracy of $\delta\mathcal{A}_{CP} = 0.15$. The maximum value of $\mathcal{A}_{CP}$ in this decay can reach the level of 50%.

7 Summary

High-p_t hadron-hadron and hadron-lepton triggers considerably enhance the HERA-B physics reach. The physics program of the high-p_t trigger includes

[c]We took the CLEO result[6] $\mathcal{B}(B^0 \to \rho^\pm \pi^\mp) = (35^{+11}_{-10} \pm 5) \times 10^{-6}$ and assumed $\mathcal{B}(B^0 \to \rho^\pm \pi^\mp) = \mathcal{B}(B^0 \to a_1^- \pi^+) + \mathcal{B}(B^0 \to a_1^+ \pi^-)$.

the measurements of CP asymmetry in $B^0 \to K^+\pi^-(K^-\pi^+)$ decay with the accuracy of $\delta\mathcal{A}_{CP} = 0.10$, sin2$\beta$ measurement using $B^0 \to D^{(*)+}D^{(*)-}$ decays with the accuracy of $\delta\text{sin}2\beta = 0.18$, sin2$\alpha$ measurement using $B^0 \to a_1^\pm\pi^\mp$ decays with the accuracy of $\delta\sin(2\alpha) = 0.19$ and CP asymmetry in $B^\pm \to \rho^0\pi^\pm$ decay with the error of $\delta\mathcal{A}_{CP} = 0.15$. It also includes $B_s - \overline{B}_s$ mixing measurement up to $x_s = 35$. All these measurements will allow HERA-B to contribute to the task of the Unitarity Triangle overconstraining.

Acknowledgments

This work is partly supported by Russian Foundation for Basic Research (grant 99-02-16878) and by Bundesministerium für Bildung und Forschung (Germany) via the Max Planck Research Award.

References

1. O.Steinkamp, these proceedings.
2. HERA-B Technical Design Report, *DESY-PRC* **95/01**, 59 (1995).
3. V.Balagura *et al*, *Nucl. Instrum. Methods* A **368**, 252 (1995);
   V.Balagura *et al*, *Nucl. Instrum. Methods* A **379**, 404 (1996);
   V.Balagura *et al*, *Nucl.Phys.B(Proc.Suppl.)* **75B**, 359 (1999);
   V.Balagura *et al*, *High-P_t Trigger for HERA-B Experiment*, to be published in the Proceedings of VII Int. Conf. on Instrumentation for Colliding Beam Physics, Hamamatsu, Japan, 1999, *Nucl.Instrum.Methods.*
4. V.Balagura *et al*, *ITEP* **3-99**, 1 (1999).
5. CLEO Collaboration, *hep-ex/0001010.*
6. CLEO Collaboration, *hep-ex/9908018.*
7. F.Parodi, P.Roudeau, A.Stocchi, *hep-ex/9903063.*
   See also
   A.Ali, D.London, *Eur.Phys.J.* **C9**, 687(1999);
   M.Ciuchini *et al*, *hep-ph/9910236*;
   S.Plaszczynski, M.-H.Schune, *hep-ph/9911280.*

NONPERTURBATIVE DESCRIPTION OF B-MESONS

WEI-MIN ZHANG
Department of Physics, National Cheng-Kung University
Tainan, Taiwan 701,ROC
E-mail: wzhang@mail.ncku.edu.tw

The B meson bound state equation, as a covariant extension of the light-front bound state equation in light-front QCD, is derived from the heavy quark equation of motion in the heavy quark effective theory (HQET) together with a heavy meson effective theory we developed recently. Solutions to the covariant B meson wave functions are then solved by minimizing the binding energy $\overline{\Lambda}$ from the bound state equation. Subsequently, the basic HQET parameters λ_1, λ_2, and the heavy quark mass m_b, m_c are consistently calculated.

1 Overview

Measurements in any B-factory must involve hadronic structure of B hadrons. While, hadronic structure is governed by nonperturbative QCD dynamics, which is still unsolved. As a result, those experimental data extracted from measurements in various B-factories certainly contain uncertainty arisen from the unknown hadronic structure. This results in uncontrollable ambiguous in searching a clear signature of CP violation or new physics from B-physics. No doubted, nonperturbatively solving the bound states of B-hadrons is one of the most essential problems in understanding B-physics.

As it is well-known, the main problem in solving hadronic structure lies at the central issue of nonperturbative QCD: how quarks are confined in hadrons and how the low energy dynamics of hadrons are dominated by the chiral symmetry breaking. Attempts in the past 25 years made no much achievement in solving directly from QCD these two difficult problems. However, for heavy hadrons, especially for B mesons, chiral symmetry may play a less important role. Hence, besides the constraint of heavy quark symmetry (HQS), B-meson bound states should be dominated alone by the confinement between the b-quark and the light constituents. This simplifies a lot the nonperturbative QCD dynamics of the B meson structure.

In this talk, I will report the recent results we obtained from a nonperturbative description of B-mesons. Our description to B mesons is based on the heavy quark equation of motion in HQET, together with a heavy meson effective field theory we developed very recently. Before I talk about our results, it may be helpful to give a brief analysis on various descriptions of the B-meson structure in the past decade. The widely used description of hadronic structure is the constituent quark model. However, the quark model is only applicable for physical processes with small momentum transfer (~ 1 GeV). While, in B-physics, most of processes involve a momentum transfer over $0 \sim 25$ GeV^2, which makes the quark model fail in a large range of the allowed momentum region. Another well-established approach is the QCD sum rule. The QCD sum rule catches the essential feature of chiral symmetry breaking effect (i.e. quark condensation) to determine the pseudo-scalar mesonic wave functions. However, chiral symmetry in B-physics is less important. Therefore, the QCD sum rule should loss its power in describing B meson structure. On the other hand, since q^2 in B physics can be very large (~ 25 GeV^2), the perturbative QCD may have some application. But the application should also be very limited since the B meson structure is still dominated by the low q^2 (about $1 \sim 4$ GeV^2) where the perturbative QCD should fail too. There is another new approach, called relativistic quark model or light-front quark model. This approach has no limitation of the constituent quark model, namely it can describe physical processes over all the region of the allowed momentum transfer. But as we have shown[1], this approach breaks the covariance, and therefore it may cause severe ambiguous in the calculations of various the B meson form factors.

To overcome the above mentioned problems, we propose recently an effective field theory to describe B meson structure[2]. Then we use the heavy quark equation of motion in HQET to determine the B meson wave functions[3]. This approach has incorporated in HQET, and is thus consistent with HQS. In this new approach, B-mesons are considered as composite particles consisting of a reduced b-quark (heavy quark in the $m_Q \to \infty$ limit) coupled with the light degrees of freedom. A structure function $\Psi(v \cdot p_q)$, corresponding to the wave function of B-meson bound state, is introduced to describe the quark-meson vertex in the effective Lagrangian. Meanwhile, this structure wave function satisfies a QCD-based bound state equation obtained from the heavy quark equation of motion. The heavy meson bound state equation is indeed a covariant extension of the light-front bound state equation I obtained recently from light-front QCD and HQET[4]. Thus, we provide a complete frame to determine B-meson structure.

2 Basic Formulation

Explicitly, the effective Lagrangian we have constructed to describe the low energy dynamics of heavy mesons reads[2,3]:

$$\mathcal{L}_{\rm eff} = \overline{h}_v iv \cdot \partial h_v + \overline{q}(i\not{D} - m_q)q - \frac{1}{4}F_a^{\mu\nu}F_{a\mu\nu} + P_v^\dagger\Big(iv\cdot \overleftrightarrow{\partial} - 2\overline{\Lambda}\Big)P_v$$

$$-V_v^{\mu\dagger}\left(iv\cdot \overleftrightarrow{\partial} -2\overline{\Lambda}\right)V_{v\mu} + \left(\overline{h}_v i\gamma_5 q_v P_v - \overline{h}_v\gamma_\mu q_v V_v^\mu + h.c.\right),$$
$$+\sum_{n=1}^{\infty}\left(\frac{1}{2m_Q}\right)^n \overline{h}_v(i\not{D}_\perp)(-iv\cdot D)^{n-1}(i\not{D}_\perp)h_v = \mathcal{L}_{\rm eff}^{QqH} + \mathcal{L}_{m_Q}\,, \quad (1)$$

where P_v and V_v^μ represent respectively the composite pseudo-scalar and vector heavy meson fields, $\overline{\Lambda} \equiv \lim_{m_Q\to\infty}(M_H - m_Q)$ is their common residual mass in the heavy quark limit, $q_v = G\Psi(v\cdot p_q)q$ represents collectively the light degrees of freedom in a heavy meson, where $\Psi(v\cdot p)$ is the heavy meson wave function mentioned earlier, p_q is a light antiquark momentum, and G a normalization constant. In Eq.(1), $\mathcal{L}_{\rm eff}^{QqH}$ obeys heavy quark symmetry and defines the nonperturbative effective coupling of heavy mesons with their constituent heavy and light quarks, while $\mathcal{L}_{m_Q}$, which contains all the $1/m_Q$ corrections, is considered as a perturbation to $\mathcal{L}_{\rm eff}^{QqH}$.

On the other hand, the reduced heavy quark field h_v in HQET obeys the following equation of motion $iv\cdot Dh_v = 0$, where $D^\mu = \partial^\mu - igA^\mu$ is the QCD covariant derivative. To ensure that our effective field theory is consistent with low energy QCD dynamics of HQET, we require that in the heavy quark limit, the heavy meson state $|H(v)\rangle$ satisfies

$$\langle H_v|\overline{h}_v iv\cdot Dh_v|H_v\rangle = 0. \quad (2)$$

By setting the heavy meson wave function $\Psi(v\cdot p_q)$ as $\Psi(v\cdot p_q) = (\overline{\Lambda} - v\cdot p_q)\varphi(v\cdot p_q)$, Eq. (2) can be reduced to an equivalent bound state equation for heavy mesons[3]:

$$(\overline{\Lambda} - v\cdot p_q)\varphi(v\cdot p_q) = C_f\int\frac{d^4p_q'}{(2\pi)^4}\left(\frac{i4\pi\alpha_s}{(p_q-p_q')^2} - 4\pi^2\Lambda_c^2\delta^4(p_q-p_q')\right)$$
$$\times\frac{(v\cdot p_q+m_q)(v\cdot p_q'+m_q)+v\cdot p_q v\cdot p_q' - p_q\cdot p_q'}{(v\cdot p_q+m_q)(p_q'^2-m_q^2)}\varphi(v\cdot p_q'), \quad (3)$$

which is indeed a covariant generalization of the light-front bound state equation for heavy mesons[4]. Also note that the interaction contains the color Coulomb interaction ($\sim \alpha_s$), plus a confining interaction ($\sim \Lambda_c^2$).

3 Numerical results

Now, we can determine the heavy meson wave function $\varphi(v\cdot p_q)$ through a variation procedure. As discussed in Ref. [2,3], the wave function should be relatively well behaved, namely, when continued into the complex p_q plane, $\varphi(v\cdot p_q)$ is analytic except for isolated singularities, and it vanishes "fast enough" as $|v\cdot p_q| \to \infty$. The wave function that satisfies these restrictions has a form

$$\varphi_n(v \cdot p_q) = \frac{1}{(v \cdot p_q + \omega - i\epsilon)^n}. \tag{4}$$

Substituting this wave function into the bound state equation, and taking a minimization to $\bar{\Lambda}$ with respect to ω, we find that for $n = 6 \sim 8$, with $m_q = 0.22$ GeV, $\alpha_s = 0.44$ and the confining scale parameter $\Lambda_c = 0.68 \pm 0.02$ GeV, the resulting basic quantities that characterize the B-meson structure are listed below:

	$\bar{\Lambda}$	$-\lambda_1$	λ_2	m_b	m_c
Theo.	0.44 ± 0.01	0.25 ± 0.02	0.11	4.85 ± 0.01	1.56 ± 0.01
Exp.	0.39 ± 0.11[5]	0.19 ± 0.1 [5]	0.11[6]	4.89 ± 0.05[7]	1.59 ± 0.02[7]

As we can see the HQET parameters determined from our theory are in good agreement with those extracted from experiments.

Acknowledgments

This work is in collaboration with H. Y. Cheng and C. Y. Cheung and is supported in part by National Science Council of the Republic of China under Grants NSC89-2112-M006-004 and NSC89-2112-M006-026.

References

1. H. Y. Cheng, C. Y. Cheung, C. W. Hwang, and W. M. Zhang, *Phys. Rev.* D **57**, 5598 (1998).
2. H. Y. Cheng, C. Y. Cheung, and W. M. Zhang, *Phys. Rev.* D **58**, 74003 (1998).
3. C. Y. Cheung and W. M. Zhang, *Phys. Rev.* D **60**, 014017 (1999).
4. W. M. Zhang, *Phys. Rev.* D **56**, 1528 (1997).
5. M. Gremm et al., *Phys. Rev. Lett.* **77**, 20 (1996).
6. In the literature, the so-called "experimental" value $\lambda_2 \simeq 0.12$ GeV^2 is obtained by assuming $m_b = M_B$. One would get $\lambda_2 \simeq 0.11$ GeV^2 if the pole-mass $m_b^{\rm pole} = 4.89$ GeV is used for $\Delta M_{B^*B} = 45.7$ MeV.
7. There are no experimental data for constituent quark masses m_b and m_c. Here we displace the pole masses from H. Fusaoka and Y. Koide, *Phys. Rev.* D **57**, 3886 (1998).

PREASYMPTOTIC EFFECTS IN b-DECAYS*

B. GUBERINA, B. MELIĆ, AND H. ŠTEFANČIĆ

Theoretical Physics Division, Rudjer Bošković Institute, P.O.Box 1016, HR-10001 Zagreb, Croatia

E-mails: guberina@thphys.irb.hr, melic@thphys.irb.hr, shrvoje@thphys.irb.hr

Large preasymptotic effects in beauty decays have been found using heavy-quark and SU(3) symmetry, as well as experimental data on charmed hyperons. Contrary to rather uniform beauty-meson lifetimes, a much larger spread of beauty-baryon lifetimes is predicted. However, it is highly unlikely that, theoretically, the $\tau(\Lambda_b)/\tau(B_d^0)$ ratio, which at present deviates more than 1σ from the experimental result, can be lowered below 0.9.

The saga of beauty never comes to an end[1]. During the last decade the general belief has been that decays of beauty quarks should be very well described theoretically in the framework of the Operator Product Expansion (OPE) and the Heavy-Quark Effective Theory (HQET). The mass of the beauty quark, being of the order of 5 GeV, appears to be heavy enough to ensure the fast convergence of the $1/m_b$ expansion. The diversity of lifetimes of 'beautiful' mesons and baryons is expected first at the subleading level in the $1/m_b$ expansion. The lifetimes of beauty mesons follow this simple theoretical prediction within $5 - 10\%$ in the $m_b \to \infty$ limit:

$$\tau(B^+) = \tau(B_d^0) = \tau(B_s^0) \,. \tag{1}$$

The only measured baryon lifetime $\tau(\Lambda_b)$ appears to be smaller by $15 - 25\%$; experimentally, it follows that

$$\frac{\tau(\Lambda_b)}{\tau(B_d^0)} = 0.81 \pm 0.05 \,(\mathrm{PDG}) \,, \tag{2}$$

while the theoretical value is about 0.98. This discrepancy raises doubt that the quark-hadron duality might be, *horribile dictu*, severely flawed.

The rate of the beauty-hadron decay is given as a sum over matrix elements of D-dimensional operators. The sum starts with the operator of dimension D=3 and shows the fast convergence in the $1/m_b$ expansion. First corrections proportional to the operator of D=5 are controllable and by taking them into account we obtain the difference of $2 - 3\%$ in the lifetimes of beauty hadrons. Therefore, the only hope to come closer to the ratio (2) is to look for

*TALK GIVEN BY B. MELIĆ AT THE 3RD INTERNATIONAL CONFERENCE ON B PHYSICS AND CP VIOLATION, TAIPEI, TAIWAN, DECEMBER 3-7, 1999

the possible larger contributions coming from the operators of dimension D=6 or higher. These operators are known to play an important role in charmed-meson decays[2] and their effects are even more pronounced in charmed-baryon decays[3]. Recently, the analysis of singly charmed baryons[4] has been extended to doubly charmed baryons[5], where the idea of 'meson-like' baryons, in which heavy quarks form a diquark, was applied. A possible physical interpretation of this philosophy is the existence of hadronic supersymmetry[6], which appears whenever the diquark is physically realizable.

However, the calculation of the contribution of D=6 operators suffers from the strong model-dependence in the evaluation of their matrix elements. Recently, Voloshin[7] has proposed the way of reducing the model dependence. His main assumption is that, owing to SU(3)-flavor and heavy-quark symmetry, the matrix elements of four-quark operators extracted from experimental data on charmed hyperons can be used for calculation in the beauty sector, provided that they are normalized at the low energy scale $\mu \sim 1$ GeV. The result of these assumptions is that lifetime differences between beauty hyperons (Λ_b, Ξ_b) can be expressed through the (measured) lifetime differences between charmed hyperons (Λ_c, Ξ_c) without invoking an explicit model-dependent calculation of four-quark operators. The possible uncertainty of this approach is of the order $\mathcal{O}(m_c^2/m_b^2) \sim 0.1$.

We have extended[8] Voloshin's analysis by introducing a rather modest model dependence in order to obtain more predictive power, i.e., the lifetimes of the (Λ_b, Ξ_b) hyperon triplet and the lifetime of Ω_b. Basically, we express the decay rates in terms of the baryon wave function squared $|\Psi(0)|^2$, which parametrizes the four-quark operator contributions and is usually given by the nonrelativistic relation $|\Psi(0)|^2 \sim f_B^2$, f_B being the meson decay constant. We use Voloshin's approach to determine the value of the $|\Psi(0)|^2$. From the lifetime differences between members of beauty hyperons we are able to extract an effective decay constant, $|\Psi(0)|^2 \sim (F_B^{eff})^2$,

$$F_B^{eff} = (0.441 \pm 0.026)\,\text{GeV}\,. \tag{3}$$

It is instructive to compare this value with the value of f_B, $f_B = 0.16 - 0.17\,$GeV, and then the ratio of the baryon over meson wave function, which is, according to our model, about 7-8 times larger than when we apply the nonrelativistic model: $|\Psi_{\Lambda_b}(0)|^2/|\Psi_B(0)|^2 \sim 4.2$. This effectively means that, in our approach, four-quark contributions in beauty-baryon lifetimes are enhanced by almost an order of magnitude.

Owing to ambiguities in the determination of the bottom quark mass, we concentrate mainly on the lifetime ratios. We follow the philosophy of using the running quark mass $\overline{m}_b$ to avoid the renormalon ambiguities and therefore have to choose $\overline{m}_b(1\,\text{GeV}) = 4.7\,\text{GeV}$.

The predicted hierarchy in the sector of beauty baryons is then

$$\tau(\Lambda_b) \simeq \tau(\Xi_b^0) < \tau(\Xi_b^-) < \tau(\Omega_b)\,, \tag{4}$$

and the obtained lifetime ratios are

$$\tau(\Xi_b^-)/\tau(\Lambda_b) \simeq 1.20\,, \qquad \tau(\Omega_b)/\tau(\Lambda_b) \simeq 1.30\,. \tag{5}$$

These ratios are much larger than those predicted by the standard nonrelativistic model.

Let us now briefly discuss the problem of the $\tau(\Lambda_b)$ over $\tau(B)$ ratio. The problem is the clear discrepancy between the experimental ratio (2) and the theory which gives

$$\tau(\Lambda_b)/\tau(B_d^0) = 0.97 + \mathcal{O}(1/m_b^3) \tag{6}$$

just by taking the first, $\mathcal{O}(1/m_b^2)$ corrections into account. Owing to the fast convergence of the $1/m_b$ expansion and because the vacuum saturation approximation for mesons works rather well, it seems that the decay rate of the B-meson cannot be significantly smaller to lower the $\tau(\Lambda_b)/\tau(B)$ ratio. The only hope then persists in the enlargement of four-quark contributions in the Λ_b-decay, but these effects also cannot be pushed over some limit. So the question is: can we accommodate the theoretical prediction on the $\tau(\Lambda_b)/\tau(B_d^0)$ ratio to the experimental result, using our enhancement of the four-quark contributions in the Λ_b-decay ?

Clearly, we need the smaller m_b to obtain the larger preasymptotic effects, but there is a competition between the $\mathcal{O}(1/m_b^2)$ effects in mesons and the $\mathcal{O}(1/m_b^3)$ effects in Λ_b. The net result is then the stable $\tau(\Lambda_b)/\tau(B)$ ratio:

$$\tau(\Lambda_b)/\tau(B_d^0) \sim 0.90 \pm 0.01 \qquad \text{for} \qquad m_b = 4.4 - 4.8\,\text{GeV}\,. \tag{7}$$

We have also checked the result against the deviation from the valence quark approximation (VQA)[9], which equals the deviation of the $\overline{B}$-parameter from one:

$$x = -\overline{B}y\,. \tag{8}$$

Here $x \sim \langle \bar{b}\Gamma_\mu b\,\bar{q}\Gamma^\mu q\rangle$ and $y \sim \langle \bar{b}^i\Gamma_\mu b^j\,\bar{q}^j\Gamma^\mu q^i\rangle$, $\Gamma_\mu = (V-A)_\mu$, and q's denote light quarks in a baryon.

We have extracted this ratio to be $|x/y| \approx 1.8 \pm 1.0$ at $\mu = 1\,\text{GeV}$, which is consistent with the result[9] of Voloshin. Owing to the large error in its determination, it is difficult to give a definite conclusion about the validity of the VQA and therefore we prefer to use $\overline{B} = 1$ in our predictions. However, it is also clear that the VQA cannot be generally valid, because of the fact that y is a μ dependent quantity, while x is not, and the result that $|x/y|$ is significantly larger than 1 might prove at the end to be both correct and fundamental. But, even if the VQA is heavily broken by almost 100%, the $\tau(\Lambda_b)/\tau(B)$ ratio *cannot* be lowered below 0.9 value.

To conclude, although there might still be some place for nonfactorizable contributions in mesons to play some role, it is highly unlikely that the lifetimes of Λ_b and the B-meson can be split by more than 10% (7). To reach the experimental value for the $\tau(\Lambda_b)/\tau(B)$ ratio (2) would require $F_B^{eff}|_{\rm fit} \sim 0.720\,{\rm GeV}$, which can hardly be accommodated in the present theory. Should future data maintain the $\tau(\Lambda_b)/\tau(B)$ ratio well below 0.9, that would indicate the violation of some of underlying concepts of the present theory, such as the quark-hadron duality. One of the tests will also be the experimental check of the predicted spread of beauty-baryon lifetimes of the order of 20% in the $\tau(\Xi_b^-)/\tau(\Lambda_b)$ ratio and of 30% in the $\tau(\Omega_b)/\tau(\Lambda_b)$ ratio.

Acknowledgments

B.M. would like to thank H.Y. Cheng and W.S. Hou for invitation to participate in this very stimulating conference. This work was supported by the Ministry of Science and Technology of the Republic of Croatia under Contract No. 00980102.

References

1. I Bigi, hep-ph/0001003, these Proceedings
2. B Guberina, S Nussinov, R Peccei and R Rückl, *Phys. Lett.* B **89**, 111 (1979); N Bilić, B Guberina and J Trampetić, *Nucl. Phys.* B **248**, 261 (1984); M Shifman, M B Voloshin, *Sov. J. Nucl. Phys.* **41**, 120 (1985).
3. M Shifman, M B Voloshin, *JETP* **64**, 698 (1986); B Guberina, R Rückl, and J Trampetić, *Z. Phys.* C **33**, 297 (1986).
4. B Guberina, B Melić, *Eur. Phys. J. C* **2**, 697 (1998).
5. B Guberina, B Melić and H Štefančić, *Eur. Phys. J. C* **9**, 213 (1999); V V Kiselev, A K Likhoded and A I Onishchenko, *Phys. Rev.* D **60**, 014007 (1999).
6. D B Lichtenberg, hep-ph/9912280.
7. M B Voloshin, hep-ph/9901445.
8. B Guberina, B Melić and H Štefančić, *Phys. Lett.* B **469**, 253 (1999).
9. M B Voloshin, hep-ph/9908455.

NON-FACTORIZABLE CONTRIBUTIONS IN B DECAYS

K. TERASAKI

Yukawa Institute for Theoretical Physics, Kyoto University,
Kyoto 606-8502, Japan
E-mail: terasaki@yukawa.kyoto-u.ac.jp

$\bar{B} \to D\pi$, $D^*\pi$, $J/\psi\bar{K}$ and $J/\psi\pi$ decays are studied by assuming that their amplitude can be decomposed into a sum of factorizable and non-factorizable ones. Non-factorizable contributions are rather small in color favored $\bar{B} \to D^{[*]}\pi$ decays while they are important in $\bar{B} \to J/\psi\bar{K}$ and $J/\psi\pi$ decays.

Factorization prescription[1,2], which has been supported by two different arguments, the large N_c argument[3] and the argument of color transparency[4], has been applied extensively to hadronic weak processes. If the large N_c argument works well in B decays, it will work well again in charm decays since it is independent of flavors. In this case, factorized amplitudes for two body decays of charm mesons which are given by products of current matrix elements should be real. However, the observed branching ratios[5] for $D \to \bar{K}\pi$ decays imply that their amplitudes have large phases. To get rid of this problem, the factorized amplitudes are usually implemented[2] by multiplying phase factors arising from final state interactions (FSI). Amplitudes with FSI, however, are given by non-leading (NL) terms in the large N_c expansion and should be discarded in the $N_c \to \infty$ limit. On the other hand, the color transparency implies that the factorization can work well under a limited kinematical condition that a heavy quark decays into another heavy quark plus a pair of light quark and anti-quark with sufficiently high energies. In this case, the factorization can work well in $\bar{B} \to D\pi$ and $D^*\pi$ decays but cannot well in $\bar{B} \to J/\psi\bar{K}$ and $J/\psi\pi$. It will be checked by comparing non-factorizable contributions with factorized ones.

We now parameterize amplitudes for $\bar{B} \to D\pi$ [and $D^*\pi$] decays using the isospin eigen amplitudes, $A_{2I}^{[*]}$, and their phases, $\delta_{2I}^{[*]}$, with isospin I, and define the ratios of decay rates, $R_{0-}^{[*]} = \Gamma(\bar{B}^0 \to D^{[*]+}\pi^-)/\Gamma(B^- \to D^{[*]0}\pi^-)$ and $R_{00}^{[*]} = \Gamma(\bar{B}^0 \to D^{[*]0}\pi^0)/\Gamma(\bar{B}^0 \to D^{[*]+}\pi^-)$. Then, we obtain

$$\cos(\delta^{[*]}) = \Big(\frac{9R_{0-}^{[*]} - 1}{4}\Big)r^{[*]} - \frac{1}{r^{[*]}}, \tag{1}$$

where $\delta^{[*]} = \delta_1^{[*]} - \delta_3^{[*]}$ and $r^{[*]} = A_3^{[*]}/A_1^{[*]} = \{3R_{00}^{[*]}R_{0-}^{[*]} + 3R_{0-}^{[*]} - 1\}^{-1/2}$. From the observed branching ratios[5], $B_{\rm exp}$, listed in Table 2, we extract phenomenologically allowed values, $B_{\rm ph}$, which satisfy the right-hand-side of Eq.(1) ≤ 1, where the values with (†) are taken as the input data. In this way, we obtain approximately $0.96 \leq \cos(\delta) \leq 1$ and $0.70 \leq \cos(\delta^*) \leq 1$, i.e., the phases of these amplitudes are very close to each other (an overall phase).

Now we study non-factorizable contributions in the $\bar{B} \to D^{[*]}\pi$ decays assuming that their amplitude is given by a sum of factorizable and non-factorizable ones, i.e., $M_{\rm tot} \simeq M_{\rm FA} + M_{\rm NF}$. Using the naive factorization, we estimate $M_{\rm FA}$ for which the BSW Hamiltonian $H_w^{\rm BSW}$ is responsible[2]. When we obtain $H_w^{\rm BSW}$ from the effective weak Hamiltonian using the Fierz reordering, we have an extra term $\tilde{H}_w$ which is given by a sum of colored current products. Matrix elements of $\tilde{H}_w$ in which soft gluon(s) should be exchanged between quarks belonging to different external meson states correspond to NL terms in the large N_c expansion and are non-factorizable. Amplitudes with FSI which are given by dynamical contributions of hadrons are also described by NL diagrams similar to the above ones in the sense of the large N_c. Therefore, $M_{\rm NF}$ for which $\tilde{H}_w$ is responsible will be controlled by hadron dynamics and can be calculated approximately by using a hard pion technique in the infinite momentum frame[6]. In this approximation, $M_{\rm NF}$ is given by a sum of equal-time-commutator term ($M_{\rm ETC}$) and surface term ($M_{\rm S}$). By the way, an amplitude for dynamical hadronic process is given by a sum of the Born term and continuum contribution, and the latter can develop a phase relative to the former. In the present case, $M_{\rm S}$ is given by a sum of all possible pole amplitudes so that $M_{\rm ETC}$ will correspond to the continuum contribution[7] and can have a phase relative to $M_{\rm S}$. Neglecting small contributions of excited meson poles, we can describe $M_{\rm NF}$ by *asymptotic* matrix elements of $\tilde{H}_w$ and axial counterpart A_π of isospin (taken between single hadron states with infinite momentum) in addition to phases (restricted by $-\pi/2 < \tilde{\delta}_{2I} < \pi/2$) between $M_{\rm ETC}$ and $M_{\rm S}$. However, we do not need to worry about $\tilde{\delta}^*_{2I}$ since $M_{\rm ETC}(\bar{B} \to D^*\pi)$ is proportional to $\langle D^{*0}|\tilde{H}_w|\bar{B}^0\rangle$ which should vanish because of spin conservation[8].

To compare our result on branching ratios with $B_{\rm ph}$ in Table 2, we need to estimate values of asymptotic matrix elements of A_π, form factors and decay constants involved. Reasonable values of them obtained from $\Gamma_{\rm exp}(\rho \to \pi\pi)$, the heavy quark effective theory and related models, and the lattice QCD lead to the approximate amplitudes in Table 1, where a_1 and a_2, $\tilde{\delta}_1$ and $\tilde{\delta}_3$, and B_H are the coefficients in $H_w^{\rm BSW}$, the phases of $M_{\rm ETC}$ and the size of $\langle D^{[*]0}|\tilde{H}_w|\bar{B}^{[*]0}\rangle$ in unit of $\langle D^{[*]0}|H_w^{\rm BSW}|\bar{B}^{[*]0}\rangle$, respectively, and the CKM matrix elements are factored out. Although the values of a_1 and a_2 have been calculated perturbatively[9], the latter is still unstable in contrast with the former. Therefore, we take $a_1 \simeq 1.024$ with the leading order QCD

Table 1. Total amplitudes in unit of 10^{-5} GeV.

Decay	$A_{\rm FA}$	+	$A_{\rm NF}$
$\bar{B}^0 \to D^+\pi^-$	$1.54\, a_1\left\{1 - 0.11\left(\frac{a_2}{a_1}\right)\right\}$	$-$	$3.70 a_2 B_H \left\{\left[\frac{4}{3}e^{i\tilde{\delta}_1} - \frac{1}{3}e^{i\tilde{\delta}_3}\right]\right\}$
$\bar{B}^0 \to D^0\pi^0$	$-1.29\, a_2\left\{\frac{f_D}{0.211\ \text{GeV}}\right\}$	$-$	$3.70 a_2 B_H \left\{\frac{\sqrt{2}}{3}\left[2\, e^{i\tilde{\delta}_1} + e^{i\tilde{\delta}_3}\right]\right\}$
$B^- \to D^0\pi^-$	$1.54\, a_1\left\{1 + 1.18\left(\frac{a_2}{a_1}\right)\right\}$	$+$	$3.70 a_2 B_H \left\{ e^{i\tilde{\delta}_3}\right\}$
$\bar{B}^0 \to D^{*+}\pi^-$	$-1.53\, a_1\left\{1 - 0.28\left(\frac{a_2}{a_1}\right)\right\}$	$-$	$3.70 a_2 B_H \times 0.694$
$\bar{B}^0 \to D^{*0}\pi^0$	$1.10\, a_2\left\{\frac{f_{D^*}}{0.211\ \text{GeV}}\right\}$	$+$	$3.70 a_2 B_H \times 0.00135$
$B^- \to D^{*0}\pi^-$	$-1.53\, a_1\left\{1 + 1.02\left(\frac{a_2}{a_1}\right)\right\}$	$-$	$3.70 a_2 B_H \times 0.696$

corrections but treat a_2 as an adjustable parameter. We now look for values of the parameters, a_2, B_H and $\tilde{\delta}_{2I}$, to reproduce $B_{\rm ph}$. The values, $0.12 \le a_2 \le 0.25$ (which should be compared with the results from the perturbative calculations[9]), $0.1 \le B_H \le 0.2$, $60° < \tilde{\delta}_1 < 90°$ and rather small $|\tilde{\delta}_3|$ (but less important), are favored. We list a typical example in Table 2.

In the same way, we obtain the following approximate amplitudes,

$$A_{\rm tot}(\bar{B} \to J/\psi\bar{K}) \simeq -i\{5.73 F_1^{\bar{K}B}(m_\psi^2) + 5.16 B_H'\} a_2 \times 10^{-5}\ \text{GeV}, \quad (2)$$

$$A_{\rm tot}(B^- \to J/\psi\pi^-) \simeq -i\{5.73 F_1^{\pi B}(m_\psi^2) + 5.46 B_H'\} a_2 \times 10^{-5}\ \text{GeV}, \quad (3)$$

where B_H' provides the size of asymptotic matrix elements, $\langle\psi|\tilde{H}_w|\bar{B}_s^{*0}\rangle$ and $\langle\psi|\tilde{H}_w|\bar{B}^{*0}\rangle$, in unit of the corresponding factorized ones of $H_w^{\rm BSW}$. If we assume flavor $SU_f(3)$ symmetry, i.e., $F_1^{\bar{K}B}(m_\psi^2) \simeq F_1^{\pi B}(m_\psi^2)$, we obtain $R = B_{\rm tot}(B^- \to J/\psi\pi^-)/B_{\rm tot}(B^- \to J/\psi K^-) \simeq 0.05$ which is compatible with the existing data[5], $R_{\rm exp} = 0.051 \pm 0.016$. Estimated values of $F_1^{\bar{K}B}(m_\psi^2)$ are approximately between 0.55 and 0.85 (but model dependent). Here we take tentatively $F_1^{\bar{K}B}(m_\psi^2) \simeq 0.7$ in addition to the values $a_2 = 0.19$ and $B_H' = B_H = 0.15$ as before. Then we obtain $B_{\rm FA} \simeq 0.05\,\%$ and $B_{\rm tot} \simeq 0.08\,\%$ which should be compared with the data[5], $B_{\rm exp}(B^- \to J/\psi K^-) = (0.099 \pm 0.010)\%$ and $B_{\rm exp}(\bar{B}^0 \to J/\psi\bar{K}^0) = (0.089 \pm 0.012)\%$. The non-factorizable contributions improve drastically the result from the factorization.

In summary, we have studied hadronic B decays assuming that their amplitude is given by a sum of factorizable and non-factorizable ones, and have observed that the latter is small in color favored $\bar{B} \to D\pi$ and $D^*\pi$ decays while it is important in the $\bar{B} \to J/\psi\bar{K}$ and $J/\psi\pi$ decays. The above result supports the argument of color transparency but does not the large N_c argu-

Table 2. Branching ratios (%) for the $\bar{B} \to D^{[*]}\pi$ decays. In $B_{\rm FA}$ from the factorized amplitudes and $B_{\rm tot}$ from the total amplitudes, $a_1 = 1.02$, $a_2 = 0.19$, $B_H = 0.15$, $\tilde{\delta}_1 = 85°$ and $\tilde{\delta}_3 = -5°$ are taken.

Decays	$B_{\rm FA}$	$B_{\rm tot}$	$B_{\rm ph}$	$B_{\rm expt}$ (*)
$\bar{B}^0 \to D^+\pi^-$	0.27	0.28	$0.28 - 0.34$	0.30 ± 0.04
$\bar{B}^0 \to D^0\pi^0$	0.007	0.012	$0.006 - 0.012$	< 0.012
$B^- \to D^0\pi^-$	0.46	0.51	0.53 ± 0.05(†)	0.53 ± 0.05
$\bar{B}^0 \to D^{*+}\pi^-$	0.25	0.27	0.276 ± 0.021	0.276 ± 0.021
$\bar{B}^0 \to D^{*0}\pi^0$	0.005	0.005	$0.004 - 0.044$	< 0.044
$B^- \to D^{*0}\pi^-$	0.42	0.45	0.46 ± 0.04(†)	0.46 ± 0.04

(*) The data values are taken from Ref.5.
(†) The values of $B_{\rm ph}$ with (†) are taken as the input data.

ment concerning with the factorization. Therefore, we need to treat carefully non-factorizable contributions when we study hadronic weak interactions.

Acknowledgments

This work was supported in part by Grant-in-Aid from the Ministry of Education, Science and Culture.

References

1. S. Oneda and A. Wakasa, *Nucl. Phys.* **1**, 445 (1956).
2. M. Bauer, B. Stech and M. Wirbel, *Z. Phys.* C **34**, 103 (1987); M. Neubelt, V. Rieckert, B. Stech, and Q. P. Xu, in *Heavy Flavors*, edited by A. J. Buras and H. Lindner (World Scientific, Singapore, 1992).
3. M. Fukugita, T. Inami, N. Sakai and S. Yazaki, *Phys. Lett.* B **72**, 237 (1977); A. J. Buras, J. -M. Gerard, and R. Rückl, Nucl. Phys. **B268**, 16 (1986).
4. J. D. Bjorken, *Nucl. Phys.* B (Proc. Suppl.) **11**, 325 (1989); M. J. Dugan and B. Grinstein, *Phys. Lett.* B **255**, 583 (1991).
5. Particle Data Group, C. Caso *et al.*, *Eur. Phys. J.* C **3**, 1 (1998).
6. K. Terasaki, S. Oneda, and T. Tanuma, Phys. Rev. D **29**, 456 (1984).
7. V. S. Mathur and L. K. Pandit, *Advances in Particle Physics*, edited by R. L. Cool and R. E. Marshak (Interscience Publishers, New York, 1968), Vol. 2, p. 383.
8. S. Pakvasa, private communication.
9. G. Buchalla, A. J. Buras and M. E. Lautenbacher, *Rev. Mod. Phys.* **68**, 1125 (1996).

SPECTROSCOPY AND LIFETIMES OF BOTTOM AND CHARM HADRONS

Fumihiko Ukegawa
(CDF Collaboration)
Institute of Physics, University of Tsukuba
Tsukuba, Ibaraki 305-8571 Japan
E-mail: ukegawa@physics.px.tsukuba.ac.jp

We review recent experimental results on spectroscopy and lifetimes of bottom and charm hadrons.

1 Introduction

There are several motivations for studying masses and lifetimes of the hadrons containing a heavy quark, either the bottom or the charm quark. First, the mass and the lifetime are fundamental properties of an elementary particle. Second, the spectroscopy of hadrons gives insights into the QCD potential between quarks. In particular, a symmetry exists [1] for heavy hadrons when the heavy quark mass is taken to be infinite, providing a powerful tool to predict and understand properties of those heavy hadrons. Third, studies of the lifetimes of heavy hadrons probe their decay mechanisms. A measurement of the lifetime, or the total decay width, is necessary when we extract magnitudes of elements of the Kobayashi-Maskawa matrix [2]. Again, in the limit of an infinite heavy quark mass things become simple and the decay of a heavy hadron should be the decay of the heavy quark Q. This leads to a prediction that all hadrons containing the heavy quark Q should have the same lifetime, that of the quark Q. This is far from reality in the case of charm hadrons, where the D^+ meson lifetime is about 2.5 times longer than the D^0 meson lifetime. Perhaps the charm quark is not heavy enough. The simple quark decay picture should be a better approximation for the bottom hadrons because of the larger b quark mass.

On the experimental side, the measurements and knowledge of the heavy hadrons (in particular bottom hadrons) have significantly improved over the last decade, thanks to high statistics data accumulated by various experiments. We shall review recent developments in these studies in the remainder of this manuscript.

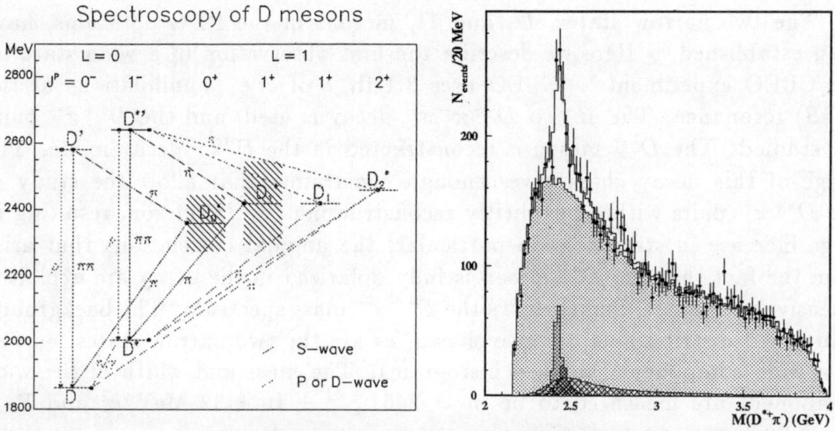

Figure 1: Left: Spectrum of charm mesons. Right: $D^{*+}\pi^-$ mass spectrum by CLEO.

2 Charm Hadron Spectroscopy

Let us consider mesons consisting of a charm quark c and a light antiquark $\bar{q}$ (Figure 1(left)). The ground states are the pseudoscalar (D) and the vector (D^*) mesons, which have long been established. They have zero orbital angular momentum ($L = 0$) between the quarks, and the total angular momentum of the mesons are given by addition of two spin 1/2 particles, namely $J = S = 0$ or 1.

If we allow for one unit of orbital angular momentum ($L = 1$) and combine it with the total spin S of the $c\bar{q}$ system, there will be four states, $J^P = 1^+$, 0^+, 1^+ and 2^+. These states are sometimes called the D^{**} mesons. They are expected to decay predominantly to $D^{(*)}\pi$ pairs via the strong interaction. Angular momentum and parity conservation restrict the decay of the 0^+ meson to the $D\pi$ final state. Similarly, the 1^+ meson decays only to $D^*\pi$ pairs. The 2^+ meson can decay either to $D^*\pi$ or $D\pi$.

In the limit of an infinite heavy quark mass, $m_Q = \infty$, the light degree of freedom is decoupled from the heavy quark, and the total angular momentum of the light antiquark $j = L + s_{\bar{q}}$ is a good quantum number. In this case the four states above form two doublets, each having $j = 1/2$ ($J = 0, 1$) or $j = 3/2$ ($J = 1, 2$), where two members of each doublet are degenerate in mass, width and other quantum numbers. The decay of the $j = 3/2$ doublet proceeds through a D-wave, while the decay of the $j = 1/2$ doublet proceeds through an S-wave. Therefore, the $j = 3/2$ doublet is expected to be narrow and the $j = 1/2$ doublet to be wide.

The two narrow states, D_1 and D_2^* mesons in the PDG notations, have been established [3]. Here we describe the first observation of a wide state by the CLEO experiment [4]. CLEO uses 3.1 fb^{-1} of e^+e^- annihilations at the $\Upsilon(4S)$ resonance. The $B^- \to D^{*+}\pi^-\pi^-$ decay is used, and the $D^{*+}\pi^-$ pairs are studied. The D^{*+} meson is reconstructed in the $D^0\pi^+$ decay mode. The usage of this decay chain gives enough constraints that allow the study of the $D^{*+}\pi^-$ pairs without explicitly reconstructing the D^0 meson, resulting in large increase in statistics. In particular, the angular correlations that arise from the fact that the D^{**} meson is fully polarized in the decay are exploited intensively. Figure 1(right) shows the $D^{*+}\pi^-$ mass spectrum. The background subtracted distributions are also shown, as are the two narrow states, as well as a wide component (hatched histogram). The mass and width of the wide component are measured to be $m = 2461\,^{+41}_{-34} \pm 10 \pm 32$ MeV/c^2 and $\Gamma = 290\,^{+101}_{-79} \pm 26 \pm 36$ MeV/c^2. Also the branching fraction is measured to be $\mathcal{B}(B^- \to D_1^{*0}\pi^-) \cdot \mathcal{B}(D_1^{*0} \to D^{*+}\pi^-) = (10.6 \pm 1.9 \pm 1.7 \pm 2.3) \times 10^{-4}$.

The DELPHI experiment reported [5] evidence for a radial excitation state $D^{*\prime +}$ using the $D^{*+}\pi^+\pi^-$ final state. However, neither OPAL nor CLEO has been able to confirm it [6].

CLEO has found [7] a new charmed baryon state that decays through $\Xi_c^*\pi$ to $\Xi_c\pi^+\pi^-$. Both charged and neutral states are observed as a peak in the mass difference $\Delta m \equiv m(\Xi_c\pi^+\pi^-) - m(\Xi_c)$, at 348.6 ± 0.6 (charged) and 347.2 ± 0.7 MeV/c^2 (neutral). An upper limit on the width is placed to be $\Gamma < 3.5$ (charged) and < 6.5 (neutral) MeV/c^2. They are interpreted as the $J^P = \frac{3}{2}^-$ Ξ_{c1} isospin doublet, namely the $L = 1$ orbital excitation of the Ξ_c baryon.

3 Bottom Hadron Spectroscopy

A lot of progress has been made in the spectroscopy of bottom hadrons in the past few years as well. All ground states have been established, and even the B_c^- meson, a bound state of the two different kinds of heavy quarks, has been observed [8]. The B^* mesons have also been established, at a mass about 50 MeV/c^2 above B. The mass splitting is smaller than in the charm mesons (~ 150 MeV/c^2), certainly consistent with expectations from the heavy quark symmetry.

The B^{**} states have also been observed. Here we describe an L3 analysis [9] as a recent example. The B mesons are reconstructed inclusively using secondary vertices and are combined with pion candidates coming from the primary vertex. Figure 2(left) shows the $B\pi$ mass distribution, where an enhancement is observed around 5.7 GeV/c^2. A fit for the masses and widths of

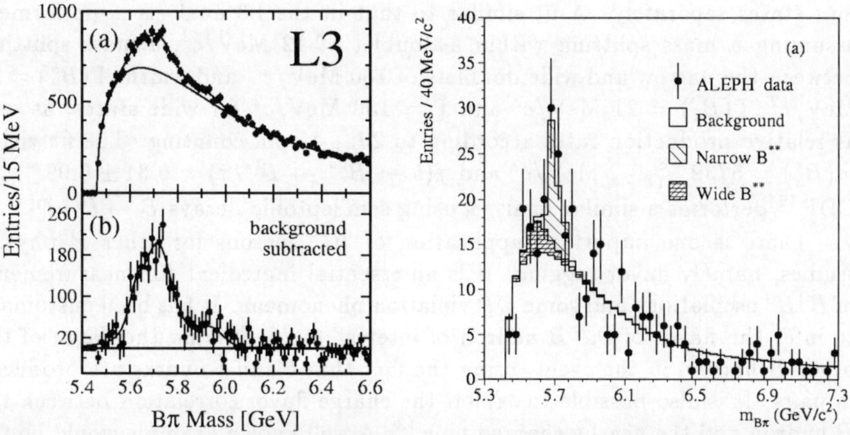

Figure 2: B^{**} analyses by L3 (Left) and ALEPH (Right). $B\pi$ mass spectra are shown.

the narrow and the wide states is performed, assuming, within each doublet, the mass splitting of 12 MeV/c^2 and equal widths, and relative production rates according to $2J+1$ spin counting. This yields $m = 5768 \pm 5 \pm 6$ MeV/c^2 and $\Gamma = 24 \pm 19 \pm 24$ MeV/c^2 for the B_2^* state, and $m = 5670 \pm 10 \pm 13$ MeV/$c^2$and $\Gamma = 70 \pm 21 \pm 25$ MeV/c^2 for the B_1^* state. The fraction of the B^{**} production is extracted to be $f(b \to B^{**} \to B^{(*)}\pi) = 0.32 \pm 0.03 \pm 0.06$. The fit actually includes a fifth component, radial excitation states B', to account for an enhancement near 5.9 GeV, giving $m = 5937 \pm 21 \pm 4$ MeV/c^2, $\sigma = 50 \pm 22 \pm 5$ MeV/c^2, and $f(b \to B' \to B^{(*)}\pi) = 0.034 \pm 0.011 \pm 0.008$.

Although the B^{**} mesons are collectively observed, their individual states have not been separated. This is because (a) typical B reconstruction relies on inclusive techniques and thus suffers from a relatively poor mass resolution, (b) the photon in the $B^* \to B\gamma$ decay is not usually included and thus $B^*\pi$ and $B\pi$ states are not separated, and (c) all B hadron species are mixed together.

OPAL [10] attempted to remedy the point (b) above, with a statistical separation of $B^*\pi$ and $B\pi$ states by looking for the photon from the B^* decay. Inclusively reconstructed B decay events are separated into B^*-enriched and B^*-depleted samples, and the B^{**} branching fraction of $\mathcal{B}(B^{**} \to B^*\pi) = 0.85\,^{+0.26}_{-0.27} \pm 0.12$ is extracted.

ALEPH [11] uses fully reconstructed decays $B \to \overline{D}^{(*)} n\pi$ and $J/\psi K^{(*)}$, solving difficulties (a) and (c). Signals of about 200 B^+ and 140 B^0 are observed. The $B\pi$ mass spectrum is shown in Figure 2(right). The limited statistics, however, prevent the determination of the masses and widths of all four states separately. A fit similar to that in the L3 analysis is performed, assuming a mass splitting within a doublet of 12 MeV/c^2, a mass splitting between the narrow and wide doublets of 100 MeV/c^2, and widths $\Gamma(B_2^*) = 25$ MeV/c^2, $\Gamma(B_1) = 21$ MeV/c^2 and $\Gamma = 150$ MeV/c^2 for wide states, as well as relative production rates according to $2J+1$ spin counting. The fit yields $m(B_2^*) = 5739\,^{+8}_{-11}\,^{+6}_{-4}$ MeV/c^2 and $f(b \to B^{**} \to B^{(*)}\pi) = 0.31 \pm 0.09\,^{+0.06}_{-0.05}$. CDF [12] performs a similar analysis using semileptonic decays $\overline{B} \to \ell^- \bar{\nu} D^{(*)} X$.

There is one important application of B^{**} mesons for other B physics studies, namely flavor tagging. It is an essential ingredient for measurements of $B^0\overline{B}^0$ oscillations and some CP violation phenomena. It has been customary to infer the flavor of the B hadron of interest by identifying the flavor of the other B hadron in the event, using the fact that b and $\bar{b}$ quarks are produced in pairs. It is also possible to exploit the charge-flavor correlation between the B hadron and the nearby charged pion [13]. A well known example would be the decay $D^{*+} \to D^0\pi^+$ when you look for $D^0\overline{D}^0$ mixing. If it is accompanied by π^+, it must have been produced as D^0, not $\overline{D}^0$. The B^* mesons cannot produce a pion kinematically. Therefore, B^{**} mesons are the main resonant state that can produce correlated pions. They can also be produced in the fragmentation (non-resonant) processes. The charge correlations are the same whether it is resonant or non-resonant. The method has been applied successfully to $B^0\overline{B}^0$ oscillation measurements [14]. The ALEPH analysis above also performs tagging studies and finds similar tagging effectiveness. Tag purity could be improved further if the $B\pi$ mass resonant regions are selected.

4 Charm Hadron Lifetimes

Recent developments in charm lifetime measurements include (a) a new precision in the D_s^+ lifetime (E791, CLEO) and (b) startup of new experiments (FOCUS, SELEX). Here we describe the D_s^+ lifetime measurements.

E791 [15] uses data from 1991 - 92 with 500 GeV/c π^- beam on foil targets. The D_s^+ meson is reconstructed with the $\phi\pi^+$, $\phi \to K^+K^-$ mode. Figure 3(left) shows the $\phi\pi^+$ mass spectrum. A signal of 1662 ± 56 is estimated. The peak around 1860 MeV/c^2 is from the Cabibbo-suppressed D^+ decay. The dashed line represents combinatorial background, where a discontinuity arises because the $D_s^+ \to \phi\pi^+$ candidates were required to be inconsistent with the $D^+ \to K^+\pi^-\pi^+$ hypothesis. Mass and decay time distributions are fit simultaneously and the D_s^+ lifetime is extracted to be $\tau(D_s^+) = 518 \pm 14 \pm 7$ fs. Using the D^0 lifetime of 415 ± 4 fs (PDG 1998), it is found that $\tau(D_s^+)/\tau(D^0) = 1.25 \pm 0.04$.

CLEO [16] uses 3.7 fb^{-1} of e^+e^- annihilation near the $\Upsilon(4S)$. The D_s^+, as

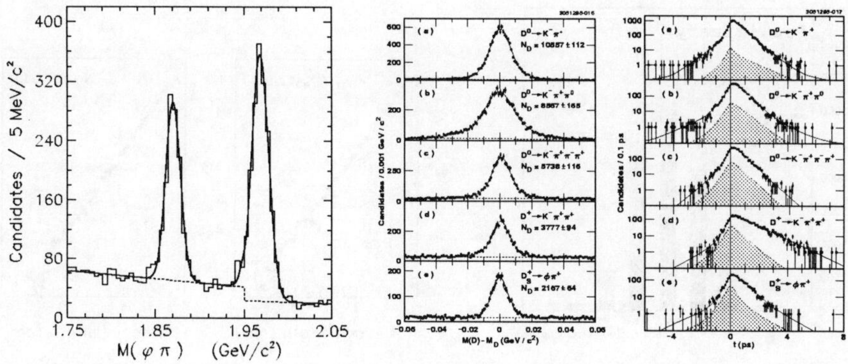

Figure 3: Left: D_s^+ lifetime measurements by E791. Right: Charm lifetime measurements by CLEO.

well as D^0 and D^+ mesons are reconstructed. Figure 3(right) shows the reconstructed mass (difference from the nominal mass) and decay time distributions. They find $\tau(D^0) = 408.5 \pm 4.1\,^{+3.5}_{-3.4}$ fs, $\tau(D^+) = 1033.6 \pm 22.1\,^{+9.9}_{-12.7}$ fs, $\tau(D_s^+) = 486.3 \pm 15.0\,^{+4.9}_{-5.1}$ fs, and the lifetime ratio of $\tau(D_s^+)/\tau(D^0) = 1.19 \pm 0.04$.

Each result establishes for the first time that the D_s^+ lifetime is significantly longer than the D^0 lifetime.

5 Bottom Hadron Lifetimes

The bottom hadrons are expected to have smaller lifetime differences among the different hadron species, of order 10% at most [17]. This poses a challenge to experiments, because it requires great precision to find and establish such small differences. The B hadron lifetime measurements performed thus far can be classified into three broad classes, using (a) inclusive B reconstruction, (b) partial reconstruction (e.g. semileptonic decays) and (c) full reconstruction (e.g. $J/\psi K^{(*)}$, $D^{(*)}n\pi$). Method (a) gives the largest statistics, but samples are less pure in terms of isolating the various B hadron species, and the selection procedure introduces a bias in the proper time distribution of the candidates that needs to be carefully accounted for. Method (b) gives respectable statistics and sample purity. Method (c) gives clean signature, perfect purity, and very precise momentum estimate on an event-by-event basis, but suffers from small branching fractions and thus low statistics. Here we describe two measurements, using method (a) and (b), respectively.

SLD [18] uses inclusively reconstructed B decays among 550 k hadronic Z^0 decays. The net charge of tracks attached to the secondary vertices is used to

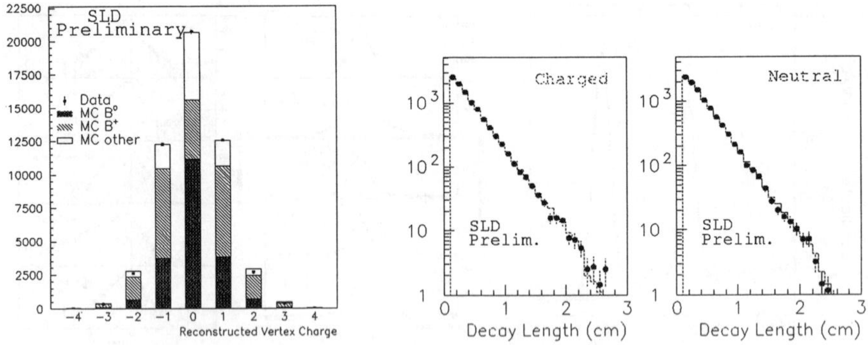

Figure 4: B^+ and B^0 meson lifetime measurement by SLD. Left: vertex charge distribution. Right: decay length distributions for charged and neutral B hadron samples. The plots show the 1997 – 98 data corresponding to 350 k hadronic Z^0 events.

estimate and separate the charge of parent B hadrons. The purity is enhanced further with information from the vertex mass, the beam polarization and the opposite hemisphere jet charge. Figure 4 shows the distributions of the vertex charge and the decay lengths for samples enriched in charged and neutral B hadrons. They find $\tau(B^+) = 1.623 \pm 0.020 \pm 0.034$ ps, $\tau(B^0) = 1.565 \pm 0.021 \pm 0.043$ ps, and the lifetime ratio of $\tau(B^+)/\tau(B^0) = 1.037\,^{+0.025}_{-0.024} \pm 0.024$.

ALEPH [19] has re-analyzed 4 M hadronic Z^0 decays taken in 1991 – 95 and measured the B^+ and B^0 meson lifetimes using semileptonic decays $\overline{B} \to \ell^- \bar{\nu} D^{*+} X$ (mostly $\overline{B}^0$) and $\overline{B} \to \ell^- \bar{\nu} D^0 X$ (mostly B^-). The D^{*+} meson is reconstructed in the $D^0\pi^+$ mode, and the D^0 meson is reconstructed in $K^-\pi^+$, $K^-\pi^+\pi^+\pi^-$, $K^-\pi^+\pi^0$ or $K^0_S\pi^+\pi^-$ mode. The charm signals are shown in Figure 5. For the $\ell^- D^0$ pairs, the D^0 meson is reconstructed in $K^-\pi^+$, $K^-\pi^+\pi^0$ or $K^0_S\pi^+\pi^-$ mode, and decays coming from the D^{*+} meson are excluded. Similar signals are observed. Distributions of the B meson proper decay time are also shown in Figure 5. The measured lifetimes are $\tau(B^+) = 1.646 \pm 0.056\,^{+0.036}_{-0.034}$ ps, $\tau(B^0) = 1.524 \pm 0.053\,^{+0.035}_{-0.032}$ ps, and the lifetime ratio is $\tau(B^+)/\tau(B^0) = 1.080 \pm 0.062 \pm 0.024$.

These two results are shown as recent representative measurements. There are many other measurements from various experiments, as compiled by the LEP B lifetime working group, and more information can be found in Ref. [20]. As of October 1999, the world average lifetimes of B^+ and B^0 mesons are $\tau(B^+) = 1.639 \pm 0.025$ ps, $\tau(B^0) = 1.553 \pm 0.029$ ps, and $\tau(B^+)/\tau(B^0) = 1.066 \pm 0.024$. The value at the time of the Hawaii Conference (March 1997) was $\tau(B^+)/\tau(B^0) = 1.06 \pm 0.04$. Namely, the precision is still improving, and

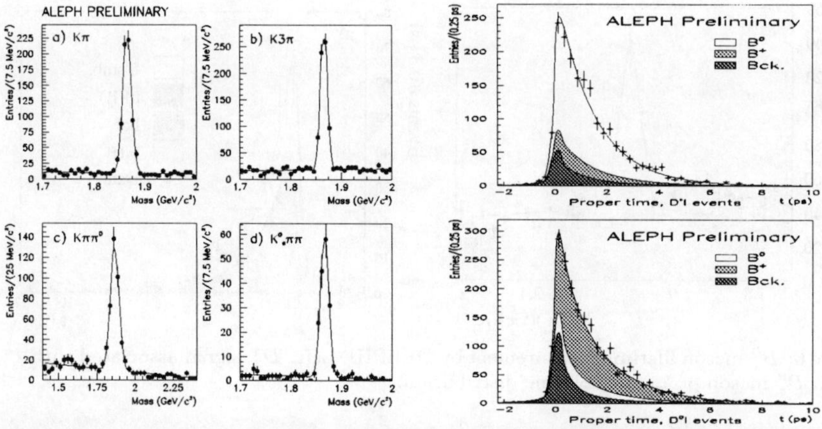

Figure 5: B^+ and B^0 meson lifetime measurement by ALEPH. Left: D^0 signal from $D^{*+} \to D^0\pi^+$ associated with ℓ^-. Right: B meson proper decay time distributions for $\ell^- D^{*+}$ and $\ell^- D^0$ samples.

we may be on the verge of observing a finite lifetime difference.

For the B_s^0 meson, a very small (of order 1%) lifetime difference from the B^0 meson is expected. The current world average value is $\tau(B_s^0) = 1.460 \pm 0.056$ ps, and $\tau(B_s^0)/\tau(B^0) = 0.94 \pm 0.07$. More precise measurements are necessary in order to see whether there is a lifetime difference of order 5%, or whether the lifetimes agree at 1% level.

Another interesting piece of physics is expected in the B_s^0 meson system. The B_s^0 meson should mix with its antiparticle to form the mass eigenstates $B_{s,L}^0$ and $B_{s,H}^0$, where the subscripts L and H denote light and heavy states. These two mass eigenstates could also have a substantial width difference $\Delta\Gamma$ of order 10%. A sizable width difference is interesting for several reasons: (a) we should see it if it indeed exists, (b) the ratio $\Delta m_s/\Delta\Gamma$ is independent of CKM matrix elements and is estimated to be large (~ -180), providing an indirect information on Δm_s, (c) it may allow CP studies using untagged decays [21], and (d) physics beyond the Standard Model should make $\Delta\Gamma$ smaller.

The width difference $\Delta\Gamma$ manifests itself as a difference in lifetimes when measured with flavor eigenstates (e.g. semileptonic decays) and CP eigenstates. Or, the decay time distributions of flavor eigenstates can be fitted with two (long and short) components to look for a width difference.

DELPHI [22] uses 3.5 M Z^0 decays and the semileptonic decay $\overline{B}_s^0 \to \ell^-\bar{\nu}D_s^+X$. The D_s^+ meson is reconstructed in $\phi\pi^+$, $\overline{K}^{*0}K^{(*)+}$, $K_S^0K^+$, $\phi\pi^+\pi^+\pi^-$ or $\phi\pi^+\pi^0$ mode, as well as the semileptonic mode $\phi\ell^+\nu$. A total signal of 290

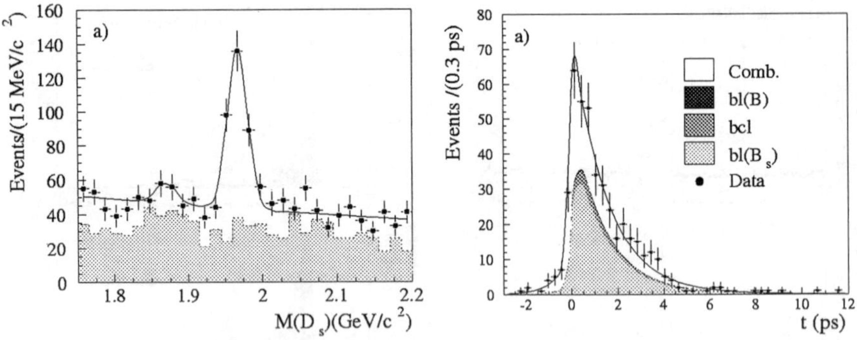

Figure 6: B_s^0 meson lifetime measurement by DELPHI. Left: D_s^+ signal associated with ℓ^-. Right: B_s^0 meson proper decay time distribution.

events is observed. Figure 6 shows the combined D_s^+ signal (excluding the semileptonic mode) and the B_s^0 decay time distribution. They find $\tau(B_s^0) = 1.42\,^{+0.14}_{-0.13} \pm 0.03$ ps, and $\Delta\Gamma/\Gamma < 0.46$. DELPHI also uses $\overline{B}_s^0 \to D_s^+ h^- X$ mode and find $\tau(B_s^0) = 1.49\,^{+0.16}_{-0.15}\,^{+0.07}_{-0.08}$ ps and $\Delta\Gamma/\Gamma < 0.58$. Other $\Delta\Gamma$ searches include an L3 analysis [23] using inclusive B reconstruction, yielding $\Delta\Gamma/\Gamma < 0.67$ and a CDF analysis [24] using $\sim$ 600 $\ell^- D_s^+$ decays, yielding $\tau(B_s^0) = 1.36 \pm 0.09\,^{+0.06}_{-0.05}$ ps and $\Delta\Gamma/\Gamma < 0.83$. These limits on the width difference, all at 95% CL, have been obtained by examining decay time distributions and looking for two components.

The B_s^0 lifetime is also measured using modes that are nearly CP eigenstates. ALEPH [25] uses the $B_s^0/\overline{B}_s^0 \to D_s^{(*)+} D_s^{(*)-} \to \phi\phi X$ signature with 32 ± 17 signal events, and finds $\tau(B_s^0) = 1.42 \pm 0.23 \pm 0.16$ ps. CDF [26] uses $B_s^0/\overline{B}_s^0 \to J/\psi\,\phi$ decays with 58 ± 12 signal events, and finds $\tau(B_s^0) = 1.34\,^{+0.23}_{-0.19} \pm 0.05$ ps. These decays are not in general pure CP eigenstates. CDF performs a transversity analysis [27] of the $J/\psi\,\phi$ decays and finds the fraction of the P-wave decays (= CP odd) to be $\Gamma_\perp/\Gamma = 0.229 \pm 0.188 \pm 0.038$. For a similar decay mode $B^0 \to J/\psi K^{*0}$, CDF [27] and CLEO [28] find the fraction to be $0.126\,^{+0.121}_{-0.098} \pm 0.028$ and $0.16 \pm 0.08 \pm 0.04$, respectively.

The above lifetime values may be compared with the world average B_s^0 lifetime measured with flavor eigenstates, $\tau(B_s^0) = 1.467 \pm 0.058$ ps. This is smaller, albeit large uncertainties, as expected for CP-even states. The ALEPH analysis quotes $\Delta\Gamma/\Gamma = 0.24 \pm 0.35$ assuming pure CP-even composition with $\tau(B_s^0) = 1.61 \pm 0.10$ ps (PDG 1996).

Since I have reached the page limit, let me conclude here by saying that new data expected in the near future from CLEO-III, Belle, BaBar and Tevatron Run-II should vastly improve the kinds of measurements described here.

Acknowledgments

I wish to thank Professor George Hou, my old friend Augustine Chen, and the other members of the local organizing committee for a very nice conference. Also, I would like to thank the experiments for their physics results contributing to this talk. In particular, the following people have provided me with great help in preparing the talk and this manuscript: D. Jaffe, J. Slaughter, K. Baird, D. Jackson, D. Abbaneo, P. Gagnon, R. Hawkings, S. Gentile, S. Blyth and A. B. Wicklund. Professor Joseph Kroll (Penn) has provided brilliant ideas and incredible inspiration as always.

References

1. N. Isgur and M. Wise, Phys. Rev. Lett. **66**, 1130 (1991), J. L. Rosner, Comm. Nucl. Part. Phys. **16**, 109 (1986).
2. M. Kobayashi and T. Maskawa, Prog. Theor. Phys. **49**, 652 (1973).
3. C. Caso *et al.*, Euro. Phys. J. C **3**, 1 (1998).
4. CLEO collaboration, CLEO CONF 99-6 (1999).
5. DELPHI collaboration, Phys. Lett. B **426**, 231 (1998).
6. OPAL collaboration, OPAL PN 352 (1998); CLEO collaboration, hep-ex/9901008 (1999).
7. CLEO collaboration, Phys. Rev. Lett. **83**, 3390 (1999).
8. CDF collaboration, Phys. Rev. Lett. **81**, 2432 (1998); Phys. Rev. D **58**, 112004 (1998).
9. L3 collaboration, Phys. Lett. B **465**, 323 (1999).
10. OPAL collaboration, OPAL PN 400 (1999).
11. ALEPH collaboration, Phys. Lett. B **425**, 215 (1998).
12. CDF collaboration, Fermilab-Pub-99/330-E, submitted to Phys. Rev. D.
13. M. Gronau, A. Nippe and J. L. Rosner, Phys. Rev. D **47**, 1988 (1993); M. Gronau and J. L. Rosner, Phys. Rev. D **49**, 254 (1994).
14. OPAL collaboration, Phys. Lett. B **327**, 411 (1994); CDF collaboration, Phys. Rev. Lett. **80**, 2057 (1998); Phys. Rev. D **59**, 032001 (1999).
15. E791 collaboration, Phys. Lett. B **327**, 449 (1999).
16. CLEO collaboration, Phys. Rev. Lett. **82**, 4586 (1999).
17. I. I. Bigi, *These Proceedings* (hep-ph/0001003) (2000).
18. SLD collaboration, SLAC-PUB-8206 (1999).
19. ALEPH collaboration, ALEPH 99-005 (1999).
20. LEP *B* lifetime working group, http://home.cern.ch/claires/lepblife.html.
21. I. Dunietz, Phys. Rev. D **52**, 3048 (1995); R. Fleischer and I. Dunietz, Phys. Lett. B **387**, 361 (1996); Phys. Rev. D **55**, 259 (1997).

22. DELPHI collaboration, DELPHI 99-109 CONF 296 (1999).
23. L3 collaboration, Phys. Lett. B **438**, 417 (1999).
24. CDF collaboration, Phys. Rev. D **59**, 032004 (1999).
25. ALEPH collaboration, ALEPH 98-064 (1998); hep-ex/9905017 (1999).
26. CDF collaboration, Phys. Rev. Lett. **77**, 1945 (1996).
27. CDF collaboration,
http://www-cdf.fnal.gov/physics/new/bottom/cdf4672/cdf4672.html.
28. CLEO collaboration, Phys. Rev. Lett. **79**, 4533 (1997).

THE LIFETIMES OF HEAVY FLAVOUR HADRONS - A CASE STUDY IN QUARK-HADRON DUALITY

I. I. BIGI

Physics Dept., Univ. of Notre Dame du Lac
Notre Dame, IN 46556, U.S.A.
E-mail: bigi@undhep.hep.nd.edu

The status of heavy quark expansions for charm and beauty lifetime ratios is reviewed. Taking note of the surprising semiquantitative success of this description for charm hadrons I interprete the new data on $\tau(D_s)$ and re-iterate the call for more precise measurements of $\tau(\Xi_c^{0,+})$ and $\tau(\Omega_c)$. A slightly larger B^- than B_d lifetime is starting to emerge as predicted; the largest lifetime difference in the beauty sector, namely in $\tau(B_c)$ vs. $\tau(B)$ has correctly been predicted; the problem posed by the short Λ_b lifetime remains. The need for more accurate data also on $\tau(B_s)$ and $\tau(\Xi_b^{-,0})$ is emphasized. I discuss quark-hadron duality as the central theoretical issue at stake here.

1 Introduction

The lifetime of a hadron represents an observable of fundamental as well as practical importance: (i) Its magnitude reveals whether the decay is driven by strong, electromagnetic or weak forces; (ii) it constitutes an essential engineering number for translating measured branching ratios into widths. Yet a strong motivation to *measure* a quantity does not necessarily imply a need for a precise *theoretical* description. Furthermore we all understand that nothing that is going to happen or not happen in the theory of weak lifetimes will make anybody abandon QCD as the theory of the strong interactions. After all, it is the only game in town after string theory has raised its ambition to become the theory of everything rather than merely the theory of the strong forces where it had first emerged.

The central theme of my talk will be that developing such a theory represents a forum for addressing the next frontier in QCD, namely *quark-hadron duality* or duality for short. The concept of duality constitutes an essential element in any QCD based description and it has been invoked since the early days of the quark model. For a long time little progress happened in this area; for a violation of duality can be discussed in a meaningful way only if one has a *reliable* theoretical treatment of *nonperturbative* effects. Let me illustrate that through an example. A priori it would be quite reasonable to assume that relating the weak decay width of a heavy flavour hadron to the fifth power of *its* mass rather than the heavy *quark* mass – $\Gamma(H_Q) \propto M^5(H_Q)$ – would incorporate boundstate effects as the leading nonperturbative corrections (and that is indeed what we originally did [1]). Only after developing a consistent theory for the weak decays of such hadrons through the operator product expansion did we realize that such an ansatz would violate duality. For it leads to a large correction of order $1/m_Q$ – $M^5(H_Q) \simeq (m_Q + \bar{\Lambda})^5 \simeq m_Q^5(1 + 5\bar{\Lambda}/m_Q)$ – which is anathema to the OPE [2]! This example already indicates that the study of heavy flavour decays had given new impetus to addressing duality: it has provided us with new theoretical tools, and it has re-emphasized the need to understand the limitations to duality since one aims at extracting fundamental KM parameters with high numerical accuracy from semileptonic decays.

Nonleptonic transitions provide a rich and multilayered lab to analyze duality and its limitations; they can act as a microscope exactly because they are thought of containing larger duality violations than semileptonic reactions.

2 Heavy Quark Expansions

The weak decay width for a heavy flavour hadron H_Q into an inclusive final state f can be expressed through an operator product expansion (OPE) [2,3]:

$$\Gamma(H_Q \to f) = \frac{G_F^2 m_Q^5}{192\pi^3}|V_{CKM}|^2 \left[c_3^{(f)} \langle H_Q|\bar{Q}Q|H_Q\rangle + c_5^{(f)} \frac{\mu_G^2(H_Q)}{m_Q^2} + \right.$$

$$\left. + \sum_i c_{6,i}^{(f)} \cdot \frac{\langle H_Q|(\bar{Q}\Gamma_i q)(\bar{q}\Gamma_i Q)|H_Q\rangle}{m_Q^3} + \mathcal{O}(1/m_Q^4) \right], \quad (1)$$

where $\mu_G^2(H_Q) \equiv \langle H_Q|\bar{Q}\frac{i}{2}\sigma \cdot GQ|H_Q\rangle$. Eq.(1) exhibits the following important features:

(A) The expansion involves (i) c-number coefficients $c_i^{(f)}$ calculable within short-distance physics; (ii) expectation values of local operators given by long distance physics; their values can be inferred from symmetry arguments, other observables, QCD sum rules, lattice studies and quark models; (iii) inverse powers of the heavy quark mass m_Q . The nonperturbative effects on the decay width – a dynamical quantity – can thus be expressed through expectation values and quark masses. Those being static quantities can be calculated with decent reliability.

(B) The free quark model expression emerges asymptotically (for $m_Q \to \infty$) from the $\bar{Q}Q$ operator since $\langle H_Q|\bar{Q}Q|H_Q\rangle = 1 + \mathcal{O}(1/m_Q^2)$.

(C) *No* $\mathcal{O}(1/m_Q)$ contribution can arise in the OPE since there is no independant dimension four operator (with colour described by a local gauge theory) [a]. This has two important consequences: (i) With the *leading* non-

[a] The operator $\bar{Q}i\not{D}Q$ can be reduced to the leading operator $\bar{Q}Q$ through the equation of motion.

perturbative corrections arising at order $1/m_Q^2$, their size is typically around 5 % in beauty decays. They had not been anticipated in the phenomenological descriptions of the 1980's. (ii) A $1/m_Q$ contribution can arise only due to a massive duality violation. Thus one should set a rather high threshold before accepting such a statement.

(D) Pauli Interference (PI) [4], Weak Annihilation (WA) [5] for mesons and W-scattering (WS) for baryons arise unambiguously and naturally in order $1/m_Q^3$ with WA being helicity suppressed [1]. They mainly drive the differences in the lifetimes of the various hadrons of a given heavy flavour.

The largest uncertainties enter in the expectation values for the dimension-six four-fermion operators in order $1/m_Q^3$. For *mesons* I will invoke approximate *factorization at a low scale of around 1 GeV*. For *baryons* there is no concept of factorization, and we have to rely on quark model results.

Below I will discuss mainly hadron-*specific* duality violations affecting the *ratios* between different hadrons of a given heavy flavour.

3 Lifetimes of Charm Hadrons

One rough measure for the numerical stability of the $1/m_c$ expansion is provided by $\sqrt{\mu_G^2(D)/m_c^2} \simeq 0.5$ as an effectice expansion parameter which is not small compared to one. Obviously one can expect – at best – a semiquantitative description. The mesonic four-quark matrix elements are calibrated by $f_D \sim 200$ MeV and $f_{D_s}/f_D \simeq 1.1 - 1.2$.

On general grounds one expects the following hierarchy in lifetimes [6,7]:

$$\tau(D^+) > \tau(D^0) \sim \tau(D_s^+) \geq \tau(\Xi_c^+) > \tau(\Lambda_c^+) > \tau(\Xi_c^0) > \tau(\Omega_c) \qquad (2)$$

Table 1 shows the predictions and data. A few comments are in order here:

- You apply the $1/m_c$ expansion at your own risk. It is easy to list reasons why it should fail to reproduce even the qualitative pattern expressed in Eq.(2). However comparing the data with the expectations shows agreement even on the semiquantitative level. This could be accidental; yet I will explore the possibility that it is not. One should note that the longest and shortest lifetimes differ by a factor of about twenty!

- PI is the main engine driving the $D^+ - D^0$ lifetime difference as already anticipated in the 'old' analysis of Guberina et al. [4]; the main impact of the HQE for this point was to show that WA cannot constitute the leading effect and that $BR_{SL}(D^0) \simeq 7\%$ is consistent with PI being the

	$1/m_c$ expect. [7]	theory comments	data
$\frac{\tau(D^+)}{\tau(D^0)}$	~ 2	PI dominant	2.55 ± 0.034 (updated)
$\frac{\tau(D_s^+)}{\tau(D^0)}$	1.0 - 1.07 0.9 - 1.3 1.08 ± 0.04	*without* WA [8] *with* WA [8] QCD sum rules [9]	1.125 ± 0.042 PDG '98 1.211 ± 0.017 E791, CLEO, FOCUS
$\frac{\tau(\Lambda_c^+)}{\tau(D^0)}$	~ 0.5	quark model matrix elements	0.489 ± 0.008 (updated)
$\frac{\tau(\Xi_c^+)}{\tau(\Lambda_c^+)}$	~ 1.3	ditto	1.75 ± 0.36 PDG '98
$\frac{\tau(\Xi_c^+)}{\tau(\Xi_c^0)}$	~ 2.8	ditto	3.57 ± 0.91 PDG '98
$\frac{\tau(\Xi_c^+)}{\tau(\Omega_c)}$	~ 4	ditto	3.9 ± 1.7 PDG '98
$y = \frac{\Delta\Gamma}{2\Gamma}\|_{D^0}$	$\leq \mathcal{O}(1\%)$	test bed for duality	$-6\% \leq y \leq 0.3\%$ CLEO

Table 1: Lifetime ratios in the charm sector

leading effect, see below [b]. In quoting a lifetime ratio of ~ 2 I am well aware that the measured value is different from two. Yet that numerical difference is within the theoretical noise level: one could use $f_D = 220$ MeV rather than 200 MeV and WA, which has been ignored here, could account for 10 - 20 % of the D^0 width.

- Since $\tau(D_s)/\tau(D^0) \simeq 1.07$ can be generated *without* WA [8], the 'old' data on $\tau(D_s)/\tau(D^0)$ had provided an independant test for WA *not* being the leading source for $\tau(D^0) \neq \tau(D^+)$; it actually allowed for it being quite irrelevant. The 'new' data reconfirm the first conclusion; at the same time they point to WA as a still significant process. A recent analysis of WA relying on QCD sum rules [9] is not quite able to reproduce the observed lifetime ratio; further analysis along these lines is called for.

- The $1/m_c^2$ contribution controlled by $\mu_G^2(D)$ *reduces* the semileptonic width common to D^0 and D^+ mesons; this brings the *absolute* values observed for $BR_{SL}(D^0)$ and $BR_{SL}(D^+)$ in line with what is expected when it is mainly PI that generates the $D^+ - D^0$ lifetime difference.

- The description of the *baryonic* lifetimes is helped by the forgiving experimental errors. More accurate measurements of $\tau(\Xi_c^{+,0}, \Omega_c)$ are needed. They might well exhibit deficiencies in the theoretical description.

[b] $\Gamma(D^+)$ is guaranteed to remain positive if the range in momentum over which PI can occur is properly evaluated. To put it differently: while one cannot count on obtaining a reliable value for $\Gamma(D^+)$, a nonsensical result will arise only if one makes a mistake.

	$1/m_b$ expect.	theory comments	data
$\frac{\tau(B^+)}{\tau(B_d)}$	$1+0.05\left(\frac{f_B}{200\text{ MeV}}\right)^2$	PI in $\tau(B^+)$ factorization at 1 GeV	1.066 ± 0.024
$\frac{\bar{\tau}(B_s)}{\tau(B_d)}$	$1.0 \pm \mathcal{O}(0.01)$		0.94 ± 0.04
$\frac{\Delta\Gamma(B_s)}{\Gamma(B_s)}$	$0.18 \cdot \left(\frac{f_{B_s}}{200\text{ MeV}}\right)^2$	first calculated in Ref. [11]	≤ 0.46 (95 % C.L.)
$\tau(B_c)$	~ 0.5 psec	largest lifetime difference!	0.46 ± 0.17 psec
$\frac{\tau(\Lambda_b)}{\tau(B_d)}$	0.9 - 1.0	quark model matrix elements	0.79 ± 0.05

Table 2: Lifetime ratios in the beauty sector

- *Non*universal semileptonic widths – $\Gamma_{SL}(D) \neq \Gamma_{SL}(\Lambda_c) \neq \Gamma_{SL}(\Xi_c) \neq \Gamma_{SL}(\Omega_c)$ – are predicted with the main effect being *constructive* PI in Ξ_c and Ω_c decays; the lifetime ratios among the baryons will thus not get reflected in their semileptonic branching ratios; one estimates [10]

$$BR_{SL}(\Xi_c^0) \sim BR_{SL}(\Lambda_c) \leftrightarrow \tau(\Xi_c^0) \sim 0.5 \cdot \tau(\Lambda_c) \tag{3}$$

$$BR_{SL}(\Xi_c^+) \sim 2.5 \cdot BR_{SL}(\Lambda_c) \leftrightarrow \tau(\Xi_c^+) \sim 1.3 \cdot \tau(\Lambda_c) \tag{4}$$

$$BR_{SL}(\Omega_c) < 15\ \% \tag{5}$$

- On general grounds one expects $\Delta\Gamma(D^0)/\Gamma(D^0) \leq \mathrm{tg}^2\theta_C \cdot SU(3)_{Fl}$ breaking $\leq \mathcal{O}(0.01)$. If the data show that the lifetime difference for the two neutral D mass eigenstates is significantly below this bound, one would have learned an intriguing lesson on duality.

4 Lifetimes of Beauty Hadrons

The numerical stability of the $1/m_b$ expansion is characterised by $\sqrt{\mu_G^2(B)/m_b^2} \simeq 0.13 \ll 1$; i.e. such an expansion should yield rather reliable numerical results. Merely reproducing the qualitative pattern would be quite unsatisfactory. I will also use $f_B \sim 200$ MeV and $f_{B_s}/f_B \simeq 1.1 - 1.2$. Table 2 shows predictions [12,7] and data. Again some comments to elucidate these findings:

- The original predictions for the *meson* lifetimes, which had encountered theoretical criticism [13], are on the mark. (i) A recent lattice study [14] finds a result quite consistent with the original work based on factorization [1]:

$$\frac{\tau(B^+)}{\tau(B_d)} = 1.03 \pm 0.02 \pm 0.03 \tag{6}$$

(ii) Sceptics will argue that predicting lifetime ratios close to unity is not overly impressive. In response one should point out that the largest lifetime difference by far – $\frac{\tau(B_c)}{\tau(B_d)} \simeq \frac{1}{3}$ – has been predicted correctly and that the absence of contributions $\sim \mathcal{O}(1/m_Q)$ had been crucial there!

- A serious challenge arises from the 'short' baryon lifetime. In terms of $\Delta \equiv 1 - \tau(\Lambda_b)/\tau(B_d)$ the data can be expressed by

$$\Delta_{\mathrm{experim}} = 0.21 \pm 0.05 \tag{7}$$

A detailed analysis of quark model calculations [15] finds however

$$\Delta_{\mathrm{theor.}} = 0.03 - 0.12 \tag{8}$$

Reanalyses by other authors agree with Eq.(8) [16] – as does a *pilot* lattice study [17]: $\Delta_{\mathrm{lattice}} = (0.07 - 0.09)$. A recent analysis based on QCD sum rules arrives at a significantly larger value: $\Delta_{\mathrm{QCDSR}} = 0.13 - 0.21$ [18]. If true it would remove the problem. However, I would like to understand better how the sum rules analysis can differ so much from other studies given that it still uses the valence quark approximation.

- An essential question for future studies concerns the lifetimes of the beauty hyperons $\Xi_b^{-,0}$. On general grounds one expects $\tau(\Xi_b^-) > \tau(\Lambda_b)$, $\tau(\Xi_b^0)$ [7]. More specifically, using the observed charm hyperon lifetimes and $SU(3)$ symmetry a very sizeable effect has been predicted [16]:

$$\frac{\tau(\Xi_b^-) - \tau(\Lambda_b)}{\tau(\Lambda_b)} \sim 0.14 - 0.21 \tag{9}$$

As said before, the ansatz $\Gamma(H_Q) \propto M(H_Q)^5$ which would yield $\tau(\Lambda_b)/\tau(B_d) \simeq 0.75$ and therefore has been re-surrected [19] is anathema to the OPE since it would imply the nonperturbative corrections to be of order $1/m_Q$! [c]

Notwithstanding my employer I am willing to consider heresy, though, since it makes some further prediction that differ from the OPE findings:

$$\frac{\bar{\tau}(B_s)}{\tau(B_d)} = \left(\frac{M(B_d)}{M(B_s)}\right)^5 \simeq 0.94 \tag{10}$$

$$\tau(\Lambda_b)/\tau(\Xi_b^0)/\tau(\Xi_b^-) \simeq 1/0.85/0.85\,; \tag{11}$$

the *expectation* $M(\Xi_b) - M(\Lambda_b) \simeq M(\Xi_c) - M(\Lambda_c)$ has been used in Eq.(11). On the down side I do not see how such an ansatz can yield a correct prediction for $\tau(B_c)$ in a natural way.

[c] It has been shown [20] (at least for semileptonic transitions) that duality is implemented as follows: 'quark phase space + nonperturbative corrections $\hat{=}$ hadronic phase space + boundstate effects'!

5 On Quark-Hadron Duality

Duality has been an early and somewhat fuzzy element of quark model arguments. It can be expressed as follows: "Rates evaluated on the *parton* level 'approximate' observable rates summed over a 'sufficient number' of *hadronic* channels." It was never stated clearly how good an approximation it provided and what is meant by 'sufficient number'; it was thought, though, that this number had to be larger than of order unity.

Heavy quark theory has opened up new theoretical tools as well as perspectives onto duality; it demonstrated that duality can hold even with one or two channels dominating – if an additional feature like heavy quark symmetry intervenes [21]. This has been demonstrated for semileptonic $b \to c$ decays. The goal is to understand better the origins of limitations to duality and to develop some *quantitative* measures for it. The new tools that are being brought to bear on this problem are (a) the OPE; (b) the so-called small velocity sum rules [22] and (c) the 't Hooft model.

The results obtained so far show there are different categories of duality – local vs. global etc. duality – depending on the amount of averaging or 'smearing' that is involved and that duality can neither be universal nor exact.

Duality is typically based on dispersion relations expressing observable rates averaged over some kinematical variables through an OPE constructed in the Euclidean region. There are natural limitations to the accuracy of such an expansion; among other things it will have to be truncated. In any case, such a power expansion will exhibit *no* sensitivity to a term like $exp(-m_Q/\Lambda_{QCD})$. Yet upon analytical continuation from the Euclidean to the Minskowskian domain this exponentially suppressed term turns into $\sin(m_Q/\Lambda_{QCD})$ – which is not surpressed at all! I.e., the OPE *cannot* account for such terms that could become quite relevant in Minkowski space and duality violations can thus enter through these 'oscillating' terms; the opening of thresholds provides a model for such a scenario. Actually they will be surpressed somewhat like $(1/m_Q^k)\sin(m_Q/\Lambda_{QCD})$ with the power k depending on the dynamics in general and the reaction in particular. This could produce also a 'heretical' $1/m_Q$ contribution from a dimension-five operator:

$$\frac{1}{m_Q^2}\sin\left(\frac{m_Q}{\Lambda_{QCD}}\right) = \mathcal{O}(1/m_Q) \tag{12}$$

The colour flow in semileptonic as well as nonleptonic spectator decays and in WA is such that duality can arise naturally; i.e., nature had to be malicious to create sizeable duality violations. Yet in PI – because it is an interference phenomenon – the situation is much more complex leading to serious concerns about the accuracy with which duality can apply here.

The most relevant features of the 't Hooft model are: (1) QCD in 1+1 dimensions obviously confines. (2) It is solvable for $N_C \to \infty$: its spectrum of narrow resonances can be calculated as can their wavefunctions.

Duality can then be probed directly by comparing the width evaluated through the OPE with a sum over the 'hadronic' resonances appropriately smeared over the threshold region:

$$\Gamma_{OPE}(H_Q) \quad \leftrightarrow \quad \sum_n \Gamma(H_Q \to f_n) \tag{13}$$

Such a program has been first pursued using *numerical* methods; it lead to claims that duality violations arise in the total width through a $1/m_Q$ term [23] and quantitatively more massively in WA [24].

However an analytical study has shown that neither of these claims is correct: perfect matching of the OPE expression with the sum over the hadronic resonances was found through high order in $1/m_Q$ [25,26]. The same result was obtained for the more intriguing case of PI [26,27].

6 Summary and Outlook

A mature formalism genuinely based on QCD has been developed for describing inclusive nonleptonic heavy flavour decays. It can tackle questions that could not be addressed before in a meaningful way; even failures can teach us valuable lessons on nonperturbative aspects of QCD, namely on limitations to duality.

A fairly successful semiquantitative picture has emerged for the lifetimes of *charm hadrons* considering the wide span characterised by $\tau(D^+)/\tau(\Omega_c) \sim 20$; while this might be a coincidence, it should be noted:

- PI provides the leading effect driving the $D^+ - D^0$ lifetime difference; this conclusion is fully consistent with the absolute value for $BR_{SL}(D^0)$.
- This year's precise new experimental result

$$\frac{\tau(D_s)}{\tau(D^0)} = 1.211 \pm 0.017 \tag{14}$$

  confirms this picture: WA is nonleading, though still significant. It represents an interesting challemge to find the footprint of WA in some classes of exclusive final states.
- More precise data on the $\Xi_c^{0,+}$ and Ω_c lifetimes are very much needed. Those might reveal serious deficiencies in the predictions. It should be noted that the semileptonic widths for baryons are *not* universal!

The scorecard for beauty lifetimes looks as follows:

- The predictions for $\tau(B^-)/\tau(B_d)$ and even more impressively for $\tau(B_c)$ appear to be borne out by experiment.
- The jury is still out on $\tau(B_s)$.
- The Λ_b lifetime provides a stiff challenge to theory. It should be noted that most authors have shown a remarkable lack of flexibility in accommodating $\tau(\Lambda_b)/\tau(B_d) < 0.9$, which is quite unusual in this line of work. Maybe experiment will show more flexibility.
- Accurate data on $\tau(\Xi_b^{-,0})$ will be essential to celebrate success or diagnose failure.

Having developed a theoretical framework for treating nonperturbative effects, we can address the issue of duality violations. While we have begun to understand better their origins, we have not (yet) found a model *theory* that could explain the short Λ_b lifetime as a duality limitation.

There are, of course, different layers of failure conceivable and the lessons one would have to draw:

- A refusal by the data to move up the value of $\tau(\Lambda_b)$ *could* be interpreted as showing that the quark model provides a very inadequate tool to estimate baryonic expectation values.
- A low value of the average B_s lifetime – say $\bar{\tau}(B_s) < 0.96\tau(B_d)$ – had to be seen as a very significant limitation to duality.

Clearly there is a lot we will learn from future data on lifetimes and other inclusive rates – one way or the other.

Acknowledgements

This truly inspiring and enjoyable meeting organized by Profs. H.-Y. Cheng and G. Hou clearly wetted the appetite of participants for the next incarnation of this series, namely BCP 4. I am grateful to my collaborators Profs. M. Shifman, N. Uraltsev and A. Vainshtein for generously sharing their insights with me. This work has been supported by the NSF under the grant PHY 96-0508.

1. I. Big, N. Uraltsev, *Phys. Lett.* **B 280** (1992) 271.
2. For a review with references to the earlier literature, see: I.I. Bigi, M. Shifman, N.G. Uraltsev, *Annu. Rev. Nucl. Part. Sci.* **47** (1997) 591.

3. For a somewhat different approach, see: V. Chernyak, *Nucl. Phys.* **B 457** (1995) 96.
4. B. Guberina, S. Nussinov, R. Peccei, R. Rückl, *Phys. Lett.* **B 89** (1979) 261.
5. M. Bander, D. Silvermann, A. Soni, *Phys. Rev. Lett.* **44** (1980) 7; H. Fritzsch, P. Minkowski, *Phys. Lett.* **B 90** (1980) 455; W. Bernreuther, O. Nachtmann, B. Stech, *Z. Phys.* **C 4** (1980) 257; I. Bigi, *Z. Phys.* **C 5** (1980) 313.
6. N. Bilic, B. Guberina et al., J. Trampetic, *Nucl. Phys.* **B 248** (1984) 33; M. Shifman, M. Voloshin, *Sov. J. Nucl. Phys.* **41** (1985) 120.
7. For a review with references to the earlier literature, see: B. Bellini, I. Bigi and P. Dornan, *Phys. Rep.* **289** (1997) 1.
8. I.I. Bigi, N.G. Uraltsev, *Z. Phys.* **C 62** (1994) 623.
9. H.-Y. Cheng, K.-C. Yang, *Phys. Rev.* **D 61** (2000) 014008.
10. M. Voloshin, *Phys. Lett.* **B 385** (1996) 369.
11. M. Voloshin et al., *Sov.J.Nucl.Phys.* **46** (1987) 112
12. I.I. Bigi, B. Blok, M. Shifman, N. Uraltsev, A. Vainshtein, in: '*B* Decays', S. Stone (ed.), World Scientific, 1994, Revised 2nd Ed., p. 132.
13. M.Neubert, C. Sachrajda, *Nucl. Phys.* **B 483** (1997) 339.
14. M. Di Pierro, C. Sachrajda, *Nucl. Phys.* **B 534** (1998) 373.
15. N.G. Uraltsev, *Phys. Lett.* **B 376** (1996) 303.
16. M. Voloshin, preprint hep-ph/9901445; B. Guberina, B. Melic, H. Stefancic, preprint hep-ph/9907468; B. Melic, these Proceed.
17. M. Di Pierro, C. Sachrajda, C. Michael, preprint hep-lat/9906031.
18. C.-S. Huang, C. Liu, S.-L. Zhu, preprint hep-ph/9906300.
19. G. Altarelli, G. Martinelli, S. Petrarca, F. Rapuano, *Phys. Lett.* **B 382** (1996) 409.
20. I. Bigi, M. Shifman, N. Uraltsev, A. Vainshtein, *Phys. Rev.* **D56** (1997) 4017.
21. M. Voloshin, M. Shifman, *Sov. J. Nucl. Phys.* **47** (1988) 511.
22. I. Bigi, M. Shifman, N. Uraltsev, A. Vainshtein, *Phys. Rev.* **D52** (1995) 196.
23. B. Grinstein, R. Lebed, *Phys. Rev.* **D 57** (1998) 1366.
24. B. Grinstein, R. Lebed, *Phys. Rev.* **D 59** (1999) 054022.
25. I. Bigi, M. Shifman, N. Uraltsev, A. Vainshtein, *Phys. Rev.* **D59** (1999) 054011.
26. I.I. Bigi, N.G. Uraltsev, *Phys. Rev.* **D60** (1999) 114034.
27. I.I. Bigi, N.G. Uraltsev, *Phys.Lett.* **B 457** (1999) 163.

THEORY OF HEAVY BARYON DECAY

S. GROOTE AND J.G. KÖRNER

Institut für Physik, Johannes Gutenberg-Universität,
Staudinger Weg 7, D-55099 Mainz, Germany

We discuss various topics in the theory of heavy baryon decays. Among these are recent applications of the Relativistic Three Quark Model to semileptonic, non-leptonic, one-pion and one-photon transitions among heavy baryons, new higher order perturbative results on the correlator of two heavy baryon currents and on the semi-inclusive decay $\Lambda_b \to X_c + D_s^{(\star)-}$.

1 Introduction

This review contains four different topics on heavy baryon decays. First we discuss some recent theoretical determinations of the quasielastic $\Lambda_b \to \Lambda_c$ form factor in HQET where there are many different results in the literature. The quasielastic $\Lambda_b \to \Lambda_c$ form factor has also been calculated in the Relativistic Three Quark Model (RTQM) which is an all-encompassing tool for the description of exclusive heavy baryon decays. We briefly describe the RTQM model and discuss various applications of the RTQM. As concerns QCD sum rules we describe some recent three-loop results on the finite mass baryon current-current correlator at $\mathcal{O}(\alpha_s)$ which is a new result important for QCD sum rule calculations. Finally we discuss the semi-inclusive decays $\Lambda_b \to X_c + D_s^{(\star)-}$ where $\mathcal{O}(\alpha_s)$ and $O(1/m_b^2)$ have been recently calculated. This is an important mode for Λ_b-decays with an expected branching ratio of $\approx 10\%$.

2 The quasielastic $\Lambda_b \to \Lambda_c$ form factor

There exist many different results on the the quasielastic $\Lambda_b \to \Lambda_c$ form factor in the literature. The predicted slope values of the form factor range from $\rho^2 = 0.33$ to $\rho^2 = 2.35$. In the simplest approach one takes a heavy quark – light diquark model and describes the transition by a one-loop Feynman diagram [1,2,3]. One takes $M_Q = m_Q$ and local point coupling factors g_1 and g_2 for the quark-diquark-baryon vertices whose strengths are fixed by the compositeness condition. The compositeness condition is nothing but the field theoretic equivalent of the familiar quantum mechanical concept of wave function normalization. In the heavy quark limit the result of such a calculation is given by the form factor $\Phi(\omega) = (\omega^2 - 1)^{-1/2} \ln(\omega + \sqrt{\omega^2 - 1})$ which is familiar from the ω-dependent renormalization of the heavy quark current. An expansion in terms of powers of $(\omega - 1)$ shows that the form factor is correctly normalized at the zero recoil point $\omega = 1$, has a slope of $\rho^2 = 1/3$ and a convexity of $c = 2/15$. The form factor $\Phi(\omega)$ is rather flat when e.g. compared to the heavy meson form factor where experimentally one finds slope values of ≈ 1. $\Phi(\omega)$ lies within the inclusive HQET sum rule bounds derived by Chiang [4] but must nevertheless be discarded since it oversaturates the semileptonic inclusive rate $\Lambda_b \to X_c + l^- + \bar{\nu}_l$ as recently shown in [5].

Improvements on this simplest approach lead one to the Relativistic Three Quark Model (RTQM) [6]. In the improvements one incorporates binding effects by replacing $M_Q = m_Q$ by $M_Q = m_Q + \bar{\Lambda}$. The vertex is softened through introduction of a nonlocal vertex and one introduces a true three-quark structure by replacing the (ud)-diquark by single u, d quarks.

3 The Relativistic Three Quark Model

According to the changes mentioned above the RTQM treats the decay $\Lambda_b \to \Lambda_c + W^-_{\text{off-shell}}$ in terms of a two-loop Feynman diagram with nonlocal vertices including binding effects. The result is that the $\Lambda_b \to \Lambda_c$ form factor becomes steeper. Depending on the choice of spin vertex structure for the heavy baryons the slope increases from the aforementioned $\rho^2 = 1/3$ to $\rho^2 = 0.75 \div 1.35$ depending on the choice of spin vertex structure to be discussed later on.

The RTQM is an all-encompassing and versatile tool for the description of heavy baryon decays in terms of a Feynman diagram description. The number of parameters associated with the nonlocality of the vertices, the binding effects and the values of the constituent quark masses is reasonably small and their values lie within common expectations. Many of the parameters are already fixed from light baryon decays where the RTQM also applies. For the loop integrations one uses the α-parametrization in its exponential form. This introduces n α-parameters $\alpha_1, \ldots, \alpha_n$ for n propagators and consequently n integrations. One introduces a Laplace transform to facilitate the vertex form factor integration which is left to the very end. The exponential α-parametrization allows one to do the tensor loop integrals directly through differentiation, i.e. without use of the Passarino-Veltman expansion. One transforms to spherical type variables which leaves one with one radial type integration and $(n-1)$ angular type integrations. All $(n+1)$ numerical integrations including the Laplace transform can be done with ease. In fact the spherical integrations can also be done analytically but the ease of the numerical integration does not warrant this effort. We shall now discuss several applications of the RTQM to heavy baryon decays.

The dependence of the Isgur-Wise function on the choice of the vertex spin structure for the heavy baryons was investigated in [7]. For the Λ_Q-type baryons both the effective couplings

$$J^1_{\Lambda_Q} = \bar{\psi}_Q \psi^T_{[u} C\gamma_5 \psi_{d]} \qquad J^2_{\Lambda_Q} = \psi_Q \psi^T_{[u} C\gamma_5 \not{v} \psi_{d]} \tag{1}$$

correctly describe the coupling of the Λ_Q to a heavy on-shell quark and two light off-shell quarks in the limit of Heavy Quark Symmetry (HQS). When inserted into the relevant two-loop diagram both coupling structures reproduce the required leading order HQET form factor structure including the unit normalization at zero recoil. However, one finds that the slope of the Isgur-Wise function depends on the choice of vertex structure. For the three choices $J^1_{\Lambda_Q}$, $\frac{1}{2}(J^1_{\Lambda_Q} + J^2_{\Lambda_Q})$ and $J^2_{\Lambda_Q}$ one finds slope values of $\rho^2 = 1.35, 1.05$ and 0.75. The fact that $\rho^2(J^1_{\Lambda_Q}) > \rho^2(J^2_{\Lambda_Q})$ agrees with the sum rule analysis of [8] although the difference in slope values in [8] is not as large. Note that taking the geometric mean of the two currents leads to a constituent type vertex structure where the projector $(\not{v}+1)/2$ projects onto the large components of the light quark fields.

Finite mass effects in heavy $\Lambda_{Q_1} \to \Lambda_{Q_2}$ transitions were analyzed in [9] by replacing the heavy quark propagators in the Feynman diagrams by the full propagator. This was effected by the replacement

$$\frac{i}{l \cdot v + \bar{\Lambda}} \frac{\not{v}+1}{2} \to \frac{i(\not{P} + \not{l} + m_Q)}{(P+l)^2 - m_Q^2} \tag{2}$$

where $(P+l)$ is the momentum of the heavy quark and l is a loop momentum. For $\Lambda_b \to \Lambda_c$ the rate is decreased by 9.3% in qualitative agreement with the findings of [10]. The decrease was found to be even larger for $\Lambda_c \to \Lambda_s$ where the strange quark was treated as a heavy quark in the reference rate for the sake of comparison. It is clear that an expansion of the full propagator in terms of powers of $1/m_Q$ would allow one to systematically explore higher order $1/m_Q$-effects in these transitions as e.g. in the zero recoil normalization of the relevant zero recoil $\Lambda_{Q_1} \to \Lambda_{Q_2}$ form factor.

In [11] the RTQM was used to calculate exclusive nonleptonic decays of heavy baryons. There are so-called factorizing and nonfactorizing contributions to these decays. The nonfactorizing contributions had never been calculated before. In the Feynman diagram approach they involve a genuine three-loop calculation which was done in [11]. As a sample result one finds that in the nonleptonic decays $\Lambda_b \to \Lambda_c + \pi^-$ the nonfactorizing contributions amount to -20% and -28% in the parity violating and parity conserving amplitudes, repectively, with an ensuing reduction in rate of $\approx 40\%$. The nonfactorizing contributions are therefore not negligible. A multitude of exclusive nonleptonic decays have been calculated within the RTQM model involving $(\bar{s}c)(\bar{u}d)$, $(\bar{b}c)(\bar{u}d)$ and $(\bar{b}c)(\bar{c}s)$ transitions [11].

The RTQM model is also well suited for heavy flavour-conserving one-pion and one-photon transitions between heavy baryons. The one-pion transitions are described by two-loop diagrams where the pion couples to a single light quark line. The $1/f_\pi$ coupling of chiral perturbation theory effectively appears through the quark level Goldberger-Treiman relation $g_\pi = 2m_q/f_\pi$. Many one-pion transitions have been calculated including transitions from excited states [12,14]. The results are remarkably close to the results of using the constituent quark model [15,16,17] for the light quarks even though the light quarks are fully off-shell in the RTQM model.

For one-photon transitions the transverse on-shell photon couples only to the light quarks in the leading order of the heavy quark expansion with a coupling strength given by the light quark charge. In addition one has to include contact graphs to assure gauge invariance of the one-photon transitions. These are generated according to the path integral formalism of Mandelstam. Again the results of the RTQM model [13,14] are remarkably close to the constituent quark model calculation [18,19]. The relation of the RTQM description of one-photon transitions to the chiral approach remains to be explored, in particular to the recent calculation of [20] which contains also chiral loops.

What we have discussed so far are some basic applications of the RTQM in the heavy baryon sector. Further work is in progress on the decays of double heavy baryons, on magnetic moments of heavy baryons and on heavy flavour-conserving nonleptonic charm and bottom baryon decays.

4 Finite mass baryonic current-current correlator at $\mathcal{O}(\alpha_s)$

The calculation of the spectral density associated with the baryonic current-current correlator is important for QCD sum rule applications. We want to report on some advances we have made in the calculation of the $\mathcal{O}(\alpha_s)$ radiative corrections to the spectral density with one finite mass quark mass and two zero quark masses [21]. The calculation involves the evaluation of two-scale three-loop Feynman diagrams which only became possible due to recent technical advances [22] in three-loop technology. Taking the appropiate limits we recover previous results derived for the zero mass case [23] and for the infinite mass case [24]. In the mesonic case the corresponding calculation has been done some time ago showing that radiative corrections to the spectral density can become quite important [25].

The basic object of study is the vacuum expectation value of the time-ordered product of two baryonic currents $< TJ(x)J(0) >$. In spinor space its Fourier transform is expanded along the spinor matrix structures $\not q$ and m with coefficients $\pi_q(q^2)$ and $\pi_m(q^2)$. We concentrate on the invariant $\pi_m(q^2)$ which has associated with it a spectral density $\rho_m(s)$ for which we shall present two- and three-loop results. Using the simplest possible current $J = \Psi(u^T Cd)$ and writing

$$\rho_m(s) = \frac{1}{128\pi^4} s^2 \left\{ \rho_0(s) \left(1 + \frac{\alpha_s}{\pi} \ln \left(\frac{\mu^2}{m^2} \right) \right) + \frac{\alpha_s}{\pi} \rho_1(s) \right\} \tag{3}$$

we obtain the Born term two-loop contribution $\rho_0(q^2) = 1 + 9z - 9z^2 - z^3 + 6z(1+z)\ln z$ where $z = m^2/q^2$. In the $\overline{\rm MS}$ scheme the radiative three-loop contribution is given by

$$\begin{aligned}
\rho_1(s) &= 9 + \frac{665}{9} z - \frac{665}{9} z^2 - 9z^3 - \left(\frac{58}{9} + 42z - 42z^2 - \frac{58}{9} z^3 \right) \ln(1-z) \\
&+ \left(2 + \frac{154}{3} z - \frac{22}{3} z^2 - \frac{58}{9} z^3 \right) \ln z + 4 \left(\frac{1}{3} + 3z - 3z^2 - \frac{1}{3} z^3 \right) \ln(1-z) \ln z \\
&+ 12z \left(2 + 3z + \frac{1}{9} z^2 \right) \left(\frac{1}{2} \ln^2 z - \zeta(2) \right) + 4 \left(\frac{2}{3} + 12z + 3z^2 - \frac{1}{3} z^3 \right) \mathrm{Li}_2(z) \\
&+ 24z(1+z) \left(\mathrm{Li}_3(z) - \zeta(3) - \frac{1}{3} \mathrm{Li}_2(z) \ln z \right)
\end{aligned} \tag{4}$$

Writing $q^2 = (m+E)^2$ we can perform a threshold expansion of the spectral density in terms of powers of E/m. We write the leading order result in a factorized form in order to facilitate comparison with HQET. One has

$$\begin{aligned}
m\rho_m \overset{m\to\infty}{=} & \frac{1}{128\pi^4} E^5 \left\{ 1 + \frac{\alpha_s}{\pi} \left(\frac{1}{2} \ln \left(\frac{m^2}{\mu^2} \right) - \frac{2}{3} \right) \right\}^2 \times \\
& \times \left\{ 1 + \frac{\alpha_s}{\pi} \left(4 \ln \left(\frac{\mu}{2E} \right) + \frac{2}{45} \left(10\pi^2 + 273 \right) \right) \right\}.
\end{aligned} \tag{5}$$

The first bracket is the square of the appropiate HQET matching coefficient $C(m/\mu, \alpha_s)$ first derived in [8] and the second bracket is the appropiate result for the leading order HQET spectral density $\rho^{\rm HQET}(E,\mu)$ first derived in [24]. We have checked that the zero mass limit of the general spectral density reproduces the result of [23]. Work is in progress on the momentum spectral density $\rho_q(s)$ and on correlators of baryonic currents with arbitrary spin structure and on sum rule applications of the spectral densities.

5 The semi-inclusive decay $\Lambda_b \to X_c + D_s^{(*)-}$

Following the analysis of the semi-inclusive B-meson decays $\bar{B} \to X_c + D_s^{(*)-}$ in [26,27] we looked at the corresponding semi-inclusive Λ_b-decays. The Λ_b decays are potentially more interesting because of the possibility to observe Λ_b polarization effects in this decay. At the leading order of α_s and the heavy quark mass expansion one expects branching ratios of $BR(\Lambda_b \to X_c + D_s^-) =$ 3.2% and $BR(\Lambda_b \to X_c + D_s^{*-}) = (4.4(L)+2.4(T))\%$ where we have separately listed the longitudinal (L) and transverse component (T) of the spin 1 D_s^{*-}. The two components can be separately measured by an angular analysis of the subsequent decays $D_s^{*-} \to D_s^- + \gamma$ and $D_s^{*-} \to D_s^- + \pi^0$. The same holds true for the measurement of polarization effects [28] which will not be discussed here.

In [28] we calculated the perturbative $\mathcal{O}(\alpha_s)$ and the nonperturbative corrections to these decays using the factorization hypothesis. Numerically one finds:

$$\begin{aligned}
\Lambda_b \to X_c + D_s^- \;:\quad & \hat{\Gamma}_S = (1 - 0.096 - 0.013) \\
\Lambda_b \to X_c + D_s^{*-} \;:\quad & \hat{\Gamma}_L = 0.65(1 - 0.110 - 0.034) \\
& \hat{\Gamma}_T = 0.35(1 - 0.108 + 0.026) \\
& \hat{\Gamma}_{L+T} = (1 - 0.096 - 0.009).
\end{aligned} \tag{6}$$

In order to clearly exhibit the percentage changes the rates have been normalized to their respective Born term rates. The second and third figures in the round brackets of Eq.(6) refer to the perturbative $\mathcal{O}(\alpha_s)$ corrections and the nonperturbative kinetic energy correction, respectively. The perturbative corrections are negative and quite uniform. They amount to $\approx 10\%$. The nonperturbative corrections range from 0.9% to 3.4% with differing signs. The longitudinal mode dominates the rate into D_s^{*-}'s. The L/T rate ratio Γ_L/Γ_T decreases by 6.8% from 1.86 to 1.73 after applying the perturbative and nonperturbative corrections.

The corresponding semi-inclusive $b \to u$ decays $\Lambda_b \to X_u + D_s^{(*)-}$ are suppressed due to the smallness of V_{bu}. They are nevertheless of interest for the analysis of so-called wrong sign $D_s^{(*)-}$'s [26]. Numerically one finds [28]

$$\begin{aligned}
\Lambda_b \to X_u + D_s^- \;:\quad & \hat{\Gamma}_S = (1 - 0.169 - 0..013) \\
\Lambda_b \to X_u + D_s^{*-} \;:\quad & \hat{\Gamma}_L = 0.73(1 - 0.178 - 0.029) \\
& \hat{\Gamma}_T = 0.27(1 - 0.115 + 0.030) \\
& \hat{\Gamma}_{L+T} = (1 - 0.161 - 0.001).
\end{aligned} \tag{7}$$

The dominance of the longitudinal mode in the decay $\Lambda_b \to X_u + D_s^{*-}$ is now more pronounced. Also the radiative corrections are no longer uniform leading to a substantial 13.3% change in the L/T rate ratio due to the perturbative and nonperturbative corrections. It would be interesting to study these semi-inclusive Λ_b decay modes including the L/T composition of the D_s^{*-}'s at future colliders.

References

1. G.V. Efimov et al., Z. Phys. **C54** (1992) 349
2. D. Ebert et al., Z. Phys. **C71** (1996) 329
3. M. Haghighat, Isfahan University of Technology Preprint IUT-PHYS/99-23
4. C.W. Chiang, Phys. Rev. **D61** (2000) 016005
5. J.G. Körner, B. Melić, to be published
6. M.A. Ivanov et al., Phys. Rev. **D56** (1997) 348
7. M.A. Ivanov et al., hep-ph/9911425
8. A.G. Grozin, O.I. Yakovlev, Phys. Lett. **285 B** (1992) 254
9. M.A. Ivanov et al., hep-ph/9910342
10. R.S. Marques de Carvalho et al., Phys. Rev. **D60** (1999) 034009
11. M.A. Ivanov et al., Phys. Rev. **D57** (1997) 5632
12. M.A. Ivanov et al., Phys. Lett. **442 B** (1998) 435
13. M.A. Ivanov et al., Phys. Lett. **448 B** (1999) 143
14. M.A. Ivanov et al., Phys. Rev. **D60** (1999) 094002
15. T.M. Yan et al., Phys. Rev. **D46** (1992) 1148
16. D. Pirjol, T.M. Yan, Phys. Rev. **D56** (1997) 5483
17. F. Hussain, J.G. Körner, S. Tawfiq, hep-ph/9909278
18. H.Y. Cheng et al., Phys. Rev. **D47** (1993) 1030
19. S. Tawfiq, J.G. Körner, P.J. O'Donnell, hep-ph/9909444
20. M.C. Banuls, A. Pich, I. Scimemi, hep-ph/9911502
21. S. Groote, J.G. Körner, A.A. Pivovarov, hep-ph/9911393
22. S. Groote, J.G. Körner, A.A. Pivovarov, Nucl. Phys. **B542** (1999) 515; Phys. Rev. **D60** (1999) 061701
23. A.A. Ovchinnikov et al., Sov. J. Nucl. Phys. **48** (1988) 358
24. S. Groote, J.G. Körner, O. Yakovlev, Phys. Rev. **D55** (1997) 358
25. S.C. Generalis, J. Phys. **G16** (1990) 367
26. R. Aleksan et al., hep-ph/9906504 and hep-ph/9906505
27. M. Fischer et al., hep-ph/9911448
28. M. Fischer et al., to be published

SEMILEPTONIC B DECAYS

JON J. THALER
University of Illinois, 1110 W. Green Street, Urbana IL61801, USA
E-mail: jjt@uiuc.edu

I review recent results in the study of semileptonic B meson decays. I discuss the measurement of CKM matrix elements, tests of HQET, and prospects for the future.

1 Introduction

Semileptonic decays of B mesons provide a clean method to measure the CKM matrix elements. Only one Feynman diagram contributes to each process, and there are no final state interactions to complicate the issue. One can use $B \to X_c \ell \nu$ for V_{cb} and $B \to X_u \ell \nu$ for V_{ub}.

Calculation of the CKM matrix elements from semileptonic decays requires the use of heavy quark effective field theory (HQET). It is thus important to verify that the phenomenology is valid. Semileptonic decays are also useful in the measurement of $B\overline{B}$ mixing, because the sign of the lepton unambiguously tags the flavor of the parent meson. In this talk, I describe some recent. They help to measure the parts of the unitarity triangle as labeled in the diagram below.

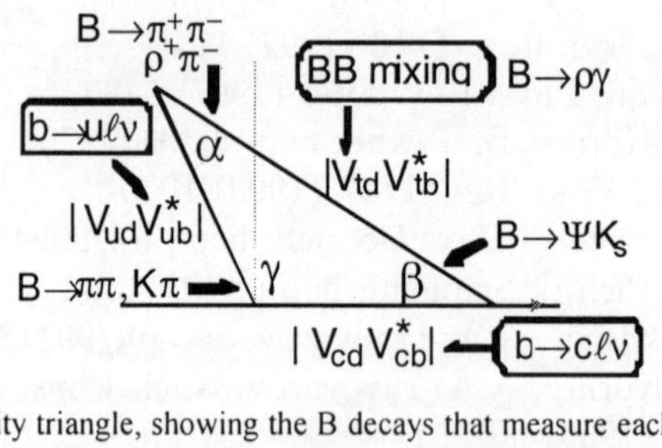

Figure 1. CKM unitarity triangle, showing the B decays that measure each part.

2 HQET

HQET is applicable to meson decay when the quarks involved in the decay are much more massive than the other mass scales in the problem. In that case, spin dependent interactions become small, and the heavy quarks are nearly at rest both before and after the decay. Measuring the decay near the maximum momentum transfer, as shown in the figure below, can minimize finite mass.

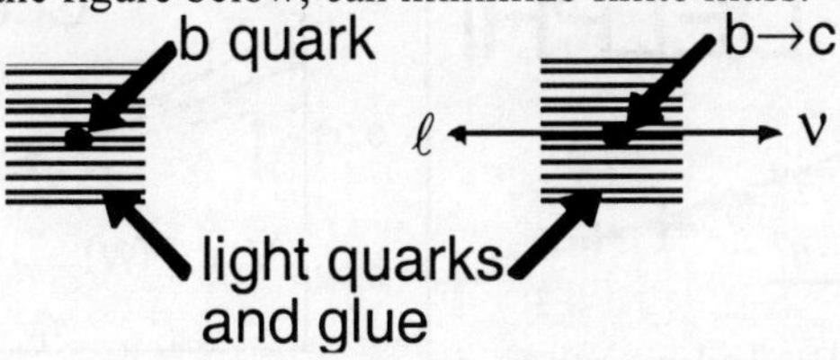

Figure 2. HQET kinematics. On the left, a meson contains a heavy quark, which is nearly at rest. On the right, the heavy quark has decayed to another heavy quark at the zero recoil (maximum q^2) kinematic point.

Both B→Dℓν and B→D*ℓν can be used to measure Vcb. B→D*ℓν is more accessible experimentally, because of the larger rate and easier D* detection. QCD radiative corrections and finite quark mass effects are accommodated by making a power series expansion in α_s and $1/m_b$. The result is [1]:

$$|V_{cb}| = 0.0411\sqrt{\frac{BR(B_{sl})}{0.105}}\sqrt{\frac{1.55\text{ps}}{\tau_B}} \times \left(1 - 0.25\frac{\mu_\pi^2 - 0.5\,GeV^2}{0.2\,GeV^2}\right) \times \left(1 \pm 0.01_{m_b} \pm 0.01_{QCD} \pm 0.015_{1/m_b^3}\right)$$

where μ_π is one of the three HQET parameters (discussed below) and the last parentheses show the various theory uncertainties.

Decays to light final states, $B \to X_u \ell\nu$, are not so easy to interpret, because HQET is not applicable. The results are much more model dependent.

3 B→D*ℓν and V_{cb}

DELPHI reported [2] a new measurement of B→D*ℓν. Their data is shown in figure 4. DELPHI obtains $BR(B^o \to D^*\ell\nu) = (5.22 \pm 0.12 \pm 0.55)\%$ and $|V_{cb}|\,A_1(1) = 0.0380 \pm 0.0013 \pm 0.0016$. In order to extract V_{cb}, they must correct for the fact that even at q^2_{max} the heavy quarks are not quite at rest. This is $A_1(1) \approx F_{D^*}(1) = 0.91 \pm 0.03$. So, $|V_{cb}| = 0.0417 \pm 0.0015 \pm 0.0017 \pm 0.0014$.

CLEO reports [3] new results on B→Dℓν. This is an important complementary measurement, because HQET predicts that the q^2 dependence of the decays should be similar. Their data, shown in figure 3, are consistent with older D*ℓν [4]. CLEO obtains $|V_{cb}| = 0.0383 \pm 0.0021 \pm 0.0020 \pm 0.0011$ from the D* mode and $|V_{cb}| = 0.042 \pm 0.005 \pm 0.004 \pm 0.004$ from the D mode. The reported theoretical uncertainty on the Dℓν result is too large; it uses $FD(1) = 1.0 \pm 0.1$. Recent theoretical calculations [5] give, 1.06 ± 0.03. This means that improved experimental results can yield an improved value for $|V_{cb}|$.

CLEO recently reported preliminary D* results [6] with smaller statistical

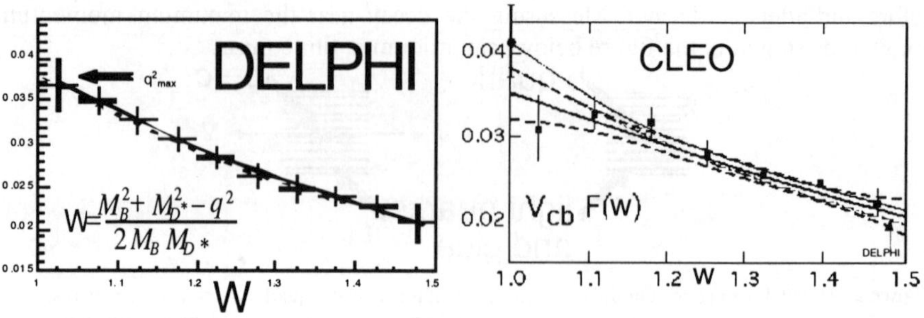

Figure 3. DELPHI and CLEO measurements of B→D*ℓν and B→Dℓν.

errors: BR($B^o \to D^*\ell\nu$) = (4.91 ± 0.17 ± 0.42)% and BR($B^- \to D^*\ell\nu$) = (5.13 ± 0.54 ± 0.64)%. The systematic error and form factors remain under study. They dominate the V_{cb} uncertainty at this time.

Inclusive decays have better statistics but more model dependence. Exclusive decays have less model dependence, by measuring the form factor at q^2_{max}.

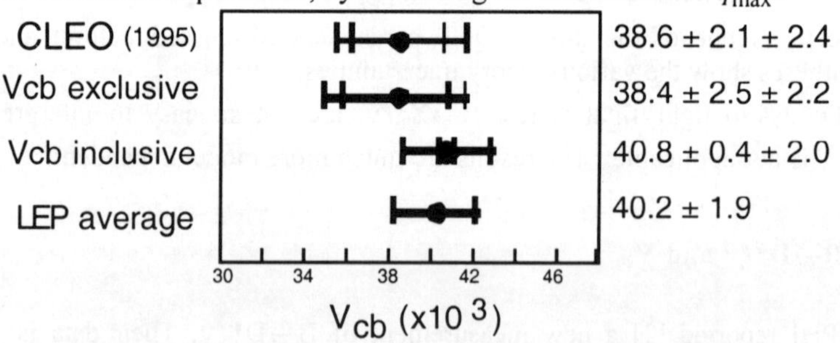

Figure 4. Combined LEP B→D*ℓν results compared with the previous CLEO result.

4 B→$X_u\ell\nu$ and V_{ub}

CLEO measures exclusive final states (ρℓν and πℓν), while the LEP experiments measure an inclusive branching ratio. Figure 5 shows CLEO's ρℓν data [7]. To improve the signal to background ratio, a lepton energy requirement of 2.3 GeV has been imposed to obtain the results shown. Unfortunately, this renders the branching ratio calculation sensitive to models of the decay form factor. The high lepton energy region does not distinguish these models. CLEO's results are:

$$BR(B \to \rho\ell\nu) = \left(2.57 \pm 0.29^{+0.33}_{-0.46} \pm 0.41\right) \times 10^{-4}$$

$$|V_{ub}| = \left(3.25 \pm 0.14^{+0.21}_{-0.29} \pm 0.55\right) \times 10^{-3} \quad \rho\ell\nu \text{ and } \pi\ell\nu \text{ combined}$$

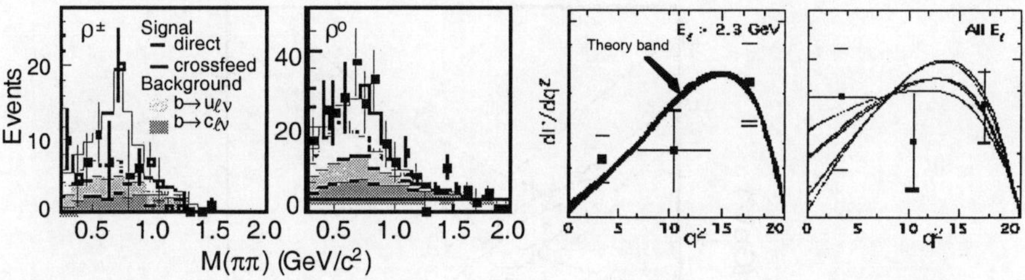

Figure 5. CLEO measurement of B→ρℓν.

LEP (ALEPH, DELPHI, and L3) [8] measure an inclusive branching ratio. The uncertainty is dominated by their knowledge of the charm background, which they must subtract. The combined LEP result is [9]:

$$BR(b \to X_u \ell \nu) = (1.67 \pm 0.35 \pm 0.38 \pm 0.20) \times 10^{-3}$$

$$|V_{ub}| = \left(4.05^{+0.39+0.43+0.23}_{-0.46-0.51-0.27} \pm 0.02 \pm 0.16\right) \times 10^{-3}$$

5 HQET parameters

Is HQET consistent? Finite mass effects can be described by three parameters.

- λ_1 measures the average kinetic energy of the b quark in the B meson. (Also called $-\mu_\pi^2$.)
- λ_2 measures the spin energy (known from $\Delta m(B^*\text{-}B)$).
- $\overline{\Lambda}$ converts from m_b to m_B.

CLEO has determined λ_1 and $\overline{\Lambda}$ by measuring [10] moments of the lepton energy and hadron mass spectra in $B \to X_c \ell \nu$.. If HQET is correct, these measurements should all be consistent. The results are shown in figure 6. There is a discrepancy at about the 2σ level. The accuracy of the method is shown by the result from the M_x moments: $\lambda_1 = -0.13 \pm 0.01 \pm 0.06$ GeV^2 and $\overline{\Lambda} = 0.33 \pm 0.02 \pm 0.08$ GeV.

Decays to excited charm states also test HQET. In the infinite mass limit, all decays should be to D and D*. At finite m_B, there is a significant D** contribution. Empirically, $BR(B \to X_c \ell \nu) = (10.55 \pm 0.13)\%$ (world average), and $BR(B \to D\text{or}D^* \ell \nu) = (7.06 \pm 0.51)\%$ (CLEO). Can $B \to D\text{or}D^* \pi \ell \nu$ explain the difference. Two LEP experiments have measurements [11]: BR = (3.43 ± 0.57 ± 0.22)% (DELPHI), and BR = (2.26 ± 0.29 ± 0.33)% (ALEPH). Theory [12] predicts only the smaller value. This seems to be another 2σ problem.

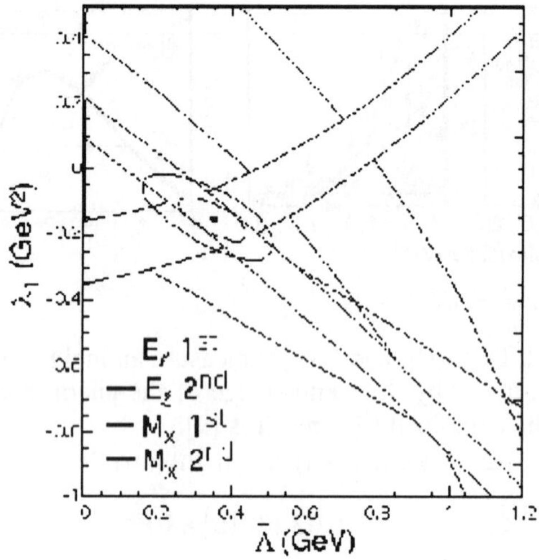

Figure 6. HQET parameters determined by lepton energy and hadron mass moments.

6 $B\overline{B}$ mixing

Mixing produces two states of different mass. The mass difference is:

$$\Delta M_d = 0.50 ps^{-1}\left[\frac{\sqrt{B_{B_d}}f_{B_d}}{200\,MeV}\right]^2\left[\frac{\overline{m}_t(m_t)}{170\,GeV}\right]^{1.52}\left[\frac{|V_{td}|}{8.8\times10^{-3}}\right]^2\left[\frac{\eta_B}{0.55}\right].$$

Knowledge of V_{td} is limited by the theoretical calculation of B_{Bd} and f_{Bd}. CLEO measures the time integrated mixing probability:

$$\chi_d = \frac{N_{B\to\overline{B}}}{N_{B\to\overline{B}} + N_{B\to B}} = \frac{\Delta M_d/\Gamma}{2\left(1+(\Delta M_d/\Gamma)^2\right)}.$$

CLEO also measures the CP asymmetry:

$$A_{CP} \equiv \frac{\chi_d(\ell^+) - \chi_d(\ell^-)}{\chi_d(\ell^+) + \chi_d(\ell^-)} \quad \text{expected to be O}(10^{-3})$$

This is less sensitive than $\frac{N_{++} - N_{--}}{N_{++} + N_{--}}$ to lepton ID systematics. CLEO observes one B via its semileptonic decay and the other B in the 2-body decay to $D^*\pi$ (or $D^*\rho$). To improve the statistical power, only the lepton, the hard π (or ρ) and the soft π from $D^*\to D\pi$, are reconstructed. Mixing causes the lepton and hard π (or ρ) to have the same charge. This method, called partial reconstruction, results in a zero-constraint fit. Background is measured by looking at known angular distributions. For example, the D* is fully aligned. Signal to background is about

2:1. The results are: $\chi_d = 0.197 \pm 0.013 \pm 0.014$, and $A_{CP} = 0.017 \pm 0.070 \pm 0.012$. This is the best single measurement of χ_d.

7 Summary

The CKM matrix element V_{cb} is now measured to 5% accuracy. A 2% measurement seems possible in the near future. Measurement of the V_{ub} matrix element, only good to 20% accuracy, will be much more difficult to improve. There are some indications, at the 2σ level, of problems with the HQET formalism. $B\overline{B}$ mixing is limited by statistics, but the V_{td} calculation is limited by theory.

References

1. I. Bigi, hep-ph/9907720; Z. Ligeti, hep-ph/9904460.
2. DELPHI 99-107 CONF 294.
3. J. Bartelt, *et al.*, *Phys. Rev. Lett.* **82**, 3746 (1999).
4. B. Barish, *et al.*, *Phys. Rev* **D51**, 1014 (1995).
5. S. Hashimoto, *et al.*, hep-ph/9907720.
6. CLEO preliminary B->D*lnu results..
7. B.H. Behrens, *et al.*, CLNS 99/1611.
8. ALEPH, *Eur. Phys. J.*, **C6**, 555 (1999); DELPHI, 98-97 CONF 165; L3, Phys. Lett., **B436**, 174 (1998).
9. LEPVUB –99/01.
10. CLEO conf 98-21.
11. DELPHI 99-103 CONF 290; Zeit. Phys. C73, 601(1997).
12. N. Isgur, hep-ph/9904398; Ebert, Faustov, & Galkin, hep-ph/9906415.

HADRONIC B DECAYS AT CLEO

ANDREAS H. WOLF
Syracuse University, Physics Bldg, Syracuse NY 13244, USA
E-mail: wolf@phy.syr.edu

CLEO has performed many analyses in the field of hadronic B decays. This talk presents two decays modes, $B \to J/\psi\phi K$ and $B \to D^{(*)}D^{(*)}$. During the first part I present a newly seen decay mode with a branching ratio of $(8.8^{+3.5}_{-3.0}[\text{stat}] \pm 1.3[\text{syst}]) \times 10^{-5}$, the second part gives an update on a study to find and analyze two-body decays involving D and D^*.

1 Introduction

With the advent of the B factory experiments we are poised to probe the CKM mechanism of the Standard Model with high precision and overconstrained parameters from hadronic B decays, some of them containing charm. (The rare and charmless decay modes are addressed by two other talks[1,2] in this conference.) Until BABAR, BELLE, and CLEO III have accumulated a sufficient amount of events, CLEO II data can provide information about these decays. In the many years of running CLEO II and CLEO II.V we have collected about 1×10^7 $B\overline{B}$ meson pairs with both detectors configurations[3,4]. The main difference between CLEO II and CLEO II.V is the addition of a silicon vertex detector.

2 $B \to J/\psi\phi K$

We reconstructed the J/ψ in its l^+l^- decay modes by identifying electron and muon tracks, and selecting J/ψ candidates, cutting on the invariant mass of the lepton pair. To combine the CLEO II and CLEO II.V data sets, this analysis employed cuts on variables that are normalized to their estimated error. ϕ candidates are found in the K^+K^- decay. We then combine the J/ψ and ϕ candidates with a charged $K^\pm$ track or a K^0_S candidate decaying into $\pi^+\pi^-$. All charged K tracks were selected using specific ionization and time-of-flight information. Finally, $B \to J/\psi\phi K$ candidates were found using two observables: First, the difference between the candidates energy and the nominal B energy (= beam energy), ΔE. Second, the "beam-constrained mass" B candidate mass, using the total momentum of the candidate and the beam energy, $M(B)$. They give a resolution of 6 MeV and 2.7 MeV/c^2, respectively. Fig. 1. I,II show the distribution of candidate events in these two

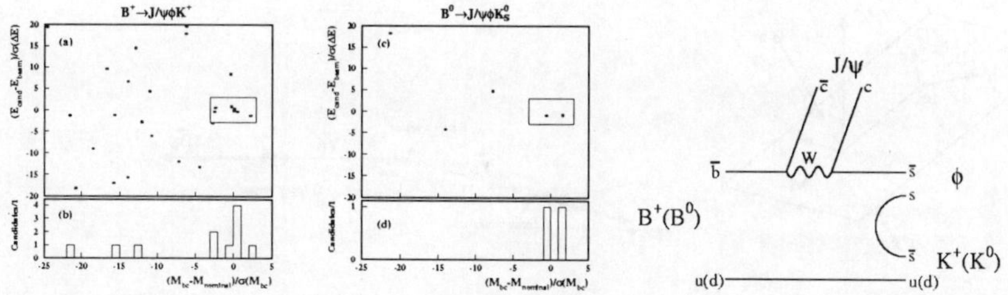

Figure 1. Energy difference and beam-constrained mass of B candidates decaying into $J/\psi\phi K$. The plots are: (a) 2-D plot with signal box, (b) Projection onto M(B) after ΔE cut, for I: $B \to J/\psi\phi K^{\pm}$ and II: $B \to J/\psi\phi K^0_S$. III: Most likely graph for $B \to J/\psi\phi K$

variables, the signal box, and the projection onto $M(B)$ after cutting on ΔE. Combining the two K modes, we find 10 signal events for $B \to J/\psi\phi K$.

Using Monte-Carlo[5] based simulations we determined the overall efficiency for this analysis, and the level of background we expect from other decays. We predict 0.5 ± 0.2 background events in the signal box, half of it combinatorical background from other B decays, the other half from $B \to J/\psi K^+K^-K$ not involving a ϕ. The second contribution was estimated from 7 such candidate events found in data (with 2.8 background events). Overall we measure a branching fraction of $\mathcal{B}(B \to J/\psi\phi K) = (8.8^{+3.5}_{-3.0}[\text{stat}] \pm 1.3[\text{syst}]) \times 10^{-5}$

Fig. 1.III shows the most likely decay process, creating a $s\bar{s}$ quark pair. No single B meson decay chain requiring this process has been observed until now. This decay is the most compelling mode showing the need for $s\bar{s}$ popping. We also do not find any evidence for resonant substructures in this decay, which would make it the first measured decay mode not being mediated by a (short–range) B meson two-body decay and subsequent (long–range) secondary decays.

3 $B \to D^{(*)}D^{(*)}$

The decays of the type $B^0 \to D^+D^-$ offer an alternative approach to measure the CKM angle β and might be helpful in resolving the ambiguities[6] left over after measuring $\sin 2\beta$ from $B \to J/\psi K_S$. We also can use D^*D^* decay, but we must note that it consist of a mixture of S, P, and D waves. This mixture makes the final state not a CP eigenstate. It was proposed[7,8] to use an angular analysis method. (The transversity angle as defined in Fig. 2.I)

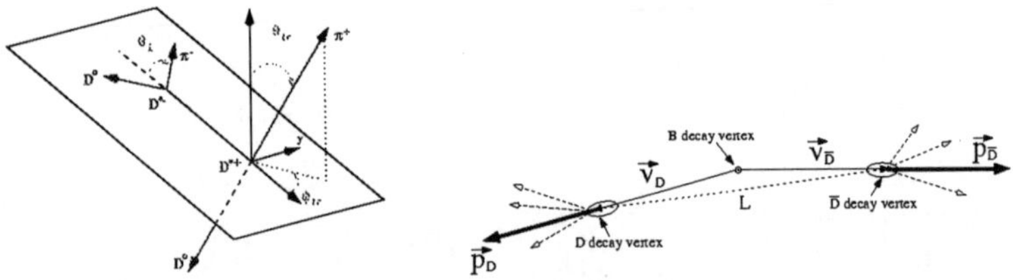

Figure 2. I. Definition of transversity angles (from the BaBar Physics Book); II. Basic geometry of lifetime analyses

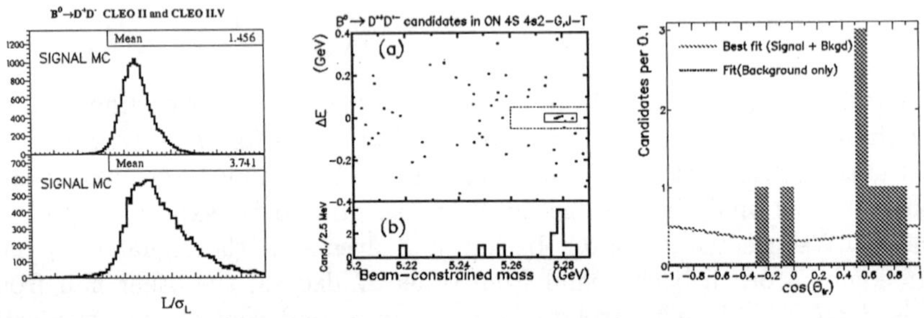

Figure 3. I: Significance of D^+D^- vertex distance for CLEO II and CLEO II.V. II. $B \to D^*D^*$ data, (a) ΔE vs $M(B)$ and signal box (inside the smaller rectangle) and the sideband definition (outside the larger rectangle), (b) Projection onto $M(B)$ after ΔE cut. III: Transversity distribution ($\cos(\theta_{tr})$) for data and the best fit.

This method tries to disentangle the CP even from the CP odd contributions. It is expected[9,10,11], that the decay is mostly CP even, which would make the analysis, even with small number of events, useful for CP measurements.

Trying to find $B^0 \to D^{*+}D^{*-}$ turns out to be easier than $B^0 \to D^+D^-$, since one can use the "slow pion" from the D^* decay to effectively "tag" D^*. When this handle becomes unavailable the best way to suppress combinatoric background is to use a lifetime or decay distance measurement. Since D mesons are relatively long lived, we can search for candidate events with a large separation between the vertices of the two D candidates (see fig. 2.II). The CLEO II.V configuration improved greatly the resolution on L after a silicon vertex detector was installed, as shown in Fig 3.I.

After the selection of D^* and/or D candidates, we again plot the energy difference and beam-constrained mass of the B candidate. Fig 3.II shows the D^*D^* mode, that has a clear signal. The extracted branching ratio is

$\mathcal{B}(B \to D^{*+}D^{*-}) = (9.9^{+4.2}_{-3.3}[\text{stat}] \pm 1.2[\text{syst}]) \times 10^{-4}$. All other related decay modes show no significant signal. We obtain $\mathcal{B}(B \to D^+D^-) < 9.4 \times 10^{-4}$ and $\mathcal{B}(B \to D^{*\pm}D^{\mp}) < 6.8 \times 10^{-4}$ at 90% CL.

We tested the transversity analysis with 8 signal events found. If the parameter A, describing the fraction of CP even components, has an extreme value of 0 or 1, we might be able to exclude some region on the other extreme end. As seen in fig. 3.III, the fit to the data results in $A = 0.49^{+0.41}_{-0.34}[\text{stat}] \pm 0.02[\text{syst}]$. We can only exclude values $A < 0.11$ with 90% confidence.

4 Summary

CLEO has found for the first time the decays $B \to J/\psi\phi K$ and $B^0 \to D^{*+}D^{*-}$, and has set upper limits on other decays with pairs of D or D^* mesons. We performed a transversity analysis and find the parameter A to be nearly unconstrained by the data due to low statistics.

Acknowledgments

We thank the CESR staff for the delivery of ground breaking luminosities to the CLEO detector and excellent running conditions over the last years. We are grateful for the support by the NSF, the DOE, and the NSERC of Canada. We also acknowledge the support from the AP Sloan Foundation, the Swiss NSF, and the A. v. Humboldt Stiftung.

References

1. J. Smith, "Direct CP Violation in B Decays" Talk given at this conference.
2. J. Alexander, "Charmless Hadronic B Decays" Talk given at this conference.
3. Y. Kubota *et al.* (CLEO Collaboration), Nucl. Instrum. Meth. **A320**, 66 (1992).
4. T.S. Hill, Nucl. Instrum. Meth. **A418**, 32 (1998).
5. R. Brun *et al.*, CERN Program Library Long Writeup W5013 (1993).
6. Y. Grossmann and H.R. Quinn, Phys. Rev. D **56**, 7259 (1997).
7. I. Dunietz Phys. Rev. D **43**, 2193 (1991).
8. P.F. Harrison and H.R. Quinn, Ed. SLAC-R-504
9. J.L. Rosner, Phys. Rev. D **42**, 3732 (1990).
10. R. Aleksan, *et al.*, Phys. Lett. B **317**, 173 (1993).
11. X.-Y. Pham and Z.-Z. Xing, Phys. Lett. B **458**, 375 (1999).

EXPERIMENTAL REVIEW OF RADIATIVE B DECAYS

T. SKWARNICKI
Syracuse University, Syracuse, NY 13244-1130, USA
E-mail:tomasz@physics.syr.edu

We review experimental results on radiative B decays with emphasis on new results from CLEO-II

1 Introduction

Interest in radiative B decays is mostly due to radiative penguins, as depicted in Figure 1. Measurements of these decays are sensitive to $V_{\rm ts}$ and $V_{\rm td}$. They provide also for sensitive probe of physics beyond the Standard Model, as non-standard particles (e.g. charged Higgs or stop and chargino) can be exchanged in the loop and compete with the heavy W and top exchange. This could alter the rate and generate sizable CP asymmetry. In the Standard Model the CP asymmetry comes from the interference of top and charm quark contributions and is expected to be small.

Loop decays can also be probed in rare kaon decays. However, the rate here is suppressed by two small CKM elements, compared to only one in B decays. Because of that, most loop decays in s quark decays, in particular radiative penguin loop, are overwhelmed by long distance interactions. Electroweak loops with emission of neutrino-antineutrino pair cannot be affected by the long distance interactions, thus experimentalists are in pursuit of such decays.

A gluon rather than a photon can also be emitted from the penguin loop. The inclusive rate is actually higher for the gluonic penguin because of the strong rather than electromagnetic coupling. However, the inclusive rate cannot be measured for the gluonic penguin because of the lack of an experimental signature and competing contributions from suppressed tree level diagrams. In contrast, high energy photons in radiative decays provide a convenient experimental signature to measure the inclusive rate, and there are no competing tree level contributions.

2 Exclusive $B \to K^*\gamma$

Radiative penguin decays were first observed experimentally when the CLEO experiment spotted exclusive $B \to K^*\gamma$ decays in the sample of $1.5 \cdot 10^6$ $B\bar{B}$ decays.[1] Updated results, based on twice the number of $B\bar{B}$ pairs, were

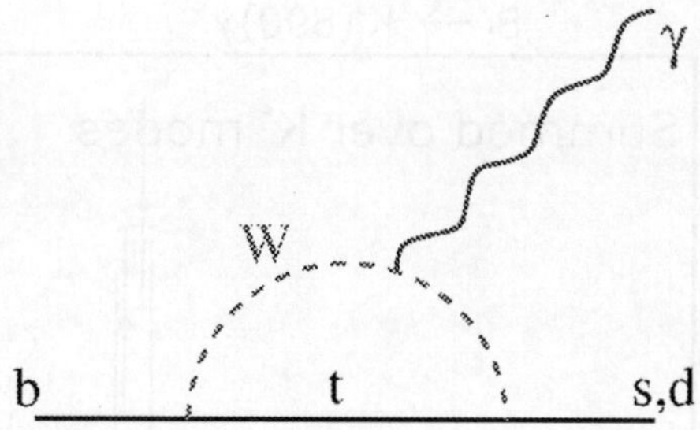

Figure 1. Penguin diagram for $b \to s(d)\gamma$ decays.

presented at the ICHEP'96 conference.[2] In this report, CLEO presents preliminary results based on $9.7 \cdot 10^6$ $B\bar{B}$ pairs which represent the full statistics collected by the CLEO II and CLEO II.V detectors.[a]

The $K^*(890)$ resonance always decays to $K\pi$. Using different combinations of charged and neutral kaons and pions CLEO employs four decay modes to reconstruct K^* ($K^\pm\pi^\mp$, $K_s^0\pi^0$, $K^\pm\pi^0$, $K_s^0\pi^\pm$). The mass of the $K\pi$ system must be consistent with the $K^*(890)$ resonance. It should be noted that there are no other K^* resonances in this mass range. Together with a hard photon, the $K\pi$ system is required to be consistent with a B meson decay. Two variables used in all B reconstructions at $\Upsilon(4S)$ decays are the energy difference, $\Delta E = E_{\text{beam}} - (E_K + E_\pi + E_\gamma)$, and the beam constrained mass, $m_B = \sqrt{E_{\text{beam}}^2 - (\vec{P}_K + \vec{P}_\pi + \vec{P}_\gamma)^2}$. For the signal events ΔE peaks at zero and m_B peaks at the nominal B meson mass, whereas the background processes produce smooth distributions extending to much broader range. CLEO uses additional background suppression cuts which are described elsewhere.[3] Figure 2 shows the beam constrained mass plotted for the selected $K\pi\gamma$ combinations. There are 125 ± 15 $B \to K^*\gamma$ events observed in the CLEO data. When separated into charged and neutral B candidates, CLEO measures the following branching ratios: $\mathcal{B}(B^0 \to K^{*0}\gamma) = (4.5 \pm 0.7 \pm 0.3) \cdot 10^{-5}$ and $\mathcal{B}(B^+ \to K^{*+}\gamma) = (3.8 \pm 0.9 \pm 0.3) \cdot 10^{-5}$. These results are consistent with the previous CLEO results based on a partial data sample.

[a] The CLEO II.V detector differs from the original CLEO II detector by use of the silicon strip detector for the inner tracking layers.

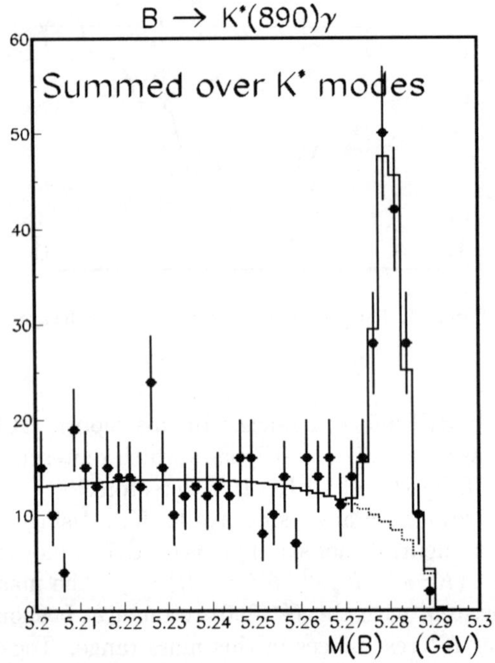

Figure 2. Beam constrained B mass for $K^*(890)\gamma$ candidates.

3 Measurement of CP Asymmetry in $B \to K^*\gamma$

With a large sample of reconstructed $B \to K^*\gamma$ decays, CLEO measures CP asymmetry in this mode for the first time. The asymmetry is defined as

$$A_{\rm CP} \equiv \frac{\mathcal{B}(b) - \mathcal{B}(\bar{b})}{\mathcal{B}(b) + \mathcal{B}(\bar{b})},$$

where $\mathcal{B}(b)$ is the branching fraction for $K^*\gamma$ decays of B^- and $\bar{B}^0$, and $\mathcal{B}(\bar{b})$ corresponds to B^+ and B^0 decays. Except for $K_s^0\pi^0\gamma$ (which CLEO excludes), the selected $K\pi\gamma$ modes are in principle self-tagging. Additional complication arises from imperfect $K^\pm/\pi^\pm$ identification in $K^\pm\pi^\mp\gamma$ mode. CLEO removes events with ambiguous flavor tags. The remaining events yield $A_{\rm CP} = +0.08 \pm 0.13 \pm 0.03$. The systematic error in the measurement of $A_{\rm CP}$ is very small, thus the increased statistics at B factory experiments (and later at hadron collider experiments) will give interesting constraints on non-SM contributions.

4 Beyond $K^*(890)$

With an increased data sample of $K\pi\gamma$ events CLEO is able, for the first time, to establish statistically significant signal for heavier K^* resonances. The $K_2^*(1430)$ resonance has a large branching ratio for $K\pi$ decays $(50 \pm 1\%)$ and small total width ($\Gamma \sim 100$ MeV), thus it is the best candidate to search for. Figure 3 shows the beam constrained mass of B candidates selected with a $K\pi$ mass consistent with the $K_2^*(1430)$. Significant B signal is observed. If all signal events are assumed to be $K_2^*(1430)\gamma$ events, CLEO obtains:

$$\mathcal{B}(B \to K_2^*(1430)\gamma) = (1.7 \pm 0.5 \pm 0.1) \cdot 10^{-5}$$

and

$$R \equiv \mathcal{B}(B \to K_2^*(1430)\gamma)/\mathcal{B}(B \to K^*(890)\gamma) = 0.4 \pm 0.1,$$

consistent with relativistic form-factor calculations by Veseli and Olsson[4] ($R = 0.4 \pm 0.2$), and inconsistent with non-relativistic form-factor results by Ali, Ohl and Mannel[5] ($R = 3.0 - 4.9$).

CLEO cannot rule out that the events seen come from non-resonant production. Further complication stems from the possible contributions from the $K^*(1410)$ resonance. This resonance has a large width $\Gamma \sim 230$ MeV and small branching ratio for $K\pi$ $(7 \pm 1\%)$. CLEO has some discrimination between these two resonances because of different spins: $J = 2$ for $K_2^*(1430)$, and $J = 1$ for $K^*(1410)$. The expected distributions of angle Θ_H between K and K^* in the K^* rest frame is shown for different spins in Figure 4a. For $J = 1$, $\cos\Theta_H$ peaks at zero, whereas $J = 2$ has a minimum at that point. In fact, CLEO excluded the region around zero in the selection of $K_2^*(1430)$ candidates. Taking this cut out and fitting for $K^*(1410)$ and $K_2^*(1430)$ simultaneously, CLEO obtains the results shown in Figure 4b. For the central value a significant $K_2^*(1430)$ contribution is found, whereas the $K^*(1410)$ rate is close to zero. On the other hand, CLEO cannot exclude even a large $\mathcal{B}(B \to K^*(1410)\gamma)$ rate due to the small $K\pi$ decay rate. The fit yields $\mathcal{B}(B \to K * (1410)\gamma) < 12.7 \cdot 10^{-5}$ at 90 % C.L. When no assumptions are made about $K^*(1410)$ rate, the $K_2^*(1430)$ signal is not significant, thus the branching ratio quoted above should be taken as the maximal possible value. The sum of the contributions from $K_2^*(1430)$ and $K^*(1410)$ resonances is statistically significant.

5 Search for $b \to d\gamma$

Measurement of $\mathcal{B}(B \to \rho\gamma)$ and $\mathcal{B}(B \to \omega\gamma)$ in comparison with $B \to K^*(890)\gamma)$ will give information on the V_{ts}/V_{td} ratio. CLEO searched for

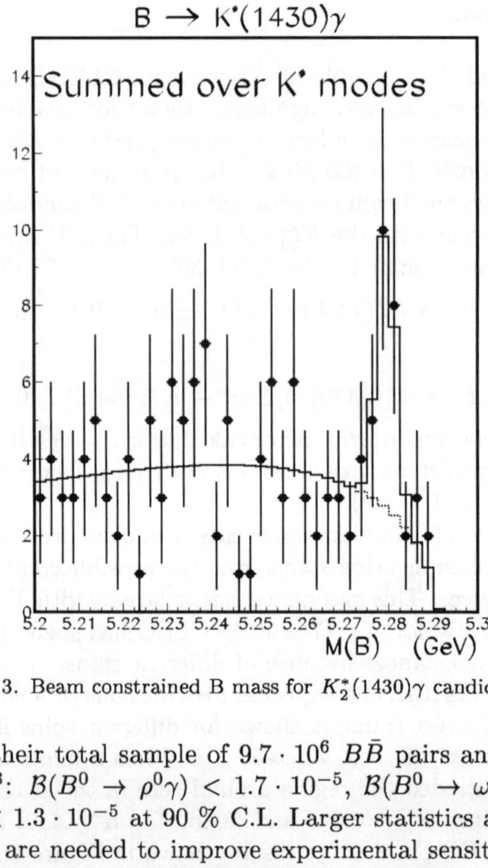

Figure 3. Beam constrained B mass for $K_2^*(1430)\gamma$ candidates.

these decays in their total sample of $9.7 \cdot 10^6$ $B\bar{B}$ pairs and set the following upper limits[3]: $\mathcal{B}(B^0 \to \rho^0\gamma) < 1.7 \cdot 10^{-5}$, $\mathcal{B}(B^0 \to \omega\gamma) < 0.9 \cdot 10^{-5}$, $\mathcal{B}(B^- \to \rho^-\gamma) < 1.3 \cdot 10^{-5}$ at 90 % C.L. Larger statistics and better particle identification are needed to improve experimental sensitivity. B-Factory experiments (BaBar, Belle and CLEO-III) will be better equipped to search for these decays. Precision measurements are likely awaiting a dedicated B experiment at hadron colliders with a good electromagnetic calorimeter, like BTeV.

6 Measurements of Inclusive $b \to s\gamma$ rate

From a theoretical point of view inclusive rates are more interesting than exclusive ones. While easier to measure experimentally, exclusive branching ratios depend on hadron form-factors, which are difficult to calculate from the first principles of the theory. When summed over all possible final states the

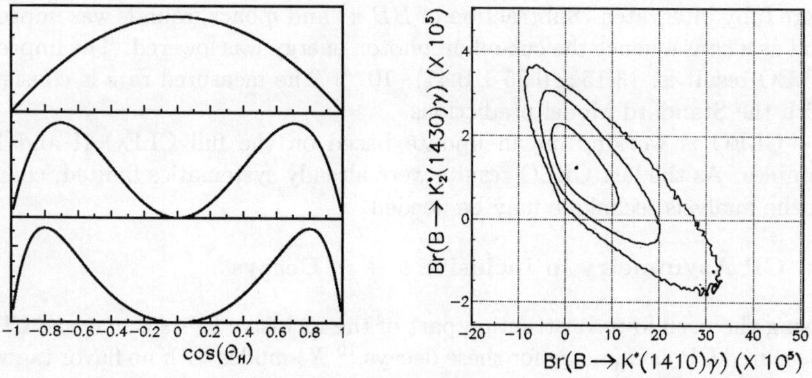

Figure 4. (a) Left plot: kaon helicity angle distribution in $K^* \to K\pi$ decays produced from $B \to K^*\gamma$ decays for various spins of K^* resonance. The top, middle, and bottom distributions correspond to the spin of 1, 2, and 3 respectively.
(b) Right plot: likelihood contours for the fit of $B \to K_2^*(1430)\gamma$ and $B \to K^*(1410)\gamma$ rates. The dot indicates location of the maximal likelihood. The contours show 1, 2, and 3 standard deviations from the maximum.

measured rates reflect short-distance quark interactions. Non-perturbative corrections occur only in the second order of expansion in $\Lambda_{\rm strong}/m_b$. The inclusive $b \to s\gamma$ rate is predicted in the Standard Model with precision of about 10%: $\mathcal{B}(B \to X_s\gamma) = (3.3 \pm 0.3) \cdot 10^{-4}$. Status of the theory is reviewed in the other contribution to this conferences.[6]

CLEO measured the inclusive $b \to s\gamma$ rate in a sample of $2.2 \cdot 10^6$ $B\bar{B}$ pairs.[7] The main background comes from continuum events, $e^+e^- \to q\bar{q}$ ($q = u, d, s, c$). CLEO used two complementary approaches to suppress continuum background: events shape cuts and pseudo-reconstruction of $B \to X_s\gamma$ decays. In the latter method, X_s is allowed to be any combination of a kaon (charged or K_s^0) and one to four pions (with at most one π^0) to cover a sufficiently broad range of possible final states.

Later, ALEPH - using its own version of pseudo-reconstruction - was able to measure the inclusive rate in much smaller sample of $b\bar{b}$ pairs ($0.8 \cdot 10^6$), which was reflected in their rather large experimental errors: $(3.11 \pm 0.80 \pm 0.72) \cdot 10^{-4}$.[8]

CLEO re-analyzed their data, adding 50% more luminosity, with improved analysis techniques.[9] The pseudo-reconstruction was improved by using information about signal and background shapes in some key variables, replacing a simple cut-off approach. Event shape and pseudo-reconstruction approaches were fully integrated. Subtraction of $B\bar{B}\ \pi^0$ and η backgrounds was improved and as a consequence the cut-off on photon energy was lowered. The improved CLEO result is: $(3.15 \pm 0.35 \pm 0.41) \cdot 10^{-4}$. The measured rate is consistent with the Standard Model predictions.

CLEO is working on an update based on the full CLEO II and II.V samples. As the last CLEO results were already systematics limited, changes in the analysis technique may be needed.

7 CP Asymmetry in Inclusive $b \to s\gamma$ Decays

Using the pseudo-reconstruction part of the inclusive $b \to s\gamma$ analysis, CLEO measured CP asymmetry for these decays.[10] X_s modes with no flavor tag were removed. In the remaining sample 8 ± 2 % of events have mistagged b flavor. After correcting for the mistag fraction, CLEO finds: $A_{\rm CP} = (+0.16 \pm 0.14 \pm 0.05) \cdot (1 \pm 0.04)$, which is consistent with zero. Analyzing full CLEO data sample will be interesting as the present results are statistics limited.

8 Non-penguin Radiative Decays

CLEO searched for $\bar{B}^0 \to D^{*0}\gamma$ decays in the full sample of $9.7 \cdot 10^6$ $B\bar{B}$ pairs.[11] Observation of these decays would provide experimental information on the importance of W exchange diagram in B decays (see Figure 5a). Using $D^{*0} \to D^0$ transitions via π^0 and γ and reconstructing D^0 in $K^-\pi^+$, $K^-\pi^+\pi^0$, and $K^-\pi^+\pi^-\pi^+$ modes, CLEO finds no evidence for these decays as illustrated in Figure 5b. The 90% C.L. upper limit is $\mathcal{B}(\bar{B}^0 \to D^{*0}\gamma) < 5 \cdot 10^{-5}$.

Some time ago CLEO also search for $B^- \to \gamma l^- \nu$ decays ($l = e$ or μ) using $2.7 \cdot 10^6$ $B\bar{B}$ decays.[12] These decays are expected to be dominated by B^- transition to B^{*-} by emission of a photon via long distance electromagnetic interactions and B^{*-} annihilation to a W. The transition to B^* removes helicity suppression and thus the expected rate is higher than in direct B^- annihilation but still small, $\mathcal{B} \sim 10^{-6}$. The search relied on a neutrino reconstruction technique via missing energy and momentum. The 90% C.L. upper limits are $\mathcal{B}(B^- \to \gamma\mu^-\nu) < 5 \cdot 10^{-5}$ and $BR(B^- \to \gamma e^-\nu) < 2 \cdot 10^{-4}$.

9 Summary

Many new measurements for $b \to s(d)\gamma$ decays reconstructed in exclusive modes are reported at this conference by CLEO which has analyzed their entire CLEO-II and CLEO-II.5 samples. Measurements of the inclusive $b \to s\gamma$ rate by CLEO are systematics dominated and no new results are reported here. CLEO also reports an upper limit on the rate for $\bar{B}^0 \to D^{*0}\gamma$.

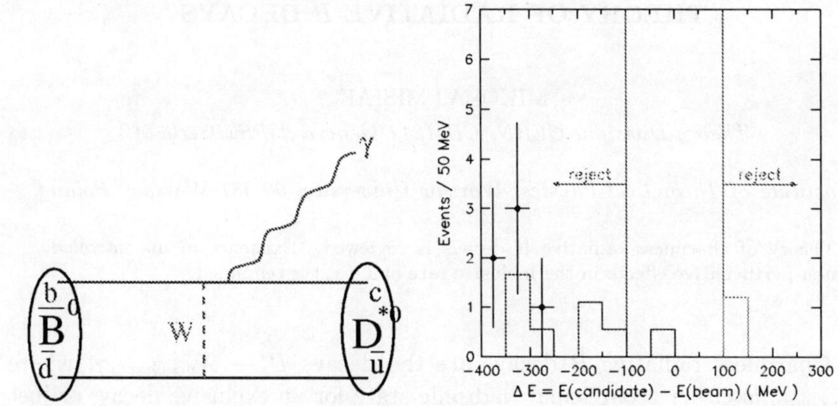

Figure 5. (a) Left plot: diagram for $\bar{B}^0 \to D^{*0}\gamma$ decays.
(b) Right plot: distribution of energies of $\bar{B}^0 \to D^{*0}\gamma$ candidate events selected by CLEO. The signal events are expected peak at zero in between the dotted lines. The data are shown by points with error bars. The histogram illustrates the expected background level as simulated in CLEO Monte Carlo.

References

1. R. Ammar et al. (CLEO Collaboration) 1993 *Phys. Rev. Lett.* **71** 674
2. R. Ammar et al. (CLEO Collaboration) 1996 *Report No. CLEO CONF 96-05*, contributed paper to *The 28th International Conference on High Energy Physics*, Warsaw, Poland.
3. T.E. Coan et al. (CLEO Collaboration) 1999 *Report No. CLNS 99/1653, CLEO 99-20 hep-ex/9912057* Submitted to *Phys. Rev. Lett.*
4. S. Veseli, M.G. Olsson 1996 *Phys. Lett.* **B367** 309
5. A. Ali, T. Ohl, T. Mannel 1993 *Phys. Lett.* **B298**, 195
6. M. Misiak, these proceedings.
7. M.S. Alam et al. (CLEO Collaboration) 1995 *Phys. Rev. Lett.* **74** 2885
8. R. Barate et al. (ALEPH Collaboration) 1998 *Phys. Lett.* **B429** 169
9. S. Glenn et al. (CLEO Collaboration) 1998 *Report No. CLEO CONF 98-17*, contributed paper to *The 29th International Conference on High Energy Physics, Vancouver, Canada*
10. S. Ahmed et al. (CLEO Collaboration) 1999 *Report No. CLEO-CONF-99-10 e-Print Archive: hep-ex/9908022*
11. M. Artuso et al. (CLEO Collaboration) 1999 *A report No. CLNS 99/1655, CLEO 99-21* Submitted to Phys. Rev. Lett.
12. T.E. Browder et al. (CLEO Collaboration) 1997 *Phys. Rev.* **D56** 11

THEORY OF RADIATIVE B DECAYS

MIKOŁAJ MISIAK
Theory Division, CERN, CH-1211 Geneva 23, Switzerland
and
Institute of Theoretical Physics, Warsaw University, 00-681 Warsaw, Poland

Theory of charmless radiative B decays is reviewed. Existence of uncontrolled non-perturbative effects in the inclusive rate at $\mathcal{O}(\alpha_s)$ is reminded.

Charmless radiative $\bar{B}$ decays are the decays $\bar{B} \to X_{\text{no charm}}\gamma$, where $X_{\text{no charm}}$ is either a particular hadronic state for an exclusive decay, or just any charmless hadronic state in the inclusive case. Such decays are generated by tree-level $b \to u$ transitions with photon radiation, loop-mediated $b \to d$ transitions, and loop-mediated $b \to s$ transitions. Examples of diagrams contributing to each of these three types of transitions are presented in Fig. 1. The loop-mediated transitions are known to be very sensitive to new physics, e.g. to existence of SUSY particles with masses below 1 TeV. In the Standard

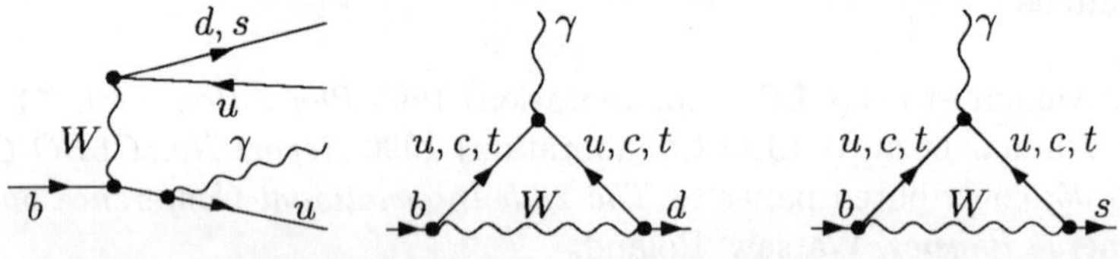

Figure 1. Examples of Feynman diagrams contributing to $\bar{B} \to X_{\text{no charm}}\gamma$.

Model, the $b \to u$ and $b \to d$ transitions are CKM-suppressed with respect to the $b \to s$ ones. The relative suppression factors are $|V_{ub}/V_{ts}|^2 \simeq 1\%$ and $|V_{td}/V_{ts}|^2 \in [2.5\%, 5\%]$, respectively. Therefore, to a good approximation, the loop-mediated $b \to s$ transitions saturate the inclusive $\bar{B} \to X_{\text{no charm}}\gamma$ branching ratio. This branching ratio is measured by CLEO[1] and ALEPH[2] at the level of around 3×10^{-4}, after the contribution from intermediate ψ states is subtracted.

$$\sum_{X_s} BR[\bar{B} \to X_s\gamma]_{\text{loop mediated}} \simeq \sum_{X_{\text{no charm}}} BR[\bar{B} \to X_{\text{no charm}}\gamma]$$
$$\simeq (3.15 \pm 0.35 \pm 0.32 \pm 0.26) \times 10^{-4} \quad \textbf{(CLEO)}^1$$
$$+BR[\bar{B} \to X^{(1)}_{\text{no charm}}\psi] \times BR[\psi \to X^{(2)}_{\text{no charm}}\gamma]. \quad (1)$$

The latter term in the above equation is the very contribution from intermediate ψ. A lower bound on its numerical size can be found by summing up the exclusive branching ratios of charmless radiative ψ decays listed in the Particle Data Book[3] (which give together around 4%) and multiplying them by $BR[\bar{B} \to X\psi]$ that is close to 1%. It follows that the intermediate ψ contribution to $BR[\bar{B} \to X_s\gamma]$ is not smaller than 4×10^{-4}, i.e. it is larger than the remainder of this branching ratio.

The CLEO result is interpreted here as the one with subtracted intermediate ψ contribution, even though no subtraction has been actually made in the measurement performed for high-energy photons.[4] However, the extrapolation to lower photon energies has been done with use of a theoretical model that did not include intermediate ψ. Photons originating from the intermediate ψ are expected to be softer than most of the other photons in $\bar{B} \to X_s\gamma$. A quantitative estimate of their softness (missing at present) is necessary to completely clarify this point.

The intermediate ψ contribution had to be included in Eq. (1), because each $X_{\text{no charm}}$ is assumed to be a QCD-eigenstate, while ψ is not stable in QCD. In other words, the diagrams in Fig. 1, when dressed by an appropriate number of gluons, give a contribution to the intermediate ψ channel as well. Thus, a separation of the intermediate ψ contribution, which may be straightforward on the experimental side, has to be thought about on the theoretical side, too.[5,15] We shall come back to this point later.

Most of this talk will be devoted to the loop-mediated $\bar{B} \to X_s\gamma$ decay that is dominant in $\bar{B} \to X_{\text{no charm}}\gamma$. In order to make a theoretical prediction, we first need to calculate the perturbative b-quark decay amplitudes to partonic states $X_s^{(p)}$ and the photon. Later, the perturbative amplitudes will enter directly into the expressions for hadronic branching ratios. A lot of effort has been devoted in the recent years to calculating the b-quark decay amplitudes with better than 10% accuracy. Single gluon corrections (Fig. 2b) to the one-loop $b \to s\gamma$ diagrams (Fig. 2a) increase the predicted amplitude by around 50%, and the branching ratio by around 100%. This effect is so large because the logarithm $\ln \frac{M_W^2}{m_b^2}$ is big and because the one-loop result is accidentally quite small - it gives only about $\frac{1}{5}$ of what is naively expected. In order to achieve better than 10% accuracy, one needs to include the NLO QCD corrections (Figs. 2c and 2d), i.e. non-logarithmic parts of two-loop diagrams[6,7] and logarithmic parts of three-loop diagrams.[8] The NLO corrections further increase the predicted branching ratio by around 20%. Both the LO and the NLO calculations include resummation of large logarithms $\ln \frac{M_W^2}{m_b^2}$ from all orders of the perturbation series.

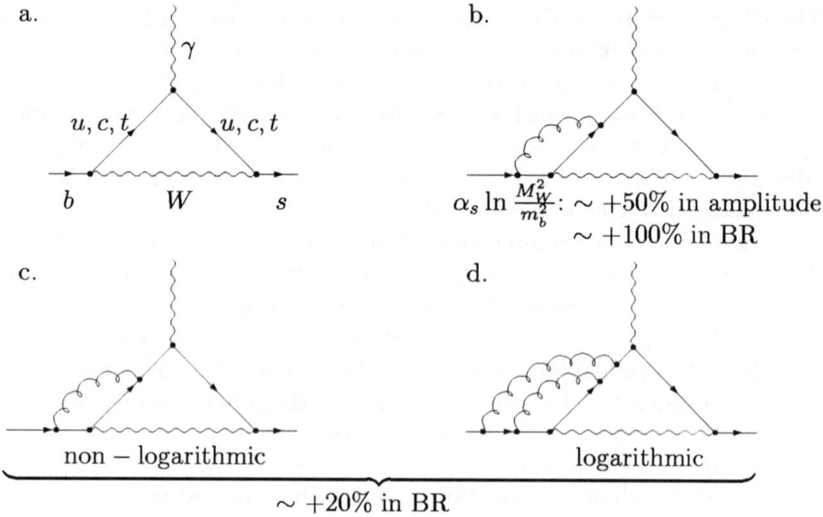

Figure 2. Examples of Feynman diagrams contributing to $b \to s\gamma$ at various orders in the renormalization-group-improved perturbation theory.

Resummation of large logarithms as well as further calculation of hadronic decay rates is most conveniently performed in the framework of an effective theory obtained from the SM by decoupling the heavy electroweak bosons and the top quark. The Lagrangian of the effective theory reads

$$\mathcal{L} = \mathcal{L}_{QCD \times QED}(u, d, s, c, b) + \frac{4G_F}{\sqrt{2}} V_{ts}^* V_{tb} \sum_{i=1}^{8} C_i(\mu) O_i, \tag{2}$$

where the first term is just the QCD×QED Lagrangian for the light quarks, and the second term contains flavour-changing local interactions O_i of either 4 quarks or 2 quarks and gauge bosons.

$$O_i = \begin{cases} (\bar{s}\Gamma_i c)(\bar{c}\Gamma'_i b), & i = 1, 2, & |C_i(\mu_b)| \sim 1, \\ (\bar{s}\Gamma_i b) \sum_q (\bar{q}\Gamma'_i q), & i = 3, 4, 5, 6, & |C_i(\mu_b)| < 0.07, \\ \frac{em_b}{16\pi^2} \bar{s}_L \sigma^{\mu\nu} b_R F_{\mu\nu}, & i = 7, & |C_7(\mu_b)| \sim 0.3, \\ \frac{gm_b}{16\pi^2} \bar{s}_L \sigma^{\mu\nu} T^a b_R G^a_{\mu\nu}, & i = 8, & |C_8(\mu_b)| \sim 0.15. \end{cases} \tag{3}$$

The symbols Γ_i and Γ'_i in O_1,..., O_6 stand for various products of the Dirac and colour matrices. The $\overline{MS}$-renormalized couplings C_i at the scale $\mu_b \sim m_b$ are known nowadays up to (and including) the following terms in their perturbative expansion:[8,9]

$$C_i(\mu_b) = C_i^{(0)}(\mu_b) + \frac{\alpha_{em}}{\alpha_s(\mu_b)} C_i^{(0)em}(\mu_b) + \frac{\alpha_s(\mu_b)}{4\pi} C_i^{(1)}(\mu_b) + \dots \tag{4}$$

Once the Wilson coefficients $C_i(\mu_b)$ are known, the b-quark decay amplitudes are given by Feynman diagrams with single insertions of the flavour-changing interactions, i.e. by matrix elements of the operators O_i between the appropriate partonic states.

For the exclusive decay $\bar{B} \to \bar{K}^*\gamma$, one needs to know matrix elements of those operators between the relevant hadronic states: $\langle \bar{K}^*\gamma|O_i|\bar{B}\rangle$. There have been many attempts to calculate these matrix elements using quark models, QCD sum rules, lattice and heavy quark symmetries (see e.g.[10]). The history of published predictions is briefly summarized in Fig. 3. The (blue) thin bars and dots are the theoretical predictions, while the (red) thick bars denote the CLEO measurements. Many recent theoretical papers on $\bar{B} \to \bar{K}^*\gamma$ are not included in the plot because only form-factors are discussed there, and no explicit number for the decay rate is given. A general conclusion one

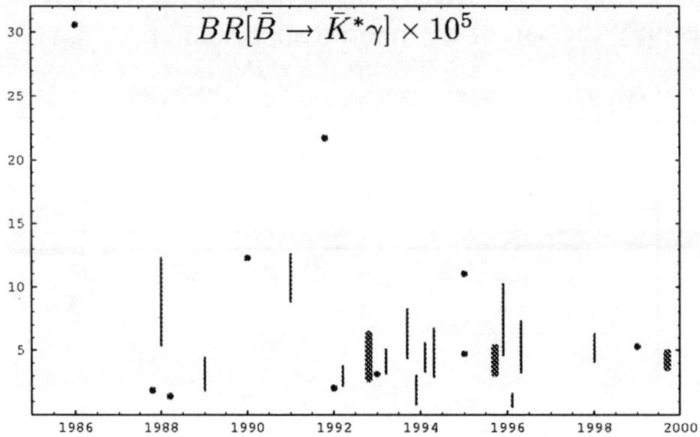

Figure 3. Brief history of published predictions and experimental results for $BR[\bar{B} \to \bar{K}^*\gamma]$.

can derive from the $\bar{B} \to \bar{K}^*\gamma$ studies is that they can help us in understanding non-perturbative QCD, but $BR[\bar{B} \to \bar{K}^*\gamma]$ is not a good place to look for new physics, given the present experimental and theoretical results for the inclusive mode $\bar{B} \to X_s\gamma$. However, observation of a sizable CP-asymmetry in the exclusive mode would be a clear signal of new physics.[11]

For the inclusive decay $\bar{B} \to X_s\gamma$, the theoretical prediction for the branching ratio can be made more precise by using the Operator Product Expansion (OPE) within the Heavy Quark Effective Theory (HQET).[12] We need to calculate

$$\sum_{X_s} \left| C_7 \langle X_s \gamma | O_7 | \bar{B} \rangle + C_2 \langle X_s \gamma | O_2 | \bar{B} \rangle + \ldots \right|^2 , \tag{5}$$

where dots stand for matrix elements of other operators that are numerically less important. Let us first look at the "77" term, i.e. the term proportional to $|C_7|^2$ in Eq. (5). This term dominates in the perturbative calculation of the b-quark decay. In full analogy to the semileptonic B-meson decay, we relate it via optical theorem to the imaginary part of the elastic forward scattering amplitude

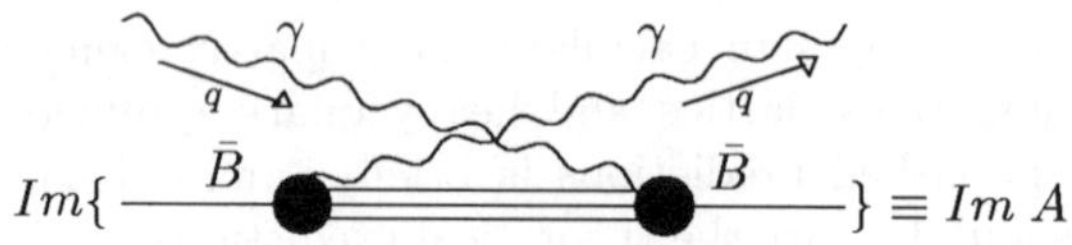

(6)

In this amplitude, we can perform OPE when the photon energies E_γ in the $\bar{B}$-meson rest frame are far from the endpoint, i.e. when $|m_B - 2E_\gamma| >> \Lambda_{QCD}$. Most of the photons in $\bar{B} \to X_s \gamma$ have energies close to the endpoint $E_\gamma^{\max} \simeq \frac{1}{2} m_b$, so they do not satisfy this requirement. Thus, at this first step, OPE gives us only the tail of the photon spectrum.

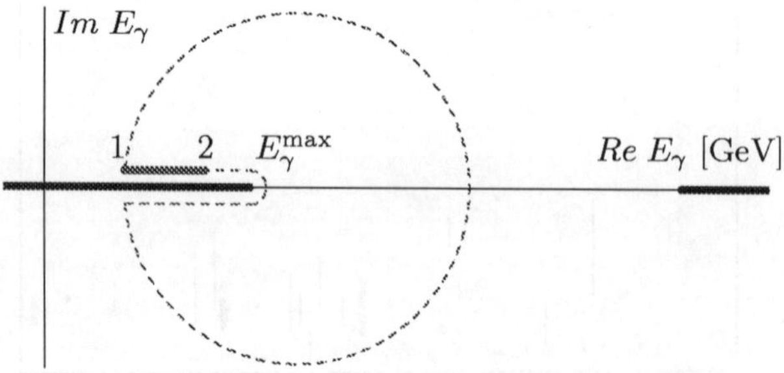

Figure 4. Physical cuts in $A(E_\gamma)$ (thick lines on the real axis) and the relevant contour.

Fortunately, we know analytic properties of the amplitude A when E_γ is formally treated as complex. We know that the discontinuity of A on the real axis is equal to ImA. Thus, if we want to know the integral of ImA from, say, 1 GeV to the endpoint, we can find it by performing the integral of A around the big circle in Fig. 4, where the condition for OPE is always fulfilled.

$$\int_{1\ \mathrm{GeV}}^{E_\gamma^{\max}} dE_\gamma \, E_\gamma^n \, Im\, A(E_\gamma) \sim \oint_{\text{big circle}} dE_\gamma \, E_\gamma^n \, A(E_\gamma), \qquad n = 0, 1, 2, \ldots \tag{7}$$

We should better not go with E_γ much below 1 GeV. There is no problem in doing so for the "77" term, but there are problems with other operators.[5,13] Anyway, the region below 1 GeV is hardly accessible experimentally, because of the $b \to c$ background.

The conclusion at this point is that we can predict the photon spectrum for not too small and not too big energies, and moments of the photon spectrum from not too small energies to the endpoint. Making such predictions requires calculating matrix elements of various local operators among $\bar{B}$-meson states. Matrix elements of higher-dimensional operators are suppressed by higher powers of Λ/m_B. Therefore, we can write a double expansion for the "77" term, i.e. we can write

$$\sum_{X_s} BR[\bar{B} \to X_s\gamma]_{E_\gamma > 1\text{ GeV}} = \left[a_{00} + a_{02}\left(\frac{\Lambda}{m_B}\right)^2 + \ldots \right]$$
$$+ \frac{\alpha_s(m_b)}{\pi}\left[a_{10} + a_{12}\left(\frac{\Lambda}{m_B}\right)^2 + \ldots \right] + \mathcal{O}\left[\left(\frac{\alpha_s(m_b)}{\pi}\right)^2\right]$$
$$+ [\text{ Contributions other than the "77" term}]. \tag{8}$$

Here, the two terms not suppressed by Λ/m_B are simply those already found in the perturbative calculation of the b-quark decay. The term proportional to a_{02} turns out to give only an around -3% contribution.[12] The remaining terms in the first two lines of Eq. (8) have stronger suppression factors, which makes them negligible.

However, we need to ask whether a similar expansion can be written for the third line of Eq. (8), i.e. for the contributions other than the "77" term. The answer to this question is *no*. These remaining contributions contain, for instance, the huge effect from intermediate ψ that has been mentioned in the beginning of this talk. The intermediate ψ becomes important either due to non-perturbative effects or because of big contributions at high orders of perturbation theory that need to be resummed. Most probably, both mechanisms are at work.

The intermediate ψ contribution can be just subtracted from both the experimental data and the theoretical predictions, using the narrow peak approximation as in Eq. (1). Other narrow $c\bar{c}$ resonances hardly ever decay radiatively to charmless states, so similar contributions from them are negligible.

Suppose we subtract the intermediate ψ contribution. Does the sum of the remaining contributions take the form of a power series as in the first two lines of Eq. (8), with the perturbatively calculable leading term?

It does not.[5,15] However, it is hard to identify any obvious source of a big non-perturbative effect in it. Operators containing no charm quark are suppressed by their small Wilson coefficients. As far as the operators containing the charm quark are concerned, we know that their contribution at the leading order in α_s can be expressed as a power series

$\langle \bar{B}|$ 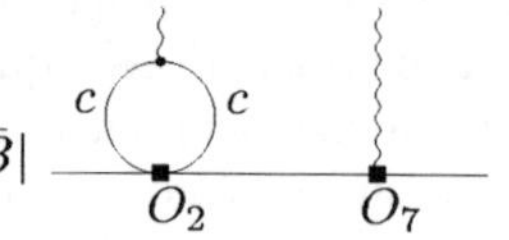

$$|\bar{B}\rangle = (\text{perturbative } 0) + \frac{\Lambda^2}{m_c^2} \sum_{n=0}^{\infty} b_n \left(\frac{m_b \Lambda}{m_c^2} \right)^n , \quad (9)$$

which can be truncated to the leading $n = 0$ term, because the coefficients b_n decrease fast with n. The calculable[5,14] $n = 0$ term makes $BR[\bar{B} \to X_s \gamma]$ increase by around 3%.

However, an analysis of non-perturbative effects in the matrix elements of O_1 and O_2 at $\mathcal{O}(\alpha_s)$ is missing. For instance,

$\langle \bar{B}|$ 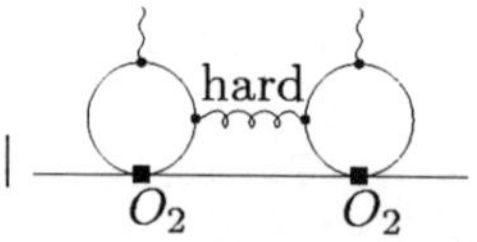

$$|\bar{B}\rangle = A_{\text{one-loop}} + B_{\psi} + C_{\text{unknown}}, \quad (10)$$

where $A_{\text{one-loop}}$ stands for the very small (less than 1% in BR) perturbative contribution from the gluon bremsstrahlung at one loop, B_{ψ} is a part of the (huge) intermediate ψ contribution, and C_{unknown} denotes the remaining non-perturbative terms. Those remaining terms would not be numerically important if they were either suppressed by $\Lambda/m_{c,b}$, or small for other reasons, or could be absorbed into the intermediate ψ contribution. Unfortunately, I am not aware of any sufficiently precise argument that any of these three possibilities is realized.

In the following, I shall assume that one of these three possibilities is realized. In such a case, the hadronic decay rate is indeed well-approximated by the partonic decay rate, up to small non-perturbative corrections

$$\frac{\Gamma[\bar{B} \to X_s \gamma]^{\text{subtracted } \psi}_{E_\gamma > E_{\text{cut}}}}{\Gamma[\bar{B} \to X_c e \bar{\nu}_e]} \simeq \frac{\Gamma[b \to X_s \gamma]^{\text{perturbative NLO}}_{E_\gamma > E_{\text{cut}}}}{\Gamma[b \to X_c e \bar{\nu}_e]^{\text{perturbative NLO}}} \times$$
$$\times \left[1 + (\mathcal{O}(\Lambda^2/m_b^2) \simeq 1\%) + (\mathcal{O}(\Lambda^2/m_c^2) \simeq 3\%) \right] . \quad (11)$$

The normalization to the semileptonic rate has been used here to cancel uncertainties due to m_b^5, CKM-angles and some of the non-perturbative corrections. One has to remember that Eq. (11) becomes a bad approximation for $E_\gamma^{\text{cut}} << 1$ GeV, and that non-perturbative corrections grow dramatically[17] when $E_\gamma^{\text{cut}} > 2$ GeV.

For $E_{\text{cut}} = 1$ GeV, Eq. (11) gives

$$BR[\bar{B} \to X_s\gamma]^{\text{subtracted } \psi}_{E_\gamma > E_{\text{cut}}} = (3.29 \pm 0.33) \times 10^{-4}, \tag{12}$$

where the dominant uncertainties originate from the uncalculated $\mathcal{O}(\alpha_s^2)$ effects and from the ratio m_c/m_b in the semileptonic decay (around 7% each).

Unfortunately, $E_{\text{cut}} = 1$ GeV is not accessible experimentally at present. We need some prediction for the photon spectrum. The solid line in Fig. 5 describes the photon spectrum[15] in the region where the theoretical HQET prediction is solid. For larger energies, the less solid the line becomes, the less solid the prediction is. In the peak region, it is simply an "artist view" of how the spectrum could look like. However, its normalization is fixed by Eq. (12), and the size of the visible $\bar{K}^*(892)$ peak is adjusted to the value measured by CLEO.[16]

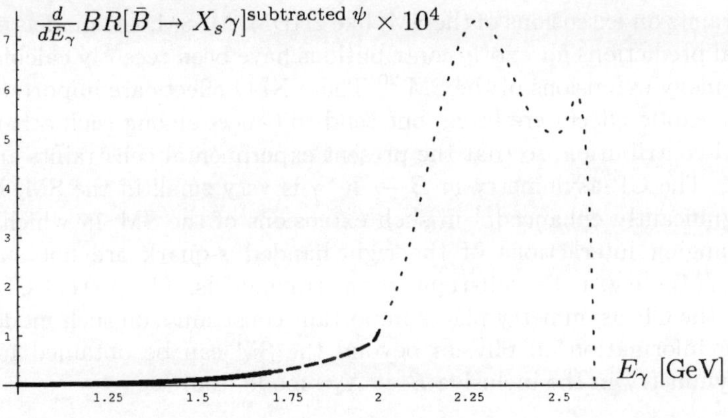

Figure 5. An "artist view" of $\frac{d}{dE_\gamma} BR[\bar{B} \to X_s\gamma]^{\text{subtracted } \psi}$.

If we knew the shape function of the $\bar{B}$ meson exactly, we could make a solid prediction for the photon spectrum in the peak region, too.[17] Unfortunately, only models for the shape function are available at present. Therefore, an optimal scenario for a comparison between theory and experiment would be measuring the photon spectrum above ~ 2 GeV without relying on any theoretical prediction for its shape, and then extrapolating in a simple manner to the predicted spectrum below ~ 2 GeV. The measured spectrum above 2 GeV would provide[18] important information for extracting V_{ub} from $\bar{B} \to X_u l\nu$.

The decay $\bar{B} \to X_d\,\gamma$ is theoretically more difficult than $\bar{B} \to X_s\,\gamma$, because diagrams with up-quark loops are no longer CKM-suppressed with respect to the remaining ones. Thus, the theoretical accuracy is at best ±30%, even for fixed Wolfenstein parameters ρ and η.

When the non-perturbative effects in up-quark loops are assumed to be small, one obtains[19]

$$\frac{1}{2}\{BR[\bar{B} \to X_d\,\gamma] + BR[B \to \bar{X}_d\,\gamma]\} \simeq 2.43[(1-\bar{\rho})^2 + \bar{\eta}^2 - 0.35(1-\bar{\rho})$$
$$+0.07] \times 10^{-5} = 1.61 \times 10^{-5} \quad (\text{for } \bar{\rho} = 0.11 \text{ and } \bar{\eta} = 0.32), \tag{13}$$

while the direct CP-asymmetries range from 7% to 35%. Estimates for the exclusive channels are: $BR[B \to \rho^{\pm}\gamma] \in [1,4] \times 10^{-6}$ and $BR[B \to (\rho^0, \omega)\gamma] \in [0.5, 2] \times 10^{-6}$.

The present experimental results for $\bar{B} \to X_s\gamma$ already place severe constraints on extensions of the SM, like 2HDM, MSSM, LR-models etc. Theoretical predictions for exotic contributions have been recently calculated at NLO in many extensions of the SM.[20] These NLO effects are important only when the exotic effects are large, but tend to cancel among each other and/or the SM contribution, so that the present experimental constraints are satisfied.

The CP-asymmetry in $B \to K^*\gamma$ is very small in the SM, but could be significantly enhanced[11] in such extensions of the SM, in which the flavour-changing interactions of the right-handed s-quark are not suppressed by m_s/M_W, e.g. in the left-right symmetric models. The present CLEO bound[16] on the CP-asymmetry places important constraints on such models. Interesting information on physics beyond the SM can be obtained from the CP-asymmetry in the inclusive $\bar{B} \to X_s\gamma$ mode, too.[21]

To conclude:

- The present theoretical prediction for $BR[\bar{B} \to X_s\gamma]$ in the SM agrees very well with the measurements of CLEO and ALEPH. However, an analysis of non-perturbative effects at order $\mathcal{O}(\alpha_s)$ is necessary in order to make sure that the theoretical uncertainties are indeed around 10%.
- Future measurements of $BR[\bar{B} \to X_s\gamma]$ should rely as little as possible on theoretical predictions for the precise shape of the photon spectrum above ~ 2 GeV.

Acknowledgements: I would like to thank P. Ball, M. Beneke, G. Buchalla, A. Hoang, M. Neubert and T. Skwarnicki for helpful discussions, and the organizers of the BCP3 conference in Taipei for the invitation and warm hospitality.

References

1. S. Ahmed *et al.* (CLEO Collaboration), hep-ex/9908022.
2. R. Barate *et al.* (ALEPH Collaboration), Phys. Lett. **B429** (1998) 169.
3. Particle Data Group, Eur. Phys. J. **C3** (1998) 1.
4. T. Skwarnicki, private communication.
5. Z. Ligeti, L. Randall and M.B. Wise, Phys. Lett. **B402** (1997) 178.
6. K. Adel and Y.P. Yao, Phys. Rev. **D49** (1994) 4945;
   C. Greub and T. Hurth, Phys. Rev. **D56** (1997) 2934;
   A.J. Buras, A. Kwiatkowski and N. Pott, Nucl. Phys. **B517** (1998) 353.
7. C. Greub, T. Hurth and D. Wyler, Phys. Rev. **D54** (1996) 3350.
8. K. Chetyrkin, M. Misiak and M. Münz, Phys. Lett. **B400** (1997) 206, **B425** (1998) 414 (E).
9. K. Baranowski and M. Misiak, hep-ph/9907427.
10. C. Bernard, P. Hsieh and A. Soni, Phys. Rev. Lett. **72** (1994) 1402;
    P. Ball and V.M. Braun, Phys. Rev. **D58** (1998) 094016;
    Z. Ligeti and M.B. Wise, Phys. Rev. **D60** (1999) 117506.
11. D. Atwood, M. Gronau and A. Soni, Phys. Rev. Lett. **79** (1997) 185.
12. A.F. Falk, M. Luke and M.J. Savage, Phys. Rev. **D49** (1994) 3367.
13. A. Kapustin, Z. Ligeti and H.D. Politzer, Phys. Lett. **B357** (1995) 653.
14. M.B. Voloshin, Phys. Lett. **B397** (1997) 275;
    A. Khodjamirian *et al.*, Phys. Lett. **B402** (1997) 167;
    A.K. Grant *et al.*, Phys. Rev. **D56** (1997) 3151;
    G. Buchalla, G. Isidori and S.J. Rey, Nucl. Phys. **B511** (1998) 594.
15. Z. Ligeti *et al.*, Phys. Rev. **D60** (1999) 034019.
16. T.E. Coan *et al.* (CLEO Collaboration), hep-ex/9912057.
17. A.L. Kagan and M. Neubert, Eur. Phys. J. **C7** (1999) 5.
18. M. Neubert, Phys. Rev. **D49** (1994) 4623.
19. A. Ali, H. Asatrian and C. Greub, Phys. Lett. **B429** (1998) 87.
20. M. Ciuchini, G. Degrassi, P. Gambino and G.F. Giudice, Nucl. Phys. **B527** (1998) 21, Nucl. Phys. **B534** (1998) 3;
    F. Borzumati and C. Greub, Phys. Rev. **D58** (1998) 074004;
    C. Bobeth, M. Misiak and J. Urban, hep-ph/9904413.
21. A.L. Kagan and M. Neubert, Phys. Rev. **D58** (1998) 094012.

CHARMLESS HADRONIC B DECAYS WITH THE FULL CLEO DATA SAMPLE

J. G. SMITH

Department of Physics, University of Colorado
Boulder, CO 80309, USA

We report the observation of 13 charmless hadronic B decays with the full CLEO data sample. Branching fractions are measured for these decays and upper limits are determined for 50 related decays. We show that these measurements are generally in agreement with theoretical expectations and that more precise determinations of the branching fractions could lead to a determination of the the CKM angle γ when improved theoretical models are available.

1 Introduction

Charmless hadronic B decays are expected to proceed primarily through $b \to s$ loop ("penguin") diagrams and $b \to u$ tree diagrams. In Fig. 1 we show four such diagrams for three of the $K\pi$ modes discussed in this paper. We also show the diagrams expected to dominate the final states with an isoscalar meson and a K or K^* meson. Interchange of d and u spectator quarks will generally provide the diagrams for both B^+ and B^0 decays. Diagrams 1c, 1d, and 1g are Cabibbo suppressed. The un-suppressed versions of these diagrams and the CKM [1] suppressed versions of the penguin diagrams lead to final states such as $\pi\pi$, $\eta'\pi$, $\eta\rho$, and $\omega\pi$.

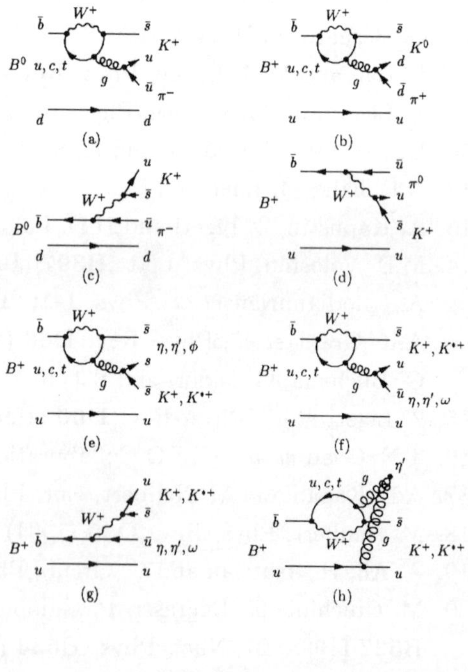

Figure 1. Feynman diagrams for some of the penguin and tree processes which are expected to be dominant for the modes described in this paper.

These decays have received a

great deal of attention because interference among penguin and tree diagrams that lead to the same final state can produce (direct) CP violation, and, for the B^0 system, interference between final states reached directly or via B-$\bar{B}$ mixing, can generate (indirect) CP violation.[2] Detectors such as BABAR and Belle, or others at hadron colliders, will attempt to measure an oscillation in the time evolution of certain B^0 decays, which is sensitive to the value of some of the CKM angles. Other approaches have been suggested, most recently a series of papers [3] which have suggested that CKM angles can be extracted with modest model dependence from ratios of certain $K\pi$ branching fractions.

In subsequent sections, we will report the results of many analyses from the CLEO experiment, most of which use the full CLEO II/II.V data sample. We conclude with interpretations of these results.

2 Analysis details

The results in this paper are based on data collected with the CLEO II detector [4] at the Cornell Electron Storage Ring (CESR). The data sample corresponds to an integrated luminosity of 9.1 fb^{-1} taken on the Υ(4S) resonance and 4.4 fb^{-1} taken slightly below. The on-resonance data sample contains 9.7×10^6 $B\overline{B}$ pairs. Approximately two-thirds of the sample has the CLEO II.V configuration, with a silicon vertex detector and a helium-propane gas mixture in the main drift chamber. Resonance states are reconstructed from charged tracks and photons with the decay channels: $\eta' \to \eta\pi^+\pi^-$, $\eta' \to \rho^0\gamma$, K^0 via $K_S \to \pi^+\pi^-$, $\rho^0 \to \pi^+\pi^-$, $\pi^0 \to \gamma\gamma$, $\eta \to \gamma\gamma$, $\eta \to \pi^+\pi^-\pi^0$, $\omega \to \pi^+\pi^-\pi^0$, $\phi \to K^+K^-$, $K^{*0} \to K^+\pi^-$, $K^{*0} \to K^0\pi^0$, $K^{*+} \to K^+\pi^0$, and $K^{*+} \to K^0\pi^+$. Charge conjugate decays are implied throughout this paper.

Candidate charged tracks are required to pass quality cuts and have specific ionization (dE/dx) consistent with that of a pion and kaon. Such tracks must not be electrons or muons, identified by calorimetry and depth of penetration of an iron muon stack, respectively. $K_S^0 \to \pi^+\pi^-$ candidates are accepted only if they are displaced from the primary interaction point by at least 3 mm. Photon candidates are isolated calorimeter showers with a measured energy of at least 30 (50) MeV in the central (end cap) region of the calorimeter. The momentum of charged tracks and photon pairs is required to be greater than 100 MeV/c to reduce combinatoric background. Photon pairs and vees are fit kinematically to the appropriate combined mass hypothesis to obtain the meson momentum vectors. Resolutions on the reconstructed masses prior to the constraint are about 5-10 MeV for $\pi^0 \to \gamma\gamma$, 12 MeV for $\eta_{\gamma\gamma}$, and 3 MeV for $K_S^0 \to \pi^+\pi^-$.

The primary means of identification of B meson candidates is through their measured mass and energy. The resolution for $\Delta E \equiv E_1 + E_2 - E_b$ (E_1 and E_2 are the energy of the two daughter particles of the B and E_b is the beam energy) is typically 20-50 MeV. The resolution for $M \equiv \sqrt{E_b^2 - \mathbf{p}_B^2}$ ($\mathbf{p}_B$ is the reconstructed B momentum) is 2.5-3.0 MeV, dominated by the uncertainty in $\mathbf{p}_B$. Signals are identified with the use of resonance masses and, in the case of vector-pseudoscalar decays and the $\eta' \to \rho^0\gamma$ channel, a variable $\mathcal{H}$ sensitive to the helicity distribution of the decay. For modes in which one daughter is a single charged track, or is a resonance pairing a charged track with a π^0, the dE/dx variables S_K and S_π are used. The latter are defined as the deviations from nominal energy loss for the indicated particle hypotheses measured in standard deviations. Studies of D^{*+} tagged $D^0 \to K^-\pi^+$ decays find a K-π separation of about 1.7 (2.0) standard deviations near 2.5 GeV/c for the CLEO II (II.V) configuration.

The large background from continuum quark production ($e^+e^- \to q\bar{q}$) is reduced with the use of event shape information. One such quantity is θ_{BB}, the angle between the thrust axis of the candidate B and that of the rest of the event (sphericity is used instead of thrust for the $K\pi$ and $\pi\pi$ analyses). Since B mesons are produced nearly at rest, there is little correlation between the two thrust axes, while candidates extracted from continuum $q\bar{q}$ events tend to be strongly correlated by the jet-like nature of the events. This difference is exploited by requiring $|\cos\theta_{BB}| < 0.9$. A multivariate discriminant $\mathcal{F}$ is also employed, with the primary inputs being the energy deposition in nine cones concentric with the thrust or sphericity axis of the candidates tracks. Monte Carlo studies indicate that backgrounds from other B decay modes are usually negligible, though they are accounted for in the few cases where they are significant.

In order to extract event yields, an unbinned extended-maximum-likelihood (ML) fit [5] is performed, which includes sidebands about the expected mass and energy peaks, of a superposition of expected signal and background distributions:

$$\mathcal{L} = e^{-(N_S+N_B)} \prod_{i=1}^{N} \{N_S \mathcal{P}_{S_i}(f_1, ..., f_m; x_1, ...x_p) + N_B \mathcal{P}_{B_i}(g_1, ..., g_m; x_1, ...x_p)\},$$

where $\mathcal{P}_{S_i}$ and $\mathcal{P}_{B_i}$ are the probabilities for event i to be signal and continuum background, respectively. The probabilities are a function of the values of the variables x used in the fit for each event, and of the parameters f and g used to describe the signal and background shapes for each variable. The variables used are ΔE, M, $\mathcal{F}$, and, where applicable, resonance masses, $\mathcal{H}$, S_K, and S_π. N_S and N_B, the free parameters of the fit, are the (positive-definite) number of signal and continuum background events in the fitted data sample, respectively. Sample sizes for these fits range from ~100 to about 30,000 events.

The signal probability distribution functions (PDFs) for each of the input variables are determined from fits to independent data samples or Monte Carlo events that simulate the response of the CLEO detector to each decay mode investigated. The GEANT[6] based simulation is tuned to reproduce detector resolution and efficiencies for a variety of benchmark processes. The parameters of the background PDFs are determined with similar fits to a side-band region of data defined by $|\Delta E| < 0.2$ GeV and $5.20 < M < 5.27$ GeV/c^2. The signal shapes used are Gaussian, double Gaussian, and Breit-Wigner as appropriate for ΔE and mass peaks. For background, resonance masses are fit to the sum of a smooth polynomial and the signal shape, to account for the component of real resonance as well as the combinatoric background. Shapes used for ΔE and M background are, respectively, a first-degree polynomial and the empirical shape[7] $f(z) \propto z\sqrt{1-z^2}\exp(-\xi(1-z^2))$, where $z \equiv M/E_b$ and ξ is a parameter to be fit. Finally, for $\mathcal{F}$, S_K, and S_π, bifurcated Gaussians are used for both signal and background.

3 $B \to K\pi$ and $B \to \pi\pi$ decays

The results of these fits for the $B \to K\pi$ and $B \to \pi\pi$ decays are given in Fig. 2 and Table 1. The figure shows the ML contour for the fit to the $K\pi/\pi\pi$ sample and projection of the fit onto the M and ΔE axes, after a cut on an event likelihood with the variable

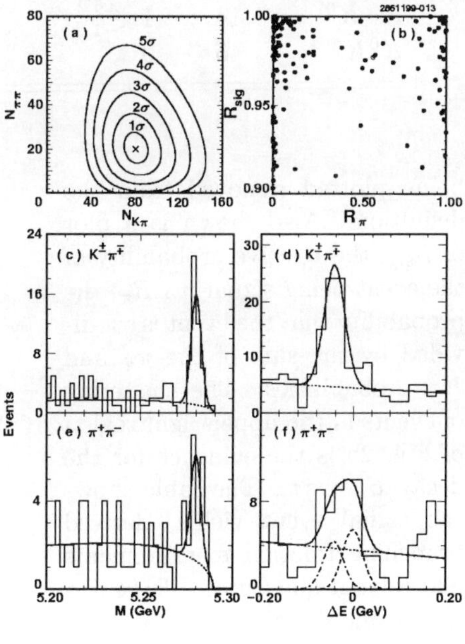

Figure 2. Results for $B \to K^\pm\pi^\mp$ and $B \to \pi^+\pi^-$. (a) Contours of the likelihood function versus $K\pi$ and $\pi\pi$ event yield; (b) Likelihood ratios – see text for explanation; (c) M for $K\pi$-like events; (d) ΔE for $K\pi$-like events; (e) M for $\pi\pi$-like events; (f) ΔE for $\pi\pi$-like events.

Table 1. Combined results for $K\pi$ and $\pi\pi$ decay modes with the efficiency, yield, significance and expectations from previously published theoretical models. The fit value $\mathcal{B}_{\rm fit}$ is given with statistical and systematic errors. The last column quotes the result as a 90% confidence level upper limit in the cases where the result is not statistically significant. The results for the modes with asterisks do not include the full CLEO data sample.

Mode	Eff (%)	Yield	$\mathcal{B}_{\rm fit}(10^{-6})$	Signif.	B(10^{-6})
$K^\pm\pi^\mp$	48	$80.2^{+11.8}_{-11.0}$	$17.2^{+2.5}_{-2.4}\pm 1.2$	11.7σ	see $\mathcal{B}_{\rm fit}$
$K^\pm\pi^0$	38	$42.1^{+10.9}_{-9.9}$	$11.6^{+3.0+1.4}_{-2.7-1.3}$	6.1σ	see $\mathcal{B}_{\rm fit}$
$K^0\pi^\pm$	14	$25.2^{+6.4}_{-5.6}$	$18.2^{+4.6}_{-4.0}\pm 1.6$	7.6σ	see $\mathcal{B}_{\rm fit}$
$K^0\pi^0$	11	$16.1^{+5.9}_{-5.0}$	$14.6^{+5.9+2.4}_{-5.1-3.3}$	4.9σ	see $\mathcal{B}_{\rm fit}$
$\pi^\pm\pi^\mp$	48	$20.0^{+7.6}_{-6.5}$	$4.3^{+1.6}_{-1.4}\pm 0.5$	4.2σ	see $\mathcal{B}_{\rm fit}$
$\pi^\pm\pi^0$	39	$21.3^{+9.7}_{-8.5}$	$5.6^{+2.6+1.4}_{-2.3-1.3}$	3.2σ	<12.7
$\pi^0\pi^0$ *	29	$2.7^{+2.7}_{-1.7}$		2.4σ	<9.3
$K^\pm K^\mp$	48	$0.7^{+3.4}_{-0.7}$		0.0σ	<1.9
$K^\pm K^0$	14	$1.4^{+2.4}_{-1.3}$		1.1σ	<5.1
$K^0\bar{K^0}$ *	5	0			<17

being plotted removed from the definition. Also shown is a plot of $R_{\rm sig}$, the relative probability of the event being signal, *vs* R_π, the probability that the event is $\pi\pi$ divided by the sum of the $\pi\pi$ and $K\pi$ probabilities. The clustering of events in the upper-right region of Fig. 2b is the evidence for the decay $B\to\pi\pi$. The table shows the signal event yields, derived branching fraction measurements and related quantities. There is a significant signal for $B^0\to\pi^+\pi^-$ and each of the four $B\to K\pi$ charge states. In Fig. 3, projection plots for the other three $K\pi$ channels are shown. Further details can be found in Ref. 8.

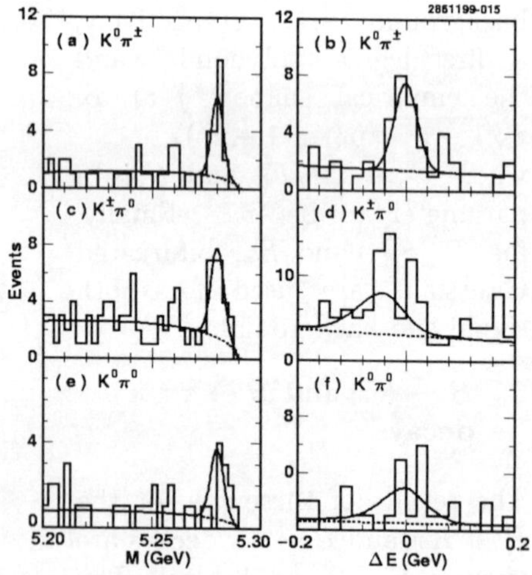

Figure 3. M and ΔE plots for $B\to K^0\pi^\pm$(a-b), $B\to K^\pm\pi^0$(c-d), and $B\to K^0\pi^0$(e-f).

4 ηX and $\eta' X$ final states

The results of the ML fits for the η and η' analyses are summarized in Table 2. This table includes the result of combining individaual analyses for the two η', two η, and two K^* decay channels. The strong signal for the $B \to \eta' K$ decay is shown in Fig. 4 for both charge states. We also observe for the first time the decay $B \to \eta K^*$, also for both charge states. The projection plots for these fits are also shown in Fig. 4. Further details concerning the η and η' modes can be found in Ref. 9.

Table 2. Combined results for η and η' decay modes with the efficiency times brancing fraction ($\epsilon\mathcal{B}_s$), significance and expectations from previously published theoretical models. The statistical and systematic errors are given for $\mathcal{B}_{\rm fit}$ except where the result is not statistically significant, in which case they are combined and the final result is quoted as a 90% confidence level upper limit.

Decay mode	$\epsilon\mathcal{B}_s$ (%)	$\mathcal{B}_{\rm fit}(10^{-6})$	Signif. (σ)	$\mathcal{B}(10^{-6})$	Theory $\mathcal{B}(10^{-6})$
$B^+ \to \eta' K^+$	13	$80^{+10}_{-9} \pm 7$	16.8	$80^{+10}_{-9} \pm 7$	7–65
$B^0 \to \eta' K^0$	4	$89^{+18}_{-16} \pm 9$	11.7	$89^{+18}_{-16} \pm 9$	9–59
$B^+ \to \eta' \pi^+$	14	$1.0^{+5.8}_{-1.0}$	0.0	< 12	1–23
$B^0 \to \eta' \pi^0$	12	$0.0^{+1.8}_{-0.0}$	0.0	< 5.7	0.1–14
$B^+ \to \eta' K^{*+}$	4	$11.1^{+12.7}_{-8.0}$	1.8	< 35	1–3.7
$B^0 \to \eta' K^{*0}$	6	$7.8^{+7.7}_{-5.7}$	1.8	< 24	1–8.0
$B^+ \to \eta' \rho^+$	5	$11.2^{+11.9}_{-7.0}$	2.4	< 33	3–24
$B^0 \to \eta' \rho^0$	8	$0.0^{+5.8}_{-0.0}$	0.0	< 12	0.1–11
$B^+ \to \eta K^+$	24	$2.2^{+2.8}_{-2.2}$	0.8	< 6.9	0.2–5.0
$B^0 \to \eta K^0$	7	$0.0^{+3.2}_{-0.0}$	0.0	< 9.3	0.1–3.0
$B^+ \to \eta \pi^+$	25	$1.2^{+2.8}_{-1.2}$	0.6	< 5.7	1.9–7.4
$B^0 \to \eta \pi^0$	18	$0.0^{+0.8}_{-0.0}$	0.0	< 2.9	0.2–4.3
$B^+ \to \eta K^{*+}$	7	$26.4^{+9.6}_{-8.2} \pm 3.3$	4.8	$26.4^{+9.6}_{-8.2} \pm 3.3$	0.2–8.2
$B^0 \to \eta K^{*0}$	12	$13.8^{+5.5}_{-4.6} \pm 1.6$	5.1	$13.8^{+5.5}_{-4.6} \pm 1.6$	0.1–8.9
$B^+ \to \eta \rho^+$	12	$4.3^{+5.2}_{-3.8}$	1.3	< 15	4–17
$B^0 \to \eta \rho^0$	15	$2.6^{+3.2}_{-2.6}$	1.3	< 10	0.1–6.5

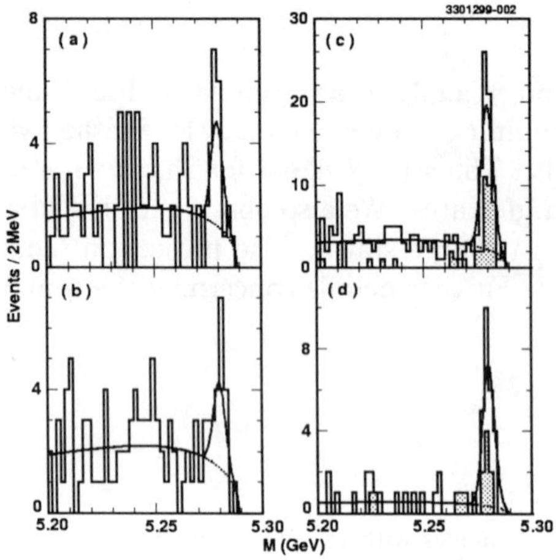

Figure 4. Projections onto the M axis. The histograms show (a) $B^+ \to \eta K^{*+}$; (b) $B^0 \to \eta K^{*0}$; (c) $B \to \eta' K$; (d) $B^0 \to \eta' K^0$. In (c) and (d) the shaded histograms correspond to the $\eta' \to \eta\pi\pi$, $\eta \to \gamma\gamma$ decay chain, while the unshaded histograms correspond to the $\eta' \to \rho\gamma$ channel. The solid (dashed) line shows the projection for the full fit (background only) with a cut on the signal likelihood (M has been removed from $\mathcal{L}$ for the purposes of this calculation.

5 Decays with an ω, ϕ, ρ or K^*

Finally, we give in Table 3 results for the ML fits for the ω, ϕ, ρ and K^* final states. Significant signals are obtained only for the $B^+ \to \rho^0\pi^+$, $B^0 \to \rho^\pm\pi^\mp$, $B^+ \to \omega\pi^+$, and $B^0 \to K^{*+}\pi^-$ decays. The projection plots for the first of these decays are shown in Fig. 5. Further details concerning these modes can be found in Ref. 10.

6 Discussion and conclusions

CLEO has observed 13 charmless hadronic B decay modes. The measured branching fractions are in the range (4–89)$\times 10^{-6}$. While the early observations were mostly of the penguin decays involving K or K^* mesons, we report here first observation of $B^0 \to \pi^+\pi^-$, $B^+ \to \omega\pi^+$ and two channels of $B \to \rho\pi$, which are presumably dominated by $b \to u$ tree diagrams.

The branching fraction for both $B \to \eta' K$ channels is several times larger than other charmless hadronic B decays. This was unexpected, though it had been pointed out by Lipkin [11] that interference effects between the two penguin diagrams, Fig. 1e and 1f, enhance the $\eta' K$ rate and suppress ηK^+. This interference flips sign for the decays involving K^* mesons, so it is $B \to \eta K^*$ which is expected to be enhanced. The branching fractions and upper limits given in Table 2 clearly exhibit this pattern.

There have been a variety of effective-Hamiltonian calculations [12,13,14] which calculate branching fractions for processes measured here; they generally employ factorization [15] approximations. In order to explain the large value of the $B \to \eta' K$ branching fractions, these and other calculations [16]

Table 3. Combined results for ω and ϕ decay modes and expectations from previously published theoretical models. See previous table for definition of symbols. The results for the modes with asterisks do not include the full CLEO data sample.

Decay mode		$\epsilon\mathcal{B}_s$ (%)	$\mathcal{B}_{\rm fit}(10^{-6})$	Signif. (σ)	$\mathcal{B}(10^{-6})$	Theory $\mathcal{B}(10^{-6})$
$B^+ \to \omega\pi^+$		26	$11.3^{+3.3}_{-2.9} \pm 1.4$	6.2	$11.3^{+3.3}_{-2.9} \pm 1.4$	$0.6 - 11$
$B^0 \to \omega\pi^0$		19	$0.8^{+1.9}_{-0.8} \pm 0.9$	0.6	< 5.5	$0.01 - 12$
$B^+ \to \omega K^+$		26	$3.2^{+2.4}_{-1.9} \pm 0.8$	2.1	< 7.9	$0.2 - 13$
$B^0 \to \omega K^0$		7	$10.0^{+5.4}_{-4.2} \pm 1.4$	3.9	< 21	$0.02 - 10$
$B^+ \to \omega\rho^+$		6	$18^{+11}_{-9} \pm 6$	2.8	< 47	$7 - 28$
$B^0 \to \omega\rho^0$		16	$0.0^{+5.7}_{-0.0}{}^{+2.9}_{-0.0}$	0.9	< 11	$0.005 - 0.4$
$B^+ \to \omega K^{*+}$		2	$5^{+19}_{-5} \pm 6$	0.3	< 52	$0.9 - 15$
$B^0 \to \omega K^{*0}$		8	$9.1^{+6.7}_{-5.1} \pm 1.9$	2.3	< 19	$0.3 - 12$
$B^+ \to \phi\pi^+$	*	27		0.0	< 4.0	$0.001 - 0.4$
$B^0 \to \phi\pi^0$	*	17		0.0	< 5.4	$0.0004 - 0.2$
$B^+ \to \phi K^+$	*	26	$1.6^{+1.9}_{-1.2} \pm 0.2$	1.3	< 5.9	$0.3 - 18$
$B^0 \to \phi K^0$	*	7	$10.7^{+7.8}_{-5.7} \pm 1.1$	2.6	< 28	$0.3 - 18$
$B^+ \to \phi K^{*+}$	*	3	$21.6^{+15.6}_{-10.4} \pm 2.6$	2.4	< 41	$0.2 - 31$
$B^0 \to \phi K^{*0}$	*	7	$8.2^{+7.7}_{-5.1} \pm 0.8$	1.9	< 21	$0.2 - 31$
$B^+ \to \rho^0\pi^+$		30	$10.4^{+3.3}_{-3.4} \pm 2.1$	5.4	$10.4^{+3.3}_{-3.4} \pm 2.1$	$0.4 - 8$
$B^0 \to \rho^\pm\pi^\mp$		12	$27.6^{+8.4}_{-7.4} \pm 4.2$	5.6	$27.6^{+8.4}_{-7.4} \pm 4.2$	$26 - 52$
$B^+ \to \rho^+\pi^0$	*	10		5.1	< 43	$11 - 27$
$B^0 \to \rho^0\pi^0$		34	$1.6^{+2.0}_{-1.4} \pm 0.8$	1.2	< 5.5	$0.9 - 2.3$
$B^+ \to \rho^0 K^+$		28	$8.4^{+4.0}_{-3.4} \pm 1.8$	3.7	< 17	$0.1 - 1.7$
$B^+ \to \rho^+ K^0$	*				< 64	$0.01 - 1.3$
$B^0 \to \rho^- K^+$		11	$16.0^{+7.6}_{-6.4} \pm 2.8$	3.5	< 32	$0.2 - 2.5$
$B^0 \to \rho^0 K^0$	*	10		2.7	< 27	$0.04 - 1.7$
$B^+ \to K^{*0}\pi^+$		18	$7.6^{+3.5}_{-3.0} \pm 1.6$	3.6	< 16	$4 - 12$
$B^0 \to K^{*+}\pi^-$	*		$22^{+8}_{-6}{}^{+4}_{-5}$	5.9	$22^{+8}_{-6}{}^{+4}_{-5}$	$1.2 - 19$
$B^0 \to K^{*0}\pi^0$		25	$0.0^{+1.3}_{-0.0} \pm 0.5$	0.0	< 3.6	$1.1 - 5$
$B^+ \to K^{*+}\pi^0$	*	4			< 80	$0.5 - 9$
$B^0 \to K^{*+}K^-$	*				< 6	
$B^+ \to K^{*0}K^+$		17	$0.0^{+1.3}_{-0.0} \pm 0.6$	0.0	< 5.3	$0.2 - 1$

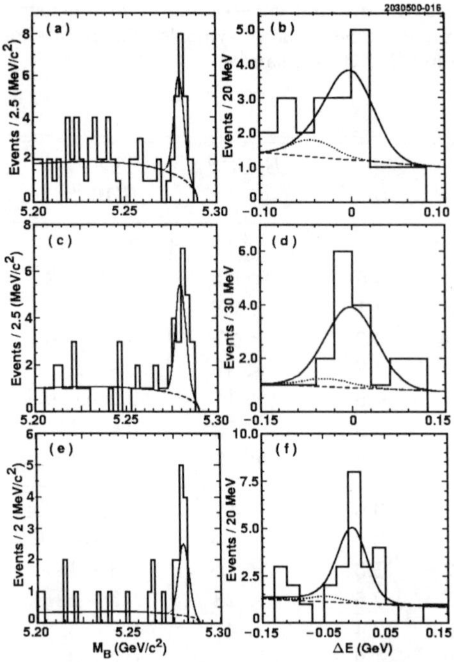

Figure 5. M and ΔE projection plots for $B^+ \to \rho^0\pi^+$ (top), $B^0 \to \rho^\pm\pi^\mp$ (middle) and $B^+ \to \omega\pi^+$ (bottom). The histograms show the data while the solid lines represent the overall fit to the data scaled to account for the extra requirement on event probability applied to make the projection, and the dashed lines represent the continuum and the other B component in the simultaneous fits.

have suggested enhancements from larger form factors, smaller strange-quark mass, and variation of the effective number of colors. Others have suggested a substantial flavor-singlet amplitude (Fig. 1h) in constructive interference with the penguins. These calculations typically fall somewhat short of the large measured branching fraction, though there are as yet no reliable calculations involving the flavor-single component. The enhancement may well be related to the large observed branching fraction for the inclusive $B \to \eta' X_s$ process.[17] The connection is not at all clear and the inclusive process is not understood quantitatively, though a qualitative suggestion involving the so-called QCD anomaly seems promising.[18]

The remaining nine observed decays and the various upper limits are generally in agreement with theoretical predictions, with the possible exception of $B^0 \to K^{*+}\pi^-$. However it was noticed about a year ago[19] that there were trends in the amount of disagreement with theory which could be reduced if the CKM angle γ were taken to be $\gtrsim 90°$ rather than the generally accepted value of 60–70°. More recently, a fit for γ has been performed[20] to 14 CLEO measurements (not all of which are statistically significant). The only modes excluded from this fit were poorly understood modes, such as the η and η' channels just discussed, and those for which a central value with statistical and systematic errors was not available. The authors fit the measured branching fractions to theoretical expressions derived using effective-theory calculations with the assumption of factorization and neglect of final-state interactions and strong phases. They vary, in addition to γ, the CKM element

V_{ub} within its experimental errors, two form factors and a parameter related to the strange quark mass. The result of the fit, shown in Fig. 6, is $\gamma = 114^{+25}_{-21}$ degrees. The good χ^2 of the fit is taken as a validation of the assumptions going into the analysis (at the current level of experimental precision), though more experimental and theoretical work is needed before such fits to charmless hadronic decays can provide useful measurements of γ.

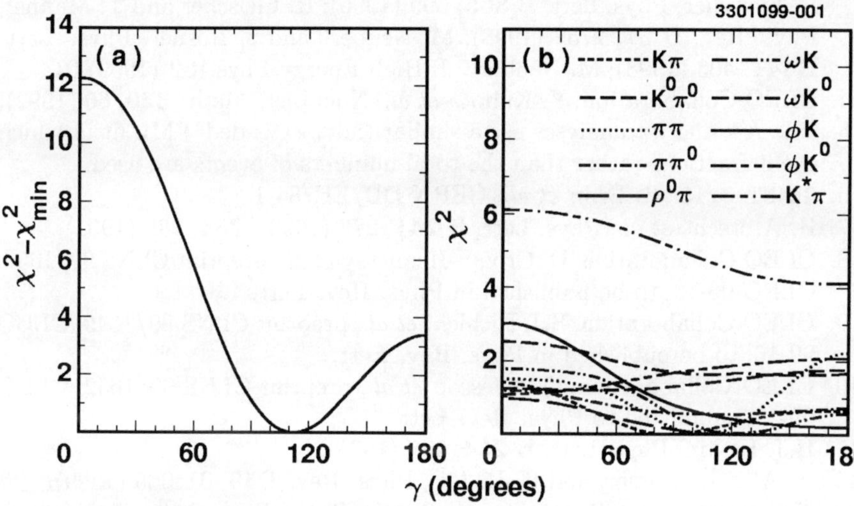

Figure 6. (a) $\chi^2 - \chi^2_{\rm min}$ vs. γ for nominal fit, where $\chi^2_{\rm min} = 11.3$ for 10 DOF; (b) Major χ^2 contributions from individual measurements.

A search for direct CP violation in these decays has been performed and is discussed elsewhere. [21]

Acknowledgments

I would like to thank H.-Y. Cheng and W.-S. Hou for their hospitality and the excellent organization of the conference. I'm grateful to dozens of my CLEO colleagues for their assistance and helpful discussions. Finally I would like to thank T. Sanda and other theorists for their encouragement in pursuing the fits to extract the CKM angle γ from the data.

References

1. M. Kobayashi and T. Maskawa, Prog. Theor. Phys. **49**, 652 (1973).
2. For an excellent recent review of the subject, see A. J. Buras and R. Fleischer, *Heavy Flavours II*, p.65, World Scientific (1997), eds. A.J. Buras and M. Lindner.
3. M. Gronau, J.L. Rosner, and D. London, Phys. Rev. Lett. **73**, 21 (1994); R. Fleischer, Phys. Lett. B **365**, 399 (1996); R. Fleischer and T. Mannel, Phys. Rev. D **57**, 2752 (1998); M. Neubert and J. Rosner, Phys. Lett. **B441**, 403 (1998); M. Neubert, J. High Energy Phys. **02** (1999) 014.
4. CLEO Collaboration, Y. Kubota *et al.*, Nucl Inst. Meth. **320**, 66 (1992).
5. The $K\pi$ and $\pi\pi$ analyses use a similar (not "extended") ML fit in which event fractions rather than the total numbers of events are used.
6. `GEANT 3.15`: R. Brun *et al.*, CERN DD/EE/84-1.
7. H. Albrecht *et al.*, Phys. Lett. B **241**, 278 (1990); **254**, 288 (1991).
8. CLEO Collaboration, D. Cronin-Hennessy *et al.*, preprint CLNS 99/1650, CLEO 99-18, to be published in Phys. Rev. Lett.
9. CLEO Collaboration, S.J. Richichi *et al.*, preprint CLNS 99/1649, CLEO 99-16, to be published in Phys. Rev. Lett.
10. CLEO Collaboration, C.P. Jessop *et al.*, preprint CLNS 99/1652, CLEO 99-19, submitted to Phys. Rev. Lett.
11. H.J. Lipkin, Phys. Lett. B **254**, 247 (1991).
12. A. Ali, G. Kramer, and C.-D. Lu , Phys. Rev. D**59**, 014005 (1999).
13. N.G. Deshpande, B. Dutta, and S. Oh, Phys. Rev. D**57**, 5723 (1998); Phys. Lett. B **473**, 141 (2000).
14. Y.-H. Chen, H.-Y. Cheng, B. Tseng, and K.-C. Yang, Phys. Rev. **D60**, 094014 (1999). See also paper by H.-Y. Cheng, hep-ph/9912372 and in these proceedings.
15. M. Bauer, B. Stech, and M. Wirbel, Z. Phys. C **43**, 103 (1987).
16. For a more complete list of the relevant theoretical papers, see for instance K. Lingel, T. Skwarnicki and J.G. Smith, Annu. Rev. Nucl. Part. Sci. **48**, 253 (1998).
17. CLEO Collaboration, T.E. Browder *et al.*, Phys. Rev. Lett. **81**, 1786 (1998).
18. D. Atwood and A. Soni, Phys. Lett. B **405**, 150 (1997).
19. X.-G. He, W.-S. Hou, and K.-C. Yang, Phys. Rev. Lett. **83**, 1100 (1999).
20. W.-S. Hou, J.G. Smith, and F. Würthwein, hep-ex/9910014, 1999.
21. CLEO Collaboration, S. Chen *et al.*, preprint CLNS 99/1651, CLEO 99-17, to be published in Phys. Rev. Lett. See also paper by J.P. Alexander in these proceedings.

HADRONIC TWO-BODY CHARMLESS B DECAYS

HAI-YANG CHENG
Institute of Physics, Academia Sinica, Taipei, Taiwan 115, Republic of China

Implications of recent CLEO measurements of hadronic charmless B decays are discussed.

1 Effective Wilson Coefficients

In the absence of first-principles calculations for hadronic matrix elements, it is customary to evaluate the matrix elements under the factorization hypothesis so that $\langle O(\mu)\rangle$ is factorized into the product of two matrix elements of single currents, governed by decay constants and form factors. However, the naive factorized amplitude is not renormalization scale- and γ_5 scheme- independent as the scale and scheme dependence of Wilson coefficients are not compensated by that of the factorized hadronic matrix elements. In principle, the scale and scheme problems with naive factorization will not occur in the full amplitude since $\langle O(\mu)\rangle$ involves vertex-type and penguin-type corrections to the hadronic matrix elements of the 4-quark operator renormalized at the scale μ. Formally, one can write

$$\langle O(\mu)\rangle = g(\mu,\mu_f)\langle O(\mu_f)\rangle, \tag{1}$$

where μ_f is a factorization scale, and $g(\mu,\mu_f)$ is an evolution factor running from the scale μ to μ_f which is calculable because the infrared structure of the amplitude, if any, is absorbed into $\langle O(\mu_f)\rangle$. Writing

$$c^{\rm eff}(\mu_f) = c(\mu)g(\mu,\mu_f), \tag{2}$$

the effective Wilson coefficients are formally scheme and μ-scale independent. The factorization approximation is then applied to $\langle O(\mu_f)\rangle$ afterwards.

In principle, one can work with any quark configuration, on-shell or off-shell, to compute the decay amplitude. If the off-shell quark momentum is chosen as the infrared cutoff, $g(\mu,\mu_f)$ will depend on the gauge of the gluon field. This is all right as the gauge dependence belongs to the infrared structure of the wave function so that the physical amplitude is gauge independent [1]. However, if factorization is applied to $\langle O(\mu_f)\rangle$, information of gauge dependence characterized by the wave function will be lost, a difficulty pointed out by Buras and Silvestrini [2]: $c^{\rm eff}$ will depend on the choice of gauge. As pointed out in [1], within the factorization framework, one must work in the on-shell quark scheme to obtain gauge invariant and infrared finite $c_i^{\rm eff}$ and infrared poles are absorbed into universal bound-state wave functions. It should be stressed that the constant matrix $\hat{r}_V$ arising from vertex-like corrections is *not* arbitrary due to the infrared finiteness of vertex-like diagrams: The infrared divergences in individual vertex-type diagrams cancel in their sum. The gauge-invariant $\hat{r}_V$ matrix in naive dimension regularization (NDR) and 't Hooft-Veltman (HV) renormalization schemes is first given in [3] and [4].

2 Nonfactorized Effects

It is known that the effective Wilson coefficients appear in the factorizable decay amplitudes in the combinations $a_{2i} = c_{2i}^{\rm eff} + \frac{1}{N_c} c_{2i-1}^{\rm eff}$ and $a_{2i-1} = c_{2i-1}^{\rm eff} + \frac{1}{N_c} c_{2i}^{\rm eff}$ $(i = 1, \cdots, 5)$. Phenomenologically, the number of colors N_c is often treated as a free parameter to model the nonfactorizable contribution to hadronic matrix elements and its value can be extracted from the data of two-body nonleptonic decays. Nonfactorizable effects in the decay amplitudes of $B \to PP$, VP can be absorbed into the parameters $a_i^{\rm eff}$. This amounts to replacing N_c in $a_i^{\rm eff}$ by $(N_c^{\rm eff})_i$. Explicitly,

$$a_{2i}^{\rm eff} = c_{2i}^{\rm eff} + \frac{1}{(N_c^{\rm eff})_{2i}} c_{2i-1}^{\rm eff}, \qquad a_{2i-1}^{\rm eff} = c_{2i-1}^{\rm eff} + \frac{1}{(N_c^{\rm eff})_{2i-1}} c_{2i}^{\rm eff}, \tag{3}$$

where $i = 1, \cdots, 5$ and

$$(1/N_c^{\rm eff})_i \equiv (1/N_c) + \chi_i \,, \tag{4}$$

with χ_i being the nonfactorizable terms, which receive contributions from nonfactorized vertex-type, penguin-type and spectator corrections. In general, χ_i and $(N_c^{\rm eff})_i$ are complex.

Naive factorization does not work in the presence of nonfactorized contributions. Nevertheless, if χ_i are universal (i.e. process independent), then we still have generalized factorization, which is likely to be justified for hadronic charmless B decays due to their large energy release. Since the Fierz transformation of $(V-A)(V+A)$ operators is quite different from that of $(V-A)(V-A)$ operators, we shall assume that $\chi_{LR} \neq \chi_{LL}$, where $\chi_{LL} \equiv \chi_{1,2,3,4,9,10}$ and $\chi_{LR} \equiv \chi_{5,6,7,8}$, or equivalently $N_c^{\rm eff}(LR) \neq N_c^{\rm eff}(LL)$ with $N_c^{\rm eff}(LL) \equiv \left(N_c^{\rm eff}\right)_{1,2,3,4,9,10}$ and $N_c^{\rm eff}(LR) \equiv \left(N_c^{\rm eff}\right)_{5,6,7,8}$. As shown in [3], the data analysis and the theoretical study of nonleptonic rare B decays all indicate that $N_c^{\rm eff}(LR) > 3 > N_c^{\rm eff}(LL)$. In principle, $N_c^{\rm eff}$ can vary from channel to channel, as in the case of charm decay. However, in the energetic two-body B decays, $N_c^{\rm eff}$ is expected to be process insensitive as supported by the data [3].

The observation $N_c^{\rm eff}(LL) < 3 < N_c^{\rm eff}(LR)$ is theoretically justified by a recent perturbative QCD calculation of $B \to \pi\pi$ decays in the heavy quark limit. As pointed out in [5] that, in the heavy quark limit, all nonfactorizable diagrams are dominated by hard gluon exchange, while soft gluon effects are suppressed by factors of $\Lambda_{\rm QCD}/m_b$. In other words, the nonfactorized term is calculable as expansion in α_s in the heavy quark limit. Following the notation of [5], we find the nonfactorized terms:

$$\chi_{1,2,3,4} = -\chi_{5,6} = \frac{\alpha_s}{4\pi}\frac{C_F}{N_c}(f^{\rm I} + f^{\rm II}), \tag{5}$$

It follows that

$$\chi_{LR} = -\chi_{LL} = -\frac{\alpha_s}{4\pi}\frac{C_F}{N_c}(f^{\rm I} + f^{\rm II}). \tag{6}$$

Several remarks are in order. (i) Since $f^{\rm I}$ is complex due to final-state interactions via hard gluon exchnge [5], so are χ_i and $N_c^{\rm eff}(LL)$ and $N_c^{\rm eff}(LR)$. Nevertheless, the complex phases of χ_i are in general small. (ii) Because ${\rm Re}\,\chi_{LL} > 0$, it is obvious that $N_c^{\rm eff}(LL) < 3$ and $N_c^{\rm eff}(LR) > 3$. Furthermore, $N_c^{\rm eff}(LL) \sim 2$ implies $N_c^{\rm eff}(LR) \sim 6$. Therefore, the empirical observation $N_c^{\rm eff}(LR) > 3 > N_c^{\rm eff}(LL)$ shown in [3] gets a firm justification from the perturbative QCD calculation.

3 Tree-dominated charmless B Decays

CLEO has observed several tree-dominated charmless B decays which proceed at the tree level through the b quark decay $b \to u\bar{u}d$ and at the loop level via the $b \to d$ penguin diagrams: $B \to \pi^+\pi^-$, $\rho^0\pi^\pm$, $\omega\pi^\pm$, $\rho^\pm\pi^\mp$. The updated branching ratios are [6]:

$$\begin{aligned}
\mathcal{B}(B^\pm \to \rho^0\pi^\pm) &= (10.4^{+3.3}_{-3.4} \pm 2.1) \times 10^{-6}, \\
\mathcal{B}(B^\pm \to \omega\pi^\pm) &= (11.3^{+3.3}_{-2.9} \pm 1.5) \times 10^{-6}, \\
\mathcal{B}(B^0 \to \rho^\pm\pi^\mp) &= (27.6^{+8.4}_{-7.4} \pm 4.2) \times 10^{-6}.
\end{aligned} \tag{7}$$

These decays are sensitive to the form factors $A_0^{B\rho}$, $A_0^{B\omega}$ and to the value of $N_c^{\rm eff}(LL)$. We consider two different form-factor models for heavy-to-light transitions: the BSW model [7] and the light-cone sum rule (LCSR) [8] and find $1.1 \le N_c^{\rm eff}(LL) \le 2.6$ from $\rho^0\pi^\pm$ and $\omega\pi^\pm$ modes. This is indeed what expected since the effective number of colors, $N_c^{\rm eff}(LL)$, inferred from the Cabibbo-allowed decays $B \to (D, D^*)(\pi, \rho)$ is in the vicinity of 2 (see [3]) and since the energy released in the energetic two-body charmless B decays is in general slightly larger than that in $B \to D\pi$ decays, it is thus anticipated that

$$|\chi(\text{two}-\text{body rare B decay})| \lesssim |\chi(B \to D\pi)|, \tag{8}$$

and hence $N_c^{\rm eff}(LL) \approx N_c^{\rm eff}(B \to D\pi) \sim 2$.

Note that the branching ratio of $\rho^0\pi^\pm$ is sensitive to the change of the unitarity angle γ, while $\omega\pi^\pm$ is not. For example, we have $\mathcal{B}(B^\pm \to \rho^0\pi^\pm) \sim \mathcal{B}(B^\pm \to \omega\pi^\pm)$ for $\gamma \sim 65°$, and $\mathcal{B}(B^\pm \to \rho^0\pi^\pm) > \mathcal{B}(B^\pm \to \omega\pi^\pm)$ for $\gamma > 90°$. It thus appears that a γ larger than $90°$, which is preferred by the previous measurement [9] $\mathcal{B}(B^\pm \to \rho^0\pi^\pm) = (15 \pm 5 \pm 4) \times 10^{-6}$, is no longer supported by the new data of $\rho^0\pi^\pm$.

4 $B \to \pi\pi$ and πK decays

The updated branching ratio for the decay $B \to \pi^+\pi^-$ has been reported in this conference to be [6]:

$$\mathcal{B}(B^0 \to \pi^+\pi^-) = (4.3^{+1.6}_{-1.4} \pm 0.6) \times 10^{-6}. \tag{9}$$

This decay mode puts a very stringent constraint on the form factor $F_0^{B\pi}$. Neglecting final-state interactions and employing $\gamma \equiv \text{Arg}(V_{ub}^*) = 65°$ and $|V_{ub}/V_{cb}| = 0.09$, and the effective number of colors $N_c^{\rm eff}(LL) = 2$, we find $F_0^{B\pi}(0) = 0.20 \pm 0.04$. This fairly small form factor will lead to two difficulties. First, the predicted $B \to K\pi$ rates will be too small compared to the data as their decay amplitudes are governed by the same form factor. Second, the predicted rate of $B \to K\eta'$ is also too small as the form factor $F_0^{BK}(0)$ cannot deviate too much from $F_0^{B\pi}(0)$ otherwise the SU(3)-symmetry relation $F_0^{BK} = F_0^{B\pi}$ will be badly broken.

There exist several possibilities that the $K\pi$ rates can be enhanced : (i) The CKM matrix element V_{ub} is small, say $|V_{ub}/V_{cb}| \approx 0.06$, so that the form factor $F_0^{B\pi}(0)$ can be large enough [10]. However, this CKM matrix element $|V_{ub}/V_{cb}|$ is smaller than the recent LEP average [11] $0.104^{+0.015}_{-0.018}$ and the CLEO result [12] $0.083^{+0.015}_{-0.016}$. (ii) A large nonzero isospin $\pi\pi$ phase shift difference of order $70°$ [3] can yield a substantial suppression of the $\pi^+\pi^-$ mode. However, a large $\pi\pi$ isospin phase difference seems to be very unlikely due to the large energy released in charmless B decays. Indeed, the Regge analysis of [14] indicates $\delta_{\pi\pi} = 11°$. (iii) Smaller quark masses, say $m_s(m_b) = 65$ MeV, will make the $(S-P)(S+P)$ penguin terms contributing sizably to the $K\pi$ modes but less significantly to $\pi^+\pi^-$ as the penguin effect on the latter is suppressed by the quark mixing angles. However, a rather smaller m_s is not consistent with recent lattice calculations. (iv) The unitarity angle γ larger than 90° will lead to a suppression of $B \to \pi^+\pi^-$ [13,3], which in turn implies an enhancement of $F_0^{B\pi}$ and hence $K\pi$ rates. Therefore, the last possibility appears to be more plausible.

We find that the CLEO $K\pi$ and $\pi\pi$ data can be accommodated by $\gamma = 105°$, $F_0^{B\pi}(0) = 0.28$, $F_0^{BK}(0) = 0.36$, $N_c^{\rm eff}(LL) = 2$. Note that $\gamma = (114^{+25}_{-21})°$ is obtained by Hou, Smith, Würthwein [15] under naive factorization. The calculated and experimental values are

$$\begin{aligned}
\mathcal{B}(\overline{B}^0 \to K^-\pi^+) &= 18.6 \times 10^{-6}, \qquad (17.2^{+2.5}_{-2.4} \pm 1.2) \times 10^{-6},\\
\mathcal{B}(B^- \to \overline{K}^0\pi^-) &= 17.0 \times 10^{-6}, \qquad (18.2^{+4.6}_{-4.0} \pm 1.6) \times 10^{-6},\\
\mathcal{B}(B^- \to K^-\pi^0) &= 12.6 \times 10^{-6}, \qquad (11.6^{+3.0+1.4}_{-2.7-1.3}) \times 10^{-6},\\
\mathcal{B}(\overline{B}^0 \to \overline{K}^0\pi^0) &= 6.0 \times 10^{-6}, \qquad (14.6^{+5.9+2.4}_{-5.1-3.3}) \times 10^{-6}.
\end{aligned} \tag{10}$$

It is known that $K\pi$ modes are penguin dominated. As far as the QCD penguin contributions are concerned, it will be expected that $\mathcal{B}(\overline{B}^0 \to K^-\pi^+) \sim \mathcal{B}(B^- \to \overline{K}^0\pi^-)$ and $\mathcal{B}(B^- \to K^-\pi^0) \sim \mathcal{B}(\overline{B}^0 \to \overline{K}^0\pi^0) \sim \frac{1}{2}\mathcal{B}(B \to K\pi^\pm)$. However, as pointed out in [3,13], the electroweak penguin diagram, which can be neglected in $\overline{K}^0\pi^-$ and $K^-\pi^+$, does play an essential role in the modes $K\pi^0$. With a moderate electroweak penguin contribution, the constructive (destructive) interference between electroweak and QCD penguins in $K^-\pi^0$ and $\overline{K}^0\pi^0$ renders the former greater than the latter; that is, $\mathcal{B}(B^- \to K^-\pi^0) > \frac{1}{2}\mathcal{B}(\overline{B}^0 \to \overline{K}^0\pi^-)$ and $\mathcal{B}(\overline{B}^0 \to \overline{K}^0\pi^0) < \frac{1}{2}\mathcal{B}(\overline{B}^0 \to K^-\pi^+)$ are anticipated. We see from Eq. (10) that, except for the decay $\overline{K}^0\pi^0$, the agreement of the calculated branching ratios for $K\pi$ modes with experiment is good. By contrast, the central value of $\mathcal{B}(\overline{B}^0 \to \overline{K}^0\pi^0)$ is much greater than the theoretical expectation. Since its experimental error is large, one has to await the experimental improvement to clarify the issue. The predicted pattern $K^-\pi^+ \gtrsim \overline{K}^0\pi^- \sim \frac{3}{2}K^-\pi^0 \sim 3\,\overline{K}^0\pi^0$ is consistent with experiment for the first three decays.

5 $B \to K\phi$, $K\eta'$ and $K^*\eta$ decays

The decay amplitude of the penguin-dominated mode $B \to K\phi$ is governed by $[a_3 + a_4 + a_5 - \frac{1}{2}(a_7 + a_9 + a_{10})]$, where a_3 and a_5 are sensitive to $N_c^{\rm eff}(LL)$ and $N_c^{\rm eff}(LR)$, respectively. The current limit $\mathcal{B}(B^\pm \to \phi K^\pm) < 0.59 \times 10^{-5}$ at 90% C.L. [9] implies that

$$N_c^{\rm eff}(LR) \geq \begin{cases} 5.0 & \text{BSW}, \\ 4.2 & \text{LCSR}, \end{cases} \tag{11}$$

with $N_c^{\rm eff}(LL)$ being fixed at the value of 2. Hence, we can conclude that $N_c^{\rm eff}(LR) > 3 > N_c^{\rm eff}(LL)$.

The improved measurements of the decays $B \to \eta' K$ by CLEO yield [6]

$$\mathcal{B}(B^\pm \to \eta' K^\pm) = \left(80^{+10}_{-\ 9} \pm 7\right) \times 10^{-6},$$
$$\mathcal{B}(B^0 \to \eta' K^0) = \left(89^{+18}_{-16} \pm 9\right) \times 10^{-6}. \tag{12}$$

This year CLEO has also reported the new measurement of $B \to K^*\eta$ with the branching ratios [16]

$$\mathcal{B}(B^\pm \to \eta K^{*\pm}) = \left(26.4^{+9.6}_{-8.2} \pm 3.3\right) \times 10^{-6},$$
$$\mathcal{B}(B^0 \to \eta K^{*0}) = \left(13.8^{+5.5}_{-4.6} \pm 1.6\right) \times 10^{-6}. \tag{13}$$

Theoretically, the branching ratios of $K\eta'$ ($K^*\eta$) are anticipated to be much greater than $K\pi$ ($K^*\eta'$) modes owing to the presence of constructive interference between two penguin amplitudes arising from non-strange and strange quarks of the η' or η. In general, the decay rates of $K\eta'$ increase slowly with $N_c^{\rm eff}(LR)$ if $N_c^{\rm eff}(LL)$ is treated to be the same as $N_c^{\rm eff}(LR)$, but fast enough with $N_c^{\rm eff}(LR)$ if $N_c^{\rm eff}(LL)$ is fixed at the value of 2. Evidently, the data much favor the latter case. As stressed in [3], the contribution from the η' charm content will make the theoretical prediction even worse at the small values of $1/N_c^{\rm eff}$ if $N_c^{\rm eff}(LL) = N_c^{\rm eff}(LR)$! On the contrary, if $N_c^{\rm eff}(LL) \approx 2$, the $c\bar{c}$ admixture in the η' will always lead to a constructive interference irrespective of the value of $N_c^{\rm eff}(LR)$. On the contrary, the branching ratios of $K^*\eta$ in general decrease with $N_c^{\rm eff}(LR)$ when $N_c^{\rm eff}(LL) = N_c^{\rm eff}(LR)$ but increase with $N_c^{\rm eff}(LR)$ when $N_c^{\rm eff}(LL) = 2$. Again, the latter is preferred by experiment. Hence, the data of both $K\eta'$ and $K^*\eta$ provide another strong support for a small $N_c^{\rm eff}(LL)$ and for the relation $N_c^{\rm eff}(LR) > N_c^{\rm eff}(LL)$. In other words, the nonfactorized effects due to $(V-A)(V-A)$ and $(V-A)(V+A)$ operators should be treated differently.

Several new mechanisms have been proposed in the past few years to explain the observed enormously large rate of $K\eta'$, for example, the large charm content of the η' [17] or the two-gluon fusion mechanism via the anomaly coupling of the η' with two gluons [18,19]. These mechanisms will in general predict a large rate for $K^*\eta'$ comparable to or even greater than $K\eta'$ and a very small rate for $K^*\eta$ and $K\eta$. The fact that the $K^*\eta$ modes are observed with sizeable branching ratios indicates that it is the constructive interference of two comparable penguin amplitudes rather than the mechanism specific to the η' that accounts for the bulk of $B \to \eta' K$ and ηK^* branching ratios.

Finally, we like to make a remark. As shown in [3], the charged $\eta' K^-$ mode gets enhanced when $\cos\gamma$ becomes negative while the neutral $\eta' K^0$ mode remains steady. Therefore, it is important to see if the disparity between $\eta' K^{\pm}$ and $\eta' K^0$ is confirmed when experimental errors are improved and refined in the future.

References

1. H.Y. Cheng, H-n. Li, and K.C. Yang, *Phys. Rev.* **D60**, 094005 (1999).
2. A.J. Buras and L. Silvestrini, *Nucl. Phys.* **B548**, 293 (1999).
3. Y.H. Chen, H.Y. Cheng, B. Tseng, and K.C. Yang, *Phys. Rev.* **D60**, 094014 (1999).
4. H.Y. Cheng and K.C. Yang, hep-ph/9910291.
5. M. Beneke, G. Buchalla, M. Neubert, and C.T. Sachrajda, *Phys. Rev. Lett.* **83**, 1914 (1999).
6. J. Smith in this proceedings.
7. M. Wirbel, B. Stech, and M. Bauer, *ibid.* **C29**, 637 (1985); M. Bauer, B. Stech, and M. Wirbel, *Z. Phys.* **C34**, 103 (1987).
8. P. Ball and V.M. Braun, *Phys. Rev.* **D58**, 094016 (1998); P. Ball, *J. High Energy Phys.* **9809**, 005 (1998).
9. CLEO Collaboration, M. Bishai *et al.*, CLEO CONF 99-13, hep-ex/9908018.
10. K. Agashe and N.G. Deshpande, hep-ph/9909298.
11. D. Abbaneo *et al.*, LEPVUB-99-01, 1999.
12. CLEO Collaboration, B.H. Behrens *et al.*, hep-ex/9905056.
13. X.G. He, W.S. Hou, and K.C. Yang, *Phys. Rev. Lett.* **83**, 1100 (1999).
14. J.-M. Gérard, J. Pestieau, and J. Weyers, *Phys. Lett.* **B436**, 363 (1998).
15. W.S. Hou, J.G. Smith, and F. Würthwein, hep-ph/9910014.
16. CLEO Collaboration, S.J. Richichi *et al.*, CLEO CONF 99-12, hep-ex/9908019.
17. I. Halperin and A. Zhitnitsky, *Phys. Rev.* **D56**, 7247 (1997); E.V. Shuryak and A. Zhitnitsky, *Phys. Rev.* **D57**, 2001 (1998).
18. M.R. Ahmady, E. Kou, and A. Sugamoto, *Phys. Rev.* **D58**, 014015 (1998).
19. D.S. Du, C.S. Kim, and Y.D. Yang, *Phys. Lett.* **B419**, 369 (1998); D.S. Du, Y.D. Yang, and G. Zhu, *Phys. Rev.* **D59**, 014007 (1999); *ibid.* **D60**, 054015 (1999).

SEARCH FOR DIRECT CP VIOLATION IN HADRONIC B DECAY

JIM ALEXANDER
Department of Physics, Cornell University, Ithaca, New York 14853-5001, USA
E-mail: jima@lns.cornell.edu

Using 9.7M $B\bar{B}$decays in the CLEO II and II.V datasets, we have searched for CP-violating asymmetries in six rare B meson decay modes. The statistical precision achieved varies from ±0.04 to ±0.25 depending on the mode, and all results are consistent with zero asymmetry.

1 Introduction

CP violating phenomena arise in the Standard Model because of the single complex parameter in the quark mixing matrix.[1] Such phenomena are expected to occur widely in B meson decays and are the incentive for most of the current B-physics initiatives in the world. While many of these investigations will focus on mixing-induced CP violation, direct CP violation is also anticipated to play a prominent role in the CP phenomena of B decay. CP asymmetries may show up in any mode where there are two or more participating diagrams which differ in weak and strong phases. $B \to K\pi$ modes, for instance, involve $b \to u$ tree diagrams carrying the weak phase $\text{Arg}(V_{ub}^* V_{us}) \equiv \gamma$ and $b \to s$ penguin diagrams carrying the weak phase $\text{Arg}(V_{tb}^* V_{ts}) = \pi$. Though the branching ratios are small, such cases are experimentally straightforward to search for. Rate differences between $B \to K^+\pi^-$ and $\bar{B} \to K^-\pi^+$ decays, if seen, would be unambiguous signals of direct CP violation.

We report here five searches for direct CP violation in charmless hadronic B decay modes, and a sixth search in the mode $B^\pm \to \psi^{(\prime)} K^\pm$. All results are based on the full CLEO II and CLEO II.V datasets which together comprise 9.66 million $B\bar{B}$ events. The charmless modes searched for are the three $K\pi$ modes, $K^\pm\pi^\mp$, $K^\pm\pi^0$, $K_S^0\pi^\pm$, the mode $K^\pm\eta'$, and the vector-pseudoscalar mode $\omega\pi^\pm$. These six modes are chosen both because all have robust signal yields, and because all are self-tagging in an experimentally clean way: the flavor of the parent b or $\bar{b}$ quark is tagged simply by the sign of the high momentum charged hadron; for $K^\pm\pi^\mp$ one must further identify K and π.

We define the sign of $\mathcal{A}_{\text{CP}}$ with the following convention:

$$\mathcal{A}_{\text{CP}} \equiv \frac{\mathcal{B}(b \to f) - \mathcal{B}(\bar{b} \to \bar{f})}{\mathcal{B}(b \to f) + \mathcal{B}(\bar{b} \to \bar{f})} \tag{1}$$

The statistical precision one can achieve in a measurement of $\mathcal{A}_{\rm CP}$ depends on the signal yield S, the CP-symmetric background B, any correlation ρ that might exist between the measurements of f and $\bar{f}$, and of course on $\mathcal{A}_{\rm CP}$ itself:

$$\sigma^2_{\mathcal{A}_{\rm CP}} = \frac{1-\mathcal{A}_{\rm CP}{}^2}{S}\left(1+\frac{B}{S}\left(\frac{1+\mathcal{A}_{\rm CP}{}^2}{1-\mathcal{A}_{\rm CP}{}^2}\right)-2\rho\sqrt{\frac{1-\mathcal{A}_{\rm CP}{}^2}{4}+\frac{B^2}{S^2}+\frac{B}{S}}\right) \tag{2}$$

In most cases there is no chance of confusing f and $\bar{f}$ so $\rho = 0$. However for $f = K^-\pi^+$ and $\bar{f} = K^+\pi^-$, a small degree of crossover is possible due to imperfect particle identification; we find $\rho = -0.11$ for this case. For $\rho \sim 0$, $B/S \sim 1$, and $\mathcal{A}_{\rm CP} \sim 0$ one has an easy rule of thumb, $\sigma \approx \sqrt{2/S}$. For $S \sim 100$ this means one expects statistical precision in the neighborhood of ± 0.15. For the $B^\pm \to \psi^{(\prime)}K^\pm$ mode, $\rho = B = 0$ and $S \approx 600$ so $\sigma \approx 0.04$. As will be discussed later, systematic errors are small, and consequently the precision in $\mathcal{A}_{\rm CP}$ measurements can be expected to be dominated by statistical errors for a long time to come. As can be seen in Eq. 2 the quantity in large parentheses is independent of luminosity, so the statistical error is determined by the leading $1/\sqrt{S}$ coefficient. This indicates that the only path to better $\mathcal{A}_{\rm CP}$ measurements will be more data. Improvements to analysis technique that reduce B/S or ρ have less impact.

2 Theoretical Expectations

The existence of a CP violating rate asymmetry depends on having both two different CP nonconserving weak phases and two different CP conserving strong phases. The former may arise from either the Standard Model CKM matrix or from new physics, while the latter may arise from the absorptive part of a penguin diagram or from final state interaction effects. A recent and comprehensive review of CP asymmetries under the assumption of generalized factorization has been published by Ali *et al* [2]. We quote in Table 1 their predictions for the modes examined in this paper. If final state interactions are not neglected, however, strong phases as large as 90° are not ruled out and $|\mathcal{A}_{\rm CP}|$ could reach ~ 0.5 in favorable cases.[3] In this nonperturbative regime numerical predictivity is limited, but a variety of relationships among asymmetries or $f, \bar{f}$ rate differences can be found in the literature.[4]

For the mode $B^\pm \to \psi^{(\prime)}K^\pm$ one expects only a very tiny Standard Model asymmetry arising from interference between the two possible $b \to c\bar{c}s$ diagrams, tree and penguin. The non-zero standard model weak phase entering in this case is due to the phase of V^*_{cs} which is second order in the Wolfenstein parameter λ. Thus the angle probed here is $\chi \equiv Arg(V_{cb}V_{cs}{}^*/V_{tb}V_{ts}{}^*)$, which is typically assumed to be zero when treating the analysis of $\sin 2\beta$ extracted from $B \to \psi K_S$. If one departs from the Standard Model, the CP asymmetry could in principle be tens of percent, depending on the scale of relative strong phases.[5]

3 Analysis

$\mathcal{A}_{\rm CP}$ measurements are intimately associated with the corresponding branching ratio determinations which are discussed in more detail in another talk at this conference.[6] We offer only a brief summary here.

Events are broadly selected using efficient quality cuts on tracks, electromagnetic showers, π^0's, and K_S's. Continuum backgrounds (from $e^+e^- \to q\bar{q}$) are suppressed by requiring the sphericity axis of the signal candidate be uncorrelated with the sphericity axis of the rest of the event ($|\cos\theta_S| < 0.8$ or 0.9 depending on mode). B candidate events are required to have invariant mass and energy within a large window around the corresponding B values. An unbinned maximum likelihood fit based on mass, energy, particle identification, event shape characteristics, and daughter resonance characteristics is then used to determine that yield and charge asymmetry of the signal. Because of the high momentum of the primary daughters, $b \to c$ backgrounds are kinematically excluded, and essentially all background come from continuum production. For the $\mathcal{A}_{\rm CP}$ determination, the important point is that this background is CP symmetric: $\mathcal{A}_{\rm CP} = 0$.

The mode $B^\pm \to \psi^{(\prime)}K^\pm$ is handled somewhat differently because of the near absence of any background in this mode. Standard ψ and ψ' reconstructions are coupled with a charged particle ($K\pi$ identification not required) and the total energy to be within 3σ of its expected value. The procedure yields 654 candidates with an estimated background of only 8 ± 1 events.

4 Systematics

The charge asymmetries measured in this analysis hinge primarily on the properties of high momentum tracks. The charged meson that tags the parent $b/\bar{b}$ flavor has momentum in all cases between 2.3 and 2.8 GeV/c. In independent studies using very large samples of high momentum tracks we have searched for and set stringent limits on the extent of possible charge-correlated bias in the CLEO detector and analysis chain for tracks in the $2-3$GeV range. Based on a sample of 8 million tracks, we find $\mathcal{A}_{\rm CP}$ bias introduced by differences in reconstruction efficiencies for positive and negative high momentum tracks

Table 1. Summary of results. Theory predictions are from Ref. 2, and include only Standard Model perturbative calculations. The 90% CL interval includes statistical and systematic errors added in quadrature.

Mode	Signal Yield	$\mathcal{A}_{CP}$	$\mathcal{A}_{CP}$ 90% CL	$\mathcal{A}_{CP}$ Theory
$K^{\pm}\pi^{\mp}$	80^{+12}_{-11}	-0.04 ± 0.16	$[-0.30, 0.22]$	$(+0.037, +0.106)$
$K^{\pm}\pi^{0}$	$42.1^{+10.9}_{-9.9}$	-0.29 ± 0.23	$[-0.67, 0.09]$	$(+0.026, +0.092)$
$K^0_S\pi^{\pm}$	$25.2^{+6.4}_{-5.6}$	$+0.18 \pm 0.24$	$[-0.22, 0.56]$	$+0.015$
$K^{\pm}\eta'$	100^{+13}_{-12}	$+0.03 \pm 0.12$	$[-0.17, 0.23]$	$(+0.020, +0.061)$
$\omega\pi^{\pm}$	$28.5^{+8.2}_{-7.3}$	-0.34 ± 0.25	$[-0.75, 0.07]$	$(-0.120, +0.024)$
$\psi^{(\prime)}K^{\pm}$	654 ± 26	-0.018 ± 0.039	$[-0.065, 0.035]$	$-$

passing the same track quality requirements as are imposed in this analysis is less than ± 0.002. For kaons this bias is in fact -0.002 due to the different nuclear interaction cross sections of K^+ and K^-. Since this is small compared to the statistical error, we ignore it, except in the case of $J/\psi^{(\prime)}K^{\pm}$, in which case the result has been corrected for this tiny effect. For $K^{\pm}\pi^{\mp}$ combinations where differential charge-correlated efficiencies must also be considered in correlation with K/π flavor, we use 37,000 $D^0 \to K\pi(\pi^0)$ decays and set a corresponding limit on $\mathcal{A}_{CP}$ bias at ± 0.005. These D^0 decays, together with an additional 24,000 $D^{\pm}_{(s)}$ decays, are also used to set a tight upper limit of 0.4 MeV/c on any charge-correlated or charge-strangeness-correlated bias in momentum measurement. The resulting limit on $\mathcal{A}_{CP}$ bias from this source is ± 0.002. We conclude that there is no significant $\mathcal{A}_{CP}$ bias introduced by track reconstruction or selection. We note that for each mode we crosscheck the asymmetry of the background events (normally a fairly large sample) and find results consistent with zero as anticipated. Particle identification is entirely based on dE/dx from the drift chamber, and has been checked for $\mathcal{A}_{CP}$ bias by studying $\sim 100K$ $D \to K\pi(\pi^0)$. The limit on bias from this source is ± 0.01 for $K^{\pm}\pi^{\mp}$ final states, and smaller for all other modes. We therefore assign a conservative systematic error ± 0.02 for all modes except $J/\psi^{(\prime)}K^{\pm}$ for which the systematic error is determined more closely to be ± 0.005.

5 Summary

We see no evidence for CP violation in the six modes analyzed here and set 90% CL intervals that reduce the possible range of $\mathcal{A}_{CP}$ by as much as a factor of four for the charmless modes and a factor of twenty for $\psi^{(\prime)}K$. While the sensitivity is not yet sufficient to probe the rather small $\mathcal{A}_{\rm CP}$ values predicted by factorization models, extremely large $\mathcal{A}_{\rm CP}$ values that might arise if large strong phase differences were available from final state interactions are firmly ruled out. For the cases of $K\pi$ and $\eta' K$, we can exclude $|\mathcal{A}_{\rm CP}|$ greater than 0.30 and 0.23 at 90% CL respectively.

References

1. For a recent review see "CP Violation and Rare Decays of K and B Mesons", A.J. Buras, Lectures given at the 14th Lake Louise Winter Institute, February 1999. hep-ph/9905437
2. "CP violating Asymmetries in Charmless Non-leptonic Decays $B \to PP$, PV, VV in the Factorization Approach", A. Ali, G. Kramer, and C.D. Lü; Phys. Rev. **D59** (1999) 014005 [hep-ph/9805403]
3. "Combining CP Asymmetries in $B \to K\pi$ Decays", M. Gronau and J.L. Rosner, Phys. Rev. **D59** (1999) 113002; [hep-ph/9809384]. It is worth pointing out also that even within the context of factorization where strong phases come strictly from short distance physics, asymmetries as large as 0.5 or greater are predicted in some modes (see previous reference).
4. See reference 5, above; and "The Possibility of Large Direct CP Violation in $B \to K\pi$-like Modes Due to Long Distance Rescattering Effects and Implications for the Angle γ" D. Atwood and A. Soni, Phys. Rev. **D58** (1998) 036005; [hep-ph/9712287]; as well as "Model Independent Analysis of $B \to \pi K$ Decays and Bounds on the Weak Phase γ, M. Neubert, JHEP 9902 (1999) 014; [hep-ph/9812396]
5. G.-H. Wu and A. Soni, hep-ph/9911419.
6. See contribution from J.G. Smith, these proceedings.

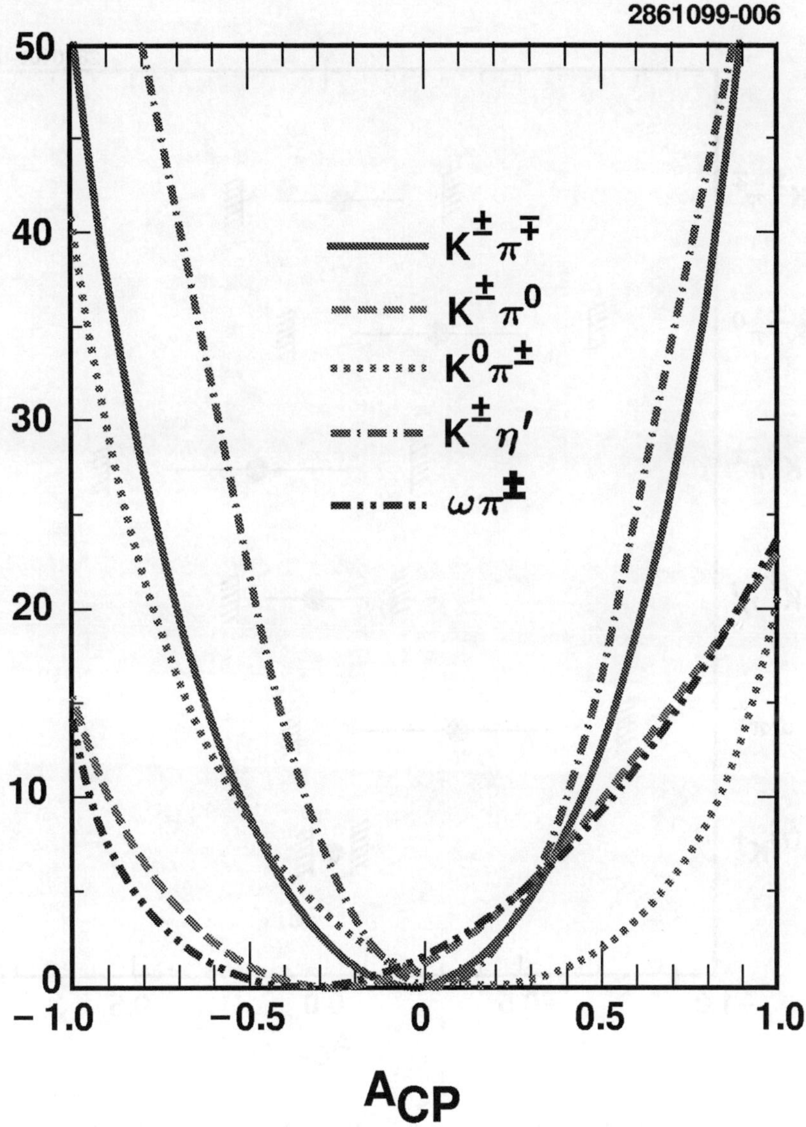

Figure 1. Likelihood functions for the five charmless modes, versus $\mathcal{A}_{CP}$.

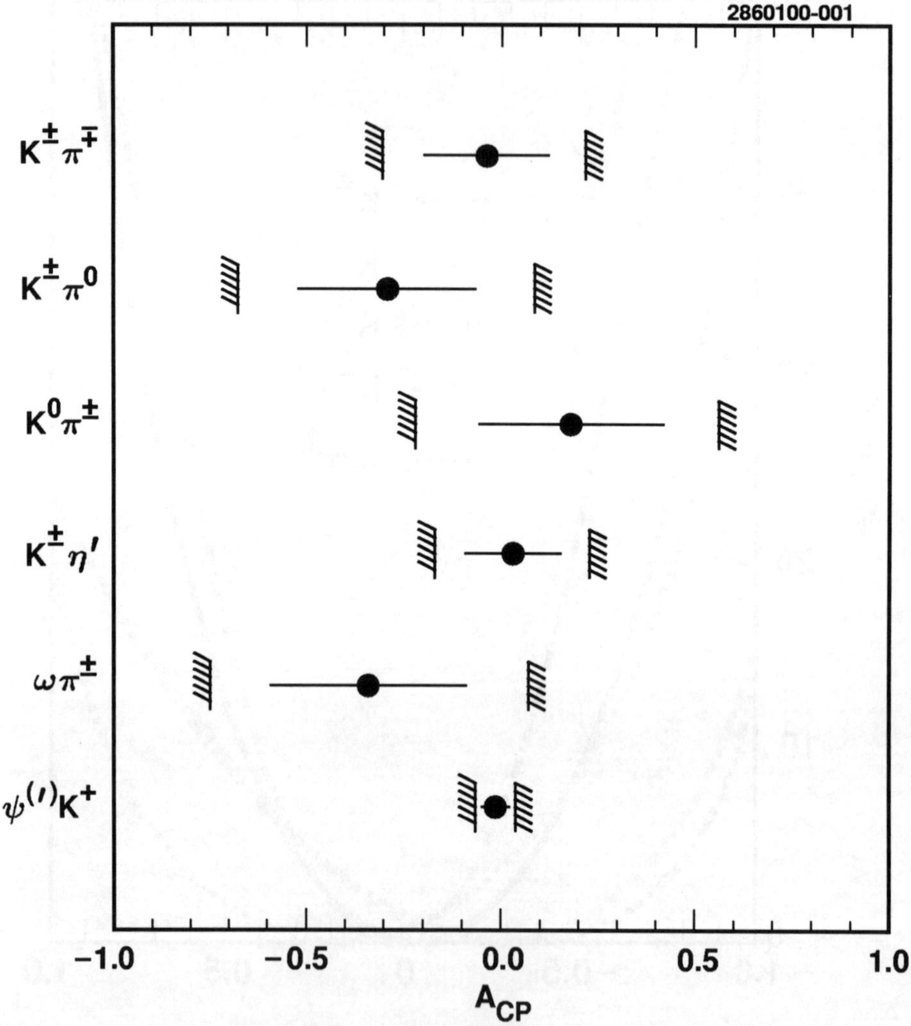

Figure 2. Central values with 1σ errors bars. Hatched exclusion markers indicate range of 90% CL interval.

CALCULATION OF DIRECT CP VIOLATION IN B DECAYS

Cai-Dian Lü
Department of Physics, Hiroshima University, Higashi-Hiroshima 739-8526, Japan
E-mail: lucd@theo.phys.sci.hiroshima-u.ac.jp

Using the generalized factorization approach, we calculate the CP asymmetries of charmless B decays. A number of decay channels has large CP asymmetries, which can be measured in the B factories.

1 Introduction

The sourse of CP violation is one of the unsolved problem in the standard model (SM). The richness of charmless hadronic decays of B meson provides a good place for study of CP violation [1]. When B^0 and $\bar{B}^0$ decay to a common final state f, B^0-$\bar{B}^0$ mixing plays a crucial role. It interferences with the direct CP asymmetries. For other decays, B^0 and $\bar{B}^0$ decay to different final states, for example $B^0 \to K^+\pi^-$, $\bar{B}^0 \to K^-\pi^+$. No mixing is involved here. They are similar to charged $B^\pm$ decays. CP asymmetry has no time dependence. The direct CP asymmetry is important even if for neutral B meson decays.

If there is only one amplitude contributing to the decay, both the strong phase and weak phase can be factored out. we have $\Gamma = \overline{\Gamma}$. So there is no direct CP asymmetry. That is the case for D meson decays and B meson going to heavy meson decays, like $B \to J/\psi K_S$. If there are two amplitudes, the decay rate of Γ and $\overline{\Gamma}$ may be different. If the strong phase difference between the two amplitudes M_1 and M_2 is not zero ($\delta_{12} \neq 0$) and the weak phase difference of the two amplitudes is also non zero ($\phi_{12} \neq 0$), we have $\Gamma \neq \overline{\Gamma}$. The direct CP asymmetry is

$$A_{CP} = \frac{2r \sin\delta_{12} \sin\phi_{12}}{1 + r^2 + 2r\cos\delta_{12}\cos\phi_{12}}, \tag{1}$$

where $r = |M_2|/|M_1|$. A_{CP} depends on $2r/(1+r^2)$, $\sin\delta_{12}$ and $\sin\phi_{12}$. If one of the three parameters is small, then A_{CP} is small. In many decays, we do not have all these conditions, then there is no sizable direct CP violation. However, most charmless decays have large values for $2r/(1+r^2)$, where M_1 is tree amplitude and M_2 is penguin amplitude. Furthermore, the CKM parameters for the tree diagram and penguin diagrams are different providing weak phase differences.

Direct CP asymmetries require an interference between two amplitudes involving both a CKM phase and a final state strong interaction phase difference. The weak phase difference arises from the superposition of penguin contributions and the tree diagrams. The strong-phase difference arises through the perturbative penguin diagrams (hard final state interaction), or non-perturbatively (soft final state interaction). The soft part is not important which is shown in some model calculations [2]. There are also some other contributions, such as annihilation diagram and Soft final state interaction. Mostly, their contributions to branching ratios are small [3]. Probably their contribution to A_{CP} is also small. This is also shown in some model calculations [2].

The method of Isospin or SU(3) symmetry [4] which requires a set of measurements to solve the uncertainties is sometimes difficult for experiments. We estimate these strong phases in specific models, such like the generalized factorization approach, which can be tested by experiments.

2 CP Violation Classification and Formulae

For charged $B^{\pm}$ decays the CP-violating asymmetries are defined as [1]

$$A_{CP} = \frac{\Gamma(B^+ \to f^+) - \Gamma(B^- \to f^-)}{\Gamma(B^+ \to f^+) + \Gamma(B^- \to f^-)}. \tag{2}$$

The charged modes are self-tagged decay channels for experiments. They are easy to be measured. For B^0 decays, more complication is from the $B^0 - \overline{B}^0$ mixing. The CP-asymmetries may be time-dependent, if the final states are the same for B^0 and $\overline{B^0}$

$$A_{CP}(t) = \frac{\Gamma(B^0(t) \to f) - \Gamma(\overline{B}^0(t) \to f)}{\Gamma(B^0(t) \to f) + \Gamma(\overline{B}^0(t) \to f)} \tag{3}$$

$$\tag{4}$$

$$\simeq a_{\epsilon'} \cos(\Delta m t) + a_{\epsilon+\epsilon'} \sin(\Delta m t). \tag{5}$$

Here the direct CP violation parameter $a_{\epsilon'}$ is defined as

$$a_{\epsilon'} = A_{CP}^{dir} = \frac{\Gamma(B^0 \to f) - \Gamma(\bar{B}^0 \to f)}{\Gamma(B^0 \to f) + \Gamma(\bar{B}^0 \to f)}, \tag{6}$$

which is the same defination as the charged B decays. And $a_{\epsilon+\epsilon'}$ is mixing-induced CP violation [1]. In this note we will concentrate on the direct CP asymmetries.

2.1 $b \to s$ ($\bar{b} \to \bar{s}$), transitions

First we parametrize the decay amplitude like this way [1]

$$\mathcal{M} = T\xi_u - P_t\xi_t e^{i\delta_t} - P_c\xi_c e^{i\delta_c} - P_u\xi_u e^{i\delta_u},$$

$$\overline{\mathcal{M}} = T\xi_u^* - P_t\xi_t^* e^{i\delta_t} - P_c\xi_c^* e^{i\delta_c} - P_u\xi_u^* e^{i\delta_u}, \tag{7}$$

where $\xi_i = V_{ib}V_{is}^*$. T and P_i are the tree and i ($i = u, c, t$) quark penguin contributions, respectively. Working in SM, we can use the unitarity relation $\xi_c = -\xi_u - \xi_t$ to simplify the above equation,

$$\begin{aligned} \mathcal{M} &= T\xi_u - P_{tc}\xi_t e^{i\delta_{tc}} - P_{uc}\xi_u e^{i\delta_{uc}}, \\ \overline{\mathcal{M}} &= T\xi_u^* - P_{tc}\xi_t^* e^{i\delta_{tc}} - P_{uc}\xi_u^* e^{i\delta_{uc}}, \end{aligned} \tag{8}$$

where we define

$$\begin{aligned} P_{tc}e^{i\delta_{tc}} &= P_t e^{i\delta_t} - P_c e^{i\delta_c}, \\ P_{uc}e^{i\delta_{uc}} &= P_u e^{i\delta_u} - P_c e^{i\delta_c}. \end{aligned} \tag{9}$$

Thus, the direct CP-violating asymmetry is

$$A_{CP} \equiv a_{\epsilon'} = \left(|\overline{\mathcal{M}}|^2 - |\mathcal{M}|^2\right) / \left(|\mathcal{M}|^2 + |\overline{\mathcal{M}}|^2\right) = A^-/A^+, \tag{10}$$

where

$$A^- = 2TP_{tc}|\xi_u^*\xi_t|\sin\phi_3\sin\delta_{tc} + 2P_{tc}P_{uc}|\xi_u^*\xi_t|\sin\phi_3\sin(\delta_{uc} - \delta_{tc}), \tag{11}$$

$$\begin{aligned} A^+ = {} & (T^2 + P_{uc}^2)|\xi_u|^2 + P_{tc}^2|\xi_t|^2 - 2P_{tc}P_{uc}|\xi_u^*\xi_t|\cos\phi_3\cos(\delta_{uc} - \delta_{tc}) \\ & -2TP_{uc}|\xi_u|^2\cos\delta_{uc} + 2TP_{tc}|\xi_u^*\xi_t|\cos\phi_3\cos\delta_{tc}. \end{aligned} \tag{12}$$

First, we note that $|\xi_u| \ll |\xi_t| \simeq |\xi_c|$, with an upper bound $|\xi_u|/|\xi_t| \leq 0.025$. In some channels, such as $B^+ \to K^+\pi^0$, $K^{*+}\pi^0$, $K^{*+}\rho^0$, $B^0 \to K^+\pi^-$, $K^{*+}\pi^-$, $K^{*+}\rho^-$, $|P_{tc}/T|$ is of $O(0.1)$, $|P_{uc}/P_{tc}| = O(0.3)$. The CP-violating asymmetry in this case is

$$A_{CP} \simeq \frac{2z_{12}\sin\delta_{tc}\sin\phi_3}{1 + 2z_{12}\cos\delta_{tc}\cos\phi_3 + z_{12}^2}, \tag{13}$$

where $z_{12} = |\xi_u/\xi_t| \times T/P_{tc}$. We show the CP asymmetry of $B \to K^{*+}\pi^-$ as an example in Figure 1. It is easy to see that, there may be large CP asymmetries in this decay channel. Besides the CKM parameter ρ and η, the CP asymmetry is also sensitive to the gluon momentum k^2, which is related to the size of strong phase. If k^2 is known, the strong phase is predictable, we may use A_{CP} to determine $\sin\phi_3$. The first 6 channels of Table 1 are this kind of decays. Two of them are reported from CLEO Collaboration with large error-bars[5]. The central values are far away from the theoretical predictions. If more data suport this, we may expect new physics signals here.

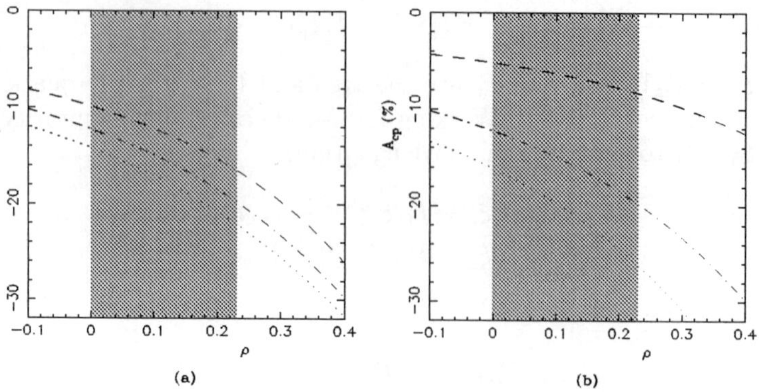

Figure 1: CP-violating Asymmetry A_{CP} in $B^0 \to K^{*+}\pi^-$ decay as a function of the CKM parameter ρ. (a) $k^2 = m_b^2/2$. The dotted, dashed-dotted and dashed curves correspond to the CKM parameter values $\eta = 0.42$, $\eta = 0.34$ and $\eta = 0.26$, respectively. (b) $\eta = 0.34$. The dotted, dashed-dotted and dashed curves correspond to $k^2 = m_b^2/2 + 2$ GeV^2, $k^2 = m_b^2/2$ and $k^2 = m_b^2/2 - 2$ GeV^2, respectively.

There are some decays with vanishing tree contributions ($T = 0$), such as $B^+ \to \pi^+ K_S^0$, $\pi^+ K^{*0}$, $\rho^+ K^{*0}$. Then for these decays, the CP-violating asymmetry is

$$A_{CP} \simeq 2\frac{P_{uc}}{P_{tc}}\left|\frac{\xi_u}{\xi_t}\right| \sin(\delta_{uc} - \delta_{tc}) \sin\phi_3. \tag{14}$$

Without the tree contribution, the suppression due to both P_{uc}/P_{tc} and $|\xi_u/\xi_t|$ is much stronger. The CP-violating asymmetries are only around $-$(1-2)%.

Some estimates of the channel $B^+ \to \pi K_S^0$ show that even including the annihilation and soft final state interaction, the CP asymmetry of this decay is still small [2]. This means that this channel is clean for new physics to show up. In table 1, we can see that CLEO's central value of this decay indicates a large CP asymmetry maybe possible.

2.2 $b \to d$ $(\bar{b} \to \bar{d})$ transitions

Similarly to the $b \to s$ transition, we can define the CP asymmetry as

$$A_{CP} = A^-/A^+, \tag{15}$$

where

$$A^- = -2TP_{tc}|\zeta_u^*\zeta_t| \sin\phi_2 \sin\delta_{tc} - 2P_{tc}P_{uc}|\zeta_u^*\zeta_t| \sin\phi_2 \sin(\delta_{uc} - \delta_{tc}), \tag{16}$$

Table 1: CP-rate asymmetries A_{CP} and branching ratios for some $B \to h_1 h_2$ decays, updated for the central values of the CKM fits $\rho = 0.20$, $\eta = 0.37$ and the factorization model parameters $\xi = 0.5$ and $k^2 = m_b^2/2 \pm 2$ GeV^2.

Decay Modes	A_{CP}-Exp.(%)	A_{CP}(%)	$BR(\times 10^{-6})$
$B^\pm \to K^\pm \pi^0$	29 ± 23	$-7.7^{-2.2}_{+4.0}$	10.0
$B^\pm \to K^{*\pm} \pi^0$	–	$-14.4^{-4.4}_{+8.2}$	4.3
$B^\pm \to K^{*\pm} \rho^0$	–	$-13.5^{-4.0}_{+7.5}$	4.8
$B^0 \to K^+ \pi^-$	4 ± 16	$-8.2^{-2.3}_{+4.3}$	14.0
$B^0 \to K^{*+} \pi^-$	–	$-17.2^{-5.5}_{+9.8}$	6.0
$B^0 \to K^{*+} \rho^-$	–	$-17.2^{-5.5}_{+9.8}$	5.4
$B^\pm \to K_S^0 \pi^\pm$	-18 ± 24	$-1.4^{-0.1}_{+0.1}$	14.0
$B^\pm \to \eta \pi^\pm$	–	$9.3^{+1.9}_{-4.1}$	5.5
$B^\pm \to \eta' \pi^\pm$	–	$9.4^{+2.1}_{-4.5}$	3.7
$B^\pm \to \eta \rho^\pm$	–	$3.1^{+0.7}_{-1.7}$	8.6
$B^\pm \to \eta' \rho^\pm$	–	$3.1^{+0.7}_{-1.8}$	6.2
$B^\pm \to \rho^\pm \omega$	–	$7.0^{+1.5}_{-3.4}$	21.0
$B^\pm \to \eta' K^\pm$	-3 ± 12	$-4.9^{-1.2}_{+2.1}$	23.0
$B^\pm \to \pi^\pm \omega$	34 ± 25	$7.7^{+1.7}_{-3.7}$	9.5

$$A^+ = (T^2 + P_{uc}^2)|\zeta_u|^2 + P_{tc}^2|\zeta_t|^2 - 2P_{tc}P_{uc}|\zeta_u^*\zeta_t| \cos\phi_2 \cos(\delta_{uc} - \delta_{tc}) \\ -2TP_{uc}|\zeta_u|^2 \cos\delta_{uc} + 2TP_{tc}|\zeta_u^*\zeta_t| \cos\phi_2 \cos\delta_{tc}, \tag{17}$$

with $\zeta_i = V_{ib}V_{id}^*$, and again we have used CKM unitarity relation $\zeta_c = -\zeta_t - \zeta_u$.

For the tree-dominated decays, such as $B^+ \to \pi^+\eta^{(\prime)}$, $\rho^+\eta^{(\prime)}$, $\rho^+\omega$, the relation $P_{uc} < P_{tc} \ll T$ holds. The CP asymmetry is

$$A_{CP} \simeq \frac{-2z_1 \sin\delta_{tc} \sin\phi_2}{1 + 2z_1 \cos\delta_{tc} \cos\phi_2}, \tag{18}$$

with $z_1 = |\zeta_t/\zeta_u| \times TP_{tc}/T'^2$, and $T'^2 \equiv T^2 - 2TP_{uc}\cos\delta_{uc}$. The CP asymmetries are proportional to $\sin\phi_2$. They are large enough for the experiments to detect them. The theoretical predictions of these decays are shown in table 1.

For the decays with a vanishing tree contribution ($T = 0$), such as $B^+ \to K^+K_S^0$, $K^+\bar{K}^{*0}$, $K^{*+}\bar{K}^{*0}$, the CP-violating asymmetry is approximately proportional to $\sin\phi_2$ again,

$$A_{CP} = \frac{-2z_3 \sin(\delta_{uc} - \delta_{tc}) \sin\phi_2}{1 - 2z_3 \cos(\delta_{uc} - \delta_{tc}) \cos\phi_2 + z_3^2}, \tag{19}$$

Table 2: Direct CP asymmetries of $B \to \pi\pi$ decays in factorization approach and perturbative QCD approach (Preliminary) [10].

Channel	A_{CP}^{direct} (Factorization)	A_{CP}^{direct} (pQCD)
$B \to \pi^+\pi^-$	$6.9\%^{+2}_{-4}$	-1.4%
$B^+ \to \pi^+\pi^0$	$0.1\%^{+0.1}_{-0.1}$	0.02%
$B^0 \to \pi^0\pi^0$	$-15\%^{-7}_{+14}$	-60%

with $z_3 = |\zeta_u/\zeta_t| \times P_{uc}/P_{tc}$. As the suppressions from $|\zeta_u/\zeta_t|$ and $|P_{uc}/P_{tc}|$ are not very big, the CP-violating asymmetry can again be of order (10-20)%. Unfortunately, these channels have smaller branching ratios [1,3].

More charmless decay channels are discussed in ref.[1]. Some of them are more complicated than the ones we discussed above. There are also some other interesting decays like $B \to K^*\gamma$, $B \to D\pi\ell\nu$, $B \to \pi\pi\ell\nu$ [6], etc. They have small CP asymmetries in SM. They are sensitive to new physics.

3 Models of Calculation

In the Factorization Approach[3,7], the two body B meson decays can be factorized as two products:

$$C_i \langle P_1 P_2 | O_i | B \rangle = C_i \langle P_1 | J_\mu | 0 \rangle \langle P_2 | J^\mu | B \rangle,$$

where C_i is the corresponding Wilson coefficients. The second factor on the right side of the equation is proportional to the meson decay constant. The last term is the corresponding form factors.

The strong-phase differences arise through Bander-Silverman-Soni Mechanism (BSS) [8]. In this picture, the perturbative penguin diagrams involving charm and up quark loops, where the light quarks can be on mass shell, providing the strong phases. They are mostly sensitive to the gluon momentum k^2. For numerical calculations, we use $k^2 = m_b^2/2 \pm 2GeV^2$.

In the perturbative QCD approach (pQCD) [9], we need one hard gluon connecting the spectator quark. Strong phases are from the non-factorizable diagram and annihilation diagram, where the innner quark or gluon propagator can be on mass shell. The pQCD approach is based on factorization, and goes one step further. In pQCD, we can calculate annihilation diagrams and also the non-factorizable contributions. The k^2 of gluon is well defined in this approach. We have calculated the $B \to \pi\pi$ decays in this approach, and the results compared with the factorization approach in Table 2.

4 Summary

The recently measured direct CP asymmetries for $B^0 \to K^+\pi^-$, $B^+ \to K^+\eta'$, $B^0 \to K^+\pi^-$, $B^+ \to \pi^+K^0$, and $B^+ \to \omega\pi^+$ are encouraging news for direct CP violation in B decays, although the signal is not significantly excess background.

CP-asymmetries of $A_{CP}(K^\pm\eta')$, $A_{CP}(\pi^\pm K^0_S)$ and $A_{CP}(\rho^\pm \overset{(-)}{K}{}^{*0})$ are small, but stable against variation in N_c, k^2 and μ. CP-asymmetries well over 10% in these decay modes will be a sign of new physics. The decay channels of $B \to K^{*\pm}\pi^\mp$, $K^{*\pm}\pi^0$, $K^{*\pm}\eta$, $K^{*\pm}\eta'$, $K^{*\pm}\rho^\mp$ and $K^{*\pm}\rho^0$, have measurably large CP-violating asymmetries. A good measurement of the CP-asymmetry in any one of these decays could be used to determine k^2. Such that the theoretical predictions of all other channels make sense. We also hope that the perturbative QCD approach could solve the remaining uncertainties in the factorization approach. With the two B factories and other hadronic machines, a number of decays is going to be measured soon.

Acknowledgments

The author thanks A. Ali and G. Kramer for collaboration on the main topic discussed in this report. We thank the organizer H.Y. Cheng and W.S. Hou for a fruitful BCP3 conference. We acknowledge the Grant-in-Aid for Scientific Research on Priority Areas (Physics of CP violation) and JSPS for support.

References

1. A. Ali, G. Kramer, C.D. Lü, Phys. Rev. D59, 014005 (1999).
2. H.Y. Jin, hep-ph/9805235; D.S. Du , X.Q. Li, Z.T. Wei, B.S. Zou, Eur. Phys. J. A4, 91 (1999); M. Neubert, Phys. Lett. B424, 152 (1998); A,N. Kamal, Phys. Rev. D60, 094018 (1999).
3. A. Ali, G. Kramer, C.D. Lü, Phys. Rev. D58, 094009 (1998); C.D. Lü, Nucl. Phys. Proc. Suppl. 74, 227 (1999).
4. M. Gronau, in these proceedings.
5. S. Chen et al., CLEO Collaboration, hep-ex/0001009.
6. C.S. Kim, J. Lee, W. Nagung, Phys. Rev. D60, 094019 (1999); Phys. Rev. D59, 114006 (1999); Phys. Rev. D59, 114005 (1999).
7. H.Y. Cheng, in these proceedings, hep-ph/9912372.
8. M. Bander, D. Silverman, A. Soni, Phys. Rev. Lett. 43, 242 (1979).
9. H.N. Li, in these proceedings.
10. C.D. Lü, K. Ukai, M.Z. Yang, in progress.

PQCD TREATMENT OF EXCLUSIVE B DECAYS

Hsiang-nan Li
Department of Physics, National Cheng-Kung University,
Tainan, Taiwan 701, Republic of China
E-mail: hnli@mail.ncku.edu.tw

We review perturbative QCD factorization theorem for exclusive B meson decays, and compute the branching ratios $B(B \to K\pi)$ using this approach. Tree, penguin and annihilation contributions, including both factorizable and nonfactorizable ones, are expressed as the convolutions of hard b quark decay amplitudes with universal meson wave functions. Strong phases arise from non-pinched singularities of quark and gluon propagators in nonfactorizable and annihilation hard amplitudes, and those from the BSS mechanism are shown to be small. We argue that final-state-interaction effects in B meson decays into two light mesons is negligible. Comparing our predictions with the CLEO data of the ratio $R = B(B_d^0 \to K^\pm\pi^\mp)/B(B^\pm \to K^0\pi^\pm)$ and of CP asymmetries, we find that the uncertainty of current data is too large to give a constraint of ϕ_3.

1 Introduction

Recently, perturbative QCD (PQCD) factorization theorem for exclusive heavy-meson decays has been developed, and applied to the semileptonic $B \to D^{(*)}(\pi)l\bar{\nu}$ decays[1,2], the nonleptonic $B \to D^{(*)}\pi(\rho)$ decays[3,4] and the penguin-induced radiative $B \to K^*\gamma$ decays[5,?]. The predictions are found to be well consistent with experimental data. The PQCD formalism has been extended to more challenging charmless decays such as $B \to \pi K(\pi)$[7]. The conventional approach to exclusive nonleptonic B meson decays relies on the factorization hypothesis[8]. Though analyses are simplified under this hypothesis, evaluations of many important ingredients, such as tree and penguin (including electroweak penguin) contributions, strong phases are not available. Final-state-interaction (FSI) effects are assumed to be absent. It will be shown that the above difficulties can be resolved in the PQCD approach.

Exclusive B meson decays involve three scale: the W boson mass M_W, at which the matching conditions of the effective weak Hamiltonian to the full Hamiltonian are defined, the typical scale t of a hard amplitude, which reflects the specific dynamics of a decay mode, and the factorization scale $1/b$, with b being the conjugate variable of parton transverse momenta. The dynamics below $1/b$ is regarded as being completely nonperturbative, and parametrized into meson wave functions $\phi(x, b)$, x being the momentum fraction. Above the factorization scale, perturbation theory is reliable, and radiative corrections produce two types of large logarithms: $\ln(M_W/t)$ and $\ln(tb)$. The former are summed by renormalization-group (RG) equations to give the evolution from M_W down to t described by the Wilson coefficients $c(t)$, while the latter are summed to give the evolution from t to $1/b$. With all the large logarithms organized, the remaining finite contributions are absorbed into the hard b quark decay amplitude $H(t)$.

Recently, Beneke *et al.* also proposed to evaluate nonfactorizable contributions to charmless B meson decays in the PQCD framework [9]. However, the momentum of the light spectator quark in the B meson has been ignored in their approach, such that quark propagators in hard decay amplitudes always remain time-like. With this approximation, leading-order information of strong phases was lost. Strong phases then arise from diagrams of the penguin topology, which, as stated above, are small compared to those from the nonfactorizable and annihilation diagrams. On the other hand, Sudakov resummation of large logarithmic corrections was not taken into account. It is then expected that higher-order corrections will be large and spoil the perturbative expansion. It has been shown [10] that the PQCD formalism without including Sudakov suppression is not applicable to exclusive processes for energy scale below 10 GeV.

B fatories at KEK and SLAC are taking data to probe the origin of CP violation. Within the KM ansatz CP violation is organized in the form of a unitarity triangle. The angle ϕ_1 can be extracted from the CP asymmetry in the $B \to J/\psi K_S$ decays, which arises from the B-$\bar{B}$ mixing. Due to similar mechanism of CP asymmetry, the decays $B \to \pi^+\pi^-$ are appropriate for the extraction of the angle ϕ_2. However, these modes contain penguin contributions such that the extraction suffers large uncertainty. Additional measurements of the decays $B^\pm \to \pi^\pm\pi^0$ and $B \to \pi^0\pi^0$ and the use of isospin symmetry may resolve the uncertainties [11]. It has been proposed that the angle ϕ_3 can be determined from the decays $B \to K\pi$, $\pi\pi$ [12,13,14,15]. The contributions to these mode involve inteference between penguin and tree amplitudes, and relevant strong phases have been formulated in terms of several independent parameters. However, without explicit calculation of these strong phases, it is still difficult to extract ϕ_3 from the currently available $B \to K\pi$ data [16].

In this talk we shall present the PQCD predictions of $B \to K\pi$ decays. Contributions from various topologies, such as tree, penguin and annihilation, including both factorizable and nonfactorizable contributions, can all be computed. That is, the factorization hypothesis is in fact not necessary. In our approach strong phases arise from non-pinched singularities of quark and gluon propagators in nonfactorizable and annihilation diagrams. Strong phases from the Bander-Silverman-Soni (BSS) mechanism [17], which is a source of strong phases in the conventional factorization approach, are of next-to-leading order and negligible.

Two-body FSI effects have been studied in various ways [18]. It was found that these effects enhance the CP asymmetry in the $B^+ \to K^0\pi^+$ mode upto order (10 - 20) %. However, Kamal has pointed out that the large CP asymmetry is due to an overestimation of FSI effects by a factor of 20 [19]. For a critic assessment of the analyses of FSI effects in the literature, refer to [19]. We adopt this viewpoint, and regard that the charge exchange in the rescattering $B^+ \to K^+\pi^0 \to K^0\pi^+$ occurs only through short-distance quark-pair annihilation, which is of higher-order and will be excluded.

As an application we evaluate the ratio R and the CP asymmetries

$$R = \frac{B(B_d^0 \to K^\pm\pi^\mp)}{B(B^\pm \to K^0\pi^\pm)}, \tag{1}$$

$$A_{CP}^0 = \frac{B(\bar{B}_d^0 \to K^-\pi^+) - B(B_d^0 \to K^+\pi^-)}{B(\bar{B}_d^0 \to K^-\pi^+) + B(B_d^0 \to K^+\pi^-)}, \tag{2}$$

$$A_{CP}^c = \frac{B(B^- \to K^0\pi^-) - B(B^+ \to K^0\pi^+)}{B(B^- \to K^0\pi^-) + B(B^+ \to K^0\pi^+)}, \tag{3}$$

using PQCD factorization theorem. In the above expression $B(B_d^0 \to K^\pm\pi^\mp)$ represents the CP average of the branching ratios $B(B_d^0 \to K^+\pi^-)$ and $B(\bar{B}_d^0 \to K^-\pi^+)$, and the definition of $B(B^\pm \to K^0\pi^\pm)$ is similar. It will be shown that the uncertainty of the data of R, A_{CP}^0 and A_{CP}^c [20],

$$R = 1.03 \pm 0.31\,, \quad A_{CP}^0 = -0.04 \pm 0.16\,, \quad A_{CP}^c = 0.17 \pm 0.24\,, \tag{4}$$

is still too large to provide useful information of ϕ_3. By fitting our predictions to the central values the CLEO data of R, we extract $\phi_3 \sim 90^o$.

2 Factorization Formulas

The decay rates of $B^\pm \to K^0\pi^\pm$ have the expressions

$$\Gamma = \frac{G_F^2 M_B^3}{128\pi}|\mathcal{A}|^2\,. \tag{5}$$

The decay amplitudes $\mathcal{A}^+$ and $\mathcal{A}^-$, corresponding to $B^+ \to K^0\pi^+$ and $B^- \to K^0\pi^-$, respectively, are given by

$$\mathcal{A}^+ = f_K V_t^* F_e^P + V_t^* \mathcal{M}_e^P + f_B V_t^* F_a^P + V_t^* \mathcal{M}_a^P - f_B V_u^* F_a - V_u^* \mathcal{M}_a\,, \tag{6}$$

$$\mathcal{A}^- = f_K V_t F_e^P + V_t \mathcal{M}_e^P + f_B V_t F_a^P + V_t \mathcal{M}_a^P - f_B V_u F_a - V_u \mathcal{M}_a\,, \tag{7}$$

with the B meson decay constant f_B, the kaon decay constant f_K and the number of colors $N_c = 3$. The notations F represent factorizable contributions (form factors), and $\mathcal{M}$ represent nonfactorizable (color suppressed) contributions. The indices a, e and P denote the annihilation, tree and penguin topologies, respectively. F_a and $\mathcal{M}_a$ are from the operators $O_{1,2}$ in the effective Hamiltonian for the $b \to s$ transition.

The decay rates of $B_d^0 \to K^\pm\pi^\mp$ have the similar expressions can be computed from Eq. (5) with amplitudes

$$\mathcal{A} = f_K V_t^* F_e^P + V_t^* \mathcal{M}_e^P + f_B V_t^* F_a^P + V_t^* \mathcal{M}_a^P - f_K V_u^* F_e - V_u^* \mathcal{M}_e \,, \quad (8)$$
$$\bar{\mathcal{A}} = f_K V_t F_e^P + V_t \mathcal{M}_e^P + f_B V_t F_a^P + V_t \mathcal{M}_a^P - f_K V_u F_e - V_u \mathcal{M}_e \,. \quad (9)$$

The notations are similar to those in Eqs. (6) and (7). F_e and $\mathcal{M}_e$ are from the operators $O_{1,2}$.

Basically, one needs to derive only the factorizaton formulas for the tree and annihilation topologies. Wilson coefficients corresponding to different operators are then inserted into the factorization formulas. The factorization formulas associated with the $B \to \pi$ form factors are written as

$$F_e^P = F_{e4}^P + F_{e6}^P \,,$$
$$\begin{aligned} F_{e4}^P = {} & 16\pi C_F M_B^2 \int_0^1 dx_1 dx_3 \int_0^\infty b_1 db_1 b_3 db_3 \alpha_s(t_e) a_4(t_e) \exp[-S_B(t_e) - S_\pi(t_e)] \\ & \times \phi_B(x_1, b_1)\{[(1+x_3)\phi_\pi(x_3) + r_\pi(1-2x_3)\phi_\pi'(x_3)]\, h_e(x_1, x_3, b_1, b_3) \\ & + 2r_\pi \phi_\pi'(x_3) h_e(x_3, x_1, b_3, b_1)\} \,, \end{aligned} \quad (10)$$
$$\begin{aligned} F_{e6}^P = {} & 16\pi C_F M_B^2 \int_0^1 dx_1 dx_3 \int_0^\infty b_1 db_1 b_3 db_3 \alpha_s(t_e) a_6(t_e) \exp[-S_B(t_e) - S_\pi(t_e)] \\ & \times \phi_B(x_1, b_1) r_K \{[2\phi_\pi(x_3) + 2r_\pi(2+x_3)\phi_\pi'(x_3)]\, h_e(x_1, x_3, b_1, b_3) \\ & + [2x_1\phi_\pi(x_3) + 4r_\pi(1-x_1)\phi_\pi'(x_3)]\, h_e(x_3, x_1, b_3, b_1)\} \,, \end{aligned} \quad (11)$$
$$F_a^P = F_{a4}^P + F_{a6}^P \,,$$
$$\begin{aligned} F_{a4}^P = {} & -16\pi C_F M_B^2 \int_0^1 dx_2 dx_3 \int_0^\infty b_2 db_2 b_3 db_3 \alpha_s(t_a) a_4(t_a) \exp[-S_K(t_a) - S_\pi(t_a)] \\ & \times \{[x_3\phi_K(x_2)\phi_\pi(x_3) + 2r_\pi r_K(1+x_3)\phi_K'(x_2)\phi_\pi'(x_3)]\, h_a(x_2, x_3, b_2, b_3) \\ & - [x_2\phi_K(x_2)\phi_\pi(x_3) + 2r_\pi r_K(1+x_2)\phi_K'(x_2)\phi_\pi'(x_3)]\, h_a(x_3, x_2, b_3, b_2)\} \,, \end{aligned} \quad (12)$$
$$\begin{aligned} F_{a6}^P = {} & 16\pi C_F M_B^2 \int_0^1 dx_2 dx_3 \int_0^\infty b_2 db_2 b_3 db_3 \alpha_s(t_a) a_6(t_a) \exp[-S_K(t_a) - S_\pi(t_a)] \\ & \times \{[2r_\pi x_3\phi_K(x_2)\phi_\pi'(x_3) + 4r_K\phi_K'(x_2)\phi_\pi(x_3)]\, h_a(x_2, x_3, b_2, b_3) \\ & + [4r_\pi\phi_K(x_2)\phi_\pi'(x_3) + 2r_K x_2\phi_K'(x_2)\phi_\pi(x_3)]\, h_a(x_3, x_2, b_3, b_2)\} \,, \end{aligned} \quad (13)$$

The expression of F_e (F_a) for $O_{1,2}$ contribution is the same as F^P_{e4} (F^P_{a4}) but with the Wilson coefficients $a_1(t_e)$ ($a_1(t_a)$). The functions h, obtained from the Fourier transformation of the lowest-order H, and the hard scales t are chosen as the maximum of the energy scales involved in b quark decay amplitudes. The Sudakov factor suppresses long-distance contributions from the large b region, and improves the applicability of PQCD to B meson decays. For the nonfactorizable amplitudes, we refer readers to [7].

In the above expressions the Wilson coefficients are defined by

$$
\begin{aligned}
a_1 &= C_2 + \frac{C_1}{N_c} , \\
a_4 &= C_4 + \frac{C_3}{N_c} + \frac{3}{2} e_q \left(C_{10} + \frac{C_9}{N_c} \right) , \\
a_6 &= C_6 + \frac{C_5}{N_c} + \frac{3}{2} e_q \left(C_8 + \frac{C_7}{N_c} \right) .
\end{aligned} \tag{14}
$$

The factors r_π and r_K,

$$
r_\pi = \frac{m_0(\pi)}{M_B} , \qquad m_0(\pi) = \frac{M_\pi^2}{m_u + m_d} , \tag{15}
$$

$$
r_K = \frac{m_0(K)}{M_B} , \qquad m_0(K) = \frac{M_K^2}{m_s + m_d} , \tag{16}
$$

with m_u, m_d, and m_s being the masses of the u, d and s quarks, respectively, are associated with the normalizations of the pseudoscalar wave functions ϕ'. The pseudovector and pseudoscalar pion wave functions ϕ_π and ϕ'_π satisfy the normalization

$$
\int_0^1 dx \phi_\pi(x) = \int_0^1 dx \phi'_\pi(x) = \frac{f_\pi}{2\sqrt{2N_c}} , \tag{17}
$$

with the pion decay constant $f_{K(\pi)}$. The kaon wave functions ϕ_K and ϕ'_K possess similar normalizations with the d quark, m_d, M_π and f_π replaced by the s quark, m_s, M_K and f_K, respectively. The B meson wave function $\phi_B(x,b)$ reduces to the standard parton model $\phi(x)$, *i.e.*, $\phi_B(x) = \phi_B(x, b=0)$, which satisfies the normalization

$$
\int_0^1 \phi_B(x) dx = \frac{f_B}{2\sqrt{2N_c}} . \tag{18}
$$

3 Numerical Analysis

For the explicit expressions of the meson wave functions and the choices of parameters, refer to [7]. We derive predictions for the ratio R and the CP asymmetries A^0_{CP} and A^c_{CP} defined in Eqs. (1) and (3) for different ϕ_3. The prediction of R increases from 0.7 to 1.3 when ϕ_3 moves from 0 to 180°. Unfortunately, the large uncertainty of the current data does not lead to a constraint of ϕ_3. Comparing our predictions with the central value of the CLEO data of R in Eq. (4), we extract $\phi_3 \sim 90^o$. The data of A_{CP} have also large uncertainties, and do not constraint ϕ_3 either. Our analysis shows that A^c_{CP} is negligible, smaller than 1%, while A^0_{CP} can reach 12%. The obtained branching ratio of each mode for $\phi_3 = 90^o$,

$$\begin{aligned} B(B^+ \to K^0\pi^+) &= 18.41 \times 10^{-6} \, , \\ B(B^- \to K^0\pi^-) &= 18.30 \times 10^{-6} \, , \\ B(B^0_d \to K^+\pi^-) &= 16.35 \times 10^{-6} \, , \\ B(\bar{B}^0_d \to K^-\pi^+) &= 20.60 \times 10^{-6} \, , \end{aligned} \tag{19}$$

are consistent with the CLEO data [20]:

$$\begin{aligned} B(B^\pm \to K^0\pi^\pm) &= (18.2^{+4.6}_{-4.0} \pm 1.6) \times 10^{-6} \, , \\ B(B^0_d \to K^\pm\pi^\mp) &= (17.2^{+2.5}_{-2.4} \pm 1.2) \times 10^{-6} \, . \end{aligned} \tag{20}$$

4 Conclusion

In this talk we have analyzed the $B \to K\pi$ decays using PQCD factorization theorem. In this approach hadronic matrix elements, including factorizable and nonfactorizable, and real and imaginary contributions can be evaluated explicitly. The universal meson wave functions have been determined from available data. The dependences of the ratio R of the neutral B decay branching ratio to the charged B decay branching ratio and of the CP asymmetries on ϕ_3 have been obtained. It has been found that the current data are not precise enough to lead to a strong constraint on ϕ_3. Fitting our predictions to the central value of the data of R, we have extracted $\phi \sim 90^o$. The strong phases arise from non-pinched singularities of quark and gluon propagators in annihilation and nonfactorizable hard amplitudes. It has been argued that strong phases from the BBS mechanism and from long-distance final state interactions are negligible. Our predictions are insensitive to the models of the meson wave functions, and can thus be regarded as being almost model-independent.

Acknowledgments

The work was supported by the National Science Council of R.O.C. under the Grant No. NSC-89-2112-M-006-004.

References

1. H-n. Li and H.L. Yu, *Phys. Rev. Lett.* **74**, 4388 (1995); *Phys. Rev.* D **53**, 2480 (1996).
2. H-n. Li, *Phys. Rev.* D **52**, 3958 (1995).
3. C.H. Chang and H-n. Li, *Phys. Rev.* D **55**, 5577 (1997).
4. T.W. Yeh and H-n. Li, *Phys. Rev.* D **56**, 1615 (1997).
5. H-n. Li and G.L. Lin, *Phys. Rev.* D **.**,
6. LM) H-n. Li and B. Melic, to appear in *Eur. Phys. J. C.*
7. Y.Y. Keum, H-n. Li and A.I. Sanda, in preparation.
8. M. Bauer, B. Stech, and M. Wirbel, *Z. Phys.* C **34**, 103 (1987); *Z. Phys.* C **29**, 637 (1985).
9. M. Beneke, G. Buchalla, M. Neubert, and C.T. Sachrajda, *Phys. Rev. Lett.* **83**, 1914 (1999).
10. N. Isgur and C.H. Llewellyn Smith, *Nucl. Phys.* B **317**, 526 (1989).
11. M. Gronau and D. London, *Phys. Rev. Lett.* **65**, 3381 (1990).
12. M. Gronau, J.L. Rosner, and D. London, *Phys. Rev. Lett.* **73**, 21 (1994); R. Fleischer, *Phys. Lett.* B **365**, 399 (1996).
13. R. Fleischer and T. Mannel, *Phys. Rev.* D **57**, 2752 (1998).
14. M. Neubert and J. Rosner, *Phys. Lett.* B **441**, 403 (1998).
15. A.J. Buras and R. Fleischer, hep-ph/9810260.
16. A.F. Falk, hep-ph/9908520, talk presented at the XIX International Symposium on Lepton and Photon Interactions at High Energies, Stanford University, 1999.
17. M Bander, D. Silverman and A. Soni, *Phys. Rev. Lett.* **43**, 242 (1979). *Z. Phys.* C **66**, 429 (1995).
18. R. Fleischer, *Eur. Phys. J. C* **6**, 451 (1999); M. Neubert and J. Rosner, *Phys. Lett.* B **441**, 403 (1998); D.S. Du, X.Q. Li, Z.T. Wei and B.S. Zou, *Eur. Phys. J. A* **4**, 91 (1999); A.F. Falk, A.L. Kagan, Y. Nir and A.A. Petrov, *Phys. Rev.* D **57**, 4290 (1998).
19. A.N. Kamal, hep-ph/9901342.
20. CLEO Coll., Y. Kwon *et al.*, hep-ex/9908039.
21. W.S. Hou and K.C. Yang, hep-ph/9908202.

FINAL STATE INTERACTION IN HEAVY HADRON DECAY

MAHIKO SUZUKI
Department of Physics and Lawrence Berkeley National Laboratory
University of California, Berkeley, CA 94720, USA
E-mail: msuzuki@lbl.gov

I present a critical account of the final-state interaction (FSI) in two-body B decays from viewpoint of the hadron picture. I emphasize that the phase and the magnitude of decay amplitude are related to each other by a dispersion relation. In a model phase of FSI motivated by experiment, I illustrate how much the magnitude of amplitude can deviate from its factorization value by the FSI.

1 Does the FSI phase diminish with the initial hadron mass ?

For a limited number of decay modes, we can extract the FSI phase directly from experiment. It is large for the D decays [1] and small for the B decays: [2]

$$\Delta\delta = 80^\circ \pm 7^\circ;\ \ D \to K^-\pi^+/\overline{K}^0\pi^+, \tag{1}$$

$$\Delta\delta < \begin{cases} 11^\circ;\ B \to D^+\pi^-/D^0\pi^- \\ 16^\circ;\ B \to D^+\rho^-/D^0\rho^- \\ 29^\circ;\ B \to D^{*+}\pi^-/D^{*0}\pi^-. \end{cases} \tag{2}$$

Though nothing is known for $B \to \pi\pi, K\pi, K\overline{K}$, we might conjecture that the FSI phases diminish as the initial mass increases. Meanwhile the relative FSI phase is known for the ggg and "γ" decay amplitudes of the J/ψ decays: [3] $\Delta\delta \approx 90^\circ$ and 75° for $J/\psi \to 0^-0^-$ and 0^-1^-, respectively. Therefore, if the FSI phases diminish with the increasing initial mass, it should start diminishing somewhere between $m_{J/\psi}$ and m_B or at higher energies.

Can we argue that short-distance (SD) QCD should dominate [4] in the FSI of B meson decays ? I am skeptical. Take the spectator decay $B^+(\overline{b}u) \to K^0(\overline{s}d)\pi^+(\overline{d}u)$. K^0 is formed with energetic $\overline{s}$ and d flying away colinearly, while $\overline{d}$ picks up the soft spectator u to form π^+. The gluons exchanged between $\overline{d}$ and $\overline{s}$ (or d) are hard. But the gluons between the soft spectator u and $\overline{s}$ (or d) are not so hard. The CM energy of the u and $\overline{s}$ (or d) grows only with $\sqrt{m_b}$, not linearly in m_b. By simple kinematics, $m_{\overline{d}u} \approx (\Lambda_{\rm QCD} m_b)^{1/2} \simeq 1.2$ GeV for $E_u \simeq \Lambda_{\rm QCD}$, which is in the middle of the resonance region of the $\overline{s}d$ channel. Then long-distance (LD) interactions cannot be ignored between K^0 and π^+. One may stick to the quark picture by invoking the local quark-hadron duality.[5] While it may be appropriate for inclusive decays, the duality is a dynamical hypothesis yet to be tested in two-body B decays.

2 Why is the FSI complicated in the B decay ?

The FSI phases of $K \to \pi\pi$ and the hyperon decay are equal to the strong interaction phase shifts by the phase theorem, since the final states are the eigenstates of S matrix. In contrast, the FSI phases of the B decay depend on both strong and weak interactions. Take the decay $B \to K\pi$. Since $K\pi$ is made of a large number of eigenchannels a of S matrix, $|K\pi^{out}\rangle = \sum_a c_a|a^{out}\rangle$, the decay amplitude through the operator $\mathcal{O}_i$ acquires an eigenphase δ_a from each eigenchannel as

$$M^{(i)}(B \to K\pi) = \sum_a c_a M_a^{(i)} e^{i\delta_a}, \quad (c_a^* = c_a; M_a^{(i)*} = M_a^{(i)}). \tag{3}$$

The net phase of $M^{(i)}$ depends on $\mathcal{O}_i$ even for the same isospin eigenstate of $K\pi$. The FSI phase has nothing to do with the phase shift of elastic $K\pi$ partial-wave amplitude, $\eta e^{i\delta_{K\pi}} \sin\delta_{K\pi} \equiv (\sum_a |c_a|^2 e^{2i\delta_a} - 1)/2i$ at the B mass.

3 Meson-meson interaction at $E_{\rm cm} \simeq 5$ GeV

To understand the FSI, we need to know how two mesons interact at $E_{\rm cm} = m_B$. The elastic meson-meson scattering amplitude can be written as

$$T(s,t) = i\sigma_{\rm tot} s e^{bt} + \sum_i \gamma_i(t) s^{\alpha_i(t)} + \cdots. \tag{4}$$

The dominant term is the Pomeron exchange, which is purely imaginary and sharply peaked in the forward direction. With a bit of theory,[6] we find $\sigma_{\rm tot} = 15 \sim 20$mb and $b \approx 2.8\text{GeV}^{-2}$ for $K\pi$. These values are much the same for other meson scatterings. The non-Pomeron terms are down by $\sqrt{s}$ at $t = 0$.

The partial-wave projection of $T(s,t)$ gives $a_0(m_B) \approx 0.17i$. It leads to a large inelasticity:

$$\sigma_{\rm inel}/\sigma_{\rm tot} = 1 - {\rm Im}a_0 \approx 0.83. \tag{5}$$

A large portion of inelasticity presumably goes into multibody channels.

What does this imply in the two-body B decays? For the factorization-allowed processes, the intermediate states would be similar to those which appear in elastic meson scattering. Among two-body intermediate states, those connected to the final state by the Pomeron exchange are the most important. For the factorization-forbidden processes in which a quark-pair annihilation must occur, the intermediate states of the Pomeron-exchange type are forbidden. Since the composition of the intermediate states are completely different, the factorization-allowed and forbidden amplitudes generally have quite different FSI phases even for the same isospin state of, say, $K\pi$.

4 Two-body intermediate approximation

In computing the FSI phases, one makes the two-body approximation for the intermediate states.[7] It is a dubious approximation when inelasticity of scattering is high. Getting a number even in this approximation is a daunting task, if one tries to do it right.

One starts with the unitarity sum of the absorptive part of decay amplitude for $B \to n$ (two-body) and keep only the two-body intermediate states:

$$\text{Im}M_n = \sum_{n'} a_0(s)_{nn'} M^*_{n'}, \quad (s = m_B^2; n' = \text{two-body}) \tag{6}$$

where $a_0(s)_{nn'} (= a_0(s)_{n'n})$ is the $J = 0$ amplitude for $n \to n'$. Let us consider only the factorization-allowed processes. The intermediate states connected to n by Pomeron exchange dominate in the sum. Therefor all $a_0(s)_{nn'}$ are purely imaginary. Then we can easily show that M_n are also all purely imaginary. This is a special case where the FSI phases do not depend on $\mathcal{O}_i$.

Phase differences of M_n arise from the non-Pomeron exchanges, ρ, $a_2, \cdots$. Computing the non-Pomeron contributions in Eq.(6) is nearly impossible: First of all, the individual terms in the unitarity sum are not real. Only the sum total is. One has to know the phases of $M_{n'}$. This is circular. To cut the circle, one might approximate $M_{n'}$ with the factorization amplitudes. That is the Born approximation valid only for small phases. After obtaining ImM_n, one has to compute ReM_n by a dispersion integral of ImM_n over all energies. Even if one could compute for all two-body intermediate states, one has no idea as to how much the multibody states modify the answer.

5 Random approximation

Computing the FSI phases is difficult because there are so many intermediate states that we know very little about. Can we take advantage of the presence of many states? A statistical approach or a random approximation comes to our mind. Such an attempt was actually made with a limited success.[6]

Start with the unitarity sum, this time including all intermediate states:

$$\text{Im}M_n = \sum_{n'} a_0(s)_{nn'} M^*_{n'}, \quad (s = m_B^2; n' = \text{all}). \tag{7}$$

Each term in the sum is a product of the pure strong-interaction amplitude and the weak decay amplitude. Let us postulate that the sign of their product is random from one intermediate state to another when there are very many states. This is a dynamical hypothesis. It is not true if SD dominates. Under this postulate one can compute the statistically likely values of the FSI phases as their standard deviations from zero. The likely value $\overline{\delta}$ of the FSI phase depends on the "favoredness" parameter ρ. It parametrizes how much favored or disfavored the decay modes are for a given decay operator: $\rho < 1$ for the favored processes such as the factorization-allowed and color-favored modes, while $\rho < 1$ for the unfavored processes such as factorization-forbidden and/or color-suppressed modes. The relation between $\overline{\delta}_n$ and ρ is given by

$$\tan^2 \overline{\delta}_n = \frac{\tau^2(\rho^2 - \tau^2)}{1 - \rho^2\tau^2}, \quad (\tau < \rho < \frac{1}{\tau}) \tag{8}$$

where τ^2 is a number determined by elasticity of the final-state rescattering ($\tau^2 \approx 0.20$ for $K\pi$). The likely phase $|\overline{\delta}_n|$ of M_n is plotted in Fig.1. It is small for the favored modes but can be large for unfavored modes. For the "typical processes" of $\rho = 1$, the value is $\simeq 20°$. It means that the FSI phases of the typical processes are most likely between $-20°$ and $20°$.

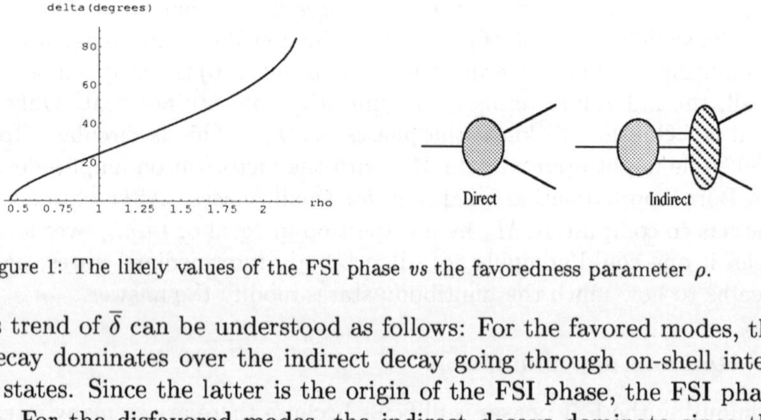

Figure 1: The likely values of the FSI phase *vs* the favoredness parameter ρ.

This trend of $\overline{\delta}$ can be understood as follows: For the favored modes, the direct decay dominates over the indirect decay going through on-shell intermediate states. Since the latter is the origin of the FSI phase, the FSI phase is small. For the disfavored modes, the indirect decays dominate since the direct decay is suppressed. The random approximation quantifies this feature in terms of the favoredness parameter ρ, given elasticity of FSI. A similar qualitative reasoning was made by Rosner[8] in the quark picture.

The limitation of the random approximation is that we cannot predict values of individual phases, but only their likely values.

6 Factorization and enhancement by FSI

6.1 Enhancement by FSI

The FSI not only generates a strong phase but also changes a magnitude of amplitude. Fermi[9] argued in potential scattering that the FSI should modify the amplitude by the wave function at origin $\varphi(0,k)$ of final two particles:

$$M \to \varphi(0,k)M. \tag{9}$$

For slow final particles, the attractive force enhances the decay by pulling them close together while the repulsive force suppresses by pushing them away. Since $\varphi(0,k)$ is equal to the Jost function $1/f(k)^*$, it also proves the phase theorem.

This theory can be extended to relativistic rescattering. In relativistic particle physics, we use the phase-amplitude dispersion relation generically referred to as the Omnès-Mushkelishvili representation[10]:

$$M(s+i\epsilon) = P(s)\exp\left(\frac{1}{\pi}\int_{s_{\rm th}}^{\infty}\frac{\delta(s')}{s'-s-i\epsilon}ds'\right), \tag{10}$$

where s is the B mass squared taken as a dispersion variable, and $P(s)$ is a polynomial arising from zeros of $M(s)$. $P(s)$ is related to the factorization amplitude since it contains no FSI. Therefore the FSI enhancement factor of the factorization amplitude is given by[a]

$$E = \exp\left(\frac{\mathcal{P}}{\pi}\int_{s_{\rm th}}^{\infty}\frac{\delta(s')}{s'-m_B^2}ds'\right). \tag{11}$$

This form of enhancement was actually proposed and studied for elastic rescattering at low energies in the late 1950's.[11] The enhancement E includes the FSI of both SD and LD. While the low energy portion of the phase integral is purely LD for an obvious reason, the high energy portion is not entirely SD.

6.2 Numerical exercise

I first illustrate the enhancement factor with the decay $K \to \pi\pi$. Since the FSI phase is equal to the scattering phase shift below the inelastic threshold, we can evaluate the low-energy portion of the phase integral up to $\simeq 1$ GeV with the $\pi\pi$ phase shifts. The result is $E = 2.2$ and 0.8 for $I = 0$ and $I = 2$ channels, respectively, which amounts to the relative enhancement of $M_{\Delta I=1/2}/M_{\Delta I=3/2} = 2.9$. This is a LD enhancement independent of $\mathcal{O}_i$. Above the inelastic threshold, $\delta(s')$ is not longer equal to the phase shift and depends on decay operators. Therefore the integral above 1 GeV provides an operator-dependent enhancement similar to the SD enhancement of RG, but not identical to it.

---

[a]For $s' > (m_{B^*}+m_\pi)^2$, $\delta(s')$ contains the initial-state interaction or dissociation, part of which enters the form factor of the factorization amplitude. It is understood that the initial-state interaction contributing only to the form factor has been removed from $\delta(s')$ of Eq.(11).

Let us turn to the B decay.

In the two-body intermediate approximation, the FSI phase approaches $\pm 90^\circ$ at high energies. Such an amplitude potentially acquires a dangerously large enhancement or suppression since the phase integral can be large. If the factorization gives a right ballpark value for a decay amplitude, $\delta(s')$ must not stay too large at high energies. In the random approximation, the sign of $\delta(s')$ may fluctuate with s'. In this case, the effect of enhancement and suppression would be much smaller.

Here I present a numerical estimate for $B \to K\pi(\pi\pi)$ with one model FSI phase motivated by experiment. Knowing that the phases are small for $D \to \overline{K}\pi$ but possibly small for $B \to K\pi$, I choose the FSI phase as shown in Fig.2: $\delta(s')$ rises to large values ($\sim 90^\circ$) around 2 GeV and falls linearly to zero at $E_{\rm max} = O(m_B)$.[b] This $\delta(s')$ suppresses the decay amplitude since

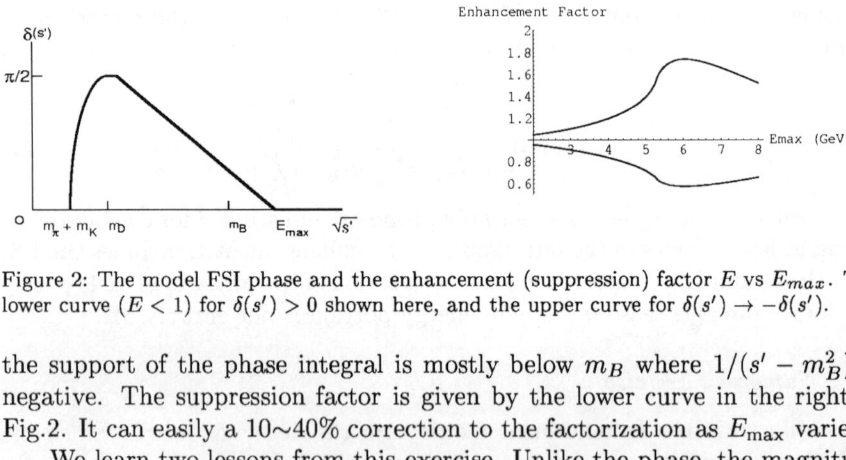

Figure 2: The model FSI phase and the enhancement (suppression) factor E vs E_{max}. The lower curve ($E < 1$) for $\delta(s') > 0$ shown here, and the upper curve for $\delta(s') \to -\delta(s')$.

the support of the phase integral is mostly below m_B where $1/(s' - m_B^2)$ is negative. The suppression factor is given by the lower curve in the right of Fig.2. It can easily a 10~40% correction to the factorization as $E_{\rm max}$ varies.

We learn two lessons from this exercise. Unlike the phase, the magnitude of amplitude receives the FSI effect from all energies. Even if the FSI phase happens to be zero at m_B, the FSI can generate a substantial correction to the factorization amplitude. Next, the low-energy (LD) contribution is power-suppressed in $1/m_B^2$, but not necessarily like $\Lambda_{\rm QCD}^2/m_B^2$. It may be more like m_D^2/m_B^2 or $m_{J/\psi}^2/m_B^2$ ($= 15 \sim 30\%$).

7 Conclusion

If LD physics is important in two-body B decays, computing individual FSI phases is nearly an impossible proposition. What we can predict at best is

[b] To be precise, this $\delta(s')$ includes part of the initial-state interaction above $s' = (m_{B^*} + m_\pi)^2$. See the footnote in the preceding page.

their statistically likely values.

The phase and the magnitude of amplitude are related by analyticity. Multiplying a large FSI phase on a factorization amplitude is very likely inconsistent with analyticity. If an amplitude is significantly enhanced or suppressed relative to its factorization value, it indicates that the FSI phase is large and LD physics is important. When we attempt to extract the CP phases from experiment, we should be apprehensive about the level of accuracy of factorization amplitudes. We must not use calculated values of FSI phases in analysis.

Acknowledgment

This work was supported by Department of Energy under contract DE-AC03-76SF00098 and National Science Foundation under grant PHY-95-14797.

References

1. M. Bishai et al., *Phys. Rev. Lett.* **78**, 3261 (1997).
2. M. Suzuki, *Phys. Rev.* D**58**, 111504 (1998).
3. G. López *et al.*, AIP Conf. Proc. **342** (AIP, New York, 1995), p.441: M. Suzuki, *Phys. Rev.* D**57**, 5717 (1998); *ibid* **60**, 051501 (1999).
4. J. D. Bjorken, *Nucl. Phys.* B(Proc. Suppl.) **11**, 325 (1991): T. W. Yeh and H. n. Li, *Phys. Rev.* D**56**, 1615 (1997): M. Beneke *et al*, *Phys. Rev. Lett.* **83**, 1914 (1999): H. n. Li, in this proceedings.
5. For earlier papers, see M. Bander *et al*, *Phys. Rev. Lett.* **43**, 242 (1979): J. Gérard and W.-S. Hou, *Phys. Rev. Lett.* D**62**, 855; *Phys. Rev.* D**43**, 2909 (1991): H. Simma *et al*, *Nucl. Phys.* B**352**, 367 (1991): N. Deshpande and X.-G. He, *Phys. Lett.* B **336**, 471 (1994).
6. M. Suzuki and L. Wolfenstein, *Phys. Rev.* D**60**, 074019 (1999).
7. H. Zhang, *Phys. Lett.* B**356**, 107 (1995): B. Blok and I. Halperin, *Phys. Lett.* B**385**, 324 (1996): J. F. Donoghue, *et al*, *Phys. Rev. Lett.* **77**, 2178 (1996): B. Blok *et al*, *Phys. Rev. Lett.* **78**, 3999 (1997): G. Nardulli and T. N. Pham, *Phys. Lett.* B**391**, 165 (1997): A. N. Kamal and C. W. Luo, *Phys. Lett.* B**398**, 151 (1997): A. F. Falk *et al*, *Phys. Rev.* D**57**, 4290 (1998).
8. J. L. Rosner, *Phys. Rev.* D**60**, 074029 (1999).
9. E. Fermi, *Suppl. Nuovo Cim.* **2**, 17 (1955). See also *Elementary Particles*, Yale University Press, New Haven, Conn., 1951, pp.58-64.
10. N.I. Muskhelishvili, *Singular Integral Equations* (Noordhoff, Gronigen, 1953), p.204: R. Omnès, *Nuovo Cim.* **8**, 316 (1958).
11. M. Jacob, G. Mahoux, and R. Omnès, *Nuovo Cim.* **23**, 838 (1962).

MEASUREMENT OF ϵ'/ϵ BY THE KTEV EXPERIMENT

AARON ROODMAN

Stanford Linear Accelerator Center, Stanford University, 2575 Sand Hill Rd., Menlo Park, California, USA

E-mail: roodman@slac.stanford.edu

The recent measurement of the Direct CP violating parameter ϵ'/ϵ by the KTeV collaboration at FNAL is presented. A value of $\epsilon'/\epsilon = 28.0 \pm 3.0(\text{stat.}) \pm 2.8(\text{syst.})$ was found using the first 20% of the KTeV 1996 and 1997 data. Special attention is given to the measurement technique and systematic errors. Finally, a comparison of this result with prior experiments and NA-48 is made.

1 KTeV Experiment

1.1 CP Violation

The primary goal of the KTeV[a] experiment at FNAL is a measurement of ϵ'/ϵ to an uncertainty approaching 10^{-4}. Recall that ϵ'/ϵ parametrizes direct CP violation in the neutral Kaon, and can be expressed as:

$$\Re(\epsilon'/\epsilon) = \frac{1}{6}\left\{1 - \frac{\Gamma(K_L \to \pi^0\pi^0)/\Gamma(K_S \to \pi^0\pi^0)}{\Gamma(K_L \to \pi^+\pi^-)/\Gamma(K_S \to \pi^+\pi^-)}\right\} \tag{1}$$

In this case, the CP violation occurs in the interference of the I = 0 and I = 2 decay amplitudes.

1.2 Experimental Method

To measure the ϵ'/ϵ double ratio with a precision of better than 10^{-3} requires exquisite control of systematic effects. To that end, KTeV uses a double-beam technique which actively exploits the double ratio. The use of simultaneous K_S and K_L beams, which alternate once per minute, nearly cancel the systematic effect of detector drifts, inefficiencies, and asymmetries. In addition, KTeV is designed to collect $K \to \pi^0\pi^0$ and $K \to \pi^+\pi^-$ decays simultaneously so that any systematics in K_S production cancel as well.

However, the K_S and K_L decays have a vastly different decay distribution, and to correct for acceptance KTeV relies on a detailed understanding of the spectrometer. It is vitally important for KTeV that the simulation of

[a] a collaboration of Arizona, Chicago, Colorado, Elmhurst, Fermilab, Osaka, Rice, Rutgers, UCLA, UCSD, Virginia, and Wisconsin

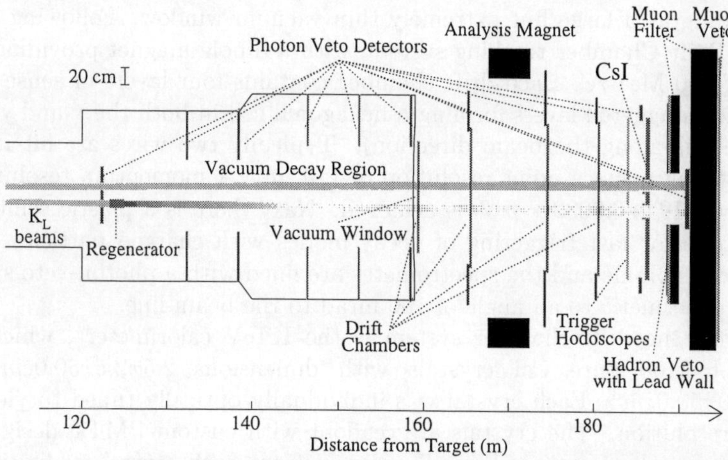

Figure 1. The KTeV Spectrometer

the spectrometer can be tested using the far more plentiful decays $K_L \to \pi^{\pm}e^{\mp}\nu$, $K_L \to \pi^+\pi^-\pi^0$ and $K_L \to \pi^0\pi^0\pi^0$. Finally, the K_S and K_L beams are sufficiently separated in space, by 15 cm, such that they can be easily separated.

1.3 KTeV Beam and Spectrometer

The two KTeV beams are produced by 800 GeV protons on a Be target, and are defined with a pair of projective collimators. Absorbers, made of Be and Pb, improve the Kaon to neutron ratio and remove photons. One of the beams is otherwise undisturbed - this is the vacuum or K_L beam. The intensity of the other beam is reduced, by a factor of roughly 3, with another Be absorber, and is incident on a 2 interaction length active regenerator, consisting of segmented plastic scintillator. The regenerator is located just after the beginning of the decay volume, as shown in Figure 1. Thus, K_S are produced by coherent regeneration, which has the advantage that the K_S beam has an identical beam size and divergence as the K_L beam. In addition, the regenerator phase and amplitude both cancel in the ϵ'/ϵ double ratio. Incoherent regeneration and inelastic scattering from the K_S beam are backgrounds, for the $K \to \pi^0\pi^0$ decays, which must be subtracted.

The KTeV spectrometer[1] consists of a 50 meter vacuum decay volume concluding at a large but extremely thin vacuum window. Following this is a four Drift Chamber tracking system, with a dipole magnet providing a P_T kick of 400 MeV/c. Each drift chamber contains four layers of sense wires, with two staggered layers forming a hexagonal cell in both the x and y views (where z is along the beam direction). Typically two wires are hit in each view, giving a space point resolution of 110 μm. A momentum resolution of $\sigma_p/p = 0.19 \oplus 0.0082 \times p$ % is achieved. Next there is a plastic scintillator hodoscope for fast triggering of decay modes with charged particles. Both the decay volume and the spectrometer are lined with a photon-veto system, which is hermetic to an angle of 100 mrad to the beam-line.

After the drift chamber system is the KTeV calorimeter[2], which consists of 3100 pure CsI crystals with dimensions $2.5x2.5x50.0cm^3$ and $5.0x5.0x50.0cm^3$. Each crystal was individually optically tuned to yield optimal resolution. The crystals are readout with custom PMTs, designed to be linear to better than 1%. The signal from each crystal is digitized at the PMT with a custom circuit utilizing an 8-range auto-ranging current splitter and an 8-bit FADC operating at 53 MHz. The calorimeter was calibrated with 200 million electrons from $K_L \to \pi^\pm e^\mp \nu$ decays; a sample which had a resolution of $\sigma(E/P) = 0.75\%$. The energy resolution could be parametrized as $\sigma(E)/E = 0.02/\sqrt{E} \oplus 0.0045$, and the position resolution as $\sigma_x = 3.9/\sqrt{E} \oplus 0.14$(mm) and $\sigma_x = 6.0/\sqrt{E}$(mm) for the 2.5 cm and 5.0 cm crystals respectively. The KTeV spectrometer ends with a hadron-veto, a beam-veto, and a muon veto.

2 $K \to \pi^+\pi^-$ and $K \to \pi^0\pi^0$

Reconstruction of the $K \to \pi^+\pi^-$ and $K \to \pi^0\pi^0$ decays is performed identically for both the K_S and K_L beams. For the case of the $K \to \pi^+\pi^-$ decays, or charged mode, we reconstruct two tracks in the drift chamber which come from a common vertex. Selection cuts are applied to the invariant mass and to the momentum transverse to the beam direction. Backgrounds from $K_L \to \pi^\pm e^\mp \nu$ and $K_L \to \pi^\pm \mu^\mp \nu$ are kept below a level of 10^{-3} by requiring that the calorimeter energy be inconsistent with an electron and by using the muon-veto system. The roughly 0.07% background from incoherent regeneration is subtracted using a detailed model of regeneration.

The reconstruction of the $K \to \pi^0\pi^0$, or neutral mode, relies entirely on the CsI calorimeter. The four photons are combined such that the inferred decay vertex of the two π^0 is consistent, where the π^0 mass has been assumed. Given the excellent resolution of the calorimeter, the probability of mis-pairing the photons is much less than 1%. Loose selection cuts are applied to the χ^2

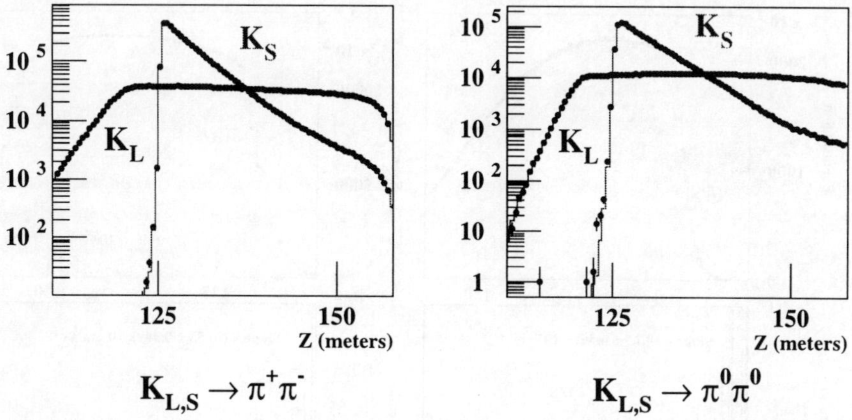

Figure 2. Decay vertex distributions for four Kaon decay modes. The points are data and the histogram is simulation

for the consistency of the two π^0 vertices, to the transverse shape of the photon's energy deposit, and to the distance between photons. In addition, the energy centroid of the photons provides a measure of the position at which the Kaon momentum vector intersects the plane of the calorimeter. For K_L decays and K_S decays from coherent regeneration this position lies inside the calorimeter beam holes. The background, under both beams, from incoherent regeneration is subtracted using an extrapolation in the energy centroid, using the detailed model of regeneration. This background amounts to 0.3% in the K_L beam, and 1.1% in the K_S beam. Background from $K \to \pi^0\pi^0\pi^0$ decays amounts to 0.3% in the K_L beam, and is removed by normalizing to the sidebands in the M_K distribution.

The distribution of the decay vertex for both K_S and K_L and both the $K \to \pi^0\pi^0$ and $K \to \pi^+\pi^-$ decay modes is shown in Figure 2 Clearly the comparison between data and simulation is excellent. Comparisons of this sort for both the 2π decay modes and the $K \to \pi^0\pi^0\pi^0$ and $K_L \to \pi^\pm e^\mp \nu$ modes provide KTeV with a way to quantify the systematic uncertainties in the acceptance correction.

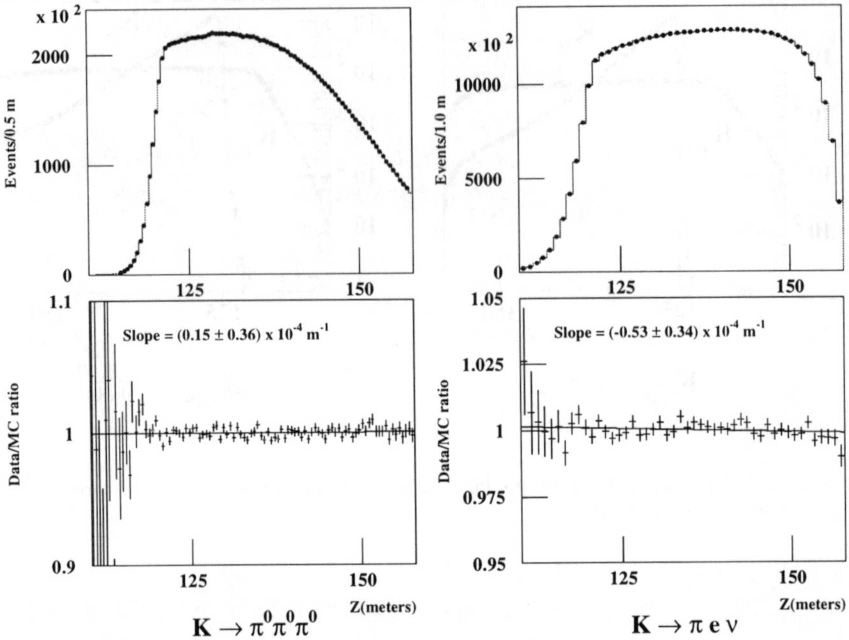

Figure 3. Decay vertex distributions for $K \to \pi^0\pi^0\pi^0$ and $K_L \to \pi^{\pm}e^{\mp}\nu$ decay modes. The points are data and the histogram is simulation

3 Systematics

The acceptance correction for the difference in decay distributions between K_S and K_L is the dominant correction in our measurement, and the slope of the data/simulation comparison of the decay vertex provides a way of quantifying the understanding of our acceptance. The comparison between data and simulation for the $K_L \to \pi^0\pi^0$ is excellent, with the slope in the ratio between data and simulation at $(0.9 \pm 1.0)10^{-4}\ m^{-1}$. The same comparison for $K_L \to \pi^+\pi^-$ shows a small slope, $(-1.7 \pm 0.63)10^{-4}\ m^{-1}$, which becomes the dominant systematic in the measurement. The decay distribution for $K \to \pi^0\pi^0\pi^0$ and $K_L \to \pi^{\pm}e^{\mp}\nu$ is shown in Figure 3, where the superior statistics allows a more accurate comparison of the acceptance. Because of the similarity of the $K_L \to \pi^{\pm}e^{\mp}\nu$ mode with $K \to \pi^+\pi^-$ and $K \to \pi^0\pi^0\pi^0$ with

$K \to \pi^0\pi^0$, any systematic effects in the acceptance for the 2π decays will be reflected in the more copious modes. Conservatively we assign a systematic from acceptance of 0.7×10^{-4} for the $K \to \pi^0\pi^0$ decays using the mean slope plus one sigma from the $K \to \pi^0\pi^0\pi^0$ comparison, and 1.6×10^{-4} for the $K \to \pi^+\pi^-$ decays using the mean slope from the $K \to \pi^+\pi^-$ comparison.

Besides acceptance the other major sources of systematic uncertainty include the $K \to \pi^0\pi^0$ backgrounds, the $K \to \pi^0\pi^0$ energy scale, and selection cuts. The systematic from $K \to \pi^0\pi^0$ backgrounds are due to the regeneration model, which in turn stem largely from a conservative systematic in the charged mode acceptance as a function of P_T. Besides acceptance, the number of decays will vary if we mis-measure the size of the decay region. This is an issue only for the $K \to \pi^0\pi^0$ decays where the decay vertex measurement is proportional to both the energy and position scale in the calorimeter. We adjust the energy scale and assess the systematic uncertainty using the position of the regenerator, where there is a sharp edge in the decay distribution. Another handle on the energy scale systematic is the position of the vacuum window, as measured with hadronic interactions yielding two π^0. The energy scale is adjusted by 0.125% with a systematic uncertainty of 0.073%. The detailed breakdown of systematics can be found in Reference [3].

This result uses data taken in 1996 for $K \to \pi^0\pi^0$ decays, and 1997 for $K \to \pi^+\pi^-$ decays. The 1996 $K \to \pi^+\pi^-$ decays were discarded due to a large inefficiency in the online trigger, which resulted from delayed hits in the drift chambers. In the 1997 data this problem persisted, although the online trigger was modified to eliminate any resulting inefficiency. However, the probability for a delayed hit was greater within the beam than outside it, and this effect required a detailed simulation using a data-derived parametrization. We are confident in the accuracy of this procedure from the agreement between data and simulation in the track occupancy at each of the drift chambers and the decay distribution.

4 Results

The value of ϵ'/ϵ is determined from a fit to the number of decays to each of the four modes, divided into 10 GeV energy bins. In addition to ϵ' the fit parameters also include the regeneration phase and amplitude and Kaon energy spectrum normalizations. The fit was performed blind to the value of ϵ'/ϵ; instead a fit parameter of the form: $\{\epsilon'/\epsilon\}^* = \left\{ \begin{matrix} +1 \\ -1 \end{matrix} \right\} \times \epsilon'/\epsilon + C$ was used, where both the value of C and the choice of $+1$ or -1 was hidden. This procedure minimizes the introduction of any experimenter's bias in the result. The true value of ϵ'/ϵ was not examined until one week prior to the announcement of the result. The adoption of such a blind analysis technique is highly recommended as a way to avoid experimenter's bias.

The value of ϵ'/ϵ was found to be[3]:

$$\Re(\epsilon'/\epsilon) = (28.0 \pm 3.0 \pm 2.8) \times 10^{-4} \tag{2}$$

which points to the presence of a direct CP violating effect at almost 7σ. When combined with the prior measurements from NA-31[4] and E-731[5], as well as the recent measurement from NA-48[6], we find an average of $\epsilon'/\epsilon = 21.2 \pm 2.8$ with a $\chi^2/\text{dof} = 2.8$. Scaling the error with the PDG procedure yields an average of $\epsilon'/\epsilon = 21.2 \pm 4.7$.

The discrepancy between the E-731 result and this KTeV result is often commented upon. However, we should consider that there is a 2.1% probability that of four measurements, with the stated uncertainties, one measurement will lie at 2.3σ or greater from the mean of all measurements.

In conclusion, KTeV has performed the most accurate measurement of ϵ'/ϵ to date. Combined with other measurements, there is now little doubt that direct CP violation has been observed. The remaining 80% of the 1997 data set is presently being analyzed, and KTeV doubled its data sample in a 1999 run, so that more precise measurements are to be expected.

Acknowledgments

This work was supported in part by the U.S. Department of Energy, the National Science Foundation, and the Ministry of Science and Education in Japan.

References

1. K. Arisaka *et.al.*, *KTeV Design Report*, Technical Report FN-580, Fermilab, 1992. and P.S. Shawhan, Ph.D. Thesis (The University of Chicago, August 1999).
2. A. Roodman, in *Proceedings of the Seventh International Conference on Calorimetry in High Energy Physics*, edited by E. Cheu *et.al.* (World Scientific, 1998), p89.
3. A.Alavi-Harati *et.al.*,, *Observation of Direct CP Violation in* $K_{S,L} \to \pi\pi$ *Decays*, *Phys. Rev. Lett.* **83**, 22 (1999).
4. G. D. Barr *et.al.*, *Phys. Lett.* B **317**, 233 (1993).
5. J. R. Patterson *et.al.*, *Phys. Rev. Lett.* **64**, 1491 (1990).
6. NA48 Collaboration (V. Fanti *et.al.*). *Phys. Lett.* B **465**, 335-348 (1999).

MEASUREMENT OF ε'/ε IN THE NA48 EXPERIMENT AT CERN

A.NAPPI

University of Perugia and INFN, Via A. Pascoli, 06100 Perugia, Italy
for the NA48 collaboration: Cagliari, Cambridge, CERN, Dubna, Edinburgh, Ferrara, Firenze, Mainz, Orsay, Perugia, Pisa, Saclay, Siegen, Torino, Vienna, Warsaw

The experiment NA48 at CERN has an ongoing program for a measurement of Re(ε'/ε), the parameter which is used to characterize CP violation in ΔS=1 $K^0 \rightarrow \pi\pi$ amplitudes, with a precision of 2×10^{-4}. The experiment is based on a novel technique to produce concurrent K_S and K_L beams. A first measurement, based on ~ 15% of the statistics available, has recently been published. This paper elaborates on salient points of the technique and of the published measurement. Information on the current status of the experiment will be provided as well.

1 Introduction

CP violation in ΔS=1 $K^0 \rightarrow \pi\pi$ decay amplitudes (*direct*) is expected if CP violation originates[1] from non-trivial phases in the CKM matrix. The effect is characterized by a non zero value of the ε' parameter. ε' is related to ε, which characterizes CP violation in K_0 $\bar{K}_0$ mixing, by the measurement of the double ratio of decay rates

$$R = \frac{\Gamma(K_L^0 \rightarrow \pi^0\pi^0)}{\Gamma(K_L^0 \rightarrow \pi^+\pi^-)} \frac{\Gamma(K_S^0 \rightarrow \pi^+\pi^-)}{\Gamma(K_s^0 \rightarrow \pi^0\pi^0)} = 1 - 6\,\mathrm{Re}\left(\frac{\varepsilon'}{\varepsilon}\right). \quad (1)$$

The first observation[2,3] of an effect, by the NA31 experiment, was not confirmed by E731[4], although the two results were only in marginal disagreement. A new round of measurements is in progress at FNAL and CERN. Both collaborations have recently published results[5,6], based on a sub-sample of the data, confirming the first observation. Other contributions to this conference discuss the interpretation of these results within the standard model[7] or in terms of new physics [8].

This paper reports on the NA48 measurement focusing on selected points of the experimental technique. For a complete account the reader is referred to [6].

2 The detector

The experiment (fig. 1a) uses a drift chamber magnetic spectrometer[9], for the measurement of charged decays, and an electromagnetic calorimeter, for photon and electron detection. The calorimeter[10] is a nearly fully active liquid krypton ionization chamber device, segmented in towers delimited by 40-μ thick longitudinal electrodes. The layout of the electrodes is designed to provide excellent space uniformity and projective geometry. Short pulse shaping time provides the desired rate capability and accurate timing information. A hodoscope made of two planes of

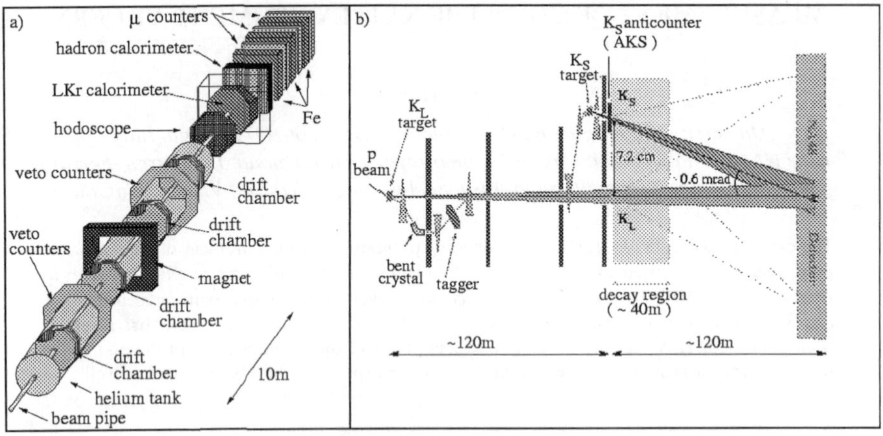

Figure 1. The NA48 apparatus: a) exploded view of the detector, b) schematics of the concurrent beams. The beam intensities on the K_L and K_S target are ~1.5×10^{12} and ~3×10^7 protons per pulse respectively

scintillator strips, 6.5 to 10 cm wide, is used for triggering and time measurement. Other components are a hadron calorimeter, used mainly for triggering purpose, and 3 planes of scintillators, placed after iron absorbers, for muon identification.

High rate capability is pursued in the design of the detectors, trigger and readout. A fully pipelined readout prevents deadtime during the trigger latency (maximum 200μs) and the readout time. Triggers based on a pipelined system, for the neutral channel, and on an array of microprocessors, for the charged one, perform simplified event reconstruction with zero or negligible dead time.

3 Experimental technique

Numbers of events can replace decay widths in (1) if a) detection efficiencies cancel between numerators and denominators, b) neutral and charged decays are collected simultaneously in the same fiducial ranges of energy and decay position, c) background is absent. Systematic errors are associated to departure from these conditions. Salient points of the technique are discussed with reference to this ideal case.

3.1 Sensitivity to detection efficiencies

Two main points are addressed in order to achieve the best possible cancellation.

Sensitivity to variations of efficiency, as a function of time or rate in the detector, is avoided by collecting data for the 4 decay channels at the same time. To this end a new technique has been developed to generate concurrent K_L and K_S beams, converging at a very small angle (0.6 mrad) towards the detector (fig. 1b). After the K_L target, a fraction of the proton beam is redirected to the K_S target, close to the decay region, by channelling in a bent crystal [11] and a magnetic channel.

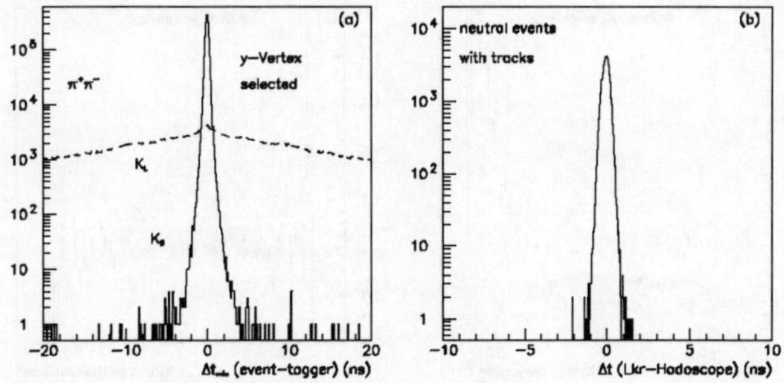

Figure 2. Histograms of (a) the minimum time difference between event and protons in the tagger for K_S and K_L charged decays identified by the vertex position; (b) difference between the "neutral" and "charged" time for neutral decays with conversions or Dalitz decays.

Identification of the beam originating the decay is performed by a time coincidence between the detected decay and a *tagger* detector placed in the proton beam that feeds the K_S target. The tagger[12] is made of thin scintillator foils (200μ in the beam core) arranged along a precisely aligned double ladder (horizontal/vertical). Double pulse separation in the order of 4ns is achieved digitizing the analog pulses of the photomultipliers by 960-MHz FADC's.

Misidentification can occur due to either inefficiency of the tagger (K_S misinterpreted as K_L) or to accidentals (K_L misinterpreted as K_S). Narrow time coincidences are needed to minimize the second effect and therefore good time resolution and absence of tails are required. The time resolution of the tagger itself is 120 ps. The detector provides multiple independent time measurements, in the liquid krypton electromagnetic calorimeter, for neutral decays, and in the hodoscope, for charged decays. This allows to apply consistency requirements, which significantly limit resolution tails. Resolutions better than 300 ps for 20 GeV photons and of the order of 200 ps per charged track are achieved. Redundant timing information is provided by the drift chambers for charged events and by a hodoscope of scintillating fibers ("*neutral*" hodoscope) immersed in the LKr for neutral decays.

The frequency of tagging errors can be directly measured only for charged events, where the vertex position identifies the parent beam (fig. 2a). The fraction of K_S events with no hit in the ±2ns coincidence window is $(1.5\pm0.1)\times10^{-4}$. As indicated by the redundant timing information, ~ 2/3 of this inefficiency is due to the tagger and appears in neutral events as well, with no bias to the double ratio. The difference of inefficiency for charged and neutral events is directly bounded by a comparison of the "neutral" and "charged" time in events with Dalitz decays or photon conversions (fig. 2b). The measured difference is $(0\pm1)\times10^{-4}$. Accidentals occur with a larger frequency (11.19±0.03)%. In first approximation this effect is the same for charged and neutral decays. A small difference is induced by rate dependent inefficiencies and is measured to be $(10\pm5)\times10^{-4}$, by comparing the frequency of hits in the side bands of the time distribution, for $\pi^+\pi^-$ and $\pi^0\pi^0$ events.

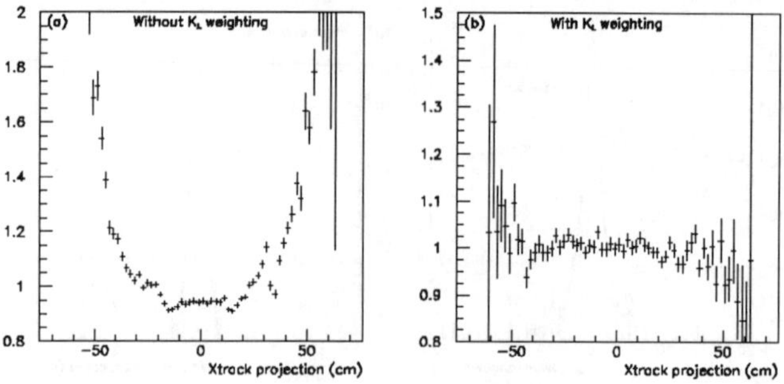

Figure 3. Ratios of the normalized distributions of track coordinates for K_S and K_L charged decays. In (b) lifetime weighting is applied to K_L events. Events in the energy range from 100 to 120 GeV are used.

Since K_L and K_S events are collected at the same time, accidental effects should be the same, except for possible effects of variations in the ratio of the beam intensities, which occur at a level of ~ 10%, and of the geometric distribution of the particles on the detectors (which, however, is very similar, as discussed below) for the two beams. In order to measure the effect, a sample of events has been overlaid with randomly triggered events, thereby artificially doubling the accidental activity. No significant difference in gains and losses are found for K_S and K_L events, but the error on the measured difference is considered a systematic error.

The second point addressed by the experiment concerns efficiencies that are functions of beam parameters, like acceptances. The difference of energy spectra is minimized by choosing a different production angle for the K_S and K_L beam and the double ratio is computed in 5 GeV energy bins. The two beams have very similar spot size at the detector and the main difference in acceptance for K_L and K_S decays is associated to the different distribution of decay lengths. In order to compensate for this effect, K_L events are given a weight proportional to the ratio of the K_S to K_L decay distributions. Acceptance corrections are thus reduced to less than 0.5%, after residual effects, related to beam divergence and finite size of the detector, are minimized by rejecting charged decays with very large momentum asymmetry of the two pions. In addition, the weighting equalizes the distribution of particles on the detector, as illustrated in fig.3, thus avoiding sensitivity to disuniformities in the detector efficiencies. Of course the price of weighting is an increase of the statistical error, since K_L events with large decay times do not contribute to the result.

3.2 Fiducial regions in energy and vertex position

The beginning of the K_S decay region is defined by an anticounter ("AKS" in fig. 1b) rather than by a cut on the reconstructed vertex. This technique avoids effects of resolution in the exponential distribution of decay times, but cannot be used in the more intense K_L beam. Thus agreement of the charged and neutral reconstruction is required for perfect cancellation of the flux normalization. For neutral events the longitudinal position of the vertex is computed, using the K^0 mass constraint, from the transverse positions and energies of the photons, whereas for charged events it is determined geometrically from the drift chambers. The transverse scales are aligned with a precision better than 300μ with electrons from K_{e3} decays. After this is done, the neutral energy scale can be adjusted in such a way that the sharp edge of the vertex distribution created by the AKS, occurs at consistent positions in charged and neutral decays. This technique ensures the agreement of the energy scales as well, since, for charged events, the K^0 momentum is computed from the opening angle and the ratio of the momenta of the two pions, using decay kinematics.

This procedure adjusts the energy scales with a precision of 5×10^{-4}. The effect on R of this uncertainty is minimized by defining the fiducial region in terms of proper time and is determined to be 5×10^{-4}. R also depends on more detailed features of the calorimetric response. Extensive checks are made using the $\pi^0\pi^0$ events themselves, electrons from K_{e3} decays, and π^0's and η's produced, during special runs, by a π^- beam hitting a thin target placed near the beginning of the decay region. The largest effect on R (9×10^{-4}) is associated to possible non-linearity effects within the estimated limit (<0.3% in the 3-100 GeV photon energy range).

3.3 Background

Apart from the tagging effects discussed in 3.1, only K_L events are affected by a relevant background, from 3-body decays. To preserve the efficiency cancellation, only variables that have the same distribution for K_L and K_S signal events are used for background rejection cuts and these are applied to K_S events also. For example, transverse momentum balance in charged decay is computed not with respect to the line joining the decay vertex with the K^0 production target, but to the line joining the target with the extrapolation of the total momentum to the center of the first drift chamber. By extrapolation from control regions in the discriminating variables, the residual background is estimated to be (0.23±0.04)% for charged events (with extrapolation factors determined using identified background events) and (0.08±0.02)% for neutral events (extrapolation factor determined by Monte-Carlo). Lifetime weighting helps reducing the neutral background, since it depletes the downstream part of the decay region, where 3-body misidentified events are moved by the vertex reconstruction algorithm.

4 Data collected during 1997

The experiment has collected the first data useful for the measurement of Re(ε'/ε) during a period of 42 days in 1997 under non-optimal conditions of the detector. The high voltage of the LKr calorimeter was not connected to a 4cm wide vertical region and it was kept at a non-optimum value of 1.5kV, due to faulty blocking capacitors. This produced a ~20% increase in electronic noise and a small space charge effect. In addition, the available charged trigger processors were saturated at the nominal beam rate. As a result of these limitations the beam intensity was kept at ~ 2/3 of its nominal value and the charged trigger was downscaled by a factor 2.

The numbers of selected events, including background, are given in table 1.

Table 1. Statistical samples (1997 data)

$K_L \to \pi^0\pi^0$	$K_S \to \pi^0\pi^0$	$K_L \to \pi^+\pi^-$	$K_S \to \pi^+\pi^-$
489×10^3	975×10^3	$1{,}071 \times 10^3$	$2{,}087 \times 10^3$

The calculation of R requires, in addition to lifetime weighting, corrections, the effects of which are given in table 2. The error associated with each correction is considered part of the systematic error, also when it is of statistical origin because the correction is measured in the data. This is true, in particular, for the first three lines in the table.

Table 2. Corrections applied to R and systematic uncertainties, in 10^{-4} units

Effect	Correction	Error
Tagging inefficiency/accidentals	+ 18	± 11
$\pi^+\pi^-$ trigger efficiency	+ 9	± 23
Accidental effects	– 2	± 14
Acceptance	+ 29	± 12
$\pi^0\pi^0$ background	– 8	± 2
$\pi^+\pi^-$ background	+ 23	± 4
Scattering/regeneration	– 12	± 3
Energy scale and linearity		± 12
Charged vertex		± 5
Total correction	+ 57	± 35

Trigger efficiency, measured using downscaled minimum bias triggers, is an important contribution in this data set. This is due to a large inefficiency (~8.3% in average but decreasing towards the end of the run) found in the charged trigger, for which a few tuning interventions (concerning time alignment and optimization of the processor time budget) occurred at different moments during the run. Although there are no a priory reasons to believe that this inefficiency is different for K_L and K_S events, the double ratio is corrected, in time periods, with the measured inefficiencies.

The final result, computed from the average of the double ratio in energy bins,

$$\mathrm{Re}(\varepsilon'/\varepsilon)=(18.5\pm4.5(\mathrm{stat})\pm5.8(\mathrm{syst}))\times10^{-4}\ ,$$

provides further confirmation of the existence of direct CP violation in K^0 decays.

5 Prospects

The previous result, based on 42 days of data taking, shows the statistical power of the experiment. More data have been collected during 110 days in 1998 and 129 days in 1999, after major repairs and improvements to the detector, which occurred mainly before the 1998 run. In particular, the LKr calorimeter has operated in stable conditions at 3kV, after the repair of the high voltage problems. A carbon fiber beam pipe was installed to reduce the multiplicity associated to accidental particles showering in the pipe. Improvements to the trigger and data acquisition system include the introduction of faster processors for the charged trigger and a new event-building system, based on a farm of PC's interconnected by a fast switch. All these changes have improved the rate capability of the experiment, which could be run at the nominal beam intensity of 1.5×10^{12} protons per pulse, without charged trigger downscaling. Less substantial changes introduced before the 1999 run have improved the efficiency and reliability of data taking. These include a new readout controller for drift chambers, a minor increase of the spill length from 2.36s to 2.58s, and online software improvements for better reliability, real time diagnostics and automatic error recovery.

The analysis of 1998 data is well advanced. The statistics of accepted events is shown in table 3. A statistical error better than 3×10^{-4} is expected and work is in progress to reduce the systematic error to a comparable level.

Table 3. Statistical samples (1998 data)

$K_L\to\pi^0\pi^0$	$K_S\to\pi^0\pi^0$	$K_L\to\pi^+\pi^-$	$K_S\to\pi^+\pi^-$
1.1×10^6	2.1×10^6	4.6×10^6	8.6×10^6

At the time of writing this contribution, the offline processing of the 1999 data has been completed. Although data quality checks are still in progress, a further factor 2 in statistics with respect to the 1998 sample can be expected.

References

1. M. Kobayashi, T. Maskawa, Prog. Theor. Phys. **49** (1973) 652
2. H. Burkhardt et al., Phys.Lett. **B206** (1988) 169
3. G.D. Barr et al., Phys. Lett. **B317** (1993) 233
4. L.K. Gibbons et al., Phys Rev. **D55** (1997) 6625
5. A. Alavi-Harati, Phys. Rev. Lett. **83** (1999) 22
6. V.Fanti et al., Phys. Lett. **B465** (1999) 335
7. S. Bertolini, these proceedings ; A.P. Soldan, these proceedings
8. A. Masiero, these proceedings; P. Ko, these proceedings
9. D.Bédérède, Nucl. Instr. and Methods **A367** (1995) 88 ; I.Augustin, Nucl. Instr. and Methods **A403** (1998) 472
10. see, e.g., M.Martini for the NA48 collaboration, Performance of the NA48 liquid krypton calorimeter for the ε'/ε measurement at the CERN SPS, in *Proceedings of the 7th International Conference on calorimetry in high energy physics* (Tucson, 1997), ed. E.Chau et al., Singapore, World Scientific 1998
11. N.Doble et al., Nucl. Instr. and Methods **B119** (1996) 181
12. P.Grafstrom et al., Nucl. Instr. and Methods **A344** (1994) 487; H.Bergauer et al., Nucl. Instr. and Methods **A419** (1998) 623

KLOE at DAΦNE
Present Status and Progress

M. Incagli
INFN, Pisa, Italy - for the KLOE collaboration [a]

The KLOE experiment at the Frascati ϕ–factory DAΦNE is fully operational and has started to collect its first data in 1999. More than 2 pb^{-1} have been collected and have been used to study the detector performances and to fully test the reconstruction and analysis programs. In the near future DAΦNE should provide KLOE with 100 pb^{-1} of data, which will allow the measurement of $\Re(\epsilon'/\epsilon)$ to a statistical accuracy 10^{-3}.

1 The DAΦNE machine

DAΦNE [1] is an e^+e^- collider optimized to work at the ϕ mass with a target luminosity L= 5×10^{32} $cm^{-2}s^{-1}$. The strategy used to reach the target luminosity is to obtain the same single bunch luminosity achieved at VEPP-2M and to increase the number of bunches (up to 120). To suppress multi-bunches instabilities the electrons and the positrons circulate in two separate storage rings and collide at two interaction points with an horizontal crossing angle of 12.5 mrad to minimize the effect of parasitic collisions. In one of the two interaction regions, equipped with two triplets of low-β quadrupoles, KLOE is installed.

In March 1999, the first test run with the KLOE experiment began. The machine has been operated in the single and multi bunch mode and an integrated luminosity of $\sim100nb^{-1}$ per day has been reached, with a peak luminosity of $\sim3\times10^{30}$ $cm^{-2}s^{-1}$.

[a] M. Adinolfi, A. Aloisio, F. Ambrosino, A. Andryakov, A. Antonelli, C. Bacci, A. Bankamp, G. Barbiellini, G. Bencivenni, S. Bertolucci, C. Bini, C. Bloise, V. Bocci, F. Bossi, P. Branchini, G. Cabibbo, R. Caloi, P. Campana, G. Capon, G. Carboni, A. Cardini, G. Cataldi, F. Ceradini, F. Cervelli, F. Cevenini, G. Chiefari, P. Ciambrone, S. Conticelli, E. De Lucia, G. De Robertis, P. De Simone, G. De Zorzi, S. Dell'Agnello, A. Denig, A. Di Domenico, S. Di Falco, A. Doria, E. Drago, O. Erriquez, A. Farilla, G. Felici, A. Ferrari, M. L. Ferrer, G. Finocchiaro, C. Forti, G. Foti, A. Franceschi, P. Franzini, M. L. Gao, P. Gauzzi, S. Giovannella, V. Golovatyuk, E. Gorini, F. Grancagnolo, E. Graziani, P. Guarnaccia, X. Huang, M. Incagli, L. Ingrosso, Y. Y. Jiang, W. Kim, W. Kluge, V. Kulikov, F. Lacava, G. Lanfranchi, J. Lee-Franzini, T. Lomtadze, C. Luisi, C. S. Mao, A. Martini, W. Mei, L. Merola, R. Messi, S. Miscetti, S. Moccia, M. Moulson, S. Mueller, F. Murtas, M. Napolitano, A. Nedosekin, L. Pacciani, P. Pagès, M. Palutan, L. Paoluzi, E. Pasqualucci, L. Passalacqua, A. Passeri, V. Patera, E. Petrolo, D. Picca, G. Pirozzi, L. Pontecorvo, M. Primavera, F. Ruggieri, P. Santangelo, E. Santovetti, G. Saracino, R. D. Schamberger, B. Sciascia, A. Sciubba, F. Scuri, I. Sfiligoi, T. Spadaro, E. Spiriti, C. Stanescu, L. Tortora, P. Valente, G. Venanzoni, S. Veneziano, Y. Wu

At present an effort is being done by the accelerator division in order to increase the peak luminosity to the run I design value of $\sim$1–3 $\times 10^{31}$ cm^{-2}s^{-1}.

2 Physics at DAΦNE/KLOE

The main interest in a ϕ–factory is that the ϕ decays at rest in a K_S, K_L pair in $\sim$34% of the times. The two kaons are monochromatic, back to back and they have $\beta \sim 0.2$, corresponding to lifetimes of 343cm and 0.6cm for the long and the short state, respectively. This allows for an efficient and clean tagging. In fact the presence of a *V-vertex* inside the beam pipe, with the mass and the momentum expected from the K_S two body decay, is the tag for a K_L in the other emisphere, ***indipendently from its decay process.*** The procedure is called a KLTAG. Similarly a KSTAG is defined as the identification of the K_L, either by its decay in a charged mode or by its interaction in the calorimeter. In such a way a pure beam of K_S is obtained, a unique possibility with respect to fixed target machines.

Therefore the KLOE experiment at DAΦNE has a wide range of possible physics items at reach, both in ***kaon physics*** (CP violation studies, rare K_S decays, CPT studies through the semileptonic asimmetry, kaon form factors) but also in *non kaon physics*: radiative ϕ decays ($\phi \rightarrow \eta\gamma, \eta'\gamma, f^0\gamma, a^0\gamma, ...$), studies of σ_{HAD} vs $\sqrt{s}$ using Initial State Radiation ($e^+e^- \rightarrow e^+e^-\gamma$ followed by $e^+e^- \rightarrow \pi^+\pi^-$), regeneration measurements at low momenta and many others. Note that part of this analysis program can be performed also at low luminosity (few pb^{-1}) and is currently undertaken.

3 The KLOE Detector

The KLOE [2,3] detector (fig.1) has been designed primarily with the goal of detecting CP violation in K^0 decays. It is a typical e^+e^- detector, of $\sim$5 m diameter and $\sim$4 m length, composed of a superconducting solenoid that provides a magnetic field of 6 KG, a central drift chamber, and an electromagnetic calorimeter installed inside the coil. The beam pipe at the interaction zone is a 10 cm ($\sim$16 λ_S) beryllium sphere 0.5 mm thick. This allows us to define a fiducial volume for K_S without complications from regeneration and to minimize multiple scattering and energy loss for charged particles. The two low-beta quadrupole triplets in the interaction region are also instrumented to improve the photon acceptance. This helps in the rejection of $K_L \rightarrow \pi^0\pi^0\pi^0$.

The performance of the KLOE detector, the final trigger configuration and the data acquisition reliability have been studied using the first data. The results obtained are well in agreement with the design specifications.

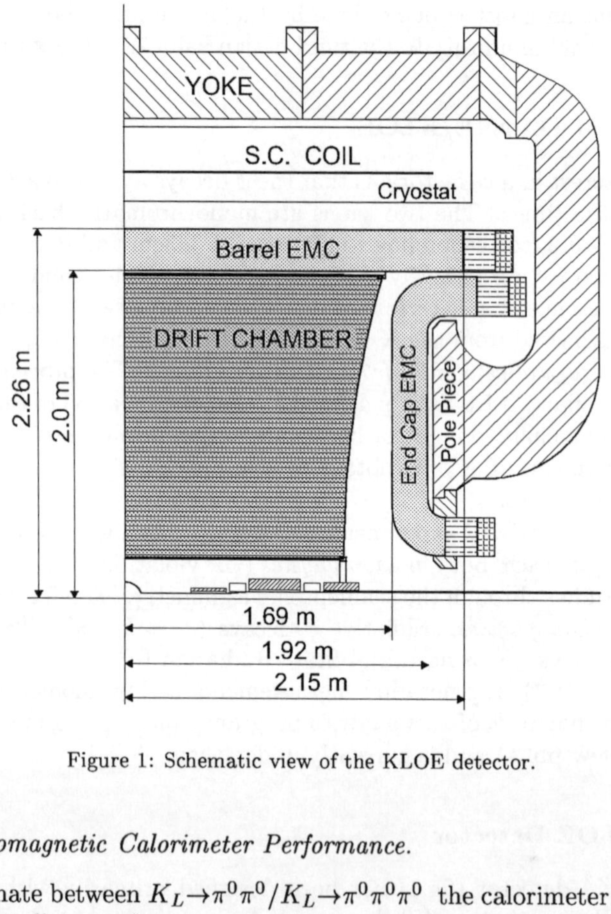

Figure 1: Schematic view of the KLOE detector.

3.1 Electromagnetic Calorimeter Performance.

To discriminate between $K_L \to \pi^0\pi^0 / K_L \to \pi^0\pi^0\pi^0$ the calorimeter should have above all full efficiency for γ's in the energy range 20–280 MeV. Of course, hermeticity and good energy resolution are also required. The calorimeter should also provide a fast trigger, should have some capability for particle identification and should have a reasonable cost. Good timing performance is also very useful for reconstruction of the K_L neutral vertex with high resolution, thanks to the method shown in fig. 2.

We have chosen to use a lead-scintillator sampling calorimeter consisting of 0.5 mm lead layer in which 1 mm diameter scintillating fibers are embedded. The energy sampling fraction is 13 %.

The calorimeter has been fully calibrated in energy and in time using cosmic rays and Bhabha events. The energy calibration is performed by equalizing the response of the photomultipliers to minimum ionizing particles

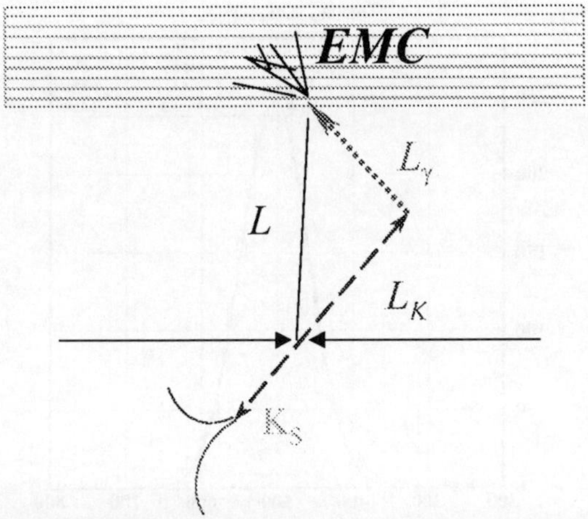

Figure 2: Neutral vertex position measurement: L is deduced from the photon conversion point, L_K is deduced from K_S direction. In addition the photon arrival time is measured and the K_L speed it is known.

and the energy scale is deduced from Bhabha and $\gamma\gamma$ events. The energy measured for bhabha events is shown in fig. 3. The energy resolution is $\sigma(E)/E$ $\sim$8% at 510 MeV which correspond to $\sigma(E)/E \sim 5.7\%/\sqrt{E\ (\mathrm{GeV})}$ in fine agreement with test beam results and the design performance. The linearity is checked with radiative Bhabhas down to 20 MeV. The time resolution is $\sigma_T \sim 80$ ps$/\sqrt{E\ (\mathrm{GeV})}$, as measured from $\gamma\gamma$ events, and it is used to recognize K_L's that interact in the calorimeter by measuring their speed. In fig. 4 the speed of the cluster found in a cone of 20 degrees around the K_L direction (obtained from the $K_S \to \pi^+\pi^-$) is shown. A clear K_L signal is seen with $\beta \sim 0.21$. As mentioned above the time information is used in KLOE to measure the neutral decay vertex position. The resolution on this measurement should be $\sim$1 cm in order to minimize the effect on the "double ratio" of event miscounting in the vicinity of the fiducial volume boundary for K_L. In particular, the relative scale for charged and neutral decays must coincide within a fraction of millimeter. The intercalibration of the scales can be obtained using events where the neutral and charged vertices coincide such as $K_L \to \pi^+\pi^-\pi^0$ or $K^+ \to \pi^+\pi^0$. A preliminary analysis of the $K_L \to \pi^+\pi^-\pi^0$ channel shows that the requested sensitivity can be reached.

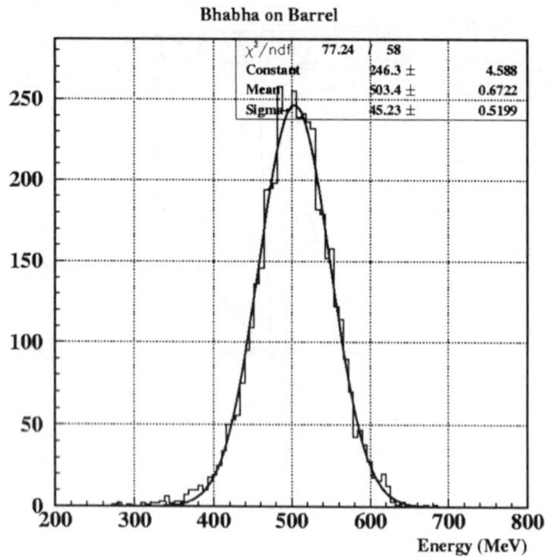

Figure 3: Calorimeter energy for Bhabha events

3.2 Tracking Chamber Performance.

The main characteristics of the tracking system are:

• Large radius and uniform sampling to collect as many K_L decays as possible, taking into account that the K_L mean decay length is 3.5 m.

• Optimized resolution for low momentum tracks by using low Z gas and thin walls to minimize multiple scattering, regeneration and γ conversion.

We have chosen a 2 m radius and 3.4 m long drift chamber [4] with a cell configuration that is almost square, with effective area of 2×2 cm^2 for the first 12 planes and 3×3 cm^2 for the outer 46 planes. The gas mixture is 90% helium and 10% isobutane for a total radiation length of ~900 m, including the ~52,000 wires. The mechanical structure is done entirely of carbon-fiber/epoxy and adds up to ~0.1 X_0.

The chamber is fully operational with very few dead channels (< 0.1 %). It is is very clean even down to low thresholds (4 mV) and the cell efficiency is ~99 %. The vertex resolution for Bhabha events is $\sigma_x \sim \sigma_y \sim$2mm, $\sigma_z \sim$5 mm in fine agreement with expectations. The same resolutions are found for lower momentum tracks such the one produced in the $\phi \rightarrow \rho\pi$ decay. The expected momentum resolution, entirely dominated by multiple scattering in our momentum range, is σ_P /p ~0.4%, and agrees with the value we find for

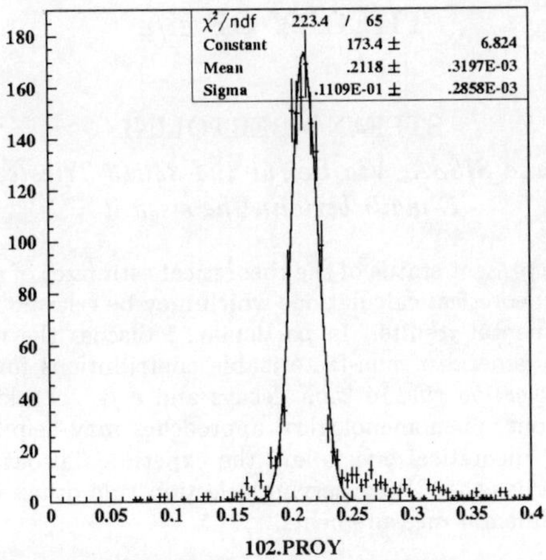

Figure 4: β for K_L interacting in the calorimeter.

Bhabha events.

4 Conclusions.

The KLOE detector performance, the DAQ and the trigger reliability, the online monitoring and the reconstruction and filtering procedures have been tested successfully with the first data collected. In the near future DAΦNE should provide KLOE with 100 pb^{-1} of data which will allow the measurement of $\Re(\epsilon'/\epsilon)$ to a statistical accuracy of 10^{-3}, the measurement of the K_{l3} form factor as well as the measurement of many radiative processes (as $\phi \rightarrow f_0\gamma$).

References

1. G. Vignola, *Workshop on Physics and Detector for DAΦNE 95*, ed. R. Baldini at al. SIS-Pubblicazioni, Frascati, 1995.
2. The KLOE Collaboration, *A general purpose detector for DAΦNE*, LNF-92/019 (1992).
3. The KLOE Collaboration, *The KLOE detector, Technical Proposal*, LNF-93/002 (1993).
4. The KLOE Collaboration, *The KLOE central drift chamber* , LNF-94/028 (1994).

THEORY OF ε'/ε

STEFANO BERTOLINI

INFN and SISSA, Via Beirut 4, I-34013 Trieste, Italy

E-mail: bertolin@he.sissa.it

I shortly review the present status of the theoretical estimates of ε'/ε . I consider a few aspects of the theoretical calculations which may be relevant in understanding the present experimental results. In particular, I discuss the role of final state interactions and in general of non-factorizable contributions for the explanation of the $\Delta I = 1/2$ selection rule in kaon decays and ε'/ε . Lacking reliable "first principle" calculations, phenomenological approaches may help in understanding correlations among theoretical effects and the experimental data. The same dynamics which underlies the CP conserving selection rule drives ε'/ε in the range of the recent experimental measurements.

The results announced this year by the KTeV[1] and NA48[2] collaborations have marked a great experimental achievement, establishing 35 years after the discovery of CP violation in the neutral kaon system[3] the existence of a much smaller violation acting directly in the decays.

While the Standard Model (SM) of strong and electroweak interactions provides an economical and elegant understanding of indirect (ε) and direct (ε') CP violation in term of a single phase, the detailed calculation of the size of these effects implies mastering strong interactions at a scale where perturbative methods break down. In addition, CP violation in $K \to \pi\pi$ decays is the result of a destructive interference between two sets of contributions, which may inflate up to an order of magnitude the uncertainties on the hadronic matrix elements of the effective four-quark operators. All that makes predicting ε'/ε a complex and subtle theoretical challenge[4].

The status of the theoretical predictions and experimental data available before the KTeV announcement early this year is summarized in Fig. 1.

The gray horizontal bands show the one-sigma average of the old (early 90's) NA31 (CERN) and E731 (Fermilab) results. The vertical lines show the ranges of the published theoretical *predictions* (before February 1999), identified with the cities where most of the group members reside. The range of the naive Vacuum Saturation Approximation (VSA) is shown for comparison.

The experimental and theoretical scenarios have changed substantially after the KTeV announcement and the subsequent NA48 result. Fig. 2 shows the present experimental world average for ε'/ε compared with the revised or new theoretical calculations that appeared during this year.

Notwithstanding the complexity of the problem, all theoretical calcula-

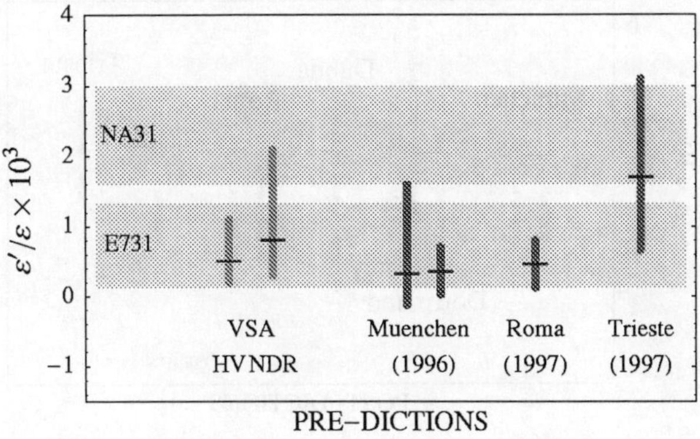

Figure 1. The 1-σ results of the NA31 and E731 Collaborations (early 90's) are shown by the gray horizontal bands. The old München, Roma and Trieste theoretical predictions for ε'/ε are depicted by the vertical bars with their central values. For comparison, the VSA estimate is shown using two renormalization schemes.

tions show a remarkable overall agreement, most of them pointing to a non-vanishing positive effect in the SM (which is by itself far from trivial).

On the other hand, if we focus our attention on the central values, many of the predictions prefer the 10^{-4} regime, whereas only a few of them stand above 10^{-3}. Is this just a "noise" in the theoretical calculations?

The answer is no. Without entering the details of the various estimates, it is possible to explain most of the abovementioned difference in terms of a single effect: the different size of the hadronic matrix element of the gluonic penguin Q_6 obtained in the various approaches.

While some of the calculations, as the early München and Rome predictions, assume for $\langle \pi\pi|Q_6|K\rangle$ values in the neighboroud of the leading $1/N$ result (naive factorization), other approaches, among which the Trieste and Dortmund calculations, find a substantial enhancement of this matrix element with respect to the simple factorization result. The bulk of such an effect is actually a global enhancement of the $I = 0$ components of the $K \to \pi\pi$ amplitudes, which affects *both* current-current *and* penguin operators, and it can be clearly understood in terms of chiral dynamics (final-state interactions).

As a matter of fact, one should in general expect an enhancement of ε'/ε with respect to the naive VSA due to final-state interactions (FSI). As Prof. Suzuki reminded us[5], in two body decays the isospin $I = 0$ final states feel an attractive interaction, of a sign opposite to that of the $I = 2$ components. This feature is at the root of the enhancement of the $I = 0$ amplitude

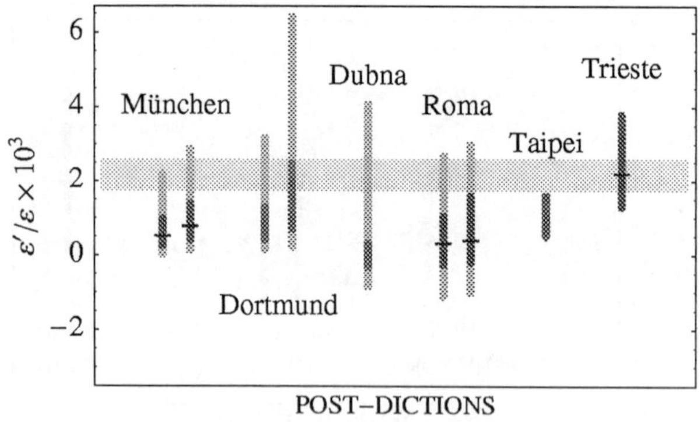

Figure 2. The recent theoretical calculations of ε'/ε are compared with the combined 1-σ average of the NA31, E731, KTeV and NA48 results ($\varepsilon'/\varepsilon = 21.2 \pm 4.6 \times 10^{-4}$), depicted by the gray horizontal band (the error is inflated according to the Particle Data Group procedure when averaging over data with substantially different central values).

over the $I = 2$ one and of the corresponding enhancement of ε'/ε beyond factorization.

The question is how to make of a qualitative statement a quantitative one, since a reliable calculation of ε'/ε must involve at the same time a convincing explanation of the $\Delta I = 1/2$ selection rule, which is still missing.

The $\Delta I = 1/2$ selection rule in $K \to \pi\pi$ decays is known since 45 years[6] and it states the experimental evidence that kaons are 400 times more likely to decay in the $I = 0$ two-pion state than in the $I = 2$ component. This rule is not justified by any general symmetry consideration and, although it is common understanding that its explanation must be rooted in the dynamics of strong interactions, there is no up to date derivation of this effect from first principle QCD.

The ratio of $I = 2$ over $I = 0$ amplitudes appears directly in the definition of ε'/ε :

$$\frac{\varepsilon'}{\varepsilon} = \frac{1}{\sqrt{2}} \left\{ \frac{\langle (\pi\pi)_{I=2} | \mathcal{H}_W | K_L \rangle}{\langle (\pi\pi)_{I=0} | \mathcal{H}_W | K_L \rangle} - \frac{\langle (\pi\pi)_{I=2} | \mathcal{H}_W | K_S \rangle}{\langle (\pi\pi)_{I=0} | \mathcal{H}_W | K_S \rangle} \right\} . \tag{1}$$

As a consequence, a consistent calculation of ε'/ε must also address the determination of the CP conserving amplitudes.

The way we approach the calculation of the hadronic $K \to \pi\pi$ transitions in gauge theories is provided by the Operator Product Expansion which allows us to write the relevant amplitudes in terms of the hadronic matrix elements of effective $\Delta S = 1$ four quark operators (at a scale μ) and of the corresponding Wilson coefficients, which encode the information about the dynamical degrees of freedom heavier than the chosen renormalization scale:

$$\mathcal{H}_{\Delta S=1} = \frac{G_F}{\sqrt{2}} V_{ud} V_{us}^* \sum_i \Big[z_i(\mu) + \tau\, y_i(\mu) \Big] Q_i(\mu)\ . \tag{2}$$

The entries V_{ij} of the 3×3 Cabibbo-Kobayashi-Maskawa matrix describe the flavour mixing in the SM and $\tau = -V_{td}V_{ts}^*/V_{ud}V_{us}^*$. For $\mu < m_c$ $(q = u, d, s)$, the relevant quark operators are:

$$\left.\begin{array}{l} Q_1 = (\bar{s}_\alpha u_\beta)_{\rm V-A}\,(\bar{u}_\beta d_\alpha)_{\rm V-A} \\ Q_2 = (\bar{s}u)_{\rm V-A}\,(\bar{u}d)_{\rm V-A} \end{array}\right\} \quad \text{Current-Current}$$

$$\left.\begin{array}{l} Q_{3,5} = (\bar{s}d)_{\rm V-A} \sum_q (\bar{q}q)_{\rm V\mp A} \\ Q_{4,6} = (\bar{s}_\alpha d_\beta)_{\rm V-A} \sum_q (\bar{q}_\beta q_\alpha)_{\rm V\mp A} \end{array}\right\} \quad \text{Gluon “penguins”} \tag{3}$$

$$\left.\begin{array}{l} Q_{7,9} \ = \frac{3}{2}\,(\bar{s}d)_{\rm V-A} \sum_q \hat{e}_q\,(\bar{q}q)_{\rm V\pm A} \\ Q_{8,10} = \frac{3}{2}\,(\bar{s}_\alpha d_\beta)_{\rm V-A} \sum_q \hat{e}_q (\bar{q}_\beta q_\alpha)_{\rm V\pm A} \end{array}\right\} \quad \text{Electroweak “penguins”}$$

Current-current operators are induced by tree-level W-exchange whereas the so-called penguin (and “box”) diagrams are generated via an electroweak loop. Only the latter “feel” all three quark families via the virtual quark exchange and are therefore sensitive to the weak CP phase. Current-current operators control instead the CP conserving transitions. This fact suggests already that the connection between ε'/ε and the $\Delta I = 1/2$ rule is by no means a straightforward one.

Using the effective $\Delta S = 1$ quark Hamiltonian we can write ε'/ε as

$$\frac{\varepsilon'}{\varepsilon} = e^{i\phi} \frac{G_F \omega}{2|\epsilon|\, {\rm Re}\, A_0}\, {\rm Im}\, \lambda_t\, \left[\Pi_0 - \frac{1}{\omega} \Pi_2 \right] \tag{4}$$

where

$$\begin{array}{l} \Pi_0 = \frac{1}{\cos\delta_0} \sum_i y_i\ {\rm Re}\langle Q_i\rangle_0\ (1 - \Omega_{\eta+\eta'}) \\ \Pi_2 = \frac{1}{\cos\delta_2} \sum_i y_i\ {\rm Re}\langle Q_i\rangle_2 \quad , \end{array} \tag{5}$$

and $\langle Q_i\rangle \equiv \langle\pi\pi|Q_i|K\rangle$. The rescattering phases $\delta_{0,2}$ can be extracted from elastic $\pi - \pi$ scattering data[7] and are such that $\cos\delta_0 \simeq 0.8$ and $\cos\delta_2 \simeq 1$. Given that the phase of ε, θ_ε, is approximately $\pi/4$, as well as $\delta_0 - \delta_2$, $\phi = \frac{\pi}{2} + \delta_2 - \delta_0 - \theta_\varepsilon$ turns out to be consistent with zero.

Two key ingredients appear in eq. 4:

1. The isospin breaking $\pi^0 - \eta - \eta'$ mixing, parametrized by $\Omega_{\eta+\eta'}$, which is estimated to give a *positive* correction to the A_2 amplitude of about

15-35%. However, it has been recently emphasized[8] that NLO chiral corrections may make $\Omega_{\eta+\eta'}$ negative and as large as -0.7, thus potentially providing a strong enhancement mechanism for ε'/ε .

2. The combination of CKM elements $\text{Im}\,\lambda_t \equiv \text{Im}(V_{ts}^* V_{td})$, which affects directly the size of ε'/ε and the range of the uncertainty. The determination of $\text{Im}\,\lambda_t$ goes through B-physics constraints and ε [9]. The latter observable depends on the theoretical determination of B_K, the $\bar{K}^0 - K^0$ hadronic parameter, which should be consistently determined within every approach. The theoretical uncertainty on B_K affects subtantially the final uncertainty on $\text{Im}\,\lambda_t$. A better determination of the unitarity triangle is expected from the B-factories and the hadronic colliders[10]. From K-physics, $K_L \to \pi^0 \nu\bar{\nu}$ gives the cleanest "theoretical" determination of $\text{Im}\,\lambda_t$, albeit representing a great experimental challenge[11].

A satisfactory approach to the calculation of ε'/ε should comply with the following requirements:

A: A consistent definition of renormalized operators leading to the correct scheme and scale matching with the short-distance perturbative analysis.

B: A self-contained calculation of *all* relevant hadronic matrix elements (including B_K).

C: A simultaneous explanation of the $\Delta I = 1/2$ selection rule and ε'/ε .

None of the available calculations satisfies all previous requirements. I summarize very briefly the various attempts to calculate ε'/ε which have appeared so far and have lead to the estimates shown in Figs. 1 and 2. A simple naive approach to the problem is the VSA, which is based on two drastic assumptions: the factorization of the four quark operators and the vacuum saturation of the completeness of the intermediate states. As an example:

$$\begin{aligned}\langle\pi^+\pi^-|Q_6|K^0\rangle &= 2\,\langle\pi^-|\bar{u}\gamma_5 d|0\rangle\langle\pi^+|\bar{s}u|K^0\rangle - 2\,\langle\pi^+\pi^-|\bar{d}d|0\rangle\langle 0|\bar{s}\gamma_5 d|K^0\rangle \\ &\quad + 2\left[\langle 0|\bar{s}s|0\rangle - \langle 0|\bar{d}d|0\rangle\right]\langle\pi^+\pi^-|\bar{s}\gamma_5 d|K^0\rangle \end{aligned} \tag{6}$$

The VSA does not allow for a consistent matching of the scale and scheme dependence of the Wilson coefficients (the HV and NDR results are shown in Fig. 1) and it carries a potentially large systematic uncertainty[4]. It is best used for LO estimates.

Generalized Factorization represents an attempt to address the issue of renormalization scheme dependence in the framework of factorization by defining effective Wilson coefficients which absorb the perturbative QCD running of the quark operators. The new coefficients, scale and scheme independent at a given order in the strong coupling expansion, are matched to the factorized matrix elements at the scale μ_F which is arbitrarily chosen in the perturbative regime. A residual scale dependence remains in the penguin matrix elements via the quark mass. Fitting the $\Delta I = 1/2$ rule and ε'/ε requires non-factorizable contributions both in the current-current and the penguin matrix elements[12].

In the München approach (phenomenological $1/N$) some of the matrix elements are obtained by fitting the $\Delta I = 1/2$ rule at $\mu = m_c = 1.3$ GeV. On the other hand, the relevant gluonic and electroweak penguin $\langle Q_6 \rangle$ and $\langle Q_8 \rangle_2$ remain undetermined and are taken around their leading $1/N$ values (which implies a scheme dependent result). In Fig. 2 the HV (left) and NDR (right) results are shown[13]. The dark range represents the result of gaussian treatment of the input parameters compared to flat scannning (complete range).

In the recent years the Dortmund group has revived and improved the approach of Bardeen, Buras and Gerard[14] based on the $1/N$ expansion. Chiral loops are regularized via a cutoff and the amplitudes are arranged in a p^{2n}/N expansion. A particular attention has been given to the matching procedure between the scale dependence of the chiral loops and that arising from the short-distance analysis[15]. The renormalization scheme dependence remains and it is included in the final uncertainty. The $\Delta I = 1/2$ rule is reproduced, but the presence of the quadratic cutoff induces a matching scale instability (which is very large for B_K). The NLO corrections to $\langle Q_6 \rangle$ induce a substantial enhancement of the matrix element (right range in Fig. 2) compared to the leading order result (left). The dark range is drawn for central values of m_s, $\Omega_{\eta+\eta'}$, $\mathrm{Im}\,\lambda_t$ and Λ_{QCD}.

In the Nambu, Jona-Lasinio (NJL) modelling of QCD[16] the Dubna group[17] has calculated ε'/ε including chiral loops up to $O(p^6)$ and the effects of scalar, vector and axial-vector resonances. Chiral loops are regularized via the heat-kernel method, which leaves unsolved the problem of the renormalization scheme dependence. A phenomenological fit of the $\Delta I = 1/2$ rule implies deviations up to a factor two on the calculated $\langle Q_6 \rangle$. The reduced (dark) range in Fig. 2 corresponds to taking the central values of the NLO chiral couplings and varying the short-distance parameters. The $\Delta I = 1/2$ rule and B_K have been studied in the NJL framework and $1/N$ expansion by Bijnens and Prades[18] showing an impressive scale stability when including vector and axial-vector resonances.

In the approach of the Trieste group, based on the Chiral Quark Model (χQM)[19], all hadronic matrix elements are computed up to $O(p^4)$ in the chiral expansion in terms of the three model parameters: the constituent quark mass, the quark condensate and the gluon condensate. These parameters are phenomenologically fixed by fitting the $\Delta I = 1/2$ rule[20]. This step is crucial in order to make the model predictive, since there is no a-priori argument for the consistency of the matching procedure. As a matter of fact, all computed observables turn out to be very weakly dependent on the scale (and the renormalization scheme) in a few hundred MeV range around the matching scale, which is taken at 0.8 GeV as a compromise between the ranges of validity of perturbation theory and chiral expansion. The $I = 0$ matrix elements are strongly enhanced by non-factorizable contributions and drive ε'/ε in the 10^{-3} regime[21].

Lattice regularization of QCD is *the* consistent approach to the problem. On the other hand, there are presently important numerical and theoretical limitations, like the quenching approximation and chiral symmetry, which may substantially affect the calculation of the weak matrix elements. In addition, chiral perturbation theory is needed in order to obtain $K \to \pi\pi$ amplitudes from the computed $K \to \pi$ transitions. As summarized by Soni at this conference[22] lattice cannot provide us at present with reliable calculations of the $I = 0$ penguin operators relevant to ε'/ε , as well as of the $I = 0$ components of the hadronic matrix elements of the current-current operators (penguin contractions), which are relevant to the $\Delta I = 1/2$ rule. This is due to large renormalization uncertainties, partly related to the breaking of chiral symmetry on the lattice. In this respect, very promising is the Domain Wall Fermion approach[23] which allows us to decouple the chiral symmetry from the continuum limit[22]. In the recent Roma re-evaluation of ε'/ε $\langle Q_6 \rangle$ is taken at the VSA value with a 100% uncertainty[24]. The result is therefore scheme dependent (the HV and NDR results are shown in Fig. 2). The dark (light) ranges correspond to Gaussian (flat) scan of the input parameters.

Other attempts to reproduce the measured ε'/ε using the linear σ-model, which include the effect of a scalar resonance with $m_\sigma \simeq 900$ MeV, obtain the needed enhancement of $\langle Q_6 \rangle$[25]. However, it is not possible to reproduce simultaneously the experimental values of ε'/ε and of the CP conserving $K \to \pi\pi$ amplitudes.

Without entering into the details of the various calculations I wish to illustrate with a simple exercise the crucial role of final-state interactions (in general of non-factorizable contributions) for ε'/ε and the $\Delta I = 1/2$ selection rule. In order to do that I focus on two semi-phenomenological approaches.

A commonly used way of comparing the estimates of hadronic matrix elements in different approaches is via the so-called B factors which represent the ratio of the model matrix elements to the corresponding VSA values.

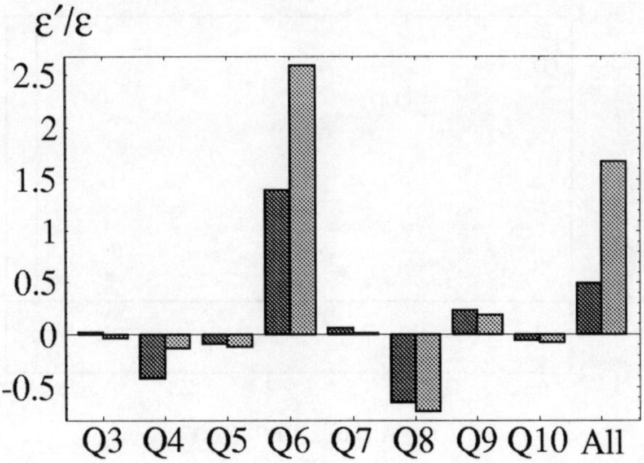

Figure 3. Predicting ε'/ε: a (Penguin) Comparative Anatomy of the München (dark gray) and Trieste (light gray) results (in units of 10^{-3}).

However, care must be taken in the comparison of different models due to the scale dependence of the B's and the values used by different groups for the parameters that enter the VSA expressions. An alternative pictorial and synthetic way of analyzing different outcomes for ε'/ε is shown in Fig. 3, where a "comparative anatomy" of the Trieste and München estimates is presented.

From the inspection of the various contributions it is apparent that the different outcome on the central value of ε'/ε is almost entirely due to the difference in the Q_6 component.

In the Münich approach[13] the $\Delta I = 1/2$ rule is used in order to determine phenomenologically the matrix elements of $Q_{1,2}$ and, via operatorial relations, some of the matrix elements of the left-handed penguins. The approach does not allow for a phenomenological determination of the matrix elements of the penguin operators which are most relevant for ε'/ε , namely the gluonic penguin Q_6 and the electroweak penguin Q_8. These matrix elements are taken around their leading $1/N$ values (factorization).

In the semi-phenomenological approach of the Trieste group the size of the effects on the $I = 0, 2$ amplitudes is controlled by the phenomenological embedding of the $\Delta I = 1/2$ selection rule which determines the ranges of the model paremeters: the constituent quark mass, the quark and the gluon condensates. In terms of these parameters all matrix elements are computed.

Fig. 4 shows an anatomy of the χQM contributions which lead to the experimental value of the $\Delta I = 1/2$ selection rule for central values of the

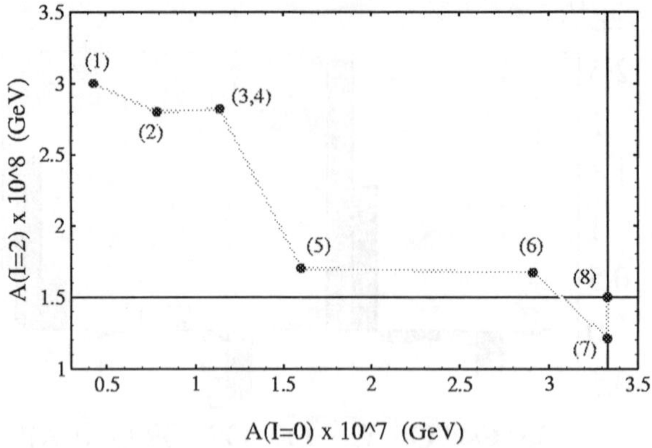

Figure 4. Anatomy of the $\Delta I = 1/2$ rule in the χQM. See the text for explanations. The cross-hairs indicate the experimental point.

input parameters.

Point (1) represents the result obtained by neglecting QCD and taking the factorized matrix element for the tree-level operator Q_2, which is the leading electroweak contribution. The ratio A_0/A_2 is thus found equal to $\sqrt{2}$: by far off the experimental point (8). Step (2) includes the effects of perturbative QCD renormalization on the operators $Q_{1,2}$[26]. Step (3) shows the effect of including the gluonic penguin operators[27]. Electroweak penguins[28] are numerically negligeable in the CP conserving amplitudes and are responsible for the very small shift in the A_2 direction. Therefore, perturbative QCD and factorization lead us from (1) to (4).

Non-factorizable gluon-condensate corrections, a crucial model dependent effect entering at the leading order in the chiral expansion, produce a substantial reduction of the A_2 amplitude (5), as it was first observed by Pich and de Rafael[29]. Moving the analysis to $O(p^4)$, the chiral loop corrections, computed on the LO chiral lagrangian via dimensional regularization and minimal subtraction, lead us from (5) to (6), while the finite parts of the NLO counterterms calculated in the χQM approach lead to the point (7). Finally, step (8) represents the inclusion of π-η-η' isospin breaking effects[30].

This model dependent anatomy shows the relevance of non-factorizable contributions and higher-order chiral corrections. The suggestion that chiral dynamics may be relevant to the understanding of the $\Delta I = 1/2$ selection rule goes back to the work of Bardeen, Buras and Gerard[14] in the $1/N$ framework

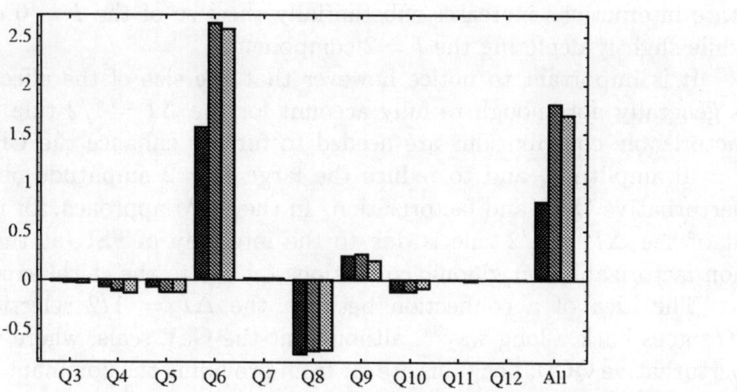

Figure 5. Anatomy of ε'/ε (in units of 10^{-3}) in the χQM approach. In black the LO results (which includes the non-factorizable gluonic corrections), in half-tone the effect of the inclusion of chiral-loop corrections and in light gray the complete $O(p^4)$ estimate.

with a cutoff regularization. A pattern similar to that shown in Fig. 4 for the chiral loop corrections to A_0 and A_2 was previously obtained in a NLO chiral lagrangian analysis, using dimensional regularization, by Missimer, Kambor and Wyler[31]. The Trieste group has extended their calculation to include the NLO contributions to the electroweak penguin matrix elements[32,21].

Fig. 5 shows the contributions to ε'/ε of the various penguin operators, providing us with a finer anatomy of the NLO chiral corrections. It is clear that chiral-loop dynamics plays a subleading role in the electroweak penguin sector (Q_{8-10}) while enhancing by 60% the gluonic penguin ($I = 0$) matrix elements. The NLO enhancement of the Q_6 matrix element is what drives ε'/ε in the χQM to the 10^{-3} ballpark.

As a consequence, the χQM analysis shows that the same dynamics that is relevant to the reproduction of the CP conserving A_0 amplitude (Fig. 4) is at work in the CP violating sector, albeit with a reduced strenght.

In order to ascertain whether these model features represent real QCD effects we must wait for future improvements in lattice calculations[22]. On the other hand, indications for such a dynamics follow from the analitic properties of the $K \to \pi\pi$ amplitudes. We should expect in general an enhancement of ε'/ε , with respect to the naive VSA estimate, due to final-state interactions. In two body decays, the $I = 0$ final states feel an attractive interaction, of a sign opposite to that of the $I = 2$ components. This feature is at the root of the enhancement of the $I = 0$ amplitude over the $I = 2$ one. Recent dispersive analyses[33] of $K \to \pi\pi$ amplitudes show how a (partial) resummation of final state interactions increases substantially the size of the $I = 0$ components, while slightly depleting the $I = 2$ components.

It is important to notice however that the size of the effect so derived is generally not enough to fully account for the $\Delta I = 1/2$ rule. Other non-factorizable contributions are needed to further enhance the CP conserving $I = 0$ amplitude, and to reduce the large $I = 2$ amplitude obtained from perturbative QCD and factorization. In the χQM approach, for instance, the fit of the $\Delta I = 1/2$ rule is due to the interplay of FSI (at the NLO) and non-factorizable soft gluonic corrections (at LO in the chiral expansion).

The idea of a connection between the $\Delta I = 1/2$ selection rule and ε'/ε goes back a long way[34], although at the GeV scale, where we can trust perturbative QCD, penguins are far from providing the dominant contribution to the CP conserving amplitudes.

I conclude by summarizing the relevant remarks:

Those semi-phenomenological approaches which reproduce the $\Delta I = 1/2$ selection rule in $K \to \pi\pi$ decays, generally agree in the pattern and size of the $I = 2$ hadronic matrix elements with the existing lattice calculations.

$I = 0$ amplitudes: the $\Delta I = 1/2$ rule forces upon us large deviations from the naive VSA: $B-$factors of $O(10)$ for $\langle Q_{1,2}\rangle_0$. Lattice calculations presently suffer from large sistematic uncertainties.

In the Trieste and Dortmund calculations, which reproduce the CP conserving $K \to \pi\pi$ amplitudes, non-factorizable effects (mainly final-state interactions) enhance the hadronic matrix elements of the gluonic penguins, and give $B_6/B_8^{(2)} \approx 2$. Similar indications stem from other $1/N$[18] and dispersive[33] approaches. Promising work is in progress on the lattice.

Theoretical error: a substantial amount of the present uncertainty in the ε'/ε prediction arises from the uncertainty in the CKM elements $\mathrm{Im}(V_{ts}^* V_{td})$ which is presently controlled by the $\Delta S = 2$ parameter B_K. A better determination of the unitarity triangle from B-physics is expected from the B-factories and hadronic colliders. From K-physics, $K_L \to \pi^0 \nu\bar{\nu}$ gives the cleanest "theoretical" determination of $\mathrm{Im}\,\lambda_t$.

New Physics: in spite of recent clever proposals (mainly SUSY[35]) it is premature to invoke physics beyond the SM in order to fit ε'/ε . A number of ungauged systematic uncertainties affect presently all theoretical estimates, and, most of all, every attempt to reproduce ε'/ε must also address the puzzle of the $\Delta I = 1/2$ rule, which is hardly affected by short-distance physics. An "anomalously" large ε'/ε is likely to be the CP violating counterpart of $A_0/A_2 \approx 22$.

Acknowledgments

My appreciation goes to the Organizers of the $\mathcal{B}_{\rm CP}$ '99 Conference for hosting a most enjoyable and stimulating meeting.

References

1. A. Alavi-Harati *et al.*, *Phys. Rev. Lett.* **83**, 22 (1999); A. Rodman, these Proceedings.
2. V. Fanti *et al.*, *Phys. Lett.* B **465**, 335 (1999); A. Nappi, these Proceedings.
3. J.H. Christenson *et al.*, *Phys. Rev. Lett.* **13**, 138 (1964).
4. S. Bertolini, J.O. Eeg and M. Fabbrichesi, *Rev. Mod. Phys.* **72**, 65 (2000).
5. M. Suzuki, these Proceedings.
6. M. Gell-Mann and A. Pais, *Proc. Glasgow Conf.*, 342 (1955).
7. J. Gasser and U.G. Meissner, *Phys. Lett.* B **258**, 219 (1991); E. Chell and M.G. Olsson, *Phys. Rev.* D **48**, 4076 (1993).
8. S. Gardner and G. Valencia, hep-ph/9909202.
9. C.S. Kim, A. Stocchi, these Proceedings.
10. R. Hughes, S. Stone, these Proceedings.
11. S. Kettel, G. D'Ambrosio, these Proceedings.
12. H.Y. Cheng, hep-ph/9906403; hep-ph/9911202.
13. A.J. Buras, Proceedings of the Kaon 99 Conference, hep-ph/9908395.
14. W.A. Bardeen, A.J. Buras and J.M. Gerard, *Nucl. Phys.* B **293**, 787 (1987).
15. T. Hambye *et al.*, *Phys. Rev.* D **58**, 014017 (1998); hep-ph/9906434; P. Soldan, these Proceedings.
16. Y. Nambu and G. Jona-Lasinio, *Phys. Rev.* **122**, 345 (1961); J. Bijnens, *Phys. Rept.* **265**, 369 (1996).
17. A.A. Bel'kov *et al.*, hep-ph/9907335;
18. J. Bijnens and J. Prades, *JHEP* **01**, 023 (1999).
19. S. Weinberg, *Physica* A **96**, 327 (1979); A. Manhoar and H. Georgi, *Nucl. Phys.* B **234**, 189 (1984).
20. S. Bertolini *et al.*, *Nucl. Phys.* B **514**, 63 (1998).
21. S. Bertolini *et al.*, *Nucl. Phys.* B **514**, 93 (1998).
22. A. Soni, these Proceedings.
23. D. Kaplan, *Phys. Lett.* B **288**, 342 (1992).
24. G. Martinelli, Proceedings of the Kaon 99 Conference, hep-ph/9910237.
25. Y.Y. Keum, U. Nierste and A.I. Sanda, *Phys. Lett.* B **457**, 157 (1999);

M. Harada *et al.*, hep-ph/9910201; J.C.R. Bloch *et al.*, nucl-th/9910029.

26. M.K. Gaillard and B. Lee, *Phys. Rev. Lett.* **33**, 108 (1974); G. Altarelli and L. Maiani, *Phys. Lett.* B **52**, 413 (1974).
27. M.A. Shifman, A.I. Vainshtein and V.I. Zacharov, *JETP Lett.* **22**, 55 (1975); *Nucl. Phys.* B **120**, 316 (1977); *JETP* **45**, 670 (1977); J. Ellis *et al.*, *Nucl. Phys.* B **131**, 285 (1977), (E) **132**, 541 (1978); F.J. Gilman and M.B. Wise, *Phys. Lett.* B **83**, 83 (1979); *Phys. Rev.* D **20**, 2392 (1979); R.S. Chivukula, J.M. Flynn and H. Georgi, *Phys. Lett.* B **171**, 453 (1986).
28. J. Bijnens and M.B. Wise, *Phys. Lett.* B **137**, 245 (1984); A.J. Buras and J.M. Gerard, *Phys. Lett.* B **192**, 156 (1987); M. Lusignoli, *Nucl. Phys.* B **325**, 33 (1989); J.M. Flynn and L. Randall, *Phys. Lett.* B **224**, 221 (1989).
29. A. Pich and E. de Rafael, *Nucl. Phys.* B **358**, 311 (1991).
30. B. Holstein, *Phys. Rev.* D **20**, 1187 (1979); J.F. Donoghue, E. Golowich, B.R. Holstein and J. Trampetic, *Phys. Lett.* B **179**, 361 (1986); A.J. Buras and J.M. Gerard, in ref. [28].
31. J. Kambor, J. Missimer and D. Wyler, *Nucl. Phys.* B **346**, 17 (1990); *Phys. Lett.* B **261**, 496 (1991).
32. V. Antonelli *et al.*, *Nucl. Phys.* B **469**, 143 (1996).
33. E. Pallante and A. Pich, hep-ph/9911233; J.F. Donoghue and E. Golowich, hep-ph/9911309; E. A. Paschos, hep-ph/9912230.
34. M.A. Shifman, A.I. Vainshtein and V.I. Zacharov, in ref. [27]; F.J. Gilman and M.B. Wise, in ref. [27].
35. A. Masiero, C.K. Chua, these Proceedings; A.I. Kagan and M. Neubert, *Phys. Rev. Lett.* **83**, 4429 (1999).

EXPERIMENTAL STUDY OF $\sin 2\beta$ AND $\sin 2\alpha$

WILLIAM TRISCHUK

University of Toronto, 60 St. George Street,
Toronto, ON, M5S 1A7, Canada
E-mail: william@physics.utoronto.ca

Detailed measurements of CP violation in B meson decay are on the horizon. Here I review the status of current measurements of $\sin 2\beta$ made at LEP and CDF. These yield an average of $\sin 2\beta = 0.82 \pm 0.39$, giving 97% confidence that β is greater than 0, evidence that CP violation occurs in B decay. I review predictions for the precision we can expect on $\sin 2\beta$ and $\sin 2\alpha$ in the next few years.

1 Introduction

Quark weak decay is governed by the unitary CKM matrix. One useful parametrisation of this 3×3 matrix, due to Wolfenstein, is:

$$V_{\text{CKM}} = \begin{pmatrix} 1-\lambda^2/2 & \lambda & A\lambda^3(\rho - i\eta) \\ -\lambda & 1-\lambda^2/2 & A\lambda^2 \\ A\lambda^3(1-\rho-i\eta) & -A\lambda^2 & 1 \end{pmatrix} + \mathcal{O}(\lambda^4)$$

where λ is the Cabbibo mixing angle in strange quark decay while A, ρ and η are three other real parameters. The unitarity of this matrix leads to one particularly interesting relationship among the elements:

$$V_{tb}^* V_{td} + V_{cb}^* V_{cd} + V_{ub}^* V_{ud} = 0. \tag{1}$$

This can be visualised as a triangle in the (ρ, η)-plane (see Fig. 1). Constraints on the CKM matrix elements from $B^0\bar{B}^0$ mixing, limits on $B_s^0\bar{B}_s^0$ mixing, ϵ_K from K meson decay, V_{ub} and V_{cb} lead to a prediction that $\beta = 0.75 \pm 0.09$ [1]. One of the most compelling reasons to measure β directly is to verify whether the CKM matrix explains CP violation in the B system.

One form of CP violation in the B system manifests itself in the decay of B^0 and $\bar{B}^0$ mesons into a common final state. The angle β can be measured from $B^0 \to J/\psi K_S^0$ decays. CP violating effects arise from the interference of direct decays to $J/\psi K_S^0$ with decays that proceed via mixing ($B^0 \to \bar{B}^0$) followed by the same decay. The measured asymmetry, $A_{CP}(t)$, can be related to the CP violating angle β by:

$$\begin{aligned} A_{CP}(t) &= \frac{\frac{dN}{dt}(\bar{B}^0 \to J/\psi K_S^0) - \frac{dN}{dt}(B^0 \to J/\psi K_S^0)}{\frac{dN}{dt}(\bar{B}^0 \to J/\psi K_S^0) + \frac{dN}{dt}(B^0 \to J/\psi K_S^0)} \\ &= \sin 2\beta \, \sin \Delta m_d t. \end{aligned} \tag{2}$$

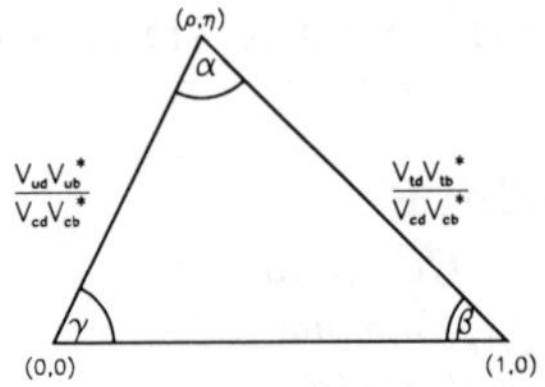

Figure 1. The unitarity triangle showing the angles α, β and γ that will be pursued in the next generation of searches for CP violation in the B sector.

There are a number of experimental complications to this clean theoretical situation. At B factories the B mesons are produced from $\Upsilon(4S)$ decay in a, J=1, coherent state. A CP asymmetry can only be observed as a function of the decay length difference between the two B mesons and the time averaged asymmetry vanishes. At hadron colliders B mesons are produced incoherently. Thus we can measure the asymmetry as a function of the single B meson decay length and the time averaged asymmetry does not vanish.

In both cases we must tag the flavour of the decaying B meson Attempts to tag the B meson are limited by the tag efficiency, ϵ, and its dilution, $\mathcal{D}$, which is a measure of the fraction, f_w, of incorrectly tagged mesons. The true CP asymmetry is reduced by the dilution:

$$A_{CP}^{obs} = \mathcal{D}A_{CP} \tag{3}$$

where $\mathcal{D}$ is defined as: $\mathcal{D} = (N_r - N_w)/(N_r + N_w) = 1 - 2f_w$, with $N_r(N_w)$ the number of right (wrong) tags. We expect a precision on sin 2β of:

$$\delta \sin\ 2\beta \propto \frac{1}{\sqrt{\epsilon\mathcal{D}^2}}\sqrt{\frac{S+B}{S^2}} \tag{4}$$

where S is the number of signal candidates and B the number of background candidates in the sample. The product $\epsilon\mathcal{D}^2$ acts as an effective tagging efficiency accounting for the number of fully tagged events. In hadron collider experiments this number can be expected to range from a few % up to 10% while at B factories it ranges from 30 – 40%.

2 Measurements of $\sin 2\beta$

Several groups have isolated $B \to J/\psi K_S^0$ samples and made measurements of the CP angle β [2,3,4]. Here I describe the most precise single measurement [5].

Table 1. Summary of CDF B meson flavour tagging efficiencies and dilutions.

Tagger	Events	efficiency (ϵ)	Dilution ($\mathcal{D}$)
SST	SVX	35.5 ± 3.7 %	16.6 ± 2.2 %
SST	non-SVX	38.1 ± 3.9 %	17.4 ± 3.6 %
SLT	all	5.6 ± 1.8 %	62.5 ± 14.6 %
JETQ	all	40.2 ± 3.9 %	23.5 ± 6.9 %

2.1 The CDF Result

From its Run I sample the CDF collaboration has selected a dataset of 400 $J/\psi K_S^0$ B meson decays. About half of these (see Fig. 2a)) have precise lifetime information allowing the extraction of $\sin 2\beta$ from an analysis of the time-dependent asymmetry (see eqn. 2). The other half (see Fig. 2b)) lack silicon information on one or more of the final state tracks. For these events we measure the time averaged asymmetry and extract β, with less precision, based on the convolution of eqn. 2 with our acceptance for B meson decay.

To extract evidence for a CP violating asymmetry we must determine the initial flavour of the B meson. The final state, a CP eigenstate, provides no clue as to whether a B^0 or $\bar{B}^0$ was responsible for the decay. We employ three methods to tag the decaying B meson's flavour. The first involves identifying a fragmentation pion near the $J/\psi K_S^0$ that results from the primary fragmentation. Its charge is related to the flavour of the B meson. We constrain the dilution of this tag with control samples of $B^{\pm}$ mesons and find that same side tagging (SST) yields a dilution of 17% (see table 1). We observe very small differences in the efficiency and dilution of SST for the data samples with/without SVX lifetime information. The other two tagging techniques involve finding another B meson in the event where a $J/\psi K_S^0$ has been reconstructed. This other B meson starts with the opposite flavour of the one that decayed into $J/\psi K_S^0$. However, either of them can mix between their production and decay. This limits the usefulness of the "opposite-side" B meson in flavour tagging. Once found we tag the away side B meson flavour in two ways. If it decays semi-leptonically then the decay lepton charge tags the flavour. This soft lepton tag (SLT) has very high dilution, exceeding 60 % (mixing limits it to a maximum of 74%) but the B meson semileptonic branching fraction (we tag both electrons and muons) results in a small efficiency. We also measure the charge of jets (JETQ) opposite the B that has decayed to $J/\psi K_S^0$. We sum the charge of tracks in a jet, weighting them by momentum and impact parameter, to obtain a measure of their parent parton charge. We do this with relatively high efficiency and find that the optimal algorithm achieves a dilution of 23%.

These tagging algorithms are applied to each decay candidate. If more than one tag is present we compute an effective dilution. If the tags agree the dilution is increased to reflect our increased confidence in the determination

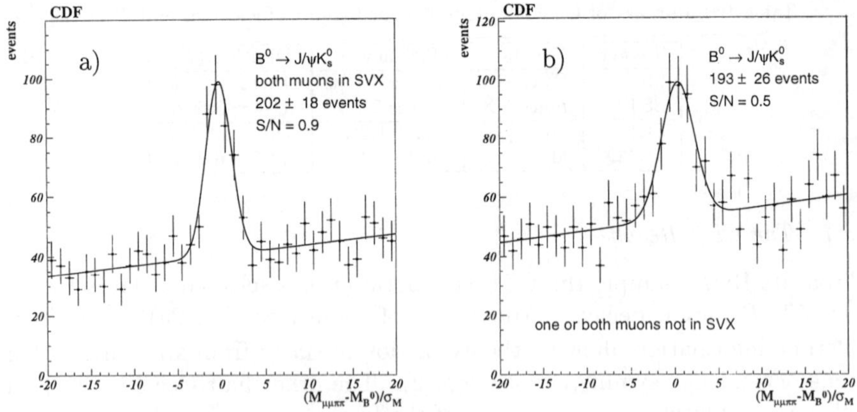

Figure 2. The CDF $J/\psi K_S^0$ data where a) both J/ψ muons have SVX information and b) where one or both J/ψ muons lack SVX information.

of the B flavour. When the tags disagree we reduce the effective dilution. Such candidates end up carrying much less weight in the final fit. Although the dilutions are computed candidate-by-candidate and included in our final fit for $\sin 2\beta$ we predict the expected statistical power of our tagging techniques. Each of the three taggers has an $\epsilon\mathcal{D}^2$ of about 2% and combining them, including correlations, we expect to have an overall $\epsilon\mathcal{D}^2$ of $(6.3 \pm 1.7)\%$. With this tagging efficiency and dilution, our 400 $J/\psi K_S^0$ candidates have the statistical power of 25 perfectly tagged events.

In performing a maximum likelihood fit, we weight each candidate not only by its predicted dilution, but also its observed decay length and mass. Our fit is shown in Fig. 3a). Events with precise lifetime information are plotted, as a function of ct, on the left. Our preferred fit is shown as a solid line. The dotted line is the result of a fit with Δm_d determined from this data. The value of Δm_d changes by one standard deviation but $\sin 2\beta$ extracted from the amplitude of these curves is unchanged. The time-averaged asymmetry of the candidates without precise ct information is shown on the right of Fig. 3a). From the combination of samples we extract $\sin 2\beta = 0.79^{+0.41}_{-0.44}$. Since we include variations of the fit inputs (ϵ, $\mathcal{D}$, Δm_d, τ_B, etc.) the quoted uncertainties include both statistical and systematic effects. Our only significant systematic is due to uncertainty on the tagging dilutions.

Figure 3b) shows a scan of our likelihood for $\sin 2\beta$. Several techniques for normalising the probability near the physical bound ($\sin 2\beta$ cannot exceed

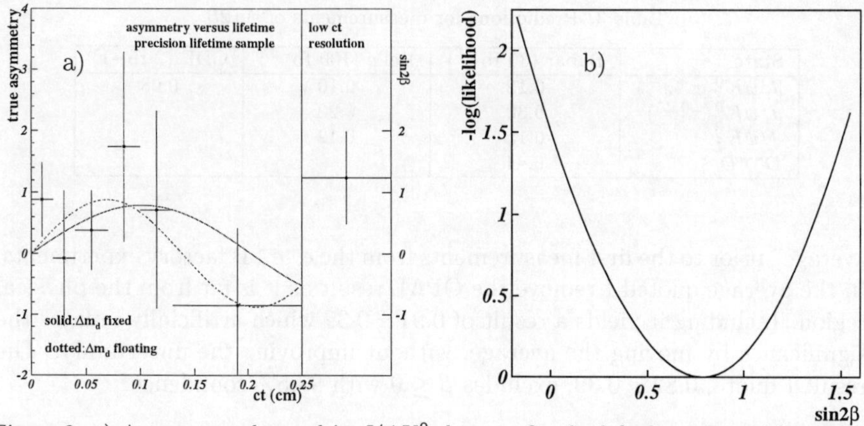

Figure 3. a) Asymmetry observed in $J/\psi K_S^0$ decays. On the left the events with precise vertex information are plotted as a function of decay time. On the right the time averaged asymmetry for events without vertex information is shown. b) A scan of the likelihood scan showing that $\sin 2\beta = 0$ is excluded with 95% CL.

Table 2. Summary of measurements of $\sin 2\beta$ and Standard Model prediction.

Experiment	$\sin 2\beta$	Reference
OPAL	$3.2^{+1.8}_{-2.0}(stat) \pm 0.5(sys)$	2
CDF (initial)	$1.8 \pm 1.1(stat) \pm 0.3(sys)$	3
CDF (update)	$0.79 \pm 0.39(stat) \pm 0.16(sys)$	5
ALEPH (prelim)	$0.93^{+0.64}_{-0.88}(stat)^{+0.36}_{-0.24}(sys)$	4
My Average	0.82 ± 0.39	
CKM model fit	0.75 ± 0.09	1

1) have been tried. The probability that $\sin 2\beta$ is actually 0, given our measurement, is excluded with 93% to 96.4% depending on the method chosen [5].

2.2 Other Results

The OPAL collaboration published a first measurement of the angle β from a sample of 24 $J/\psi K_S^0$ candidates including background [2]. CDF also published an initial determination using only the SST tagging algorithm, applied to $J/\psi K_S^0$ candidates with precise lifetime information [3]. This has now been superseded by the measurement described above [5]. Finally, the ALEPH collaboration has presented a preliminary measurement of β from 23 $J/\psi K_S^0$ candidates in their data. These results are summarised in table 2.

I have tried combining these results in several ways to produce a "world

Table 3. Predictions for measurements of $\sin 2\beta$.

State	Babar (30 fb^{-1})	Belle (100 fb^{-1})	CDF (2 fb−1)
$J/\psi K^0_S(\pi^+\pi^-)$	0.12	0.10	0.08
$J/\psi K^0_S(\pi^0\pi^0)$	0.30	0.20	–
$J/\psi K^0_L$	0.15	0.12	–
$D^{*+}D^{*-}$	0.44	–	–

average" prior to the first measurements from the e^+e^- B factory experiments. In the average quoted I remove the OPAL result as it is far from the physical region. Including it yields a result of 0.91 ± 0.39 which artificially inflates the significance by moving the average, without improving the uncertainty. The result I quote, 0.82 ± 0.39, excludes $\beta \leq 0$ with 97.5% confidence.

3 The Future of CP Violation Measurements in B Decay

The measurements made so far only hint at CP violation in B meson decay. Dedicated B factory experiments were beginning to take data as this meeting took place. CDF and D0 will resume data taking in early 2001. Later BTeV and the LHC experiments will come online. Here I focus on the "immediate" prospects for observing CP violation in B meson decay, summarising the anticipated precisions from Babar and Belle as well as the Tevatron.

3.1 Outlook for $\sin 2\beta$

The B factory experiments have predicted precisions for several different B decay modes. I summarise their results for decay modes that study β in table 3. These are quoted for one year of running at the design luminosity of their respective machines. While the $J/\psi K^0_S$ modes, already studied by OPAL, CDF and ALEPH, will likely yield the highest precision other modes will provide important cross-checks. Final states where the K^0_S decays into $\pi^0\pi^0$ (about 1/3 of the decays), will provide a first test of the B factory experiment's π^0 reconstruction capabilities. Similarly, the K^0_L decays are more challenging to reconstruct and likely to have lower precision, however they are predicted to have the opposite CP asymmetry from those K^0_S decays. A measurement of $\sin 2\beta$ from these decays that agrees (in magnitude) with that from the K^0_S decays will provide a cross-check of possible charge asymmetries or biases. Finally modes such as $D^{*+}D^{*-}$, while not likely to be statistically competitive with the J/ψ modes but will provide a first arena to study the effect of strong phases obscuring an underlying CP violating angle.

At the Tevatron we anticipate a 20-fold increase in luminosity during the first two years of running. In CDF [a] we expect to increase our $J/\psi K^0_S$ sample by improving our silicon vertex detector acceptance by a factor of 1.5. Our projected $\sin 2\beta$ precision is based on a sample of 10,000 $J/\psi K^0_S$ candidates. The control samples used to measure dilution will grow by similar factors. Neglecting the expected improvements in tagging capability we anticipate a precision of $\delta \sin 2\beta \approx 0.08$. By including a time-of-flight detector in CDF-II we should increase our effective tagging capabilities with away-side $K^\pm$ mesons. We also hope to further increase our sample size by including a trigger for $J/\psi \to e^+e^-$ decays. Predictions for similar modes from the B factories can be found in table 3.

3.2 Outlook for $\sin 2\alpha$

Attempts to study CP angles other than β will meet significant experimental and theoretical challenges. The next obvious CP eigenstate, $B^0 \to \pi^+\pi^-$, may measure $\sin 2\alpha$. However it suffers from the presence of non-CP violating phases, from penguin decays. Proposals to measure these phases are reviewed in Ref. [6].

Apart from the theoretical complication of interpreting the relative phase of $B^0 \to \pi^+\pi^-$ and $\bar{B}^0 \to \pi^+\pi^-$ there would remain the experimental problem of isolating a clean sample of these decays and measuring their asymmetry. First, and foremost, is the branching fraction $\mathcal{B}(B^0 \to \pi^+\pi^-) = (0.43 \pm 0.18) \times 10^{-5}$ [7]. For the B factories this will make the collection of a significant sample of these decays a fairly long process. The isolation of the $\pi^+\pi^-$ final state is further complicated by the presence of the much larger branching into $B^0 \to K^+\pi^-$, in experiments without particle ID.

Even with the huge B^0 production rate at the Tevatron, the absence of particle ID will hinder the separation of $\pi^+\pi^-$ candidates from $K^+\pi^-$ background. The study of $B^0 \to \pi^+\pi^-$ final states at a hadron collider also requires a trigger for B decays into two charged tracks. CDF is developing such a trigger, using information from the silicon vertex detector within 20 μs of the collision to identify pairs of tracks each having an impact parameter larger than 100 μm. This trigger will be at least 50% efficient for decays of $B \to h^+h^-$ and provide the background reduction necessary to make recording these events feasible. Using this trigger we expect ≈ 5000 $B^0 \to \pi^+\pi^-$ in 2 fb^{-1} (based on the latest branching fractions from CLEO). From this sample, despite a background of 20000 $B^0 \to K^+\pi^-$ decays, we should determine the

[a] D0 will have similar detector capabilities to CDF during run II. One would expect D0 to make B physics measurements with similar precision during this run.

$\pi^+\pi^-$ charge asymmetry with a precision of 0.13.

Similarly pragmatic studies of the $\pi^+\pi^-$ asymmetry at B factories project uncertainties of 0.15 and 0.26 for one nominal year of running at Belle and Babar, respectively. It remains to be seen whether penguin phases can be sorted out to allow a clean interpretation of this asymmetry in terms of $\sin 2\alpha$.

4 Summary

First measurements of the CKM angle, β, have been made in B meson decay but we are only beginning to study the CP violating effects predicted in the B sector by the CKM matrix. From the measurements to date it is clear that the higher B meson production cross-section in hadron colliders can be a significant advantage in the study of these effects. CDF's 400 event sample provides a more precise measurement of $\sin 2\beta$ than the ≈ 20 event samples of OPAL and ALEPH, despite the fact that the hadronic backgrounds are larger and the dilutions/efficiencies are smaller. Several measurements of $\sin 2\beta$ with precisions of $\approx$0.1 can be expected before the next meeting in this series. The tools are also in place to collect the data samples necessary for a measurement of $\sin 2\alpha$. However, the theoretical challenges of interpreting observed $\pi^+\pi^-$ asymmetries, in terms of α along with the experimental challenges of separating the $\pi^+\pi^-$ signal from $K^+\pi^-$ (and other) backgrounds will mean that these results will not be available until further into the decade.

Acknowledgments

The author acknowledges the support of the National Sciences and Engineering Research Council of Canada for his work on CDF. I thank those who provided me with information included in this review. Finally, I also thank the organisers (especially George Hou) for hosting an interesting conference.

References

1. S. Mele, *Phys. Rev. Lett.* **59**, 113011 (99);
   For similar results see also A. Stocchi, these proceedings.
2. K. Ackerstaff *et al* (OPAL Collaboration), Euro. Phys. C**5**, 379, (1998).
3. F. Abe. *et al* (CDF Collaboration), *Phys. Rev. Lett.* **81**, 5513 (1998).
4. R. Forty, these proceedings.
5. T. Affolder *et al* (CDF Collaboration), hep-ex/9909003.
6. The Babar Physics Book, Chapter 5, SLAC-R-504, (1998).
7. J. Smith, these proceedings.

PENGUINS AND MIXING DEPENDENT CP VIOLATION

N. G. DESHPANDE

Institute of Theoretical Science, University of Oregon, Eugene, OR 97403, USA
E-mail: desh@oregon.uoregon.edu

Constraints on angles of Unitarity triangle are reviewed, and in particular constraint on γ from limit on Δm_s is emphasized. Effects of penguin diagram on measurement of β and α are then reviewed. New measurements on $B \to \pi^+\pi^-$ in QCD improved factorization approximation suggest large penguin effects. It is possible to estimate the error in measurement of α as a function of γ for different $|V_{ub}/V_{cb}|$ values.

1 Constraints on Angles of Unitarity Triangle

Constraints on unitarity triangle expressed in terms of Wolfenstin parameterization are given below:

[a] From charmless semileptonic B decays [1]:

$$\left|\frac{V_{ub}}{V_{cb}}\right| = 0.08 \pm 0.02 \tag{1}$$

which yields

$$\left(\rho^2 + \eta^2\right)^{1/2} = 0.36 \pm 0.09 \tag{2}$$

[b] From $B_d - \bar{B}_d$ mixing. Error is dominated by f_B and the bag factor B_B:

$$|V_{td}| = 0.009 \pm 0.003 \tag{3}$$

which yields

$$|1 - \rho - i\eta| = 1.0 \pm 0.3 \tag{4}$$

[c] Value of ϵ in K system. Error is dominated by hadronic matrix elements:

$$\eta\,(1 - \rho + 0.35) = 0.48 \pm 0.20 \tag{5}$$

[d] Constraints from $B_s - \bar{B}_s$ [2]

$$(\Delta m_s) > 14.3 ps^{-1}\,(90\% CL) \tag{6}$$

Using the relation from Box diagrams

$$\left|\frac{V_{td}}{V_{ts}}\right| = \xi \left[\frac{m_{B_s}\Delta m_d}{m_{B_d}\Delta m_s}\right]^{1/2} \tag{7}$$

where

$$\xi = \frac{f_{Bs}}{f_{Bd}}\sqrt{\frac{B_{Bs}}{B_{Bd}}} \tag{8}$$

Lattice calculations yield for ξ the value [3]

$$\xi = 1.15 \pm 0.05 \tag{9}$$

This translates into the bound

$$|V_{td}/V_{ts}| < 0.214 \tag{10}$$

or

$$|1 - \rho - i\eta| < 0.96 \tag{11}$$

The last constraint in particular implies $\gamma < 90°$ and $75° < \alpha < 120°$. This allowed range of γ leads to unique estimate of errors in α as we shall see.

2 CP Violation Through Mixing

Strategy to measure β and α involve measuring time dependent asymmetry in B decays to CP eigenstate. Defining the long and short lived eigenstates of B as

$$|B\rangle_{L,S} = p\,|B^{\circ}\rangle \pm q\,|\bar{B}^{\circ}\rangle, \tag{12}$$

the amplitudes for decays into CP eigenstates are defined as

$$A = \langle f_{CP}|H_w|B^{\circ}\rangle \tag{13}$$

$$\bar{A} = \langle f_{CP}|H_w|\bar{B}^{\circ}\rangle\,. \tag{14}$$

The asymmetry is then defined by

$$Asy(t) = \left[\left(1 - |\lambda|^2\right)\cos\left(\Delta M t\right) - 2Im(\lambda)\sin\left(\Delta M t\right)\right] / \left[|1 + |\lambda|^2\right] \tag{15}$$

where $\lambda = (q/p)\left(\bar{A}/A\right)$. In the standard model $(q/p) = e^{-2i\beta}$. If A is dependent on a single weak phase,

$$\left(\bar{A}/A\right) = e^{-2i\phi_{\text{weak}}} \tag{16}$$

then we have the expression

$$Asy(t) = -Im(\lambda)\sin\left(\Delta M t\right) \tag{17}$$

2.1 Measurement of β

The mode that has the least theoretical uncertainty is $B \to \psi Ks$. The amplitude for this mode can be written in terms of Tree and Penguin contribution as

$$A = V_{cb}V_{cs}^*T + V_{tb}V_{ts}^*P = V_{cb}V_{cs}^*(T - P) + V_{ub}V_{us}^*P \tag{18}$$

since $|V_{ub}V_{us}^*/V_{cb}V_{cs}^*|$ is $\approx 1/50$, and the Penguin contribution has predominantly $\bar{c}c$ in a color octet state, the contribution due to penguin diagram is less than 1%.

If $B \to D^+D^-$ mode is used instead, the penguin contribution is much larger, and there is no color suppression either.

$$A = V_{cb}V_{cd}^*T + V_{tb}V_{td}^*P = V_{cb}V_{cd}^*(T - P) - V_{ub}V_{ud}^*P \tag{19}$$

The value of $|V_{ub}V_{ud}^*/V_{cb}V_{cd}^*| \approx 0.3$, and although P is suppressed compared to T due to small Wilson coefficients, one can expect a contamination due to penguin of a few percent.

2.2 Measurement of α

The mode $B^o \to \pi^+\pi^-$ lends itself to the earliest measurement of α. For this mode the amplitude is

$$A = V_{ub}V_{ud}^*T + V_{tb}V_{td}^*P = V_{ub}V_{ud}^*(T - P) + V_{cb}V_{cd}^*P \tag{20}$$

The value of $|V_{cb}V_{cd}^*/V_{ub}V_{ud}^*| \approx 3$ giving a crude estimate of around 15% for the penguin contamination. Gronau and London [4] have presented a method of extracting α from measurements of $B^o \to \pi^+\pi^-$, $\overline{B^o} \to \pi^+\pi^-$, $B^o \to \pi^o\pi^o$, $\overline{B^o} \to \pi^o\pi^o$ and $B^+ \to \pi^+\pi^o$. However, the most recent theoretical estimates of $B^o \to \pi^o\pi^o$ branching ratio are around 5×10^{-7}, making this method academic at present. However, we now discuss theoretical developments that may allow us to extract the correct α from measurements of asymmetry in $B^o \to \pi^+\pi^-$ alone.

3 Determination of α from $B^o \to \pi^+\pi^-$

This is based on recent work of Agashe and Deshpande [5]. Recently, the CLEO collaboration has reported the first observation of the decay $B \to \pi^+\pi^-$ [6].The effective Hamiltonian for B decays is:

$$\mathcal{H}_{eff} = \frac{G_F}{\sqrt{2}}\Big[V_{ub}V_{ud}^*\left(C_1O_1^u + C_2O_2^u\right)$$

$$+V_{cb}V_{cd}^*\left(C_1O_1^c + C_2O_2^c\right) - V_{tb}V_{td}^*\sum_{i=3}^{6} C_iO_i\Bigg]. \tag{21}$$

The C_i's are the Wilson coefficients (WC's). In a recent paper, Beneke *et al.* found that the matrix elements for the decays $B \to \pi\pi$, in the large m_b limit, can be written as [7]

$$\langle\pi\pi|O_i|B\rangle = \langle\pi|j_1|B\rangle\langle\pi|j_2|0\rangle \times\left[1 + \sum r_n\alpha_s^n(m_b) + O(\Lambda_{QCD}/m_b)\right], \tag{22}$$

where j_1 and j_2 are bilinear quark currents. If the radiative corrections in α_s *and* $O(\Lambda_{QCD}/m_b)$ corrections are neglected, then the matrix element on the left-hand side factorizes into a product of a form factor and a meson decay constant so that we recover the "conventional" factorization formula. These authors computed the $O(\alpha_s)$ corrections. In this approach, the strong interaction (final-state rescattering) phases are included in the radiative corrections in α_s and thus the $O(\alpha_s)$ strong interaction phases are determined [7]. The matrix element for $B \to \pi^+\pi^-$ is [7]:

$$i\bar{A}\left(\bar{B}_d \to \pi^+\pi^-\right) = \frac{G_F}{\sqrt{2}}\Bigg[V_{ub}V_{ud}^*\left(a_1 + a_4^u + a_6^u r_\chi\right) + V_{cb}V_{cd}^*\left(a_4^c + a_6^c r_\chi\right)\Bigg] \times X. \tag{23}$$

Here

$$X = f_\pi\left(m_B^2 - m_\pi^2\right)F_0^{B\to\pi^-}\left(m_\pi^2\right), \tag{24}$$

where $f_\pi = 131$ MeV is the pion decay constant and $F_0^{B\to\pi^-}$ is a form factor. In the above equations, the a_i's are (combinations of) WC's with the $O(\alpha_s)$ corrections added. The values of the a_i's are given in Table 1 [7]. The imaginary parts of a_i's are due to final-state rescattering. For the CP conjugate processes, the CKM elements have to be complex-conjugated. We discuss two values of the form factors: $F^{B\to\pi^-} = 0.27$ and 0.33. Model calculations indicate that the $SU(3)$ breaking in the form factors is given by $F^{B\to K^-} \approx 1.13\ F^{B\to\pi^-}$ [8,9]. The large measured $BR(B \to K\eta')$ requires $F^{B\to K^-} \gtrsim 0.36$ [10] which, in turn, implies a larger value of $F^{B\to\pi^-}$ (≈ 0.33). If $F^{B\to K^-} \lesssim 0.36$, then we require a "new" mechanism to account for $BR(B \to K\eta')$: high charm content of η' [11], QCD anomaly [12] or new physics. Also, if $F^{B\to\pi^-} < 0.27$, then the value of $F^{B\to K}$ is too small to explain the measured BR's for $B \to K\pi$ [13]. We use $|V_{cb}| = 0.0395$, $|V_{ud}| = 0.974$,

a_1	$1.047 + 0.033\, i$
a_2	$0.061 - 0.106\, i$
a_4^u	$-0.030 - 0.019\, i$
a_4^c	$-0.038 - 0.009\, i$
$a_6^{u,c}\, r_\chi$	-0.036

Table 1. The factorization coefficients for the renormalization scale $\mu = m_b/2$ [7].

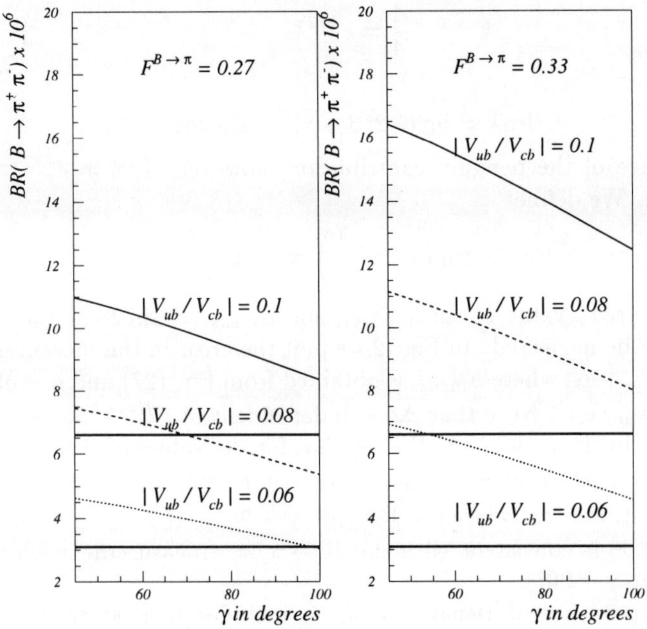

Figure 1. CP-averaged $BR\left(B \to \pi^+\pi^-\right)$ as a function of γ for $F^{B\to\pi^-} = 0.27$ (left) and 0.33 (right) and for $|V_{ub}/V_{cb}| = 0.1$ (solid curves), 0.08 (dashed curves) and 0.06 (dotted curves). The BR measured by the CLEO collaboration lies (at the 1 σ level) between the two horizontal (thicker) solid lines. The errors on the CLEO measurement have been added in quadrature to compute the 1 σ limits.

$|V_{cd}| = 0.224$, $m_B = 5.28$ GeV and $\tau_B = 1.6$ ps [1]. In Fig. 1 we show the CP-averaged BR for $B \to \pi^+\pi^-$ as a functions of γ for $F^{B\to\pi^-} = 0.33$ and 0.27 and for $|V_{ub}/V_{cb}| = 0.1$, 0.08 and 0.06.

The CLEO measurement is $B \to \pi^+\pi^- = (4.7^{+1.8}_{-1.5} \pm 0.6) \times 10^{-6}$ [6]. If $F^{B\to\pi^-} = 0.33$ and for $\gamma \lesssim 90°$, we see from the figures that smaller values of $|V_{ub}/V_{cb}| \approx 0.06$ are preferred: $|V_{ub}/V_{cb}| = 0.08$ is still allowed at the 2σ level for $\gamma \sim 100°$. The smaller value of $|V_{ub}/V_{cb}|$ leads to greater penguin contamination. However, if the smaller value of the form factor (0.27) is used, then the CLEO measurement is consistent with $|V_{ub}/V_{cb}| \approx 0.08$. We obtain similar results using "effective" WC's (C^{eff})'s and $N = 3$ in the earlier factorization framework [11].

Since the $B_d - \bar{B}_d$ mixing phase is 2β, if we neglect the (QCD) penguin operators, *i.e.*, set $a_{4,6} = 0$ in Eq. (23), we get

$$\frac{\bar{A}}{A} = e^{-i2\gamma} \tag{25}$$

and

$$\mathrm{Im}\lambda = \sin\left(-2(\beta+\gamma)\right) = \sin 2\alpha. \tag{26}$$

In the presence of the penguin contribution, however, $\bar{A}/A \neq e^{-i2\gamma}$ so that $\mathrm{Im}\lambda \neq \sin 2\alpha$. We define

$$\mathrm{Im}\lambda = \mathrm{Im}\left(e^{-i2\beta}\frac{\bar{A}}{A}\right) \equiv \sin 2\alpha_{meas.} \tag{27}$$

as the "measured" value of $\sin 2\alpha$, *i.e.*, $\sin 2\alpha_{meas.} = \sin 2\alpha$ if the penguin operators can be neglected. In Fig. 2 we plot the error in the measurement of α, $\Delta\alpha \equiv \alpha_{meas.} - \alpha$, where $\alpha_{meas.}$ is obtained from Eq. (27) and α is obtained from γ and $|V_{ub}/V_{cb}|$. Note that $\Delta\alpha$ is independent of $F^{B\to\pi^-}$ since the form factor cancels in the ratio $\bar{A}/A$. We see that for the values of $|V_{ub}/V_{cb}| \approx 0.06$ preferred by the $B \to \pi^+\pi^-$ measurement (if $F^{B\to\pi^-} \approx 0.33$), the error in the determination of α is large $\sim 15°$ (for $\gamma \sim 90°$). If $F^{B\to\pi^-} \approx 0.27$, then $|V_{ub}/V_{cb}| \approx 0.08$ is consistent with the $B \to \pi^+\pi^-$ measurement which gives $\Delta\alpha \sim 10°$ (for $\gamma \sim 90°$).

The computation of Beneke *et al.* [7] includes final state rescattering phases, *i.e.*, it is *exact* up to $O(\Lambda_{QCD}/m_b)$ and $O(\alpha_s^2)$ corrections. Thus, the value of $\sin 2\alpha$ "measured" in $B \to \pi^+\pi^-$ decays (Eq. (27)) is a known function of γ and $|V_{ub}/V_{cb}|$ only (in particular, there is no dependence on the phenomenological parameter $\xi \sim 1/N$ and strong phases are included unlike in the earlier factorization framework [11]). Since, the "true" value of α can also be expressed in terms of γ and $|V_{ub}/V_{cb}|$, we can estimate the "true" value of $\sin 2\alpha$ from the "measured" value of $\sin 2\alpha$ for a given value of $|V_{ub}/V_{cb}|$ (of course, up to $O(\Lambda_{QCD}/m_b)$ and $O(\alpha_s^2)$ corrections); this is shown in Fig. 3 where we have restricted γ to be in the range $(40°, 120°)$ as indicated by constraints on the unitarity triangle from present data. If $0° \leq \gamma \leq 180°$ is allowed, then there will be a discrete ambiguity in the determination of $\sin 2\alpha$ from $\sin 2\alpha_{meas.}$.

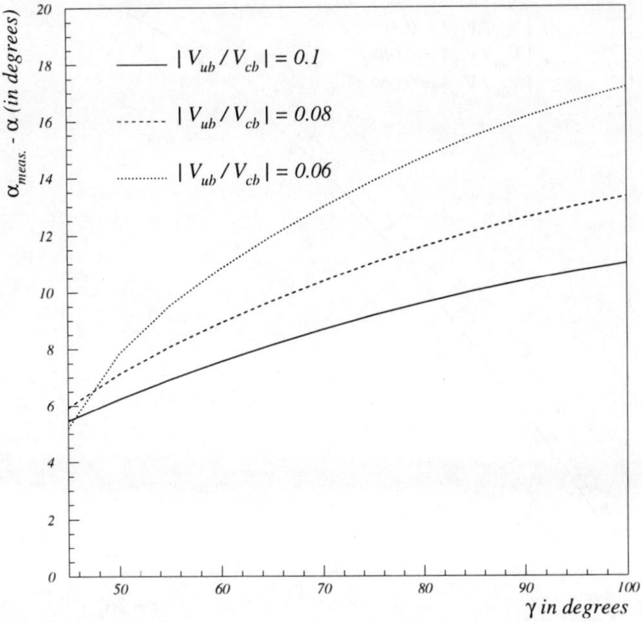

Figure 2. The error in the measurement of CKM phase α using (only) time-dependent $B \to \pi^+\pi^-$ decays as a function of γ for $|V_{ub}/V_{cb}| = 0.1$ (solid curve), 0.08 (dashed curve) and 0.06 (dotted curve).

4 Conclusions

We have shown how α can be obtained from the measured value of $\sin 2\alpha$ inspite of large penguin effects. The theoretical work can be extended to $K\pi$ modes to obtain values of γ from the measured branching ratios. Naive factorization suggest $\gamma \approx 100^\circ$ [14,13].

References

1. *Review of Particle Physics* by Particle Data Group, Eur. Phys. J. **C3** (1998) 1.
2. Talk by G. Blaylock at the XIX International Symposium on Lepton and Photon Interactions at High Energies, Stanford University, August 9-14, 1999 (http://LP99.slac.stanford.edu/db/program.asp).
3. A. J. Buras, invited talk given at the Symposium on Heavy Flavors, Santa Barbara, July 7-11, 1997, hep-ph/9711217.
4. M. Gronau, D. London, Phys. Rev. Lett. **65** (1990) 3381.

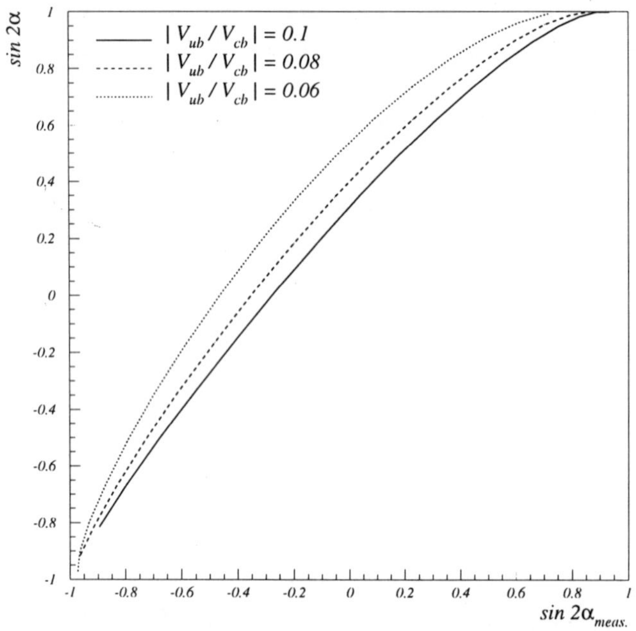

Figure 3. The "true" value of $\sin 2\alpha$ as a function of the value of $\sin 2\alpha$ "measured" in $B \to \pi^+\pi^-$ decays for $|V_{ub}/V_{cb}| = 0.1$ (solid curve), 0.08 (dashed curve) and 0.06 (dotted curve).

5. K. Agashe, N.G. Deshpande, hep-ph/9909298, to be published in Phys. Rev. **D**.
6. Talk by R. Poling at the XIX International Symposium on Lepton and Photon Interactions at High Energies, Stanford University, August 9-14, 1999 (http://LP99.slac.stanford.edu/db/program.asp).
7. M. Beneke *et al.*, Phys. Rev. Lett. **83** (1999) 1914.
8. M. Bauer, B. Stech, M. Wirbel, Z. Phys. **C34** (1987) 103.
9. P. Ball, V. M. Braun, Phys. Rev. **D58** (1998) 094016.
10. See, for example, N.G. Deshpande, B. Dutta, S. Oh, Phys. Rev. **D57** (1998) 5723.
11. See, for example, A. Ali, C. Greub, Phys. Rev. **D57** (1998) 2996.
12. P. Ball, J.M. Frere, M. Tytgat, Phys. Lett. **B365** (1996) 367; A. Ali, J. Chay, C. Greub, P. Ko, Phys. Lett. **B424** (1998) 161.
13. See, for example, B. Dutta, S. Oh, hep-ph/9911263.
14. N. G. Deshpande, X-G. He, W-S. Hou, S. Pakvasa, Phys. Rev. Lett. **82** (1999) 2240; W-S. Hou, J.G. Smith, F. Wurthwein, hep-ex/9910014.

DETERMINING THE CP VIOLATING PHASE γ

MICHAEL GRONAU
Department of Physics, Technion, Haifa 32000, Israel

We review several methods for determining the Kobayashi-Maskawa phase $\gamma = \mathrm{Arg}V_{ub}^*$ from rate and CP asymmetry measurements in hadronic B decays. We focus on the processes $B \to DK$, $B_s \to D_s K$, and on charmless decays to two light pseudoscalars and decays to a pair of a pseudoscalar and a vector meson. Theoretical uncertainties underlying these methods are discussed.

1 Introduction

B meson decays open the window into new phenomena of CP nonconservation, providing useful information about CP violating phases[1]. Phase measurements for α, β and γ, which are not independent in the Kobayashi-Maskawa framework, are important for two reasons. First, they improve to a higher precision our knowledge of the CKM mixing matrix, consisting of fundamental quark couplings in our present theory. And second, they are sensitive probes for sources of CP violation outside the CKM matrix, and for new flavor changing interactions contributing to $B^0 - \bar{B}^0$ mixing and to rare B decays.

The phase $\beta \equiv \mathrm{Arg}(-V_{cb}^* V_{cd}/V_{tb}^* V_{td})$ $(= \mathrm{Arg}V_{td}^*$ in the standard phase convention) is measured by the time-dependent CP asymmetry in $B^0(t) \to \psi K_S$ [2]. Theoretically, this measurement provides a very clean determination of $\sin(2\beta)$, since the single phase approximation holds in this case to better than 1%[3,4]. The recent CDF measurement[5], $\sin(2\beta) = 0.79^{+0.41}_{-0.44}$, is an encouraging proof of the method, although it does not yet constitute definite evidence for CP violation in B decays, as predicted in the CKM framework.

The determination of $\sin 2\alpha$ $(\alpha \equiv \mathrm{Arg}(-V_{tb}^* V_{td}/V_{ub}^* V_{ud}))$ in $B^0 \to \pi^+\pi^-$ suffers from the appearance of a second amplitude (P) due to QCD penguin operators[3,4]. The weak phase[6] of P differs from the phase of the dominant tree amplitude (T). The ratio $|P|/|T| = 0.3 \pm 0.1$, obtained[7] by comparing within broken flavor SU(3) the measured rates[8] of $B \to \pi\pi$ and $B \to K\pi$, implies a potentially large deviation of the measured asymmetry from $\sin 2\alpha$. In fact, the time-dependent asymmetry in $B^0(t) \to \pi^+\pi^-$ contains two terms [3]

$$A(t) = a_{\mathrm{dir}} \cos(\Delta m t) + \sqrt{1 - a_{\mathrm{dir}}^2} \sin 2(\alpha + \theta) \sin(\Delta m t) , \tag{1}$$

where a_{dir} and θ are both proportinal to $|P|/|T|$ and are functions of the relative strong phase between P and T. A precise knowledge of $|P|/|T|$ would permit an accurate determination of α from a measurement of the two terms in the asymmetry [3,9]. One way of calculating $|P|/|T|$ in perurbative QCD is described by Deshpande at this meeting [10]. However, the uncertainty in this calculation has not yet been shown to be under control.

As long as $|P|/|T|$ cannot be calculated reliably, a theoretically cleaner way of resolving the penguin correction will require combining the asymmetry in $B^0 \to \pi^+\pi^-$ with other measurements. A very early suggestion [11] was to also measure the rates of isospin related processes, $B^+ \to \pi^+\pi^0$ and $B^0/\bar{B}^0 \to \pi^0\pi^0$. However, it is expected that the color-suppressed rates of tagged B^0 and $\bar{B}^0$ decays to neutral pions will involve large experimental errors, at least in the first round of experiments. Alternatively, α can be determined by applying the isospin technique to $B \to \rho\pi$ decays [12], or by relating the time-dependence in $B^0(t) \to \pi^+\pi^-$ and in $B_s(t) \to K^+K^-$ using flavor U-spin symmetry [13]. Since none of these methods is expected to be both theoretically clean and experimentally accessible in the near future, some of these methods will have to be combined in order to reduce the error in α.

The phase $\gamma \equiv \text{Arg}(-V^*_{ub}V_{ud}/V^*_{cb}V_{cd})$ ($= \text{Arg}V^*_{ub}$ in the standard convention) is the relative weak phase between a CKM-favored ($b \to c$) and a CKM-suppressed ($b \to u$) decay amplitude. Therefore, it contributes to direct CP asymmetries, in which two such amplitudes interfere, without the need for neutral B mixing. Asymmetry measurements in charged B decays and in self-tagged neutral B decays have the advantage of avoiding the price of flavor tagging. However their theoretical interpretation in terms of the weak phase γ involves an unknown strong phase and an unknown ratio of two interferening hadronic matrix elements. A few methods were proposed to overcome this difficulty, often by measuring not only the asymmetry, but also the rates of certain related processes which provide information on these matrix elements. In the next few sections we describe several ways of measuring γ in four classes of processes, discussing in each case both the theoretical uncertainties and the experimental limitations.

2 $B \to DK$

A simple idea for measuring γ in $B^\pm \to DK^\pm$ was proposed some time ago [14]. Neglecting very small CP violation in $D^0 - \bar{D}^0$ mixing, one can write a triangle amplitude relation

$$\sqrt{2}A(B^+ \to D^0_1 K^+) = A(B^+ \to D^0 K^+) + A(B^+ \to \bar{D}^0 K^+) \ . \qquad (2)$$

D^0_1 is an even CP-eigenstate, decaying, for instance, to $\pi^+\pi^-$, while D^0 and $\bar{D}^0$ are two opposite flavor states. The two amplitudes on the right-hand-side contain CKM factors $V^*_{ub}V_{cs}$ and $V^*_{cb}V_{us}$, both of which are $\mathcal{O}(\lambda^3)$ [15], and have a relative weak phase γ. If the two amplitudes were of about equal magnitudes, then the triangle relation and its charge-conjugate would permit a determination of $\sin\gamma$ within certain discrete ambiguities.

At a closer examination one observes, however, that the first amplitude is suppressed relative to the second one by a CKM factor, $|V_{ub}^* V_{cs}|/|V_{cb}^* V_{us}| \approx 0.4$, and by a color-suppression factor of about 0.25. [This prediction comes from evidence for color-supression in $B \to \bar{D}\pi$ [16]]. An order of magnitude suppression of this amplitude relative to the measured amplitude [17] of $B^+ \to \bar{D}^0 K^+$, causes serious experimental difficulties in tagging the flavor of D^0, through decays such as $D^0 \to K^-\pi^+$. The very rare decay $B^+ \to D^0 K^+ \to (K^-\pi^+)K^+$ interferes strongly with the doubly Cabibbo-suppressed decay of $\bar{D}^0$ in $B^+ \to \bar{D}^0 K^+ \to (K^-\pi^+)K^+$.

Two ways were proposed for partially evading this difficulty by considering only very rare decays, typically with branching ratios of a few times 10^{-7}. Aside from the small rates, both methods have other limitations.

- Study only color-suppressed decays [18] $B^0 \to D^0(\bar{D}^0)K^{*0}$, where the flavors of neutral D and K^* are determined through $D^0 \to K^-\pi^+$, $K^{*0} \to K^+\pi^-$. Here the interference between Cabibbo-favored and Cabibbo-suppressed neutral D decays is much weaker. Still, such interference occurs and prohibits a precise determination of γ.

- Study $B^+ \to D^0(\bar{D}^0)K^+ \to fK^+$, with two neutral D decay modes, such as $f = K^-\pi^+,\ K^-\rho^+$. In this case the two interfering amplitudes have comparable magnitudes, one being color-suppressed in B decay and the other being doubly-Cabibbo-suppressed in D decay. Measurement of the rates of these two processes and their charge-conjugates would permit a determination of γ [19], provided that the two doubly Cabibbo-suppressed D decay branching ratios were known. Uncertainties in these branching ratios prohibit an accurate determination of γ. An intrinsic uncertainty [20] follows from the difficulty of disentangling doubly Cabibbo-suppressed D^0 decays from $D^0 - \bar{D}^0$ mixing [21]. The mixing may be larger than conventionally estimated [22], as anticipated recently from large resonance contributions in D^0 decays [23]. Therefore, a precise measurement of γ using this method requires knowledge of the $D^0 - \bar{D}^0$ mixing parameters.

There are also variants [24], which combine the above two schemes [14,19] based on CP and flavor states. In all cases, the very small branching ratios of the color-suppressed processes imply that such measurements cannot be carried out effectively in the first round of experiments at e^+e^- B factories, and would have to wait for hadron collider experiments providing at least 10^9 B's.

An interesting question remains [25]: What can be learned by studying only the more abundant color-allowed $B^{\pm} \to DK^{\pm}$ decays which have larger branching ratios? Considers the charge-averaged ratio of rates for neutral D mesons decaying to an even (odd) CP state and for a color-allowed flavor state

$$R_i \equiv \frac{2[\Gamma(B^+ \to D_i K^+) + \Gamma(B^- \to D_i K^-)]}{\Gamma(B^+ \to \bar{D}^0 K^+) + \Gamma(B^- \to D^0 K^-)} \ , \qquad i = 1, 2 \ . \tag{3}$$

One finds (neglecting the small $D_1^0 - D_2^0$ width-difference)

$$R_{1,2} = 1 + \bar{r}^2 \pm 2\bar{r} \cos\bar{\delta} \cos\gamma \ , \tag{4}$$

where $A(B^+ \to D^0 K^+)/A(B^+ \to \bar{D}^0 K^+) \equiv \bar{r} \exp[i(\bar{\delta} + \gamma)]$. This leads to two general inequalities

$$\sin^2\gamma \leq R_{1,2} \ , \qquad i = 1, 2 \ , \tag{5}$$

one of which *must* imply a new constraint on γ, unless $\gamma = \pi/2$.

Assuming, for instance, $\bar{r} = 0.1$, $\bar{\delta} = 0$, $\gamma = 40^\circ$, one finds $R_2 = 0.85$. With 10^8 B^+B^- pairs, and using measured B and D decay branching ratios [26], one estimates an error [25] $R_2 = 0.85 \pm 0.05$. In this case, Eq.(5) excludes the range $73^\circ < \gamma < 107^\circ$ with 90% confidence level. Including measurements of the CP asymmetries in $B^{\pm} \to D_i K^{\pm}$ permit, in principle, a determination of γ (and not only bounds on the angle). This depends, of course, on the unknown strong phase $\bar{\delta}$. For further studies of related methods, see [27].

3 $B_s^0(t) \to D_s K$

The interference between the two λ^3 subprocesses $b \to c\bar{u}s$ and $b \to u\bar{c}s$ operates also in the time-dependent decay $B_s^0 \to D_s^- K^+$, leading to a $\sin(\Delta m_s t)$ term [28]

$$\Gamma(B_s^0(t) \to D_s^- K^+) = e^{-\Gamma_s t}[|A|^2 \cos^2(\frac{\Delta m_s t}{2}) + |\bar{A}|^2 \sin^2(\frac{\Delta m_s t}{2}) - |A\bar{A}| \sin(\delta + \gamma) \sin(\Delta m_s t)] \ . \tag{6}$$

The two color-allowed amplitudes, corresponding to $b \to c\bar{u}s$ and $b \to u\bar{c}s$, have magnitudes $|A|$ and $|\bar{A}|$, and involve relative strong and weak phases, δ and γ, respectively.

These four parameters describe in a similar way the time-dependence in the three processes in which the initial and/or final states are charge-conjugated, $\bar{B}_s^0(t) \to D_s^- K^+$, $B_s^0(t) \to D_s^+ K^-$, $\bar{B}_s^0(t) \to D_s^+ K^-$. Thus, measuring the time-dependent oscillations in these four processes, all of which require tagging the flavor of initial B_s^0, permits a determination of γ [29]. It is obvious that, for an accurate measurement, one would also need to know the width-difference between the two B_s mass eigenstates, which was neglected in (6). Further studies and discussions of width-dependent effects can be found in [30].

4 $B \to PP$

We will consider B decays to two light pseudoscalars, $B \to PP$ where $P = \pi, K$, within the framework of flavor SU(3) symmetry [31,32,33,34]. Final states involving η and η' can be studied similarly [35]. Occasional reference to SU(3) breaking effects will be made. A more ambitious approach, relying on generalized factorization [10,36,37,38], has not yet been justified quantitatively to a satisfactory level.

The low energy effective weak Hamiltonian [39] governing $B \to PP$

$$\mathcal{H} = \frac{G_F}{\sqrt{2}} \sum_{q=d,s} \left(\sum_{q'=u,c} \lambda_{q'}^{(q)} [c_1 Q_1^{(q')} + c_2 Q_2^{(q')}] - \lambda_t^{(q)} \sum_{i=3}^{10} c_i Q_i^{(q)} \right) , \tag{7}$$

where $\lambda_{q'}^{(q)} = V_{q'b}^* V_{q'q}$, consists of the sum of three types of four quark operators: two $(V-A)(V-A)$ current-current operators ($Q_{1,2}$), four QCD penguin operators ($Q_{3,4,5,6}$), and four electroweak penguin (EWP) operators ($Q_{7,8,9,10}$) with different chiral structures. The EWP operators with dominant Wilson coefficients, Q_9 and Q_{10} ($|c_{7,8}|/|c_9| \approx 0.04$), have both a $(V-A)(V-A)$ structure. All the four-quark operators, $(\bar{b}q_1)(\bar{q}_2 q_3)$, can be decomposed into a sum of $\overline{\mathbf{15}}$, $\mathbf{6}$ and $\overline{\mathbf{3}}$ representations [31,33,34]. The QCD penguin operators are pure $\overline{\mathbf{3}}$.

A simple proportionality relation was noted recently [40] to hold between the dominant EWP operators Q_9 and Q_{10} and the current-current ("tree") operators Q_1 and Q_2, both trasforming as given representations

$$\mathcal{H}_{EWP}^{(q)}(\overline{\mathbf{15}}) = -\frac{3}{2}\frac{c_9 + c_{10}}{c_1 + c_2}\frac{\lambda_t^{(q)}}{\lambda_u^{(q)}}\mathcal{H}_T^{(q)}(\overline{\mathbf{15}}) , \tag{8}$$

$$\mathcal{H}_{EWP}^{(q)}(\mathbf{6}) = \frac{3}{2}\frac{c_9 - c_{10}}{c_1 - c_2}\frac{\lambda_t^{(q)}}{\lambda_u^{(q)}}\mathcal{H}_T^{(q)}(\mathbf{6}) . \tag{9}$$

The superscripts $q = d, s$ denote strangeness-conserving and strangeness-changing transitions, respectively. The two ratios of Wilson coefficients are equal within 3% [39]

$$\frac{c_9 + c_{10}}{c_1 + c_2} \approx \frac{c_9 - c_{10}}{c_1 - c_2} \approx -1.12\alpha . \tag{10}$$

As a consequence of (8) and (9), processes given by $\overline{\mathbf{15}}$ and $\mathbf{6}$ transitions contain EWP and current-current contributions which are proportional to each other. For instance, in the $\Delta S = 1$ $\overline{\mathbf{15}}$-dominated decay, $B \to (K\pi)_{I=3/2}$, where $|I = 3/2\rangle = |K^0\pi^+\rangle + \sqrt{2}|K^+\pi^0\rangle$, the ratio of these amplitudes

$$\delta_{EW} = -\frac{3}{2}\frac{c_9 + c_{10}}{c_1 + c_2}\frac{|V_{tb}^*V_{ts}|}{|V_{ub}^*V_{us}|} = 0.65 \pm 0.15 \tag{11}$$

is of order one [41], in spite of the fact that EWP amplitudes are higher order in electroweak couplings.

This simple result, obtained in the limit of flavor SU(3), was used [42] in order to obtain a bound on γ in $B^\pm \to K\pi$. Expressing $B^+ \to K\pi$ in terms of reduced SU(3) amplitudes depicted in graphical form [33], one has [43]

$$A(B^+ \to K^0\pi^+) = \lambda_t^{(s)}[P + EW] + |\lambda_u^{(s)}|e^{i\gamma}[P_{uc} + A] \ , \tag{12}$$
$$\sqrt{2}A(B^+ \to K^+\pi^0) = -\lambda_t^{(s)}[P + EW] - |\lambda_u^{(s)}|[(T + C)(e^{i\gamma} - \delta_{EW}) + P_{uc} + A] \ .$$

The dominant QCD penguin amplitudes P in the two processes, carrying a weak phase $\mathrm{Arg}\lambda_t^{(s)} = \pi$, are equal because of isospin. Unequal rates of the two processes would be evidence for interference with smaller amplitudes involving a different weak phase. Defining the charge-averaged ratio of rates [42]

$$R_*^{-1} \equiv \frac{2[B(B^+ \to K^+\pi^0) + B(B^- \to K^-\pi^0)]}{B(B^+ \to K^0\pi^+) + B(B^- \to \bar{K}^0\pi^-)} \ , \tag{13}$$

one finds, to leading order in small quantities

$$R_*^{-1} = 1 - 2\epsilon\cos\phi(\cos\gamma - \delta_{EW}) + \mathcal{O}(\epsilon^2) + \mathcal{O}(\epsilon\epsilon_A) + \mathcal{O}(\epsilon_A^2) \ . \tag{14}$$

ϵ and ϵ_A are ratios of tree-to-penguin and rescattering[44] -to-penguin amplitudes, respectively

$$\epsilon e^{i\phi} = \frac{|\lambda_u^{(s)}|}{|\lambda_t^{(s)}|}\frac{T + C}{P + EW} \ , \quad \epsilon_A = \frac{|\lambda_u^{(s)}[P_{uc} + A]|}{|\lambda_t^{(s)}[P + EW]|} \ . \tag{15}$$

The first ratio is given by [8,45]

$$\epsilon = \sqrt{2}\frac{V_{us}}{V_{ud}}\frac{f_K}{f_\pi}\frac{|A(B^+ \to \pi^0\pi^+)|}{|A(B^+ \to K^0\pi^+)|} = 0.21 \pm 0.05 \ , \tag{16}$$

while the second one is roughly [8,46]

$$\epsilon_A \sim \lambda\frac{|A(B^+ \to \bar{K}^0K^+)|}{|A(B^+ \to K^0\pi^+)|} < 0.12 \tag{17}$$

Neglecting second order terms, one obtains a bound [42]

$$|\cos\gamma - \delta_{EW}| \geq \frac{|1 - R_*^{-1}(K\pi)|}{2\epsilon}\,, \tag{18}$$

which would provide useful information about γ if a value different from one were measured for R_*^{-1}. The present value, $R_*^{-1}{}_{\rm exp} = 1.27 \pm 0.48$, is consistent with one. Further information about γ, applying also to the case $R_*^{-1} = 1$, can be obtained by measuring separately B^+ and B^- decay rates [47]. The solution obtained for γ involves uncertainties due to SU(3) breaking in subdominant amplitudes and an uncertainty in $|V_{ub}/V_{cb}|$, both of which affect the value of δ_{EW}. Combined with errors in $\epsilon \propto |A(B^+ \to \pi^+\pi^0)/A(B^+ \to K^0\pi^+)|$, and in the rescattering parameter ϵ_A, the resulting uncertainty in γ is unlikely to be smaller than 20° [47]. For other ways of studying γ in $B \to K\pi$, see [48].

5 $B \to VP$

B mesons decays to a charmless vector meson (V) and a pseudoscalar meson (P) involve a larger number of SU(3) amplitudes [49,50] than $B \to PP$. SU(3) relations between EWP and current-current amplitudes, following from (8) and (9), reduce considerably the number of independent amplitudes. In this section we describe briefly two applications [40].

5.1 $B^\pm \to \rho K$

Defining a charge-averaged ratio of rates for $B^\pm \to \rho K$

$$R_*^{-1}(\rho K) \equiv \frac{2[B(B^+ \to \rho^0 K^+) + B(B^- \to \rho^0 K^-)]}{B(B^+ \to \rho^+ K^0) + B(B^- \to \rho^- \bar{K}^0)}\,, \tag{19}$$

one obtains, with some analogy to (18), the bound

$$|\cos\gamma| \geq \frac{|1 - R_*^{-1}(\rho K)|}{2\epsilon_V} - \delta_{EW}\left(\frac{\epsilon_P}{\epsilon_V}\right)\,, \tag{20}$$

where

$$\epsilon_V = \sqrt{2}\frac{V_{us}}{V_{ud}}\frac{f_K}{f_\pi}\frac{|A(B^+ \to \rho^0\pi^+)|}{|A(B^+ \to \rho^+ K^0)|}, \quad \epsilon_P = \sqrt{2}\frac{V_{us}}{V_{ud}}\frac{f_{K^*}}{f_\rho}\frac{|A(B^+ \to \rho^+\pi^0)|}{|A(B^+ \to \rho^+ K^0)|}. \tag{21}$$

Although this constraint is weaker than (18), it shows that measuring charge-averaged ratios of rates, which differ from one, is of interest also for PV final states.

5.2 $B^0 \to K^{*\pm}\pi^{\mp}$ *vs.* $B^\pm \to \phi K^\pm$

Considering the amplitudes for $B^0 \to K^{*+}\pi^-$ and $B^+ \to \phi K^+$, both of which are expected to be QCD penguin dominated, and keeping only dominant and subdominant terms, one finds

$$\begin{aligned} A(B^0 \to K^{*+}\pi^-) &= -\lambda_t^{(s)}\, P_P - \lambda_u^{(s)}\, T_P \,, \\ A(B^+ \to \phi K^+) &= \lambda_t^{(s)}\, [P_P + EW] \,, \end{aligned} \tag{22}$$

where the suffix P denotes the pseudoscalar which includes the spectator in the graphic SU(3) amplitudes [49]. The EWP contribution is related to the tree amplitude by [40]

$$\lambda_t^{(s)} EW = \frac{1}{3}\delta_{EW}\, |\lambda_u^{(s)}| T_P \,. \tag{23}$$

Defining the ratio

$$R = \frac{|A(B^0 \to K^{*+}\pi^-)|^2 + |A(\bar{B}^0 \to K^{*-}\pi^+)|^2}{|A(B^+ \to \phi K^+)|^2 + |A(B^- \to \phi K^-)|^2} \,, \tag{24}$$

one finds

$$R = \frac{1 + r^2 - 2r\, \cos\delta \cos\gamma}{1 + (\delta_{EW}/3)^2\, r^2 - (2/3)\, \delta_{EW}\, r\, \cos\delta} \,, \tag{25}$$

where

$$r\, e^{i\delta} = \frac{|\lambda_u^{(s)}|\, T_P}{|\lambda_t^{(s)}|\, P_P} \,. \tag{26}$$

Present 90% confidence level limits [8], $\mathcal{B}(B^0 \to K^{*\pm}\pi^{\mp}) > 12 \times 10^{-6}$ and $\mathcal{B}(B^\pm \to \phi K^\pm) < 5.9 \times 10^{-6}$, imply $R > 2$, which is evidence for a nonzero contribution of T_P. In order to use this inequality for information about γ, one must include some input about r and δ. A reasonable assumption, supported both by perturbative [36] and statistical [51] calculations, is that δ does not exceed 90°. A very conservative assumption about r is $r \leq 1$. Making these two assumptions, one finds [for δ_{EW} we will use the range (11)]

$$\cos\gamma - \frac{2}{3}\delta_{EW} < \frac{-1 + r^2[1 - 2(\delta_{EW}/3)^2]}{2r} \,. \tag{27}$$

This implies $\gamma > 62^\circ$ for $r = 1$, and $\gamma > 105^\circ$ for $r = 0.5$. Indirect evidence [50] exists already for $r < 0.55$. Direct information on r will be obtained from future rate measurements of $B^+ \to K^{*0}\pi^+$ and $B^+ \to \rho^+\pi^0$ or $B^0 \to \rho^+\pi^-$.

These bounds neglect smaller terms in the amplitudes, primarily the color-suppressed tree amplitude. If this contribution is 10% (20%) of the color-favored tree amplitude T_P [36], then the limits move up or down by about

5° (10°). The bounds also assume [by SU(3)] equal QCD penguin contributions in the two processes. The constraint becomes stronger if the penguin amplitude in $B^+ \to \phi K^+$ is larger than in $B^0 \to K^{*+}\pi^-$, as predicted by factorization [52].

6 Conclusion

We discussed several ways of determining the weak phase γ. The first two schemes, based on $B \to DK$ and $B_s \to D_s K$, seem at first sight to involve no hadronic uncertainties. However, at a closer look, they require taking care of smaller effects, such as $D^0 - \bar{D}^0$ mixing and the width-difference in the B_s^0 system. Methods based on charmless decays to two pseudoscalars, and decays to a pseudoscalar and a vector meson, involve dynamical hadronic parameters, such as those describing SU(3) breaking and rescattering amplitudes. Some of these quantities are under control, and others should be studied through a dialogue between theory and experiments.

Acknowledgments

This work was supported in part by the Israel Science Foundation founded by the Israel Academy of Sciences and Humanities, by the United States – Israel Binational Science Foundation under Research Grant Agreement 98-00237, by the Technion V.P.R. fund - Harry Werksman Research Fund, and by P. and E. Natahan Research Fund.

References

1. For a recent review, see M. Gronau, hep-ph/9908343, to appear in the Proceedings of the 1999 Chicago Conference on Kaon Physics, Chicago, IL, June 1999, J. L. Rosner and B. Winstein, eds. (University of Chicago Press, 2000). For an earlier comprehensive study, see *The BaBar Physics Book*, P. F. Harrison and H. R. Quinn, eds. (SLAC Report SLAC-R-504, 1998).
2. A. B. Carter and A. I. Sanda, Phys. Rev. Lett. **45**, 952 (1980); Phys. Rev. D **23**, 1567 (1981); I. I. Bigi and A. I. Sanda, Nucl. Phys. **193**, 85 (1981).
3. M. Gronau, Phys. Rev. Lett. **63**, 1451 (1989).
4. D. London and R. D. Peccei, Phys. Lett. B **223**, 257 (1989).
5. W. Trischuk, these proceedings; CDF Collaboration, hep-ex/9909003.
6. R. Sinha, these proceedings, and Phys. Rev. D **60**, 074020 (1999).

7. A. Dighe, M. Gronau and J. L. Rosner, Phys. Rev. Lett. **79**, 4333 (1997); M. Gronau and J. L. Rosner, hep-ph/9909478, Phys. Rev. D. **61**, 0540xx (2000).
8. J. Smith, these proceedings; CLEO CONF 99-13, CLEO CONF 99-14.
9. F. DeJongh and P. Sphicas, Phys. Rev. D **53**, 4930 (1996); P. S. Marrocchesi and N. Paver, Int. J. Mod. Phys. A **13**, 251 (1998); J. Charles, Phys. Rev. D **59**, 054007 (1999).
10. N. Deshpande, these proceedings, and hep-ph/9909298.
11. M. Gronau and D. London, Phys. Rev. Lett. **65**, 3381 (1990).
12. H. J. Lipkin, Y. Nir, H. R. Quinn and A. Snyder, Phys. Rev. D **44**, 1454 (1991); M. Gronau, Phys. Lett. B **265**, 389 (1991); H. R. Quinn and A. Snyder, Phys. Rev. D **48**, 2139 (1993).
13. I. Dunietz, Proceedings of the Workshop on *B* Physics at Hadron Accelerators, Snowmass, CO, 1993, p. 83; D. Pirjol, Phys. Rev. D **60**, 54020 (1999); R. Fleischer, Phys. Lett. B **459**, 306 (1999).
14. M. Gronau and D. Wyler, Phys. Lett. B **265**, 172 (1991).
15. L. Wolfenstein, Phys. Rev. Lett. **51**, 1945 (1983).
16. T. E. Browder, K. Honscheid and D. Pedrini, Ann. Rev. Nucl. Part. Sci. **46**, 395 (1997).
17. CLEO Collaboration, M. Athanas *et al.*, Phys. Rev. Lett. **80**, 5493 (1998).
18. M. Gronau and D. London, Phys. Lett. B **253**, 483 (1991); I. Dunietz, Phys. Lett. B **270**, 75 (1991).
19. D. Atwood, I. Dunietz and A. Soni, Phys. Rev. Lett. **78**, 3257 (1997).
20. J. P. Silva and A. Soffer, hep/ph-9912242.
21. D. Asner, these proceedings; CLEO Collaboration, CLEO CONF 99-17.
22. L. Wolfenstein, Phys. Lett. B **164**, 170 (1985); J. Donoghue, E. Golowich, B. Holstein and J. Trampetic, Phys. Rev. D **33**, 179 (1986); T. A. Kaeding, Phys. Lett. B **357**, 151 (1995).
23. M. Gronau, Phys. Rev. Lett. **83**, 4005 (1999).
24. A. Soffer, Phys. Rev. D **60**, 054032 (1999).
25. M. Gronau, Phys. Rev. D **58**, 037301 (1998).
26. Particle Data Group, C. Caso, *et al.*, Zeit. Phys. C **3**, 1 (1998).
27. Z-z. Xing, Phys. Rev. D **58**, 093005 (1998); M. Gronau and J. L. Rosner, Phys. Lett. B **439**, 171 (1998); J.-H. Jang and P. Ko, Phys. Rev. D **58**, 111302 (1998); N. Sinha and R. Sinha Phys. Rev. Lett. **80**, 3706 (1998); B. Kayser and D. London, hep-ph/9909561.
28. M. Gronau, Phys. Lett. B **23**, 479 (1989).
29. R. Aleksan, I. Dunietz, B. Kayser and F. Le Diberder, Nucl. Phys. **361**, 141 (1991).

30. R. Aleksan, A. Le Yaouanc, L. Oliver, O. Pene and J.C. Raynal, Zeit. Phys. C **67**, 251 (1995); I. Dunietz, Phys. Rev. D **52**, 3048 (1995); R. Fleischer and I. Dunietz, Phys. Rev. D **55**, 259 (1997).
31. D. Zeppenfeld, Zeit. Phys. C **8**, 77 (1981).
32. L. L. Chau *et al.*, Phys. Rev. D **43**, 2176 (1991).
33. M. Gronau, O. Hernández, D. London, and J. L. Rosner, Phys. Rev. D **50**, 4529 (1994).
34. B. Grinstein and R.F. Lebed, Phys. Rev. D **53**, 6344 (1996).
35. A. S. Dighe, M. Gronau, and J. L. Rosner, Phys. Lett. B **367**, 357 (1996); **377**, 325(E) (1996); Phys. Rev. D **54**, 3309 (1996); Phys. Rev. Lett. **79**, 4333 (1997).
36. M. Beneke, G. Buchalla, M. Neubert, and C. Sachrajda, Phys. Rev. Lett. **83**, 1914 (1999).
37. H. -Y. Cheng, these proceedings, and hep-ph/9910291.
38. H. -N. Li, these proceedings, and hep-ph/9903323.
39. G. Buchalla, A.J. Buras and M.E. Lautenbacher, Rev. Mod. Phys. **68**, 1125 (1996).
40. M. Gronau, hep-ph/9911429, to be published in Phys. Rev. D.
41. R. Fleischer, Zeit. Phys. C **62**, 81 (1994); N. G. Deshpande and X.-G. He, Phys. Lett. B **336**, 471 (1994).
42. M. Neubert and J. L. Rosner, Phys. Lett. B **441**, 403 (1998).
43. M. Gronau, D. Pirjol, and T.-M. Yan, Phys. Rev. D **60**, 034021 (1999).
44. For a definition of rescattering in terms of graphical SU(3) amplitudes, see M. Gronau and J. L. Rosner, Phys. Rev. D **58**, 113005 (1998).
45. M. Gronau, J. L. Rosner, and D. London, Phys. Rev. Lett. **73**, 21 (1994).
46. A. Falk, A. L. Kagan, Y. Nir, and A. A. Petrov, Phys. Rev. D **57**, 6843 (1998).
47. M. Neubert and J. L. Rosner, Phys. Rev. Lett. **81**, 5076 (1998); M. Neubert, JHEP **9902**, 014 (1999); M. Gronau and D. Pirjol, Phys. Rev. D **61**, 013005 (2000).
48. R. Fleischer and T. Mannel, Phys. Rev. D **57**, 2752 (1998); M. Gronau and J. L. Rosner, Phys. Rev. D **57**, 6843 (1998); A. Buras and R. Fleischer, Euro. Phys. J. C **11**, 93 (1999).
49. A. S. Dighe, M. Gronau, and J. L. Rosner, Phys. Rev. D **57**, 1783 (1998).
50. M. Gronau and J. L. Rosner, hep-ph/9909478, Phys. Rev. D **61**, 0540xx (2000).
51. M. Suzuki, these proceedings, and Phys. Rev. D **60**, 074019 (1999).
52. C. D. Lu, these proceedings, and Phys. Rev. D **58**, 094009 (1998).

HADRONIC MATRIX ELEMENTS FROM THE LATTICE

AMARJIT SONI

Theory Group, Brookhaven National Laboratory, Upton, NY 11973, USA
E-mail: soni@bnl.gov

Lattice matrix elements are briefly reviewed. In the quenched approximation f_B, B_B and B_K are now under good control. Experimental hints for $f_{D_S}^{\rm expt} > f_{D_S}^{\rm QQCD}$ are noted; precise determination of f_{D_S} from experiment as well as from the lattice is strongly advocated. Lattice calculations of the form factor for $B \to \pi\ell\nu$ at relatively large value of q^2 have made good progress and should be useful in conjunction with precise measurement of the differential spectra expected from B-factories. Recent attempt at $K \to \pi\pi$ using staggered quarks is briefly discussed; use of non-perturbative renormalization, improved actions and operators with staggered quarks is emphasized. Due to the good chiral behavior of domain wall quarks it would be useful to study $K \to \pi\pi$ with this discretization.

1 Introduction

Hadronic matrix elements are of crucial importance for constraining the parameters of the Standard Model (SM) in conjunction with experimental information. Lattice approach has already attained considerable success in handling B-parameters and decay constants and to a lesser degree semi-leptonic form factors. [1] Weak decays into purely hadronic final states, of which $K \to \pi\pi$ is the simplest example, are extremely problematic. [2] After almost a decade the topic is again getting some attention due in part to important developments with regard to chiral symmetry on the lattice. [3]

2 Heavy-light decay constants: effects of quenching. [1]

Lattice methods have made considerable progress in pinning down, fairly accurately, f_B and other heavy-light decay constants, in the quenched approximation (see Table 1). Unfortunately, the effects of quenching could be substantial; current estimates place them around $\sim (20 \pm 10)\%$, seriously limiting the phenomenological applications. Since the past few years there is heightened activity to accurately ascertain the effects of quenching. For technical reasons, the dynamical simulations are with $N_f = 2$ i.e. with only two "light" sea-quarks rather than the three (u, d, s) in real life. Furthermore, the mass of the sea-quarks tends to be relatively heavy. The indications from these dynamical simulations is that $f_B^{\rm dynamical} > f_B^{\rm quenched}$.

Table 1. Heavy-light decay constants and their ratios. [1]

Quantity	Quenched ($N_f = 0$)	Partially Unquenched ($N_f = 2$)
f_B/MeV	170 ± 20	200 ± 30
f_{B_S}/MeV	190 ± 20	220 ± 30
f_D/MeV	205 ± 20	225 ± 30
f_{D_S}/MeV	225 ± 20	245 ± 30
f_{B_S}/f_B	$1.14 \pm .06$	$1.14 \pm .06$
f_{D_S}/f_D	$1.10 \pm .06$	$1.10 \pm .06$

3 Hints from Experiment: $f_{D_S}^{\text{expt}} > f_{D_S}^{\text{QQCD}}$?

Table 2 exhibits a compilation of the results for f_{D_S} from several experiments. Curiously, the central value of all but one experiment is above the value for f_{D_S} from quenched QCD simulations: $f_{D_S}^{\text{QQCD}} = 225 \pm 20$ MeV. Since the errors in the existing experimental numbers are rather large we cannot draw strong conclusions; it seems plausible, nevertheless, that these experiments are also indicating that quenched QCD tends to underestimate the heavy-light pseudoscalar decay constants.

Table 2. Experimental determinations of f_{D_S}. [4]

Experiment	Mode	f_{D_S}/MeV
S. Aoki *et al.* (WA75)	$\mu^+\nu_\mu$	$238 \pm 47 \pm 21 \pm 48$
J.Z. Bai *et al.* (BES)	"	$430^{+150}_{-130} \pm 40$
K. Kodema *et al.* (E653)	"	$190 \pm 34 \pm 20 \pm 26$
M. Chada *et al.* (CLEO)	"	$280 \pm 19 \pm 28 \pm 34$
M. Acciari *et al.* (L3)	$\tau^+\nu_\tau$	$309 \pm 58 \pm 33 \pm 38$
F. Parodi *et al.* (DELPHI)	"	330 ± 95
(ALEPH)	inclusive	284 ± 62

4 Precise studies of f_{D_S}

For experiment as well as for dynamical lattice simulations, a precise determination of $f_{B_D}(f_{B_S})$ is significantly more problematic than f_{D_S}; in fact experimentally direct determination of f_B is not of immediate reach. Therefore, it would be very useful if the experimental as well as the lattice determinations of f_{D_S} could be improved to 10–15% accuracy. A comparison between the two would then serve as a very useful guide for correcting $f_{B_d}(f_{B_S})$ from the lattice. In this regard the ratios f_{B_d}/f_{D_S} and f_{B_S}/f_{D_S} from the lattice would clearly be useful.

5 Heavy-light B parameters

Lattice has had success with heavy-light B-parameters for a long time although calculations in the static approximation are still somewhat problematic. [1] Table 3 presents a brief summary. Recall that a precise value of the ratio $f_{B_S}(B_{B_S})^{1/2}/f_{B_d}(B_{B_d})^{1/2}$ is needed in accurately deducing V_{td}/V_{ts} once B_S-$\bar{B}_S$ oscillations get experimentally measured.

Table 3. Summary of heavy-light B-parameters. [1]

	Quenched	"Unquenched"
$B_{B_d}(m_b)$	.86(4)(8)	.86(4)(8)
B_{B_S}/B_{B_d}	1.00(1)(2)	1.00(1)(2)
$f_{B_d}(\hat{B}^{nlo}_{B_d})^{1/2}$	195 ± 25 MeV	230 ± 35
$\frac{f_{B_S}(B^{nlo}_{B_S})^{1/2}}{f_{B_d}(B^{nlo}_{B_d})^{1/2}}$	$1.14 \pm .06$	$1.14 \pm .07$

In view of the importance of the aforesaid ratio it may be useful to determine it also more directly from the SU(3) breaking ratio: [5]

$$\frac{M_{B_S}(\mu)}{M_{B_d}(\mu)} = \frac{\langle \bar{B}_{B_S}|(\bar{b}\gamma_\rho(1-\gamma_5)s)^2|B_{B_S}\rangle}{\langle \bar{B}_{B_d}|(\bar{b}\gamma_\rho(1-\gamma_5)d)^2|B_{B_d}\rangle}$$

Many of the systematic errors should cancel in this ratio. However, in the attempts that have been so far made with this direct method, [5,6] the errors are not yet small enough to make this method competitive with the traditional $f_B^2 B$ method.

6 Semi-leptonic form factors, $B \to \pi(\rho)\ell\nu$.

The large b-quark mass presents a difficult computational problem. The cleanest lattice simulations are for 'rest to rest' i.e. both the initial and final meson at rest. Then the large B mass forces q^2 (q = lepton 4-momentum $= p_B - p_\pi$) to be rather large. For some phenomenological applications the value of the form factor(s) are needed near $q^2 \sim 0$. This entails large extrapolations, introducing additional errors and model dependence. However, one important phenomenological application, namely deducing the mixing angle V_{ub} from experimental data, only requires precise knowledge of the form factor at one value of q^2. This should work so long as the experiment has enough data so that even the differential rate around that region of q^2 is accurately determined, which is anticipated to be feasible at the B-factories. In this regard two recent approaches are noteworthy. Both of these efforts avoid the use of large extrapolations in q^2 and focus instead on accurate predictions in a limited region of q^2.

UKQCD focussed on near the end-point or the zero-recoil region where the lattice data tends to be cleanest. Furthermore, heavy quark symmetry also provides useful scaling relations in this region. [8] Their result for the form factor f^+ as a function of q^2 is given below.

Table 4. $f^+(q^2)$ from UKQCD. [7]

$q^2(\mathrm{GeV}^2)$	16.7	18.1	19.5	20.9	22.3
$f^+(q^2)$	0.9^{+1+2}_{-2-1}	1.1^{+2+2}_{-2-1}	1.4^{+2+3}_{-2-1}	1.8^{+2+4}_{-2-1}	2.3^{+3+6}_{-3-1}

The FNAL group [9] is also focussing on an approach towards the semi-leptonic form factors for $B \to \pi\ell\nu$, $D \to \pi(K)\ell\nu$ suitable for accurate determinations of mixing-angles in conjunction with high statistics data samples anticipated from experiments. The key idea here again is to concentrate directly on the differential decay spectrum in an interval with $0.4 \lesssim \vec{p}_\pi/\mathrm{GeV} \lesssim 0.8$ thus avoiding the need for large extrapolation in q^2. The partial width over this interval can be computed on the lattice. From the experimental point of view this should have the advantage of using a range of q^2 wherein the decay rate is not small unlike near the end-point.

7 B_K

The kaon B-parameter, B_K, has been studied most extensively on the lattice. [2] Two important limitations that lattice simulations still need to adequately address are SU(3) breaking (the 'kaon' on the lattice is a pseudoscalar made of degenerate quarks with $m_{\mathrm{pseudoscalar}} \sim m_K$) and the quenched approximation. Both of these effects are expected to be rather small $\sim$ 5–10% and we get $\hat{B}_K = .85 \pm .13$. This number is based on results of various groups [2], amongst which the one from JLQCD is the most precise. [10] The error on the lattice number contains a guess-estimate of the SU(3) breaking and quenching errors. We note, in passing, that there are some preliminary indications that unquenching will increase [11] or decrease [12] B_K by just a few percent. For now, we have not changed the central value of B_K due to this effect; only the systematic errors are increased to reflect this possibility.

8 $K \to 2\pi$ Decays and ϵ'/ϵ.

It was realized long ago that chiral perturbation theory (ChPT) can be used to simplify the problem so that $\langle \pi\pi|Q|K\rangle$ can be obtained by computing on the lattice simpler entities: $\langle \pi|Q|K\rangle$ and $\langle \text{vac}|Q|K\rangle$, where Q is a 4-quark operator. [13] The coefficients in this relation can be calculated using lowest order ChPT. Traditionally this strategy has been available for staggered fermions as they possess a remnant chiral symmetry. [14] Since Wilson fermions explicitly break chiral symmetry this approach cannot be used with this discretization. [15] The new development in this regard is that domain wall quarks (DWQ) [16] are found to be quite practical for simulating QCD and possess excellent chiral behavior. [17] Therefore, the $K \to \pi$ (and $K \to$ vac) method for dealing with $K \to 2\pi$ is amenable to this discretization as well. [3]

Final state interactions (FSI) are of course a serious limitation of the $K \to \pi$ method. However, it should still be very instructive to quantify how well this concrete approximation works. Actually direct $K \to 2\pi$ decays may also be amenable to the lattice due to an elegant application of the CPS symmetry. [18] The restrictions of the Maiani-Testa theorem [19] are bypassed here by working near the threshold. In any case the direct $K \to 2\pi$ methods are computationally extremely intensive and for now there is not much to report based on these methods.

The credit for an extensive study of the $K \to 2\pi$ in the $K \to \pi$ approach with staggered fermions goes to Pekurovsky and Kilcup (PK). [20,21] The work presents a first study of lattice spacing dependence, finite size effects as well as quenching effects. Unfortunately PK used lattice weak coupling perturbation theory [22] (LWCPT) to renormalize the operators and this seems to completely fail for staggered quarks in the case of LR operators such as Q_6 which are crucial for ϵ'/ϵ.

At $\beta = 6.0$, PK observe a significant enhancement of the $\Delta I = 1/2$ channel over the 3/2; however at $\beta = 6.2$, i.e. closer to the continuum limit (compared to $\beta = 6.0$), they find that the central value of the enhancement weakens appreciably. While the large errors at $\beta = 6.2$ do not allow a strong conclusion, more work is needed to unambiguously show that the $\Delta I = 1/2$ enhancement survives in the continuum limit.

PK's calculation of ϵ'/ϵ is seriously hampered as the one-loop LWCPT that they use for renormalization completely fails for Q_6. The perturbation theory corrections are several hundreds of percents showing extreme sensitivity to the renormalization scale as well as the quark mass (see Table 5). As PK themselves clearly emphasize their calculation of ϵ'/ϵ is extremely fragile and not at all reliable.

Table 5. $\langle Q_6 \rangle$ in arbitrary units with one-loop perturbative matching using two values of q^*; for comparison, the renormalized results ("bare") are also given (Table 2 from Ref. 20).

Quark Mass	0.01	0.02	0.03	0.04	0.05
$q^* = 1/a$	0.1 ± 1.2	-0.9 ± 0.4	-1.2 ± 0.2	-1.6 ± 0.3	-1.1 ± 0.2
$q^* = \pi/a$	-13.1 ± 1.8	-9.0 ± 0.5	-7.1 ± 0.3	-6.3 ± 0.5	-4.6 ± 0.5
Bare	-55.6 ± 5.0	-35.4 ± 1.5	-27.0 ± 0.9	-22.3 ± 1.4	-16.4 ± 1.5

It seems quite plausible that for the renormalization of the $\Delta S = 1$, 4-quark operators, as has also been known to some extent to be the case for quark bilinears, the breaking of flavor symmetry by the staggered approach is responsible for the failure of LWCPT. In any case, use of non-perturbative renormalization (NPR) [23] improved actions and/or operators with the staggered approach are highly desirable in the context of $K \to \pi\pi$ decays.

Domain wall quarks are extremely attractive as at the expense of introducing a fictitious 5th dimension they preserve the full SU(N) $\times$ SU(N) chiral symmetries of the continuum theory in the limit of an infinite 5th dimension. [16] Quenched QCD numerical simulations showed that in practice for $\beta \gtrsim 6.0$, 10–20 sites in the 5th dimension may be sufficient to render very good chiral behavior. [17] Early numerical studies also seem to indicate that the discretization errors are effectively $0(a^2)$; if substantiated this improved scaling behavior may off-set the extra cost of the 5th dimension. [17,24]

Calculation of $K \to 2\pi$ and ϵ'/ϵ in this method has been in progress for quite sometime. [3] With DWQ, considerable progress has been made in non-perturbative renormalization of quark bilinears, $\Delta S = 2$ and $\Delta S = 1$ Hamiltonians and so far the method seems promising. [25]

I thank the organizers for a very interesting workshop. This work was supported in part by the U.S. DOE contract DE-AC02-98CH10886.

References

1. For recent reviews of heavy light matrix elements see, S. Hashimoto, hep-lat/9909136; T. Draper, *Nucl. Phys. Proc. Suppl.* **73**, 43 (1999).
2. For recent reviews of kaon matrix elements see, Y. Kuramashi, hep-lat/9910032; G. Martinelli, *Nucl. Phys. Proc. Suppl.* **73**, 58 (1999).
3. T. Blum, *Nucl. Phys. Proc. Suppl.* **73**, 167 (1999).

4. S. Aoki *et al.* (WA75), *Prog Theo. Phys.* **89**, 131 (1993); J.Z. Bai *et al.* (BES), *Phys. Rev. Lett.* **74**, 4599 (1995); D. Costa *et al.* (CLEO), *Phys. Rev.* D **49**, 5690 (1994); K. Kodama *et al.* (E653), *Phys. Lett.* B **382**, 299 (1996); M Chada *et al.* (CLEO), *Phys. Rev.* D **58**, 32002 (1998); M. Acciarri *et al.* (L3), *Phys. Lett.* B **396**, 327 (1997); F. Parodi, P. Roudeau and A. Stocchi, hep-ex/9903063 and references therein; in particular, the ALEPH result is taken from this paper.
5. C. Bernard, T. Blum and A. Soni, *Phys. Rev.* D **58**, 014501 (1998).
6. L. Lellouch *et al.* (UKQCD), hep-ph/9912322; hep-ph/9909018.
7. K.C. Bowler *et al.* (UKQCD), hep-lat/9911011.
8. N. Isgur and M. Wise, *Phys. Rev.* D **42**, 2388 (1990); M. Neubert, *Phys. Rept.* **245**, 259 (1994).
9. S. Ryan *et al.*, hep-lat/9810041; hep-lat/9910010.
10. S. Aoki *et al.* (JLQCD), *Phys. Rev. Lett.* **80**, 5271 (1998).
11. G. Kilcup *et al. Nucl. Phys.* B(Proc. Suppl.) **53**, 345 (1997).
12. A. Soni, *Nucl. Phys.* B(Proc. Suppl.) **47**, 43 (1996).
13. C. Bernard *et al. Phys. Rev.* D **32**, 2343 (1985).
14. G. Kilcup and S. Sharpe, *Nucl. Phys.* B **283**, 493 (1987).
15. M. Bochicchio *et al.*, *Nucl. Phys.* B **262**, 331 (1985).
16. D. Kaplan, *Phys. Lett.* B **288**, 342 (1992); R. Narayanan and H. Neuberger, *Phys. Lett.* B **302**, 62 (1993); *Nucl. Phys.* B **412**, 574 (1994); Y. Shamir, *Nucl. Phys.* B **409**, 90 (1993) and V. Furman and Y. Shamir, *Nucl. Phys.* B **439**, 54 (1995).
17. T. Blum and A. Soni, *Phys. Rev.* D **56**, 174 (1997); *Phys. Rev. Lett.* **79**, 3595 (1997); hep-lat/9712004; T. Blum, A. Soni and M. Wingate, *Phys. Rev.* D **60**, 114507 (1999); P. Chen *et al.* hep-lat/9812011; A. Ali Khan *et al.* (CP-PACS) hep-lat/9909049; L. Wu *et al.* (RBC Collaboration), hep-lat/9909117.
18. C. Bernard *et al.*, *Nucl. Phys.*(Proc. Suppl) **4**, 483 (1988); C. Bernard and A. Soni, *Nucl. Phys.*(Proc. Suppl) **9**, 155 (1989). See also, C. Dawson, *et al.*, *Nucl. Phys.* B **514**, 313 (1998).
19. L. Maiani and M. Testa, *Phys. Lett.* B **245**, 585 (1990).
20. D. Pekurovsky and G. Kilcup, hep-lat/9812019.
21. D. Pekurovsky, hep-lat/9909141.
22. S. Sharpe and A. Patel, *Nucl. Phys.* B **417**, 307 (1994).
23. G. Martinelli *et al.*, *Nucl. Phys.* B **445**, 81 (1995).
24. Y. Kikukawa, R. Narayanan and H. Neuberger, *Phys. Lett.* B **399**, 105 (1997); J. Noaki andY. Taniguchi, *Phys. Rev.* D **61**, 054505 (2000).
25. C. Dawson (RBC Collaboration), hep-lat/9909107.

RARE KAON DECAYS

S. KETTELL
Brookhaven National Laboratory
Upton, NY 11973-5000

The current status of rare kaon decay experiments is reviewed. New limits in the search for Lepton Flavor Violation are discussed, as are new measurements of the CKM matrix.

1 Introduction

The field of rare kaon decays has a long and rich history. As the field has evolved so has the definition of 'rare'. In this review it refers to branching ratios less than $\sim 10^{-10}$. The two areas of greatest interest have been the very sensitive searches for physics beyond the Standard Model (SM) through lepton flavor violating (LFV) decays and the studies of the SM picture of CKM mixing and CP violation that have recently begun to come to fruition.

2 Lepton Flavor Violating Decays

There is solid experimental evidence for an additive quantum number associated with each family of charged leptons. While there is no SM mechanism for LFV (even if $m_\nu \neq 0$, LFV is too small to observe in the charged lepton sector), there is no underlying gauge symmetry preserving lepton flavor. Observation of LFV would be unambiguous evidence for physics beyond the SM, and is predicted by many extensions to the SM.

Due to excellent sensitivity, the mass scale probed by rare kaon decay experiments is quite large. This can be seen by comparing the $K_L^\circ \to \mu e$ decay, through a hypothetical LFV vector boson with coupling g_X and mass M_X to the conventional $K_{\mu 2}$ decay (g and M_W); a lower limit on M_X can be derived:

$$\begin{aligned} M_X &\simeq \frac{g_X}{g}(\sin^2\theta_c)^{1/4} M_W \left[\frac{\Gamma(K^+ \to \mu^+\nu)}{\Gamma(K_L^\circ \to \mu e)}\right]^{1/4} \qquad (1) \\ &\simeq 200\mathrm{TeV/c^2} \times \frac{g_X}{g} \times \left[\frac{10^{-12}}{\mathrm{B}(K_L^\circ \to \mu e)}\right]^{1/4} \end{aligned}$$

2.1 $K_L^\circ \to \mu e$

The E871 experiment at BNL has finished, with two long runs in 1995–96. No events were seen in the signal region, with an expected background of 0.1 events (see Fig. 1), and the 90% CL limit[1] is $B(K_L^\circ \to \mu e) < 4.7 \times 10^{-12}$.

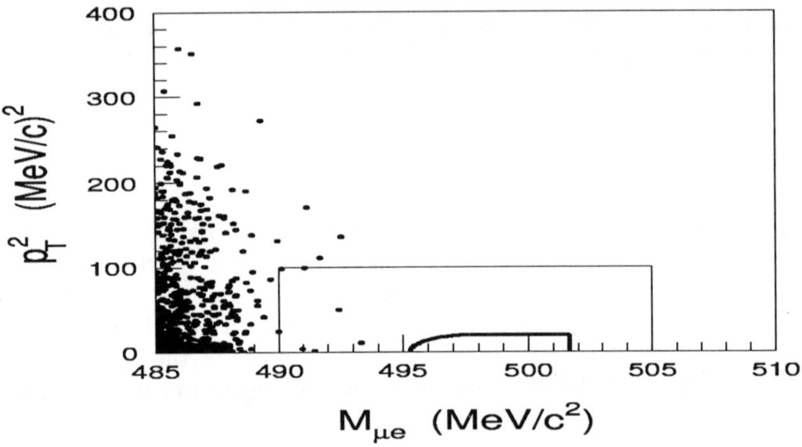

Figure 1. Final E871 data sample after all cuts, with no events in the signal region. The exclusion box was used to set cuts in an unbiased way on data far from the signal region. The shape of the signal box was optimized to maximize signal/background.

There are no near-term plans to pursue this decay further; it is expected to be limited by background from $K_L^\circ \to \pi^\pm e^\mp \nu_e$ with a $\pi \to \mu$ decay and an electron scattering in the vacuum window or first tracking chamber at $\sim 10^{-13}$.

2.2 $K^+ \to \pi^+ \mu^+ e^-$

The E865 experiment at BNL has collected data for the decay $K^+ \to \pi^+ \mu^+ e^-$ during the 1995, 1996, and 1998 runs of the AGS. The 90% CL limit on this mode from the 1995 run[2], similar in sensitivity to the predecessor experiment E777, was $B(K^+ \to \pi^+ \mu^+ e^-) < 2 \times 10^{-10}$. From the 1996 run[3] with no events above a $\pi \mu e$ likelihood $> 20\%$, the limit is $B(K^+ \to \pi^+ \mu^+ e^-) < 4.8 \times 10^{-11}$ (see Fig. 2). Combining the results from E777 and the E865 runs in 1995 and 1996 a limit of $B(K^+ \to \pi^+ \mu^+ e^-) < 3.2 \times 10^{-11}$ is obtained. The final sensitivity, including 1998 data, is expected to be ~3 times better. There are no plans to continue with this search.

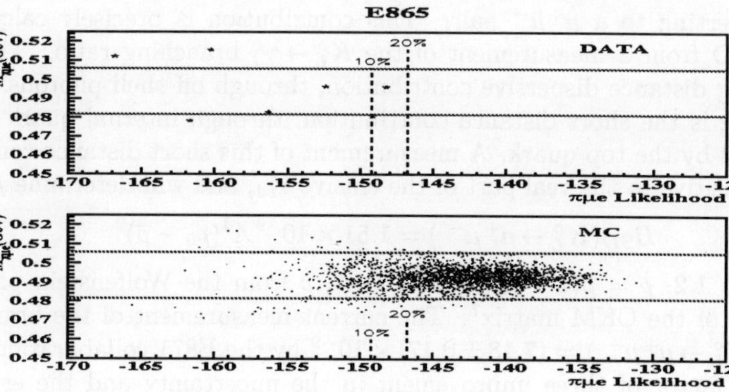

Figure 2. Final 1996 data sample after all cuts, with no events above $\pi\mu e$ likelihood of 20%. Also shown are $\pi\mu e$ Monte Carlo events passing all cuts.

2.3 $K_L^\circ \to \pi^\circ \mu e$

The current limit on $K_L^\circ \to \pi^\circ \mu e$ from E799-I[4] at FNAL, is B($K_L^\circ \to \pi^\circ \mu e$) $< 3.1 \times 10^{-9}$ (90% CL). This measurement has very small background levels, so E799-II (KTeV) will be able to substantially improve upon this limit.

3 $K_L^\circ \to \mu^+\mu^-$ and other Semi-Rare Decays

Two modes for which it may be possible to extract fundamental CKM parameters, from measurement of the short distance contributions, are $K_L^\circ \to \mu^+\mu^-$ and $K_L^\circ \to \pi^\circ \ell^+\ell^-$. The $K_L^\circ \to \pi^\circ \ell^+\ell^-$ modes are covered in detail by Taku Yamanaka elsewhere in this volume. The current 90% CL limit[5] for $K_L^\circ \to \pi^\circ e^+e^-$ is B($K_L^\circ \to \pi^\circ e^+e^-$) $< 5.6 \times 10^{-10}$ and the limit[6] for $K_L^\circ \to \pi^\circ \mu^+\mu^-$ is B($K_L^\circ \to \pi^\circ \mu^+\mu^-$) $< 3.4 \times 10^{-10}$. Both of these measurements have background which will slow improvement in these modes.

3.1 $K_L^\circ \to \mu^+\mu^-$

The mode $K_L^\circ \to \mu^+\mu^-$, which has played such an important role in the development of the SM (e.g. the GIM mechanism and the prediction of the charm quark), has now been measured to the unprecedented precision of 1.5%, with 6200 events, by E871.[8]

The decay $K_L^\circ \to \mu^+\mu^-$ is dominated by $K_L^\circ \to \gamma\gamma$ with the two real photons converting to a $\mu^+\mu^-$ pair. This contribution is precisely calculated using QED from a measurement of the $K_L^\circ \rightarrow \gamma\gamma$ branching ratio. There is also a long distance dispersive contribution, through off-shell photons. Most interesting is the short distance contribution, through internal quark loops, dominated by the top quark. A measurement of this short distance contribution is sensitive to the real part of the elusive V_{td}, and will determine ρ:

$$B_{SD}(K_L^\circ \rightarrow \mu^+\mu^-) = 1.51 \times 10^{-9} A^4 (\rho_0 - \overline{\rho})^2 \tag{2}$$

with $\rho_0 = 1.2$, $\overline{\rho} = \rho(1 - \lambda^2/2)$ and λ, A, ρ from the Wolfenstein parameterization of the CKM matrix[7]. The current measurement of the branching ratio $B(K_L^\circ \rightarrow \mu^+\mu^-) = (7.18 \pm 0.17) \times 10^{-9}$ by the E871 collaboration[8] represents a factor of three improvement in the uncertainty and the error on $B(K_L^\circ \rightarrow \mu^+\mu^-)$ no longer dominates the error on the ratio:

$$\begin{aligned}\frac{\Gamma(K_L^\circ \rightarrow \mu\mu)}{\Gamma(K_L^\circ \rightarrow \gamma\gamma)} &= \frac{B(K_L^\circ \rightarrow \mu^+\mu^-)}{B(K_L^\circ \rightarrow \pi^+\pi^-)} \left|\frac{\eta_{+-}}{\eta_{\circ\circ}}\right| \frac{B(K_S^\circ \rightarrow \pi^+\pi^-)}{B(K_S^\circ \rightarrow \pi^\circ\pi^\circ)} \frac{B(K_L^\circ \rightarrow \pi^\circ\pi^\circ)}{B(K_L^\circ \rightarrow \gamma\gamma)} \qquad (3)\\ &= [(3.474 \pm .054) \times 10^{-6}][1.004 \pm .002][2.19 \pm .03][.632 \pm .009]\\ &\quad [1.55\%][0.23\%][1.28\%][1.42\%]\\ &= (1.213 \pm 0.030) \times 10^{-5}\end{aligned}$$

This value is only slightly above the unitarity bound from the on-shell two photon contribution

$$\frac{\Gamma(K_L^\circ \rightarrow \mu^+\mu^-)}{\Gamma(K_L^\circ \rightarrow \gamma\gamma)} > 1.195 \times 10^{-5} \tag{4}$$

and leaves very little room for the short distance contribution. Currently, with conservative estimates of the long distance dispersive contribution, a limit on ρ can be extracted $\rho > -0.33$ (90% CL)[8].

Unlike $K_L^\circ \rightarrow \mu^+\mu^-$, the decay $K_L^\circ \rightarrow e^+e^-$ is predominantly through two off-shell photons, so this decay is less interesting for extracting SM parameters. However, the recent observation of four events by E871[9], with $B(K_L^\circ \rightarrow e^+e^-) = (8.7^{+5.7}_{-4.1}) \times 10^{-12}$ is consistent with Chiral Perturbation Theory (ChPT) predictions[10,11] and is the smallest branching ratio ever measured.

3.2 $K_L^\circ \rightarrow e^+e^-\gamma$, $K_L^\circ \rightarrow \mu^+\mu^-e^+e^-$

Many new measurements of radiative kaon decays have been reported recently.[12] In particular, new measurements of $K_L^\circ \rightarrow \ell^+\ell^-\gamma$ and $K_L^\circ \rightarrow \ell^+\ell^-\ell'^+\ell'^-$ have been made by NA48 and KTeV. These measurements are substantially improved over previous values and even larger improvements will be obtained when the complete data sets are analyzed. For example, the new measurement from NA48[13] $B(K_L^\circ \to e^+e^-\gamma) = (1.06\pm0.02\pm0.02\pm0.04)\times 10^{-5}$ is ~70 times more sensitive than the previous measurement and KTeV will improve this by an additional factor of ~20 beyond NA48. These modes, which are not rare, with branching ratios from 10^{-8}–10^{-5}, may help to determine the long distance dispersive contribution to $K_L^\circ \to \mu^+\mu^-$. Additional work in ChPT is needed to extract the short distance physics.[10,14]

4 $K \to \pi\nu\overline{\nu}$

The $K \to \pi\nu\overline{\nu}$ modes — $K^+ \to \pi^+\nu\overline{\nu}$ and $K_L^\circ \to \pi^\circ\nu\overline{\nu}$ — are the 'golden modes' for determining the CKM parameters ρ and η; and, along with the other golden mode $B_d^\circ \to \psi K_S^\circ$ and perhaps $\Delta_{m_{B_s}}/\Delta_{m_{B_d}}$, provide the best opportunity to over-constrain the unitary triangle and to search for new physics. The unitarity of the CKM matrix can be expressed as

$$V_{us}^* V_{ud} + V_{cs}^* V_{cd} + V_{ts}^* V_{td} = 0 \tag{5}$$

with the three vectors $V_{is}^* V_{id}$ converging to form an elongated triangle in the complex plane. The first vector $V_{us}^* V_{ud}$ is well determined from the decay $K^+ \to \pi^\circ e^+ \nu_e$. The height will be measured by $K_L^\circ \to \pi^\circ\nu\overline{\nu}$ and the third side $V_{ts}^* V_{td}$ will be measured by $K^+ \to \pi^+\nu\overline{\nu}$. The $K \to \pi\nu\overline{\nu}$ decays are sensitive to the magnitude and imaginary part of V_{td}. From these two modes the unitarity triangle can be completely determined.

The theoretical uncertainty in $K^+ \to \pi^+\nu\overline{\nu}$ (~7%) is small and even smaller in $K_L^\circ \to \pi^\circ\nu\overline{\nu}$ (~1%); the hadronic matrix element can be extracted from the well measured $K^+ \to \pi^\circ e^+ \nu_e$ (K_{e3}) decay. The branching ratios have been calculated to next-to-leading-log approximation,[15] complete with isospin violation corrections[16] and two-loop-electroweak effects.[17] Based on current understandings of SM parameters, the branching ratios can be expressed as:[18]

$$\begin{aligned} B(K^+ \to \pi^+\nu\overline{\nu}) &= \frac{\kappa_+ \alpha^2 B(K_{e3})}{2\pi^2 \sin^4\theta_W |V_{us}|^2} \sum_l |X_t^l(x_t) V_{ts}^* V_{td} + X_c^l(x_c) V_{cs}^* V_{cd}|^2 \quad (6) \\ &= 8.88 \times 10^{-11} A^4 [(\overline{\rho}_0 - \overline{\rho})^2 + (\sigma\overline{\eta})^2] \\ &= (0.82 \pm 0.32) \times 10^{-10} \\ B(K_L^\circ \to \pi^\circ\nu\overline{\nu}) &= \frac{\tau_{K_L}}{\tau_{K^+}} \frac{\kappa_L \alpha^2 B(K_{e3})}{2\pi^2 sin^4\theta_W |V_{us}|^2} \sum_l |Im(V_{ts}^* V_{td}) X(x_t)|^2 \quad (7) \\ &= 4.08 \times 10^{-10} A^4 \eta^2 \\ &= (3.1 \pm 1.3) \times 10^{-11} \end{aligned}$$

(see Ref 18 for definitions). The decay $K_L^\circ \to \pi^\circ \nu\bar{\nu}$ is direct CP violating, and offers the best opportunity for measuring the Jarlskog invariant J_{CP},[19] the fundamental measure of CP violation in the SM.

4.1 $K^+ \to \pi^+ \nu\bar{\nu}$

In E787's analysis of the 1995–97 data sample, one clean $K^+ \to \pi^+ \nu\bar{\nu}$ event lies in the signal box (see Fig. 3), with a measured background of 0.08 ± 0.02

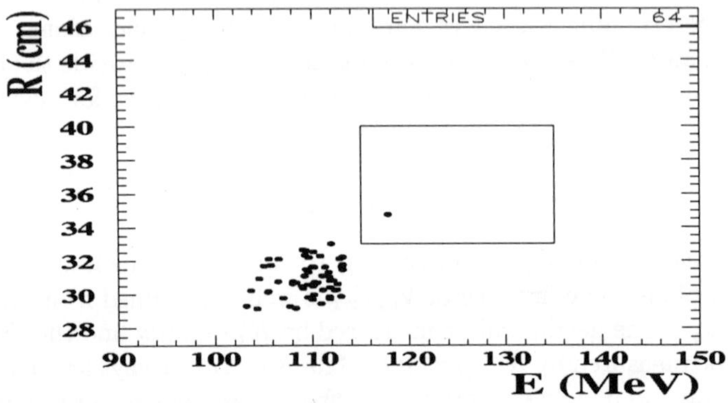

Figure 3. Final E787 data sample from 1995–97 after all cuts. One clean $K^+ \to \pi^+ \nu\bar{\nu}$ event is seen in the box. The remaining events are $K^+ \to \pi^+ \pi^\circ$ background.

events. Based on this one event[20] the branching ratio is $B(K^+ \to \pi^+ \nu\bar{\nu}) = 1.5^{+3.4}_{-1.2} \times 10^{-10}$. From this measurement, a limit of $0.002 < |V_{td}| < 0.04$ is determined; in addition, the following limits on $\lambda_t \equiv V_{ts}^* V_{td}$ can be set: $|Im\lambda_t| < 1.22 \times 10^{-3}$, $-1.10 \times 10^{-3} < Re\lambda_t < 1.39 \times 10^{-3}$, and $1.07 \times 10^{-4} < |\lambda_t| < 1.39 \times 10^{-3}$. The final sensitivity of the E787 experiment, based on data from 1995–98, should reach a factor of two further, to the SM expectation for $K^+ \to \pi^+ \nu\bar{\nu}$.

A new experiment, E949, is under construction and will run in 2001–03. Taking advantage of the entire AGS proton flux and the experience gained with the E787 detector, E949 with modest upgrades should observe 10 SM events in a two year run. The background is well-understood and is ~10% of the SM signal.

A proposal for a further factor of 10 improvement has been initiated at FNAL. The CKM experiment (E905) plans to collect 100 SM events, with B/S

$\sim$ 10%, in a two year run starting in $\sim$2005. This experiment will use a new technique, with K^+ decay-in-flight and momentum/velocity spectrometers.

4.2 $K_L^\circ \to \pi^\circ \nu\bar{\nu}$

The current best direct limit[21] on $K_L^\circ \to \pi^\circ \nu\bar{\nu}$ comes from the KTeV run in 1997: $B(K_L^\circ \to \pi^\circ \nu\bar{\nu}) < 5.9 \times 10^{-7}$ (90% CL).

An even more stringent limit can be derived in a model independent way[22] from the E787 measurement of $K^+ \to \pi^+ \nu\bar{\nu}$:

$$B(K_L^\circ \to \pi^\circ \nu\bar{\nu}) < 4.4 \times B(K^+ \to \pi^+ \nu\bar{\nu}) = 2.9 \times 10^{-9} \text{ (90\%CL)} \quad (8)$$

The next generation of $K_L^\circ \to \pi^\circ \nu\bar{\nu}$ experiments will start with E391a at KEK, which hopes to reach a sensitivity of $\sim 10^{-10}$. While this does not reach the SM level, the experiment will make a large step forward and learn more about how to do this difficult experiment. The plan would then be to move the detector to the JHF and push to a sensitivity of $\mathcal{O}(10^{-13})$.

Two other experiments propose to reach sensitivities of $\mathcal{O}(10^{-13})$ — E926 (KOPIO/RSVP) at BNL and E804 (KAMI) at FNAL. KAMI plans to reuse the excellent CsI calorimeter from KTeV and to operate at high kaon momentum to achieve good photon energy resolution and efficiency. The flux will increase substantially over KTeV due to the Main Injector. KOPIO follows a different strategy. The kaon center of mass will be reconstructed using a bunched proton beam and a very low momentum K_L beam. This gives 2 independent criteria to reject background: photon veto and kinematics — allowing background levels to be directly measured from the data — and gives further confidence in the signal by measuring the spectra. The necessary flux will by obtained using the entire AGS proton current. The low energy beam also substantially reduces backgrounds from neutrons and other sources. After three years of running, 65 SM events will be observed with a S/B $\geq$ 2:1.

5 Conclusions and Future Prospects

The unprecedented sensitivities of the rare kaon decay experiments in setting limits on LFV have constrained many extensions of the SM. The discovery of $K^+ \to \pi^+ \nu\bar{\nu}$ has opened the doors to measurements of the unitarity triangle completely within the kaon system. Significant progress in the determination of the fundamental CKM parameters will come from the generation of experiments that is starting now. Comparison with the B-system will then over-constrain the triangle and test the SM explanation of CP violation.

Acknowledgments

I would like to thank members of several experiments for access to their data and plots and for useful discussions, in particular, I would like to thank Bill Molzon, Taku Yamanaka, Ron Ray, Bob Tschirhart, Robin Appel, Hong Ma, Mike Zeller, Doug Bryman and Laurie Littenberg. This work was supported under U.S. Department of Energy contract #DE-AC02-98CH10886.

References

1. D. Ambrose *et al*, *Phys. Rev. Lett.* **81**, 5734 (1998).
2. D. Bergman, Yale University Ph.D. Thesis; S. Pislak, University of Zurich Ph.D. Thesis.
3. M.Zeller, Proc KAON99 (ed T.Quinn, University of Chicago Press, 2000).
4. K. Arisaka *et al*, *Phys. Lett.* B **432**, 230 (1998).
5. T. Yamanaka, Proc. Rencontres de Moriond (ed. C. Barthelemy, 1999).
6. J.Whitmore, Proc. KAON99 (ed T.Quinn, University of Chicago Press, 2000).
7. L. Wolfenstein, *Phys. Rev. Lett.* **51**, 1945 (1983).
8. D. Ambrose *et al*, submitted to *Phys. Rev. Lett.*, W. Molzon, Proc. Lepton Photon 99, http://arXiv.org/abs/hep-ex/0001024.
9. D. Ambrose *et al*, *Phys. Rev. Lett.* **81**, 4301 (1998).
10. G. Valencia, *Nucl. Phys.* B **517**, 339 (1998).
11. D.G. Dumm and A. Pich,, *Phys. Rev. Lett.* **80**, 4633 (1998).
12. S. Kettell, Proc. DAΦNE Workshop (ed. G. Capon *et al*, Frascati Physics Series, 2000); also BNL–67140.
13. V. Fanti *et al*, *Phys. Lett.* B **458**, 553 (1999).
14. D'Ambrosio, *et. al.*, *Phys. Lett.* B **423**, 385 (1998).
15. G. Buchalla and A. Buras, *Nucl. Phys.* B **412**, 106 (1994).
16. W.J. Marciano and Z. Parsa, *Phys. Rev.* D **53**, R1 (1996).
17. G. Bucahalla and A. Buras, *Phys. Rev.* D **57**, 216 (1998).
18. G. Bucahalla and A. Buras, *Nucl. Phys.* B **548**, 309 (1999).
19. C. Jarlskog, *Phys. Rev. Lett.* **55**, 1039 (1985).
20. S. Adler *et al*, *Phys. Rev. Lett.* **79**, 2204 (1997), S. Adler *et al*, to be submitted to *Phys. Rev. D.*
21. A. Alavi-Harati *et al*, to be published in *Phys. Rev. D.*
22. Y. Grossman and Y. Nir, *Phys. Lett.* B **398**, 163 (1997).

THEORY OF RARE KAON DECAYS

GIANCARLO D'AMBROSIO

INFN-Sezione di Napoli, Complesso Universitario Monte S. Angelo, Via Cintia 80126 Napoli Italy

E-mail: dambrosio@na.infn.it

We review some recent theoretical results on rare kaon decays. Particular attention is devoted to find Standard Model tests. This is theoretically easy in $K \to \pi\nu\overline{\nu}$, while a careful study of the long distance contributions is needed for $K_L \to \pi^0 e\overline{e}$, $K_L \to \mu\overline{\mu}$ and $K \to \pi\pi\gamma$.

1 Introduction

Historically, rare kaon decays, like $K_{L,S} \to \mu\overline{\mu}$, $K \to \pi\nu\overline{\nu}$, and $K^0 - \overline{K^0}$ mixing have been fundamental to understand weak interactions and particle physics: i) the solution to the large Flavour Changing Neutral Currents (FCNC) was GIM mechanism and ii) CP violation in the Standard Model (SM) led to three families [1,2]. Originally QCD corrections to $W-$ box, Z^0- and $\gamma-$penguin contributions were not evaluated and phenomenological models were used to estimate long distance contributions [3]. For instance for $K_S \to \gamma\gamma$, since the short distance amplitude was vanishing, the long distance one was computed by a $\pi^+ - \pi^-$ loop with a $K_S\pi\pi$ constant weak vertex from $A(K_S \to \pi\pi)_{\rm exp}$ and γ's coupled electromagnetically; $A(K_S \to \gamma\gamma)$ then is finite [3] but with at least two legittimate questions: i) what is the correct off-shell extrapolation for $K_S \to \pi\pi$ and ii) why $K_S F^{\mu\nu} F_{\mu\nu}$ local operators (and higher dimensional operators like $\partial^2 K_S F^{\mu\nu} F_{\mu\nu}$) are suppressed?

There have been two major theoretical steps in this field since the seminal paper of Gaillard-Lee: i) the OPE (Operator Product Expansion) and ii) Chiral pertubation theory (χPT). In OPE the physical processes are determined by an effective hamiltonian, written as a product of local operators O_i and (Wilson) coefficients c_i : $\mathcal{H}_{eff} = \sum_i c_i(\mu)\, O_i(\mu)$; indeed the scale dependence must cancel in the product since physical processes are $\mu-$independent. Thus OPE factorizes short ($c_i(\mu)$) and long distance contributions. Long distance matrix elements are evaluated by symmetry arguments, i.e. χPT [4,2]. Then, for instance, it is easy to understand the problems posed above for $K_S \to \gamma\gamma$: chiral symmetry constrains the momentum dependence of the vertex $K_S\pi\pi$ and imposes the $\mathcal{O}(p^4)$ $K_S F^{\mu\nu} F_{\mu\nu}$ local operator to be zero, while chiral power counting suppresses $\partial^2 K_S F^{\mu\nu} F_{\mu\nu}$ ($\mathcal{O}(p^6)$) compared to the $\mathcal{O}(p^4)$ $\pi\pi$-loop contribution [5]. This prediction is in agreement with experiment [6].

B-physics will test SM measuring the CKM triangle [7] with sizes $V_{qb}^* V_{qd}$; the area of this triangle, $J_{CP}/2$, is invariant for all CKM triangles and non-zero if CP is violated; in the Wolfenstein parametrization:

$$|J_{CP}| \overset{Wolfenstein}{\simeq} A^2\lambda^6\eta \tag{1}$$

with $V_{us} = \lambda$, $V_{cb} = A\lambda^2, \Im m(V_{td}) = -A\lambda^3\eta$. As a consequence of the improved understanding of low energy physics we can test precisely (1) in rare kaon channels[1].We can establish that $K \to \pi\nu\overline{\nu}$ is dominated by far by short distance

$$A(s \to d\nu\overline{\nu})_{SM} \sim G_F^2 \sum_{q=u,c,t} V_{qs}^* V_{qd} \quad m_q^2 \quad (\overline{s}d)_{V-A} (\overline{\nu}\nu)_{V-A} \tag{2}$$

and due to the large top mass, is very sensitive to the CKM matrix element,V_{td} [8,9]. Thus K-physics will measure J_{CP} from the triangle with sizes $V_{qs}^* V_{qd}$ and test extensions of the SM. K-physics is also an ideal window to test Flavour violation [14]: scales of 100 TeV are probed [1]. Here we first study $K \to \pi\nu\overline{\nu}$, then $K \to \pi e\overline{e}$, $K_L \to \mu\overline{\mu}$ and $K \to \pi\pi\gamma$.

2 $K \to \pi\nu\overline{\nu}$

2.1 *Standard Model*

The SM predicts the $V - A \otimes V - A$ effective hamiltonian in (2) and Wilson coefficients known at next–to–leading order [9]. $SU(2)$ isospin symmetry relates hadronic matrix elements for $K \to \pi\nu\overline{\nu}$ to $K \to \pi l\overline{\nu}$ to a very good precision [10]. QCD corrections have been evaluated at next-to-leading order[1] and the main uncertainties is due the strong corrections to the charm loop contribution.

The structure in (2) leads to a pure CP violating contribution to $K_L \to \pi^0\nu\overline{\nu}$, induced only from the top loop contribution and thus proportional to $\Im m(\lambda_t)$ and free of hadronic uncertainties. This leads to the prediction [9]

$$B(K_L \to \pi^0\nu\overline{\nu})_{SM} = 4.25 \times 10^{-10} \left[\frac{\overline{m}_t(m_t)}{170GeV}\right]^{2.3} \left[\frac{\Im m(\lambda_t)}{\lambda^5}\right]^2 . \tag{3}$$

$K^{\pm} \to \pi^{\pm}\nu\overline{\nu}$ receives CP conserving and violating contributions proportional to $\Re e(\lambda_c)$, $\Re e(\lambda_t)$ and $\Im m(\lambda_t)$. Theoretical uncertainty from the charm loop induces 5% error on the width. If one takes into account the various indirect limits, i.e.V_{ub} and ε, on CKM elements one obtains the SM values[1,9]:

$$B(K_L \to \pi^0 \nu\bar{\nu}) = (2.8 \pm 1.1)\times 10^{-11} \quad B(K^\pm \to \pi^\pm \nu\bar{\nu}) = (0.8 \pm 0.3)\times 10^{-10} \tag{4}$$

Using only the experimental knowledge on $B_{s,d}-$ mixing the more general bound [11] $B(K^\pm \to \pi^\pm \nu\bar{\nu})_{SM} < 1.67 \times 10^{-10}$ is obtained. To be compared respectively with the experimental results in [12] and in [13,14]

$$B(K_L \to \pi^0 \nu\bar{\nu}) \leq 5.9 \times 10^{-7} \qquad B(K^\pm \to \pi^\pm \nu\bar{\nu}) = \left(1.5^{+3.4}_{-1.2}\right) \times 10^{-10} \tag{5}$$

A future measurement [14] of $B(K_L \to \pi^0 \nu\bar{\nu})$ with 10 % error implies a determination of $\Im m(\lambda_t)$, as we can see from (3), and the area J in (1), with 5 % error.

2.2 New Physics

The isospin structure of any $\bar{s}d$ operator (bilinear in the quark fields) leads to the model independent relation [15] and to an interesting bound with E787 [13,14]

$$B(K_L \to \pi^0 \nu\bar{\nu}) < \frac{\tau_{K_L}}{\tau_{K^+}} B(K^\pm \to \pi^\pm \nu\bar{\nu}) \underset{E787}{<} 2.9 \cdot 10^{-9} \quad \text{at} \quad 90\% C.L.$$

Lately it has been pointed out the possibility of new physics to enhance substantially the SM predictions [16]. Also the recent value of ε'/ε [17,18,19], though not incompatible with the SM, allows large values for new sources of CP violating contributions. Adopting a supersymmetric framework (susy) where the flavour breaking scale is much larger than susy breaking scale (M_s) , which is in turn larger than the EW scale ($M_s \geq M_W$) the bigger contributions to the $K \to \pi l\bar{l}$ effective hamiltonian come from operators of dimensions four and five. Indeed once the heavy (M_s) degrees of freedom have been integrated out , $Z-$(susy)penguin effects could be parametrized by an effective dimension-4 operator $\bar{s}dZ$ with coupling Z_{ds} and with a contribution to (2) $\mathcal{O}(G_F Z_{ds})$ [16,1], while the superbox diagrams generate a suppressed $\mathcal{O}(1/M_s^2)$ effective dimension-6 operator $\bar{s}d\bar{\nu}\nu$ [20]. We can parametrize the relevant dimension-5 effective hamiltonian as [20]

$$\mathcal{H}_{eff}^{|\Delta S|=1;d=5} = \frac{4G_F}{\sqrt{2}} \left[C_\gamma^+ O_\gamma^+ + C_\gamma^- O_\gamma^- + C_g^+ O_g^+ + C_g^- O_g^- + \text{h.c.}\right], \tag{6}$$

$$O_\gamma^\pm = \frac{Q_d e}{16\pi^2}(\bar{s}_L \sigma_{\mu\nu} d_R \pm \bar{s}_R \sigma_{\mu\nu} d_L) F^{\mu\nu} \tag{7}$$

$$O_g^\pm = \frac{g}{16\pi^2}(\bar{s}_L \sigma_{\mu\nu} t^a G_a^{\mu\nu} d_R \pm \bar{s}_R \sigma_{\mu\nu} t^a G_a^{\mu\nu} d_L) \tag{8}$$

In the SM the coefficients are chirally suppressed [21], and thus extensions of SM might be interesting [22]. If we consider the bounds coming from ε'/ε and $K_L \to \mu\overline{\mu}$ [23], as we shall see, large NP enhancements can be obtained for $K_L \to \pi^0 \nu\overline{\nu}$ and for $K^{\pm} \to \pi^{\pm}\nu\overline{\nu}$ (respectively at most a factor 10 and 3 in the branching compared to the SM predictions) [20].

3 $K_L \to \pi^0 e^+ e^-$

Electromagnetic interactions in $K_L \to \pi^0 e^+ e^-$ add new structures to the direct CP violating contribution analogous to (2) (additional single photon exchange contributions are smaller): i) indirect CP –violating contribution $K_L = K_2 + \tilde{\varepsilon} K_1 \stackrel{\tilde{\varepsilon} K_1}{\to} \pi^0 \stackrel{*}{\gamma} \to \pi^0 e^+ e^-$ and ii) a $CP-$ conserving contribution: $K_L \to \pi^0 \stackrel{*}{\gamma}\stackrel{*}{\gamma} \to \pi^0 e^+ e^-$. Thus to really probe the short distance window one has to have under control i) and ii) and also we must warn about the danger of the potentially large background contribution from $K_L \to e^+ e^- \gamma\gamma$ to $K_L \to \pi^0 e^+ e^-$ [24]. The present bounds from KTeV are [14]:

$$B(K_L \to \pi^0 e^+ e^-) < 5.6 \times 10^{-10} \quad B(K_L \to \pi^0 \mu^+ \mu^-) < 3.4 \times 10^{-10} \qquad (9)$$

and the prediction for the direct CP violation contribution analogous to (3) is

$$B(K_L \to \pi^0 e^+ e^-)^{SM}_{CPV-dir} = 0.69 \times 10^{-10} \left[\frac{\overline{m}_t(m_t)}{170 GeV} \right]^2 \left[\frac{\Im m(\lambda_t)}{\lambda^5} \right]^2 ;$$

using the present constrains on $\Im m(\lambda_t)$ one obtains [1,25]

$$2.8 \times 10^{-12} \leq B(K_L \to \pi^0 e^+ e^-)^{SM}_{CPV-dir} \leq 6.5 \times 10^{-12}.$$

New physics contributions [20] from an effective dimension-4 Z_{ds} vertex and $\mathcal{H}^{|\Delta S|=1;d=5}_{eff}$ in (6) can enhance the branching up to a factor 10.

3.1 Indirect CP violation contribution, $K_S \to \pi^0 e^+ e^-$ and $K^{\pm} \to \pi^{\pm} l^+ l^-$

Electromagnetic interaction induce a new operator to the effective hamiltonian for $s \to d l^+ l^-$: $(\overline{s}d)_{V-A} (l^+ l^-)_V$, which generates a contrbution to $K \to \pi l^+ l^-$ dominated by long distance and thus studied in χPT. These decays ($K^{\pm} \to \pi^{\pm} \stackrel{*}{\gamma}$ and $K_S \to \pi^0 \stackrel{*}{\gamma}$) start at $\mathcal{O}(p^4)$ with loops (dominated by the $\pi\pi-$cut) and unknown counterterm contributions [26]. Higher order contributions ($\mathcal{O}(p^6)$) might be large, but not completely under control since new and with unknown coefficients counterterm structures will appear [27]. However we can still parameterize, quite generally the $K \to \pi \overset{*}{\gamma}(q)$ form factor as [28]

$$W_i(z) = G_F M_K^2 (a_i + b_i z) + W_i^{\pi\pi}(z) , \qquad i = \pm, S \tag{10}$$

with $z = q^2/M_K^2$, and where $W_+^{\pi\pi}(z)$ is the loop contribution, given by the $K \to \pi\pi\pi$ unitarity cut and completely known up to $\mathcal{O}(p^6)$. All our results can be expressed in terms of the unknown parameters a_i and b_i, expected of $\mathcal{O}(1)$. At the first non-trivial order, $\mathcal{O}(p^4)$, $b_i = 0$, while a_i receive counterterm contributions not determined yet. At $\mathcal{O}(p^6)$, $b_i \neq 0$, while a_i receive new counterterm contributions. Due to the generality of (10) we expect that $W_i(z)$ is a good approximation to the complete form factor.

From the $K^+ \to \pi^+ e^+ e^-$ experimental width and slope, E865 obtains [29]

$$a_+ = -0.587 \pm 0.010 \qquad b_+ = -0.655 \pm 0044 \tag{11}$$

Also the fit with (10), i.e. with the genuine chiral [29] contributions $W_i^{\pi\pi}(z)$, is better (χ^2 $/d.o.f. \sim 13.3/18$) than just a linear slope ($\chi^2/d.o.f. \sim 22.9/18$), showing the validity of the chiral expansion.Then the universality of the form factor in (10) is further tested by using (11) to predict the branching $B(K^+ \to \pi^+\mu^+\mu^-)$, which indeed perfectely agrees with the new experimental value[30] $B(K^+ \to \pi^+\mu^+\mu^-)_{\rm exp} = (9.22 \pm 0.60 \pm 0.49) \cdot 10^{-8}$. This value is however larger [28] by 3.3 σ 's than the old experimental result [31]. Also the slope in the muon channel, though with large statistical errors, is now consistent with (11).

We should stress that it is not clear at the moment the meaning of the apparent slow convergence in the chiral expansion in $K^+ \to \pi^+ l^+ l^-$, indeed the values in (11) do not respect the naive chiral dimensional analysis expectation $b_+/a_+ \sim M_K^2/M_V^2$.

There is no model independent relation among a_S and a_+ and thus a secure determination of $B(K_L \to \pi^0 e^+ e^-)_{CP-indirect}$ requires a direct measurement of $B(K_S \to \pi^0 e^+ e^-)$, possibly to be performed by KLOE at DAΦNE [28]. The dependence from b_S is very mild and thus we predict $B(K_S \to \pi^0 e^+ e^-) \simeq 5.2\, a_S^2 \times 10^{-9}$. We can take advantage of the interference term among direct and indirect the CP –violating terms [28]: for $a_S \lesssim -0.5$ or $a_S \gtrsim 1.0$ we obtain $B(K_L \to \pi^0 e^+ e^-)_{CPV} \gtrsim 10^{-11}$ and thus perform an independent measurement of $\Im m \lambda_t$, with a precision increasing with the value of $|a_S|$.

3.2 CP conserving contributions: "γγ" intermediate state contributions

The general amplitude for $K_L(p) \to \pi^0 \gamma(q_1) \gamma(q_2)$ can be decomposed as a sum of two independent Lorentz and gauge invariant amplitudes: $A(z, y)$, where the two photons are in a state of total angular momentum $J = 0$ (J, total diphoton angular momentum), and the higher angular momentum state $B(z, y)$, also chirally and kinematically suppressed [2]. The double differential rate is given by

$$\frac{\partial^2 \Gamma}{\partial y \, \partial z} = \frac{m_K}{2^9 \pi^3} [\, z^2 \,|\, A + B \,|^2 + \left(y^2 - \frac{\lambda(1, r_\pi^2, z)}{4} \right)^2 \,|\, B \,|^2 \,] \,, \tag{12}$$

where $y = p \cdot (q_1 - q_2)/m_K^2$, $z = (q_1 + q_2)^2/m_K^2$, $r_\pi = m_\pi/m_K$ and $\lambda(a, b, c)$ is the usual kinematical function. Thus in the region of small z (collinear photons) the B amplitude is dominant and can be determined separately from the A amplitude. This feature is crucial in order to disentangle the CP-conserving contribution $K_L \to \pi^0 e^+ e^-$. Here we assume, consistentely with models [27] and possibly to test in $K_L \to \pi^0 e^+ e^- \gamma$, that the dominant contribution is generated by on-shell $\gamma\gamma$ rescattering [32]. It turns out that for the $J = 0$ contribution $A(K_L \to \pi^0 e^+ e^-)_{J=0} \sim m_e$ (m_e electron mass) [33]; while the suppressed B-type amplitude, generate $A(K_L \to \pi^0 e^+ e^-)_{J \neq 0}$ competitive with the CP violating contributions [27].

The leading finite $\mathcal{O}(p^4)$ amplitudes of $K_L \to \pi^0 \gamma\gamma$ generates only the A–type amplitude in Eq. (12) [34], but underestimates the observed branching ratio [35,14], $(1.68 \pm 0.07 \pm 0.08) \times 10^{-6}$ by a large factor. The two presumably large $\mathcal{O}(p^6)$ contributions have been studied: i) $\mathcal{O}(p^6)$ unitarity corrections [36,37] enhance the $\mathcal{O}(p^4)$ branching ratio by 40%, and generate a B–type amplitude, ii) vector meson exchange contributions to $K_L \to \pi^0 \gamma\gamma$ are in general model dependent [38,39] but one can parameterize their contribution to A and B by an effective vector coupling a_V [39] . The agreement with experimental $K_L \to \pi^0 \gamma\gamma$ rate and spectrum would demand $a_V \sim -0.8$ [37,40] . We have related a_V with the linear slope, α, of the $K_L \to \gamma\gamma^*$ form factor [40], which is also generated by vector meson exchange contribution. In factorization it is possible to describe VMD contributions to these observables in terms of one free parameter k_F. The resulting phenomenology is successful and it is also suggested that very interestingly matching with short distance should be performed at the resonance scale and not at the kaon mass [40,41]. The new data from KTeV [35] confirms sharply our prediction : $a_V = -0.72 \pm 0.05 \pm 0.06$ and show a clear evidence of events at low z. This turns in a stringent determination for the CP conserving contribution to $K_L \to \pi^0 e^+ e^-$: $1. < B(K_L \to \pi^0 e^+ e^-) \cdot 10^{12} < 4$ [27,40].

4 $K_L \to l^+l^-$

To fully exploit the potential of $K_L \to \mu^+\mu^-$ in probing short–distance dynamics it is necessary to have a reliable control on its long–distance amplitude. However the dispersive contribution generated by the two–photon intermediate state cannot be calculated in a model independent way and it is subject to various uncertainties [42,43,44]. The branching ratio can be generally decomposed as $B(K_L \to \mu^+\mu^-) = |\Re e\mathcal{A}|^2 + |\Im m\mathcal{A}|^2$, and the dispersive contribution can be rewritten as $\Re e\mathcal{A} = \Re e\mathcal{A}_{long} + \Re e\mathcal{A}_{short}$. The recent measurement of $B(K_L \to \mu^+\mu^-) = (7.18 \pm 0.17) \times 10^{-9}$ [6] is almost saturated by the absorptive contribution $B(K_L \to \mu^+\mu^-)_{abs} = (7.1 \pm 0.2) \times 10^{-9}$, obtained from the experimental $K_L \to \gamma\gamma$ width. This leaves very small room for the dispersive contribution determined by $B(K_L \to \mu^+\mu^-)_{\rm exp}$, $\Re e\mathcal{A}_{exp} = (-1.0 \pm 3.7) \times 10^{-10}$ or $|\Re e\mathcal{A}_{exp}|^2 < 5.6 \times 10^{-10}$ at 90% C.L.[42].

Within the Standard Model the known[45] NLO short-distance contribution $|\Re e\mathcal{A}_{short}|^2 = (0.9 \pm 0.4) \times 10^{-9}$, gives the possibility to extract information on V_{td} and New Physics, once $\Re e\mathcal{A}_{long}$ is under control. Two methods [42,44] and a criticism [43] have been proposed to predict $\Re e\mathcal{A}_{long}$.

Large-N_c argument and $U(3)\otimes U(3)$ symmetry have been invoked in [44] for $\Re e\mathcal{A}_{long}$; however these arguments give a poor phenomenological description of the experimental $K_L \to \gamma\gamma$ width and thus caution must be used before to completely accept this result [46].

We instead have proposed [42] a low energy parameterization of the $K_L \to \gamma^*\gamma^*$ form factor that include the poles of the lowest vector meson resonances with arbitrary residues

$$f(q_1^2, q_2^2) = 1 + \alpha \left(\frac{q_1^2}{q_1^2 - m_V^2} + \frac{q_2^2}{q_2^2 - m_V^2} \right) + \beta \frac{q_1^2 q_2^2}{(q_1^2 - m_V^2)(q_2^2 - m_V^2)} \ . \tag{13}$$

The parameters α and β, expected to be $\mathcal{O}(1)$ by chiral power counting, are in principle directly accessible by experiment in $K_L \to \gamma\ell^+\ell^-$ and $K_L \to e^+e^-\mu^+\mu^-$. Now we can determine only α_{exp}, while β is constrained by QCD: the form factor defined in Eq. (13) goes as $1 + 2\alpha + \beta$ for $q_i^2 \gg m_V^2$ and one has to introduce an ultraviolet cutoff Λ; since in this region the perturbative QCD calculation of the two-photon contribution gives a small result we find $|1 + 2\alpha + \beta| \ln(\Lambda/M_V) < 0.4$, showing a mild behaviour of the form factor at large q^2 and limiting β . Thus we predict [42] $|\Re e\mathcal{A}_{long}| < 2.9 \times 10^{-5} \quad (90\% C.L.)$. These bounds could be very much improved if the α and β parameters were measured with good precision and a more stringent bound on $|\Re e\mathcal{A}_{exp}|$ is established. Recentely $K_L \to e^+e^-$ has been measured at BNL E871 [47]: $B(K_L \to e^+e^-) = (8.7^{+5.7}_{-4.1}) \times 10^{-12}$, however the theoretical prediction[43,44] for this branching is not sensitive to the slopes of the form factor.

5 $K \to \pi\pi\gamma$

The $K \to \pi\pi\gamma$ amplitude is usually decomposed also in electric (E) and the magnetic (M) terms [48,2,49]. In the electric transitions one generally separates the bremsstrahlung amplitude E_B generally enhanced from the factor $1/E^*_\gamma$ for $E^*_\gamma \to 0$, (E^*_γ is the photon energy in the kaon rest frame). Summing over photon helicities there is no interference among electric and magnetic terms. At the lowest order (p^2) in χPT one obtains only E_B. Magnetic and electric direct emission amplitudes, appearing at $\mathcal{O}(p^4)$, can be decomposed in a multipole expansion [50,49,2]. We do not discuss $K_{S,L} \to \pi^0\pi^0\gamma$ due to the small branching ratio ($< 10^{-8}$) [51] and $K_S \to \pi^+\pi^-\gamma$ [52], where no new experimental results have been reported recentely. While motivated by new results we update $K_L \to \pi^+\pi^-\gamma$ and $K^+ \to \pi^+\pi^0\gamma$.

decay	BR(bremsstrahlung)	BR(direct emission)
$K^\pm \to \pi^\pm\pi^0\gamma$ $T^*_{\pi^+} = (55-90)MeV$	$(2.57 \pm 0.16) \times 10^{-4}$ $(\Delta I = 3/2)$	$(4.72 \pm 0.77) \times 10^{-6}$ $E1, M1$
$K_L \to \pi^+\pi^-\gamma$ $E^*_\gamma > 20MeV$	$(1.49 \pm 0.08) \times 10^{-5}$ $(CP$ violation$)$	$(3.09 \pm 0.06) \times 10^{-5}$ $M1, E2$

$K_L \to \pi^+\pi^-\gamma(\gamma^*)$: bremsstrahlung ($E_B$) is suppressed by CP violation (η_{+-}) and firmly predicted theoretically [49]. This contribution has been also measured by interference with the $M1$ transition in $K_L \to \pi^+\pi^-e^+e^-$ [53,54]. Due to the large slope, KTeV parametrizes the magnetic amplitude

$$\mathcal{F} = \widetilde{g}_{M1}\left[\frac{a_1}{(M_\rho^2 - M_K^2) + 2M_K E^*_\gamma} + a_2\right] \tag{14}$$

finding $a_1/a_2 = (-0.729 \pm 0.026(\text{stat}))GeV^2$ and the branching given in the table, which fixes also $\widetilde{g}_{M1}$. Such large slope can be accomodated in various Vector dominance schemes [51,48,57], while the rate is very sensitive to $SU(3)-$breaking and unknown p^4 unknown low energy contributions and thus difficult to predict.

$K^+ \to \pi^+\pi^0\gamma$ New data from BNL E787[55] show vanishing interference among bremsstrahlung and electric transition. Thus the direct emission branching ($B(K^+ \to \pi^+\pi^0\gamma)^{DE}_{\text{exp}}$), in the table, must be interpreted as a pure magnetic transition. Theoretically one can identify two different sources for M, appearing at $\mathcal{O}(p^4)$: i) a pole diagram with a Wess-Zumino term and ii) a pure weak contact term, also generated in factorization by an anomalous current [56]. $B(K^+ \to \pi^+\pi^0\gamma)^{DE}_{\rm exp}$ is substantially smaller than previous values, but still show that this last contribution is non-vanishing.

CP Violation

Direct CP violation can be established in the width charge asymmetry in $K^\pm \to \pi^\pm\pi^0\gamma$, $\delta\Gamma/2\Gamma$ and in the interference E_B with E_1 in $K_L \to \pi^+\pi^-\gamma$ (E with M_1 in $K_L \to \pi^+\pi^-e^+e^-$); both observables are kinematically difficult since one is looking for large photon energy distribution[2]. SM charge asymmetry were looked in [58] expecting $\delta\Gamma/2\Gamma \leq 10^{-5}$. General bounds on new physics in $M1$ transitions have been studied in [59], while the effects of dimension-5 operators in (6) $E1$ transitions have been studied in [60], where for instance it has been shown that the value of $\Re\,(\varepsilon'/\varepsilon)$ allows in particular kinematical regions a factor 10 larger than SM.

6 Conclusions

The recent experimental results in $K-$decays, i.e. ε'/ε, $K \to \pi\nu\overline{\nu}$ and $K_L \to \pi^0\gamma\gamma$, let us hope that the near future will be full of exciting new results and challenges to the SM and its extensions. In particular $K \to \pi\nu\overline{\nu}$ can determine J in (1). I remind that, though not discussed here, T-invariance [54] and Flavour violation[14] are other interestiong windows to search for new physics. From the theoretical side we expect an improving in matching long and short distance contributions[46] and this would lead to an accurate determination of low energy parameters, for instance a reliable relation between a_S and a_+ in (10).

7 Acknowledgements

I am happy to thank the organizers of the Conference for the nice atmosphere and G.Isidori for many interesting discussions. This work is supported in part by TMR, EC–Contract No. ERBFMRX–CT980169 (EURODAΦNE).

8 References

References

1. G. Buchalla, hep-ph/9912369, Y. Nir, WIS-99 [hep-ph/9904271], A.J. Buras, [hep-ph/9905437] and references therein.
2. G. D'Ambrosio and G. Isidori, *Int. J. Mod. Phys.* A **13**, 1 (1998).
3. M.K. Gaillard and B.W. Lee, *Phys. Rev. Lett.* **33,** 108 (1974).

4. S. Weinberg, *Physica* A **96**, 327 (1979); J. Gasser and H. Leutwyler, *Ann. Phys.* (N.Y.) **158,** 142 (1984); A.V. Manohar and H. Georgi, *Nucl. Phys.* B **234,** 189 (1984).
5. G. D'Ambrosio and D. Espriu, *Phys. Lett.* B**175,** 237 (1986);J.L. Goity, *Z. Phys.* C **34,** 341 (1987).
6. Review of Particle Physics, C. Caso *et al.* (Particle Data Group), *Eur. Phys. J.* C **3** 1 (1998).
7. S. Stone, this proceedings.
8. G. Buchalla and G. Isidori, *Phys.Lett.* B **440,** 170 (1998) and references therein.
9. L. Littenberg, *Phys. Rev.* D **39,** 3322 (1989); J. Ellis and J.S. Hagelin *Nucl. Phys.* B **217,** (1983); G. Buchalla and A.J. Buras, *Phys. Rev.* D **54**, 6782 (1996).
10. W.Marciano and Z. Parsa, *Phys. Rev.* **53**, R1 (1996).
11. G. Buchalla and A.J. Buras, *Nucl. Phys.* B **548,** 309 (1999).
12. A.Alavi-Harati et al., KTeV collaboration, hep-ex/9907014.
13. S. Adler et al.(E787 collaboration), *Phys. Rev. Lett.* **79,** 2204 (1997); hep-ex/0002015.
14. S. Kettel, this proceedings.
15. Y. Grossman and Y. Nir, *Phys. Lett.* B **398,** 163 (1997).
16. G.Colangelo and G. Isidori, *JHEP* **09,** 009 (1998).
17. A.Alavi-Harati et al., KTeV collaboration, *Phys. Rev. Lett.* **83** , 22 (1999); A.Roodman, this proceedings.
18. V.Fanti *et al. Phys. Lett.* B **465,** 335 (1999); A. Nappi, this proceedings.
19. S. Bertolini, this proceedings.
20. A.J. Buras *et al.*, hep-ph/9908371.
21. S. Bertolini, J. Eeg and M. Fabbrichesi, *Nucl. Phys.* B **449,** 197 (1995); X.-G. He and G. Valencia, hep-ph/9909399.
22. A. Masiero and H. Murayama, *Phys. Rev. Lett.* **83**, 907 (1999); A. Masiero, these proceedings.
23. A.J. Buras and L.Silvestrini, *Nucl. Phys.* B **546,** 299 (1999).
24. H.B. Greenlee, *Phys. Rev.* D **42,** 3724 (1990).
25. A.J. Buras, M.E. Lautenbacher, M. Misiak and M. Münz, *Nucl. Phys.* B **423,** 349 (1994); S. Bosch *et al.*, hep-ph/9904408.
26. G. Ecker, A. Pich and E. de Rafael, *Nucl. Phys.* B **291,** 692 (1987).
27. J.F. Donoghue and F. Gabbiani, *Phys. Rev.* D **51,** 2187 (1995).
28. G. D'Ambrosio, G. Ecker, G. Isidori and J. Portolés, *JHEP* **08,** 004 (1998).
29. R.Appel *et al.*, E865 collaboration, *Phys.Rev.Lett.* **83**, 4482 (1999).
30. H.Ma *et al.*, E865 collaboration, hep-ex/9910047.

31. S. Adler *et al.*, *Phys. Rev. Lett.* **79,** 4756 (1997).
32. J.F. Donoghue and F. Gabbiani, *Phys. Rev.* D **56,** 1605 (1997); K. Murakami *et al.*,KEK E162, *Phys.Lett.* B **463** 333(1999).
33. J.F. Donoghue, B. Holstein and G. Valencia, *Phys. Rev.* D **35,** 2769 (1986); G. Ecker, A. Pich and E. de Rafael, *Nucl. Phys.* B **303,** 665 (1988); T. Morozumi and H. Iwasaki, *Prog. Theor. Phys.* **82,** 371 (1989).
34. G. Ecker, A. Pich and E. de Rafael, Phys. Lett. B **189,** (1987) 363; L. Cappiello and G. D'Ambrosio, Nuov. Cim. A **99,** (1988) 155.
35. A.Alavi-Harati *et al.*, KTeV collaboration, *Phys.Rev.Lett.* **83,** 917 (1999).
36. L. Cappiello, G. D'Ambrosio and M. Miragliuolo, *Phys. Lett.* B **298,** 423 (1993). J. Kambor and B.R. Holstein, *Phys. Rev.* D **49,** 2346 (1994).
37. A. G. Cohen, G. Ecker and A. Pich, *Phys. Lett.* B **304,** 347 (1993).
38. L. M. Sehgal, *Phys. Rev.* D **38,** 808 (1988); P. Heiliger and L. M. Sehgal, *Phys. Rev.* D **47,** 4920 (1993).
39. G. Ecker, A. Pich and E. de Rafael, *Phys. Lett.* B **237,** 481 (1990).
40. G. D'Ambrosio and J. Portolés, *Nucl. Phys.* B **492**, 417 (1997).
41. G. Ecker, J. Kambor and D. Wyler, *Nucl. Phys.* B **394,** 101 (1993); G. D'Ambrosio and J. Portolés, *Nucl. Phys.* B **533,** 494 (1998).
42. G. D'Ambrosio, G. Isidori and J. Portolés, *Phys.Lett.* B **423**, 385 (1998).
43. G. Valencia, *Nucl. Phys.* B **517,** 339 (1998).
44. D. Gomez Dumm and T. Pich, *Phys. Rev. Lett.* **80,** 4633 (1998).
45. G. Buchalla and A.J. Buras, *Nucl. Phys.*, B **412** (1994) 106.
46. M. Knecht, S. Peris, M. Perrotet and E. de Rafael, *Phys.Rev.Lett.* **83,** 5230 (1999); E. de Rafael, hep-ph/9909210.
47. D. Ambrose *et al.,* BNL (E871), *Phys. Rev. Lett.* **81,** 4309 (1998).
48. G. Ecker, H. Neufeld and A. Pich, *Nucl. Phys.* B **314,** 321 (1994).
49. G. D'Ambrosio *et al.*, in *The DAΦNE Physics Handbook*, eds. L. Maiani, G. Pancheri and N. Paver, LNF (1992) 231; G. D'Ambrosio *et al.*, in *The Second DAΦNE Physics Handbook*, eds. L. Maiani, G. Pancheri and N. Paver (Frascati, 1995) 265.
50. Y.-C.R. Lin and G. Valencia, *Phys. Rev.* D **37,** 143 (1988).
51. P. Heiliger and L.M. Sehgal, *Phys. Lett.* B **307,** 182 (1993)
52. G. D'Ambrosio, M. Miragliuolo and F. Sannino, *Z. Phys.* C **59,** 451 (1993). G. D'Ambrosio and G. Isidori *Z. Phys.* C **65,** 649 (1995).
53. J.Belz, KTeV collaboration, hep-ex/9903025; A.Alavi-Harati *et al.*, KTeV collaboration, *Phys.Rev.Lett.* **84**, 408 (2000).
54. T.Yamanaka, these proceedings.
55. T.K.Komatsubara, BNL E787, Talk at Kaon99.

56. H.-Y. Cheng, *Phys. Rev.* D **42,** 72 (1990); J. Bijnens, G. Ecker and A. Pich, *Phys. Lett.* B **286,** 341 (1992).
57. G. D'Ambrosio and J.Portolés, *Nucl. Phys.* B **533,** 523 (1998).
58. M. McGuigan and A.I. Sanda, *Phys. Rev.* D **36,** (1987) 1413; C.O. Dib and R.D. Peccei, *Phys. Lett.* B **249,** 325 (1990); N. Paver, Riazzudin and F. Simeoni, *Phys. Lett.* B **316,** 397 (1993); H.Y. Cheng, *Phys. Rev.* D **49,** 3771 (1994).
59. X.-G. He and G. Valencia, hep-ph/9908298.
60. G. Colangelo, G. Isidori and J.Portolés, *Phys.Lett.* B **470**,134 (1999).

RARE SEMI/LEPTONIC B DECAYS

TAKUYA MOROZUMI

Recent theoretical development on semi-leptonic Rare B decays are reviewed. The following decay modes are discussed.

1. Inclusive
   $B \to X_s l^+ l^-$, $B \to X_d l^+ l^-$
2. Exclusive
   $B \to K^* l^+ l^-$, $B \to K l^+ l^-$
   $B \to \rho l^+ l^-$, $B \to \pi l^+ l^-$

1 Missions of search for semi-leptonic Rare B decays

The branching ratios of these decays are order α small compared with $B \to K^*\gamma$. Why do we need to find them ? The reason is given as follows.

- We need to test the $\Delta B = 1$ FCNC, the part which has not been tested by $B \to K^*\gamma$ and $b \to s\gamma$. They are Z electroweak penguins and $\Delta B = 1$ Box diagrams.
- Because the top quark contributes to the process, they are useful to determine the KM matrix elements, $|V_{tb}V_{ts}|$ and $|V_{tb}V_{td}|$.
- New physics search
  The decay rate of $B \to K^*\gamma$ already constraints on $\Delta B = 1, \Delta S = 1$ FCNC. However, some new physics effect may not change the decay rate of $B \to K^*\gamma$ while it may change the other observables of rare semi-leptonic B decays. Such new physics effect may be detected with the rare semi-leptonic decays.

2 Theory

The effective interaction of FCNC at M_w scale is well understood. [1]
The coefficient of local four fermi interactions are computed up to $O(\alpha_{(MW)})$ and are given as follows,

$$\begin{aligned} \alpha G_F[C_{9MW}(\bar{s}_L\gamma_\mu b_L)(\bar{l}\gamma_\mu l) + \\ C_{10MW}(\bar{s}_L\gamma_\mu b_L)(\bar{l}\gamma_\mu\gamma_5 l) + \\ C_{qMW}O_{4quark}\cdots], \end{aligned} \tag{1}$$

where,

$$O(C_{9,10Mw}) = (\frac{Mt}{MW})^{2n},$$
$$O(C_{qMw}) = 1 + O(\alpha_s(MW)).$$

At m_b scale, there are large QED, QCD logarithimic corrections to the coefficients and they are summed up by using renormalization group,

$$O(C_{9mb}) = Log(\frac{MW}{mb}) \times [\alpha_s(MW)Log(\frac{MW}{mb})]^n$$
$$+ [\alpha_s(MW)Log(\frac{MW}{mb})]^n,$$
$$O(C_{10mb}) = O(C_{10MW}) = (\frac{Mt}{MW})^{2n}.$$

With the effective hamiltonian at m_b scale at hand, the quark level amplitude for $b \to sll$ in the standard model is given by,

$$\begin{aligned}
\mathcal{M}_{SM} &= \frac{G_F\alpha}{\sqrt{2}\pi} V_{ts}^* V_{tb} \\
&[\ (C_9^{eff} - C_{10})\ \bar{s}_L\gamma_\mu b_L\ \bar{l}_L\gamma^\mu l_L \leftarrow (V-A) \\
&+ (C_9^{eff} + C_{10})\ \bar{s}_L\gamma_\mu b_L\ \bar{l}_R\gamma^\mu l_R \leftarrow (V+A) \\
&- 2C_{7L}^{eff}\ \bar{s}i\sigma_{\mu\nu}\frac{q^\nu}{q^2} m_b Rb\ \bar{l}\gamma^\mu l \leftarrow (b \to s\gamma^*) \\
&- 2C_{7R}^{eff}\ \bar{s}i\sigma_{\mu\nu}\frac{q^\nu}{q^2} m_b Lb\ \bar{l}\gamma^\mu l\ ] \leftarrow (b \to s\gamma^*) \\
&(C_{7R} = \frac{m_s}{m_b} C_{7L}).
\end{aligned} \tag{2}$$

$q^2 = s$ Lepton invariant mass squared. C_9^{eff} depends on s and it is not merely local.

$$\begin{aligned}
&(C_9^{eff} - C_{10}) \simeq 9 >> (C_9^{eff} + C_{10}) \simeq -0.4 \\
&C_{7L}^{eff} \simeq -0.3 >> C_{7R} \simeq 6 \times 10^{-3}
\end{aligned} \tag{3}$$

The typical values for $C_9(mb)$, $C_{10}(mb)$ and $C_{7L}(mb)$ in the standard models are given in table(1). Theoretical uncetainties of the wilson coefficients are about 10% by changing $m_t = 175 \pm 9$, $\mu = [mb/2, 2mb]$.
The general features of the interaction can be summarized as follows. The non-local interaction C_{7L} is dominant for very low small $q^2(= s)$. While $q^2 < m_\Psi^2$

Table 1. Wilson coefficients at m_b

$C_9(m_b)$	4.153
$C_{10}(m_b)$	-4.546
$C_{7L}(m_b)$	-0.311
$C_{7R}(m_b)$	6×10^{-3}

which is below charmonuum thershold, all the four types of interaction can be equally important. Near the maximum of q^2, $C_9 - C_{10}$ is dominant. In the following, we study how to check the above features. Before doing this, let me summarize the experimental constraints from the existing data.

1. $B \to X_s\gamma$ and $B \to K^*\gamma$ sets a constraint. Using the inclusive decay rate, we obtain,

   $$|C_{7L}|^2 + |C_{7R}|^2 = 0.081 \pm 0.014,$$

   $$(C_{7L}, C_{7R})_{SM} \simeq (-0.311, O(10^{-2} - 10^{-3})).$$

2. The upper bound on $B \to K^*\mu\bar{\mu}$ ($Br \leq 4.0 \times 10^{-6}$) set the constraint [3].

   $$\sqrt{|C_9|^2 + |C_{10}|^2} \leq 10,$$

   $$(C_9, C_{10})_{SM} \simeq (4, -4.5).$$

The exsiting upper bound on the decay rate ($b \to see(\mu\mu) < 5.8(5.7) \times 10^{-5}$) is still too high to gives the meaning constraint.

3 How to check the Standard model

The following Measurements may be usefull to check the structure of the interaction of FCNC.

- Decay rates
- Forward-backward (Energy) asymmetry
- polarization asymmtries
- angle distributions

Table 2. **Dependence on the sign change of each Wilson coefficient, $C \to -C$**

	$Br - Br\|_{SM}$	$Asym_{LOW(S)}$	$Asym_{High(S)}$
SM	0	−	+
$-C_{7L}$	+	+	+
$-C_{10}$	0	+	−
$-C_9$	+	−	−

It can be seen that the branching ratio and forward backward asymmetry is written as follows [4],

$$\frac{dBr}{d\hat{s}} = K(1-\hat{s})^2[(|C_9^{eff}|^2 + |C_{10}|^2)(1+2\hat{s}) + 4(|C_{7L}|^2 + |C_{7R}|^2)\frac{2+\hat{s}}{\hat{s}} + 12Re(C_9^{eff*}C_{7L})],$$

$$\frac{d\bar{A}}{d\hat{s}} = \frac{-3K(1-\hat{s})^2 ReC_{10}^*(C_9^{eff}\hat{s} + 2C_{7L})}{\frac{dBr}{d\hat{s}}},$$
$$\hat{s} = q^2/mb^2, K > 0.$$

Branching ratio and Forward backward asymmetry are sensitive to the sign of $ReC_9^*C_{7L}(<0)$, $ReC_{10}^*C_9(<0)$ $ReC_{10}^*C_{7L}(>0)$. Inside the parentheses, the signs of the standard model are shown.

Summary of the changing the sign of the Wilson coefficients. Here we mean that we change the sign of Wilson coefficients at m_b. We do not change the sign of the whole C_9^{eff}.

- Changing the sign of each coefficient leads to the different patterns of the change of the three quantities. Branching ratio, Low (s) and high (s) forward backward asymmetry. Therefore measuring the quantities, we may guess the interactions. See table2.

- If the sign of the Forward and Backward asymmetry is changed from the standard model prediction in the whole region, the sign of C_{10} is flipped. If the Forward-backward asymmetry at High (s) is negative, then it may be signature of $V + A$ dominance in the leptonic current. (In the SM, it is $V - A$ dominance.) It means either sign of C_9 or that of C_{10} is flipped. Flipping the sign of C_{10} does not change the branching ratio while the sign change of C_9 may increase the branching ratio. Therefore two cases may be distinguishable.

- If branching ratio increases and the asymmetry over all region is positive, C_{7L} sign may be opposite to that of the standard model.

The other interesting cases
Q: What kind of new physics effect significantly decreases the branching ratio compared with that of the standard model ?
A: Subtract the dominant interaction in the standard model [6],

$$C_{LL}\bar{s_L}\gamma_\mu b_L \bar{l_L}\gamma^\mu l. \qquad C_{LL} < 0 \tag{4}$$

Q: If the new physics changes the coefficients of non-local operators C_{7L} and C_{7R}, can we distinguish it from the standard model ? [5]
A: There are many combinations of (C_{7L}, C_{7R}) which satisfies the constraint from the $B \to X_s\gamma$.

$$|C_{7L}|^2 + |C_{7R}|^2 \simeq 0.08 \tag{5}$$

If we attribute $B \to X_s\gamma$ to C_{7R}, both branching ratio and asymmetry increases because the negative interference terms between C_{7L} and the other coefficients vanish. (Provided only these two coefficients can be changed.)

4 Exclusive decays

By using the exclusive decays (see for example, [8]), we can perform various analysis which can not be done in the inclusive decay. Here we introduce a method for extracting the ratio C_{7R}/C_{7L} with azimuthal angle distribution of $B \to K^*(\gamma^*, Z, BOX) \to (K\pi)ll$, [7].

$$r(\phi, s) \simeq \frac{1}{2\pi}\frac{|C_{7L}\exp(i\phi) + C_{7R}\exp(-i\phi)|^2}{(|C_{7L}|^2 + |C_{7R}|^2)}$$

$r(\phi, s)$= Probability per unit radian for emitting K to the direction with azimuthal angle ϕ between $K\pi$ decay plane and l^+l^- decay plan.
$C_{7L}(C_{7R})$ = amplitude which produces the left-handed (right-handed) polarized photon (γ^*).
(The above equation approximately holds at very low invariant mass squared region of dileptons so that $B \to K^*\gamma^*$ is dominant.)

Table 3. **Angle Dependence on two typical choice of C_{7R}, C_{7L}**

$\alpha \simeq 0$	$C_{7R} \simeq 0(SM)$	No ϕ dependence
$\alpha = \pi/4$	$\|C_{7R}/C_{7L}\| = 1$	$1 - cos2(\phi - v)/2$

Table 4. **Theoretical predictions of the exclusive and inclusive decays: I=From Melikov Nikitin Simula [10], $|V_{ts}| = 0.038$, QM=Dispersion Quark Model II=From A. Ali, P.Ball, Handoko, and Hiller [3], $|V_{ts}| = 0.0385$ LCS=Light Cone QCD sum rule**

	I(QM)	II(LCS)
$B \to K^* l^+ l^- (l = e)$	1.5×10^{-6}	2.3
$B \to K^* l^+ l^- (l = \mu)$	1.15×10^{-6}	1.9
$B \to K^* l^+ l^- (l = \tau)$	1.0×10^{-7}	1.9
$B \to K^* \nu\bar{\nu}$	1.5×10^{-5}	
$B \to K l^+ l^- (l = e, \mu)$	4.4×10^{-7}	5.7
$B \to K l^+ l^- (l = \tau)$	1.0×10^{-7}	1.3
$B \to K \nu\bar{\nu}$	5.6×10^{-6}	
$B \to X_s l^+ l^- (l = e)$	$(8.4 \pm 2.3) \times 10^{-6}$	
$B \to X_s l^+ l^- (l = \mu)$	$(5.7 \pm 1.2) \times 10^{-6}$	
$B \to X_s l^+ l^- (l = \tau)$	$(2.6 \pm 0.5) \times 10^{-7}$	

Physics motivation
In the standard model C_{7L} is dominant. There are models (ex.left-right model, [2]), C_{7R} can be as large as C_{7L}. Can we distinguish the models with $O(C_{7L}/C_{7R}) = 1$? We use the following parametrization.

$$\begin{aligned} C_{7L} &= -\sqrt{|C_{7L}|^2 + |C_{7R}|^2} \cos\alpha \exp i(u+v), \\ C_{7R} &= \sqrt{|C_{7L}|^2 + |C_{7R}|^2} \sin\alpha \exp i(u-v), \\ |C_{7L}|^2 + |C_{7R}|^2 &= 0.081(from \quad B \to X_s\gamma). \end{aligned}$$

Azimutal angle distribution for very low invariant mass squared is:

$$r(\phi, s) = \frac{1}{2\pi}\left\{1 - \frac{1}{2}\sin 2\alpha \cos 2(\phi - v)\right\},$$

From table 3, we can see that the azimuthal angle distribution for the standard model and that for the case $|C_{7R}/C_{7L}| = 1$ are very different.

5 Summary

- It is shown how we can determine the various type of contribution in $b \to sll$ process by using differential branching ratio, Foward backward asymmetry. The quantity like the ratio C_{7L}/C_{7R} which can not be detemined from the $b \to s\gamma$ may be extracted by using the angle distribution of $B \to K^*(K\pi)ll$.
- They are interesting. Still much work is needed since the typical exclusive decay rates are $O(10^{-6})$ for $b \to s$ FCNC and $b \to d$ FCNC is futher down by a factor of $O(\lambda^2, \lambda = 0.22)$ smaller.[9]
- In order that the experimental discovery (with small statistics) is meaningful, we have to reduce and controle the theoretical uncertainties of form factors of exclusive processes. After this is done, various ideas on the sensitive measurements for new physics will be realistic.

Acknowledgments

T.M. likes to thank organizers of B and CP conference and Misiak for comments. and also likes to thank S. Fukae for drawing some figures. The work of T.M. is supported by Grant-in-Aid for Scientific Research on Priortty Areas (Physics of CP violation) and is partly supported by JSPS.

References

1. Misiak, *Nucl. Phys.* B **393**, 23 (1993) ,*Nucl. Phys.* B **439**, 461 (1995) Buras and Münz, *Phys. Rev.* D **52**, 186 (1995)
2. K. Fujikawa, and A. Yamada, *Phys. Rev.* D **49**, 5890 (1994),
3. A. Ali, P. Ball, G. Hiller, L. Handoko, hep-ph/9910221.
4. Ali, Mannel, and T. Morozumi,*Phys. Lett.* B **273**, 505 (1991).
5. S. Fukae, C.Kim, and T. Yoshikawa , HUPD9920.
6. S. Fukae, C. Kim, T. Yoshikawa and T. Morozumi, *Phys. Rev.* D **59**, 074013 (1999).
7. C. S. Kim, Y. G. Kim, C. Lü, and T. Morozumi , hep-ph/0001151.
8. C. Q. Geng and C. Kao, *Phys. Rev.* D **54**, 5636 (1996).
9. Krüger and Sehgal, *Phys. Rev.* D **55**, 2799 (1997), C.Kim, T. Morozumi, and A. I. Sanda, *Phys. Rev.* D **56**, 7240 (1997).
10. Melikov Nikitin Simula, *Phys. Rev.* D **57**, 6814 (1998).

$D^\circ - \bar{D}^\circ$ Mixing and Rare Decays

Hwanbae Park
Department of Physics, Korea University, Seoul, Korea
E-mail: hbpark@hep.korea.ac.kr

Current status of charm mixing search, lifetime difference measurement and rare decay search are reported. The best upper limit at 95% CL for mixing is reported from the CLEO collaboration. E791 has reported lifetime difference measurement and results of rare decay searches. Rare decay searches of the FOCUS collaboration are in progress and their sensitives to branching ratios are at the level of a few $\times 10^{-6}$.

The Standard Model (SM) predictions of charm mixing and rare decays are negligibly small and new physics effects are not hidden by the SM effects. There are new physics scenarios such as a large mass of fourth down-quark, two higgs-doublet model which gives an additional box diagram by replacing $W^{\pm}$ with the charged Higgs, flavor changing neutral Higgs (FCNC) model in which up-quark sector is treated differently from down-type sector. There are also leptoquark model in which scalar leptoquark bosons participate in a mass difference via exchange inside a box diagram, and supersymmetry model which gives an addtional box diagram contribution. [1] These scenarios predict large mixing effects and any experimental observations in searches of mixing and rare decay would indicate the existence of new physics beyond the Standard Model.

1 Mixing

Mixing occurs because two weak eigenstates are not the mass eigenstates. The probability of finding $\bar{D}^\circ$ at time t with produced D° at $t = 0$ is

$$\Gamma(D^\circ \to \bar{D}^\circ) = \frac{1}{4}\left|\frac{q}{p}\right|^2 e^{-\Gamma_+ t}[1 + e^{\Delta\Gamma t} - 2e^{(\frac{1}{2}\Delta\Gamma t)}\cos(\Delta m t)] \tag{1}$$

in which two states oscillate with rate expressed by Δm and $\Delta\Gamma$. $\Delta m \ll \Gamma$ and $\Delta\Gamma \ll \Gamma$ are good approximations for the charm particle and the mixing rate is

$$r_{\text{mix}} = \frac{\Gamma(D^\circ \to \bar{D}^\circ \to \bar{f})}{\Gamma(D^\circ \to f)} = \frac{1}{4}e^{-\Gamma t}\left|\frac{q}{p}\right|^2 (x^2 + y^2)\Gamma^2 t^2 \tag{2}$$

with following definition of the mixing parameters,

$$x = \frac{\Delta m}{\Gamma}; \quad y = \frac{\Delta\Gamma}{2\Gamma} \tag{3}$$

$\Delta\Gamma$ is expected to be of the same order of magnitude as Δm, and new physics or long distance contribution effects are sensitive to Δm.

Mixing is studied by searching for wrong sign signals or measurement of lifetime difference in CP eigenstates. The direct search is to look for the produced $D^\circ(\bar{D}^\circ)$ that decays as $\bar{D}^\circ(D^\circ)$. The flavor of the produced D is tagged by the charge of the bachelor pion in $D^{\star +} \to D^\circ\pi^+$, and D° then decays hadronically ($K^+\pi^-$, $K^+\pi^-\pi^+\pi^-$) and semileptonically ($K^+\ell^-\bar{\nu_\ell}$). Since the Doubly Cabbibo Suppressed Decay (DCSD) gives the wrong sign signals as well as mixing, and the charm mixing in the SM is expected to be small, the DCSD contribution in the hadronic modes may not be neglected. The amplitude of the mixing is given by

$$A_{\rm WS} = A_{\rm DCSD}(D^\circ \to \bar{f}) + A_{\rm mix}(D^\circ \to \bar{D}^\circ \to \bar{f}) \tag{4}$$

The wrong sign decay rate is then

$$\begin{aligned} r_{\rm WS} = \tfrac{1}{4}e^{-\Gamma t} &< \bar{f}|H|\bar{D}^\circ >^2_{\rm CF} |\tfrac{q}{p}|^2 \times \\ &[4|\lambda|^2 + (2Re(\lambda)\Delta\Gamma + 4Im(\lambda)\Delta m)t + (\Delta m^2 + \tfrac{1}{4}\Delta\Gamma^2)t^2] \end{aligned} \tag{5}$$

where

$$\lambda = \frac{p}{q}\frac{< \bar{f}|H|D^\circ >_{\rm DCSD}}{< \bar{f}|H|\bar{D}^\circ >_{\rm CF}} \tag{6}$$

The significant differences in the structure of the proper decay time for the DCSD, the interference between DCSD and mixing, and the mixing allow us to estimate their respective contributions.
The CP conjugate rate is

$$\begin{aligned} \bar{r}_{\rm WS} = \tfrac{1}{4}e^{-\Gamma t} &< f|H|D^\circ >^2_{\rm CF} |\tfrac{p}{q}|^2 \times \\ &[4|\bar{\lambda}|^2 + (2Re(\bar{\lambda})\Delta\Gamma + 4Im(\bar{\lambda})\Delta m)t + (\Delta m^2 + \tfrac{1}{4}\Delta\Gamma^2)t^2] \end{aligned} \tag{7}$$

where

$$\bar{\lambda} = \frac{q}{p}\frac{< f|H|\bar{D}^\circ >_{\rm DCSD}}{< f|H|D^\circ >_{\rm CF}} \tag{8}$$

With definition of the CP violation angle ϕ and strong phase δ:

$$e^{-i\phi} = \frac{p}{q}; \quad \sqrt{R_{\rm DCSD}} \cdot e^{-i\delta} = \frac{< \bar{f}|H|D^\circ >_{\rm DCSD}}{< \bar{f}|H|\bar{D}^\circ >_{\rm CF}} \tag{9}$$

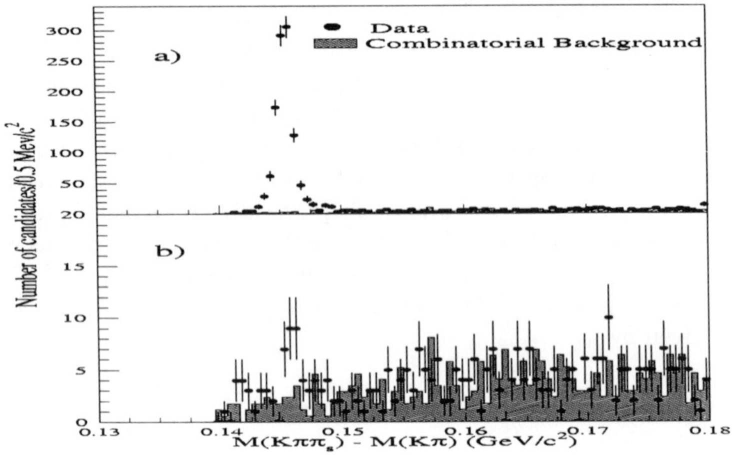

Figure 1: ALEPH mixing search in the $K\pi$ mode. (a) is the right sign and (b) is the wrong sign distributions. The hatched histogram is the combinatorial background and is assumed same shape from events in the sideband region of D° mass above $2.1\text{GeV}/c^2$, and is normlized to the number of candidates having $\Delta M > 160\text{MeV}/c^2$.

Eq. (5) is simplied

$$\begin{aligned} r_{WS} = e^{-\Gamma t} &< \bar{f}|H|\bar{D}^\circ >^2_{CF} \times \\ &[R_{DCSD} \\ &+ (y\sqrt{R_{DCSD}} \cdot \cos(\delta \pm \phi) - x\sqrt{R_{DCSD}} \cdot \sin(\delta \pm \phi))\Gamma t \\ &+ \tfrac{1}{2} R_{\text{mix}}(\Gamma t)^2] \end{aligned} \tag{10}$$

Defining $x' = x \cdot \cos\delta + y \cdot \sin\delta$ and $y' = y \cdot \cos\delta - x \cdot \sin\delta$, and assuming CP invariance ($\phi = 0$), Eq. (10) then becomes

$$r_{\text{WS}} \propto e^{-\Gamma t}[R_{\text{DCSD}} + y'\sqrt{R_{\text{DCSD}}}(\Gamma t) + \frac{1}{2}R_{\text{mix}}(\Gamma t)^2] \tag{11}$$

The dependence on x' and y' is distingushable due to the interference with direct decay amplitude, which induce the linear interference.

2 Mixing Searches

2.1 *Wrong Sign Searches*

The ALEPH has used 4×10^6 hadronic Z° events collected from 1991 to 1995 and searched the wrong sign signals in $D^{*+} \to D^\circ\pi^+$ with D° decaying to $K^+\pi^-$.[2] The right sign and wrong sign mass plots are shown in Fig. 1.

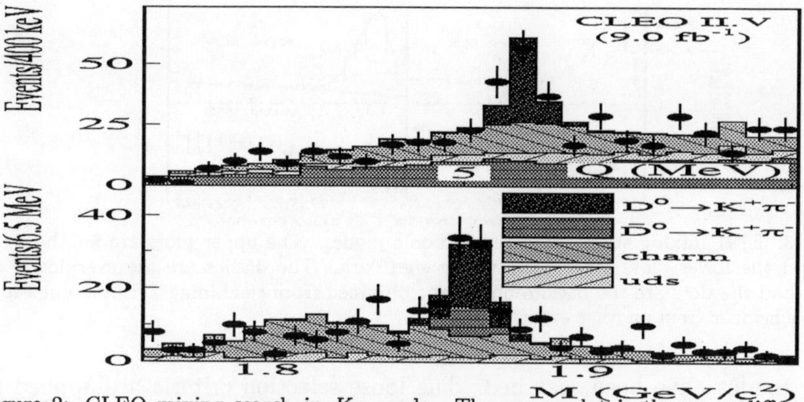

Figure 2: CLEO mixing search in $K\pi$ mode. The upper plot is the mass difference Q ($M_{K\pi\pi} - M_{K\pi} - M\pi$) and the lower plot is reconstructed D mass distributions of wrong sign data. Backgrounds from various sources are shown.

$N_{RS} = 1038.8 \pm 32.5 \pm 4.3$ and $N_{WS} = 19.1 \pm 6.1 \pm 3.5$ after combinatorial and physics backgrounds substraction are obtained. They also measured a relative branching ratio of $B(D^\circ \to K^+\pi^-)/B(D^\circ \to K^-\pi^+) = (1.84 \pm 0.59 \pm 0.34)\%$ and set $r_{\text{mix}} < 0.92\%$ at 95% CL on the mixing rate assuming no interference between DCSD and mixing.

The CLEO has used 9.0 fb^{-1} of e^+e^- data with CLEO II.V detector from 1996 to 1999 runs. The analysis takes great advantage of the silicon vertex detector. The D mass and Q ($M_{K\pi\pi} - M_{K\pi} - M\pi$) plots for the wrong sign candidates are shown in Fig. 2. The signal shape is determined by the right sign distribution and the background shapes are estimated by Monte Carlo simulation. $44.8^{+9.7}_{-8.7}$ and 13527 ± 116 for the wrong sign and the right sign events are found from fits, respectively.[3] Using these numbers, they measure

$$\frac{R_{\text{WS}}(K\pi)}{R_{\text{RS}}(K\pi)} = (0.335^{+0.064}_{-0.066} \pm 0.040)\% \tag{12}$$

From fits, they also find one-dimensional intervals at 95% CL of:

$$\frac{1}{2}x'^2 < 0.039\%; \quad -5.6\% < y' < 1.3\% \tag{13}$$

The E791 is the fixed target hadroproduction experiment at Fermilab which uses a 500 GeV/c^2 π^- beam. They logged 2×10^{10} hadronic interactions and reconstructed more than 200,000 hadronic events for 1990-1991 runs. The wrong sign signals in hadronic ($K\pi/K3\pi$) and semileptonic ($K\ell\bar{\nu}$)

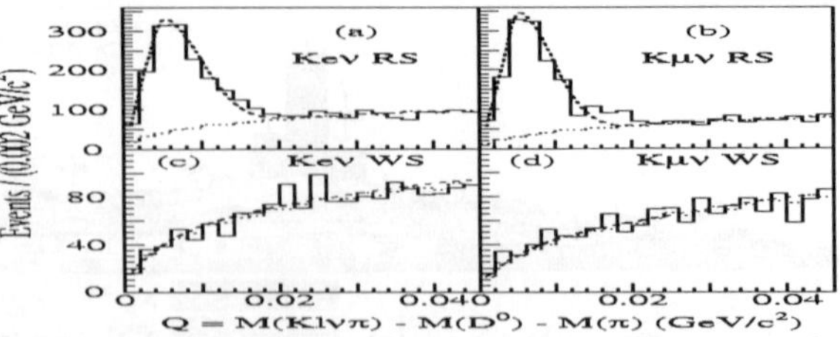

Figure 3: E791 mixing search with semileptonic modes. The upper plots are for the right sign and the lower plots are for the wrong sign data. The dashes are the overplot of fit results and the dots are the background shape obtained from combining D° from one event with bachelor π from another event.

decay modes have been searched. The loose selection criteria are applied to reduce data samples and then selection cuts are optimized with artificial neural networks. 5643 and 3469 of the right sign signals for $K\pi$ and $K3\pi$ are obtained, respectively.[4] They first performed most general fit allowing for CP violation, mixing interference and no model dependence, and set the 90% CL limits:

$$r_{\text{mix}}(D^\circ \to \bar{D}^\circ) < 1.45\%; \quad r_{\text{mix}}(\bar{D}^\circ \to D^\circ) < 0.74\% \tag{14}$$

Assuming CP violation only in the interference term, they find 90% CL limits of:

$$r_{\text{mix}} < 0.85\% \tag{15}$$

They also evaluated relative branching ratio for the DCSD modes with assumption of no mixing:

$$R_{\text{DCSD}}(K\pi) = (0.68^{+0.34}_{-0.33} \pm 0.07)\%; \quad R_{\text{DCSD}}(K3\pi) = (0.25^{+0.36}_{-0.34} \pm 0.03)\% \tag{16}$$

In the semileptonic analysis, $D^{\star +} \to D^\circ \pi^+$ with D° decaying to $K^+\ell^-\bar{\nu}_\ell$ is searched. There is a two-fold momentum ambiguity in the D° momentum due to the missing neutrino and the higher momentum solution is picked from Monte Carlo study. They fit to the mass difference and the decay time distribution with D° mass fixed. The right and wrong sign Q plots are shown in Fig. 3. $N_{\text{RS}} = 1237 \pm 45$ for $Ke\nu$ and $N_{\text{RS}} = 1267 \pm 44$ for $K\mu\nu$ are obtained.[5] After performing two-dimensional unbinned maximum likelihood fit and correcting for the different detector acceptance for the different decay time behaviors of the mixed and right sign decays, they obtained:

$$r_{\text{mix}}(Ke\nu) = (0.16^{+0.42}_{-0.37})\%; \quad r_{\text{mix}}(K\mu\nu) = (0.06^{+0.44}_{-0.40})\% \tag{17}$$

Using these numbers and they have determined the weighted average and set 90% CL limit:

$$r_{\text{mix}} = (0.11^{+0.30}_{-0.27})\%; \quad r_{\text{mix}} < 0.5\% \tag{18}$$

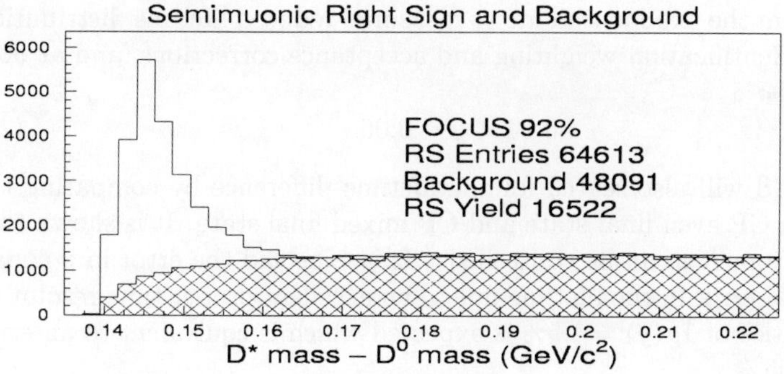

Figure 4: FOCUS preliminary mixing search in semimuonic mode. The hatched background shape is modeled by replacing pions in primary by pions from a different event.

The FOCUS is the fixed target photoproduction experiment at Fermilab and is the successor to E687. The data collected during 1996-1997 run have reconstructed 15 times more charm decays than the E687. Data is being used in search of semileptonic and hadronic wrong sign decay modes. A mass difference plot for their preliminary right sign semimuonic analysis with 92% of their total data is shown in Fig. 4. Approximately 16500 background subtracted right sign candidates are found and roughly equal numbers of events in the electron mode are obtained as well. The wrong sign signal region is not looked for until selection cuts have been optimized and backgrounds understood. r_{mix} will be extracted from two dimensional fit to the mass difference and the proper time distribution. At 90% CL limit on a sensitivity extrapolation from the preliminary studies indicate:

$$r_{\text{mix}} < 0.12\% \tag{19}$$

2.2 Lifetime Difference Searches

The lifetime difference is obtained by the measurement of the lifetimes of CP eigenstates or comparing number of decays observed at any decay time:

$$\begin{aligned} \Gamma(D^\circ \to K^+K^-) - \Gamma(D^\circ \to K^-\pi^+) &= \Gamma_+ - \tfrac{1}{2}(\Gamma_+ + \Gamma_-) = \tfrac{1}{2}\Delta\Gamma \\ \ln\frac{N(D^\circ \to K^+K^-)}{N(D^\circ \to K^-\pi^+)} &= \ln(\frac{A_{KK}}{A_{K\pi}}) - \frac{\Delta\Gamma}{2}t \end{aligned} \tag{20}$$

The E791 measured the lifetimes of the $D^\circ \to K^+K^-$ which is CP even eigenstate and $D^\circ \to K^-\pi^+$ which is CP mixed state. [6] After background subtraction 6683 ± 161 K^+K^- and 60571 ± 353 $K^-\pi^+$ events are found. The lifetime is extracted from the exponential fits to measured reduced lifetime distribution after particle identification weighting and acceptance corrections, and at 90% CL limit on y is:

$$-0.04 < y < 0.06 \tag{21}$$

The FOCUS will also search for the lifetime difference by comparing the lifetimes of the CP even final state and CP mixed final state. It is shown that the reflection from $K\pi$ can be controlled.[7] It is expected the error in y equals to the fractional error in the lifetime due to an excellent proper time resolution and a sensitivities of $1\sigma(y) = 1.5\%$ is expected which is equivalent to an error on r_{mix} of 0.005%.

3 Rare Decay Searches

Charm rare decay searches are studied in flavor changing neutral current (FC-NC), lepton flavor number violation (LFNV) and lepton number violation (L-NV) decays. The FCNC decays are highly suppressed by the Glashow, Ilopoulos, and Maiani (GIM) mechanism and branching ratios are expected to be 10^{-19}-10^{-9}. The LFNV and LNV are forbidden in the Standard Model. Any results beyond the SM predictions would be a clear evidence of new physics. Dilepton modes have experimental advantages due to clean and efficient lepton identification and small combinatoric backgrounds. To avoid any bias in selection criteria due to presence or absence of signal candidates, all events within a signal mass window are masked.

The E791 has recently published results on searches for 24 different decay modes in FCNC, LFNV and LNV modes.[8] They did not observe any signals and set 90% CL upper limit. The results of searches are compared to previous limit in Fig. 5. 8 are new measurements in D_s^+ and 14 are of significant improvements.

The FOCUS is now in process of searching for forbidden and rare dilepton modes. The blind analysis is being used in the seaches and is very similar to that of the E791. The current sensitivites for D^+ modes are at (4-6) $\times\ 10^{-6}$ level at 90% CL on upper limit.

4 Summary

In charm mixing search the best limits in the most general case comes from the CLEO measurement: $r_{\text{mix}} < 0.05\%$ at 90% CL upper limit. The E791 also has the best published limit using the semileptonic decays: $r_{\text{mix}} < 0.5\%$ at 90% CL upper limit. Analysis in progress by the FOCUS are sensitive to $r_{\text{mix}} \sim 5 \times 10^{-4}$. First results from searches for a lifetime difference $\Delta\Gamma$ have

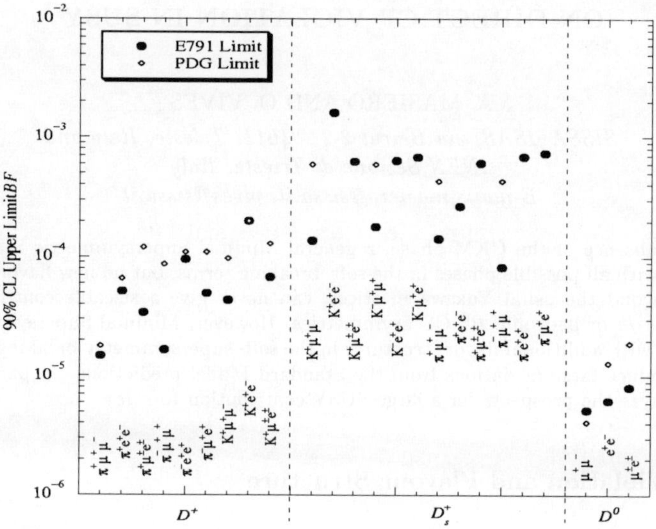

Figure 5: Comparision of branching fraction measurements at the 90% CL limit from E791 (dark circles) with previous best limits (open diamonds).

recently been published by the E791 and they find $-0.04 < y < 0.06$ at 90% CL. Work in the CLEO and the FOCUS is in progress and expected sensitivies on y to the few $\times 10^{-3}$ level in the very near future. The E791 has recently published limits on the branching ratios for 24 forbidden and rare decay modes. In general their results are new or are of significant improvements over previous limits. The rare decay searches of the FOCUS is at the early stage and their sensitivity to D^+ branching ratios is expected to be a few $\times 10^{-6}$.

References

1. J.L. Hewett, *Searching for New Physics with Charm*, hep-ph/9505246.
2. R. Barate *et al* , *Phys. Lett.* B **436**, 211 (1998).
3. D. Asner's talk in this proceeding.
4. E.M. Aitala *et al*, *Phys. Rev.* D **57**, 13 (1998).
5. E.M. Aitala *et al*, *Phys. Rev. Lett.* **77**, 2384 (1996).
6. E.M. Aitala *et al*, *Phys. Rev. Lett.* **83**, 32 (1999).
7. P. Sheldon, *Charm Mixing and Rare Decays* in Heavy Flavors 8, Southampton, UK (1999) [hep-ex/9912016].
8. E.M. Aitala *et al*, *Phys. Lett.* B **462**, 401 (1999).

ON DIRECT CP VIOLATION IN SUSY

A. MASIERO AND O. VIVES

SISSA–ISAS, via Beirut 2-4, 34014, Trieste, Italy and
INFN Sezione di Trieste, Italy
E-mail: masiero@sissa.it, vives@sissa.it

In the absence of the CKM phase, a general Minimal Supersymmetric Standard Model with all possible phases in the soft–breaking terms, but no new flavor structure beyond the usual Yukawa matrices, can never give a sizeable contribution to ε_K, ε'/ε or hadronic B^0 CP asymmetries. However, Minimal Supersymmetric models with additional flavor structures in the soft–supersymmetry breaking terms can produce large deviations from the Standard Model predictions. In particular we analyze the prospects for a large SUSY contribution to ε'/ε

1 CP Violation and Flavour Structure

The Standard Model of electroweak interactions is known to be able to accommodate the experimentally observed CP–violation through a unique phase, δ_{CKM}. However, most of the extensions of the SM include new observable phases that may significantly modify the pattern of CP violation. For instance, it is known that in the constrained minimal supersymmetric extension of the SM (CMSSM), after all possible rephasings of the parameters and fields, there remain at least two new physical phases [1]. More CP violating phases appear in unconstrained MSSM where, for instance, universality in the soft breaking terms is not imposed. In the last few years, the possibility of having non–zero SUSY phases has again attracted a great deal of attention. Several new mechanisms have been proposed to suppress supersymmetric contributions to the EDMs of the neutron and the electron below the experimental bounds while allowing SUSY phases $\mathcal{O}(1)$ [2,3,4].

We have recently shown that the presence of large susy phases is not enough to produce sizeable supersymmetric contributions to CP violation observables [5]. In fact, in the absence of the CKM phase, a general MSSM with all possible phases in the soft–breaking terms, but no new flavor structure beyond the usual Yukawa matrices, can never give a sizeable contribution to ε_K, ε'/ε or hadronic B^0 CP asymmetries. However, as recently emphasized (see next section), as soon as one introduces some new flavor structure in the soft Susy–breaking sector, even if the CP violating phases are flavor independent, it is indeed possible to get sizeable CP contribution for large Susy phases and $\delta_{CKM} = 0$. Then, we can rephrase our sentence above in a different way: *A new result in hadronic B^0 CP asymmetries in the framework of supersymme-*

try would be a direct proof of the existence of a completely new flavor structure in the soft–breaking terms. This means that B–factories may have a chance to probe the flavor structure of the supersymmetry soft–breaking terms even before the direct discovery of the supersymmetric partners .

2 Trilinear scalar terms and ε'/ε

In the CMSSM the SUSY contribution to ε'/ε is small [6]. However in a MSSM with a more general framework of flavor structure it is relatively easy to obtain larger SUSY effects to ε'/ε. In ref. [7] it was shown that such large SUSY contributions arise once one assumes that i) hierarchical quark Yukawa matrices are protected by flavor symmetry, ii) a generic dependence of Yukawa matrices on Polonyi/moduli fields is present (as expected in many supergravity/superstring theories), iii) the Cabibbo rotation originates from the down-sector and iv) the phases are of order unity. In fact in [7] it was illustrated how the observed ε'/ε could be mostly or entirely due to the SUSY contribution.

The universality of the breaking is a strong assumption and is known not to be true in many supergravity and string inspired models [8]. On the other hand the mass matrix of the left- and right-handed squarks (LR mass matrix) has the same flavor structure as the fermion Yukawa matrix and both in fact originate from the superpotential couplings. It may be appealing to invoke the presence of an underlying flavor symmetry restricting the form of the Yukawa matrices to explain their hierarchical forms. Then the LR mass matrix is expected to have a very similar form as the Yukawa matrix. Indeed, we expect the components of the LR mass matrix to be roughly the SUSY breaking scale (e.g., the gravitino mass) times the corresponding component of the quark mass matrix. However there is no reason for them to be simultaneously diagonalizable based on this general argument. It turns out that, following the simplest implementation along the lines of the above described idea, the amount of flavor changing LR mass insertion in the s and d–squark propagator results to roughly saturate the bound from ε'/ε if a SUSY phase of order unity is present [7]. Such results are obtainable even if the soft breaking A terms are real under relatively general assumptions [9]. Major implications of this approach are the predictions that the EDM of the neutron and the leptonic flavor changing phenomena, like $\mu \to e\gamma$ decay, should be within experimental reach in a near future.

There exist other ways of originating large SUSY contributions to ε'/ε [10]. The message which emerges from all these studies is that, apart from the very constrained situation of the minimal supergravity model (CMSSM), in general SUSY extensions of the SM where universality is not present there exist possibilities to obtain sizeable SUSY contributions to ε'/ε without invoking very particular or contrived circumstances.

Acknowledgments

We thank the organizers for the pleasant atmosphere in which this meeting took place, and S. Bertolini, D. Demir, G.Eyal, T. Kobayashi , S. Khalil, H. Murayama, Y. Nir and L. Silvestrini for enlightening discussions. The work of A.M. was partially supported by the European TMR Project "Beyond the Standard Model" contract N. ERBFMRX CT96 0090; O.V. acknowledges financial support from a Marie Curie EC grant (TMR-ERBFMBI CT98 3087).

References

References

1. M. Dugan, B. Grinstein and L.J. Hall, Nucl. Phys. **B 255**, 413 (1985); S. Dimopoulos and S. Thomas, Nucl. Phys. **B 465**, 23 (1996), hep-ph/9510220;
2. T. Ibrahim and P. Nath, Phys. Rev. **D 58**, 111301 (1998), hep-ph/9807501;
   M. Brhlik, G. Good and G. Kane, Phys. Rev. **D 59**, 115004 (1999), hep-ph/9810457;
   M. Brhlik, L. Everett, G.L. Kane and J. Lykken, Phys. Rev. Lett. **83**, 2124 (1999), hep-ph/9905215;
   M. Brhlik, L. Everett, G.L. Kane and J. Lykken, Fermilab preprint no. FERMILAB-PUB-99-230-T, August 1999, hep-ph/9908326;
   T. Ibrahim and P. Nath, Santa Barbara U. preprint no. NSF-ITP-99-129, October 1999, hep-ph/9910553;
   M. Brhlik, L. Everett, G.L. Kane, S.F. King and O. Lebedev, Virginia Pol. Inst. Report no. VPI/IPPAP/99/08, Sept. 1999, hep-ph/9909480.
3. S. Abel and J.M. Frere, Phys. Rev. **D 55**, 1623 (1997), hep-ph/9608251; S. Khalil, T. Kobayashi and A. Masiero, Phys. Rev. **D60**, 075003 (1999), hep-ph/9903544.
4. S. Dimopoulos and G.F. Giudice, Phys. Lett. **B357**, 573 (1995), hep-ph/9507282;
   A.G. Cohen, D.B. Kaplan and A.E. Nelson, Phys. Lett. **B388**, 588 (1996), hep-ph/9607394;
   A. Pomarol and D. Tommasini, Nucl. Phys. **B466**, 3 (1996), hep-

ph/9507462.

5. D.A. Demir, A. Masiero and O. Vives, SISSA Report no. SISSA/107/99/EP, accepted for publication in Phys. Rev. **D**, hep-ph/9909325;
   D.A. Demir, A. Masiero and O. Vives, SISSA Report no. SISSA/134/99/EP, hep-ph/9911337;
   S. Baek and P. Ko, Phys. Rev. Lett. **83**, 488, (1999), hep-ph/9812229.
6. E. Gabrielli and G. Giudice, Nucl. Phys. **B433**, 3 (1995).
7. A. Masiero and H. Murayama, Phys. Rev. Lett. **83**, 907 (1999), hep-ph/9903363.
8. See, e.g. Y. Kawamura, H. Murayama and M. Yamaguchi, Phys. Rev. **D51**, 1337 (1995);
   A. Brignole, L. Ibanez and C. Munoz, in Perspectives on Supersymmetry, ed. G. Kane, Singapore, World Scientific, 1998.
9. R. Barbieri, R. Contino and A. Strumia, Pisa U. Report no. IFUP-TH-45-99, Aug. 1999, hep-ph/9908255.
10. K. Babu, B. Dutta and R.N. Mohapatra, U. of Maryland Report no. UMD-PP-99-111, hep-ph/9905464;
    S. Khalil and T. Kobayashi, Phys. Lett. **B 460**, 341 (1999), hep-ph/9906374;
    G. Colangelo and G. Isidori, JHEP **09**, (1998), 009, hep-ph/9808487;
    A.J. Buras et al., Munich Technical University Report no. TUM-HEP-353/99, hep-ph/9908371;
    S. Baek, J.-H. Jang, P. Ko and J.H. Park, Korea Inst. Sci Report No. KAIST-99/04, hep-ph/9907572;
    G. Eyal, A. Masiero, Y. Nir and L. Sivestrini, JHEP **11**, (1999), 032, hep-ph/9908382;
    A. Kagan and M. Neubert, Phys. Rev. Lett. **83**, 4429, (1999), hep-ph/9908404.

ANALYZING ε'/ε IN THE $1/N_c$ EXPANSION[a]

T. HAMBYE
INFN - Laboratori Nazionali di Frascati, I-00044 Frascati, Italy

G.O. KÖHLER, E.A. PASCHOS, P.H. SOLDAN
Institut für Physik, Universität Dortmund, D-44221 Dortmund, Germany

We present a recent analysis of ε'/ε in the $1/N_c$ expansion. We show that the $1/N_c$ corrections to the matrix element of Q_6 are large and positive, indicating a $\Delta I = 1/2$ enhancement similar to the one of Q_1 and Q_2 which dominate the CP conserving amplitude. This enhances the CP ratio and can bring the standard model prediction close to the measured value for central values of the parameters.

1 Introduction

Direct CP violation in $K \to \pi\pi$ decays was recently observed by the KTeV and NA48 collaborations.[1,2] The present world average [2] for the parameter ε'/ε is $\mathrm{Re}\,\varepsilon'/\varepsilon = (21.2 \pm 4.6) \cdot 10^{-4}$. In the standard model CP violation originates in the CKM phase, and direct CP violation is governed by loop diagrams of the penguin type. The main source of uncertainty in the calculation of ε'/ε is the QCD non-perturbative contribution related to the hadronic nature of the $K \to \pi\pi$ decay. Using the $\Delta S = 1$ effective hamiltonian,

$$\mathcal{H}_{eff}^{\Delta S=1} = \frac{G_F}{\sqrt{2}}\,\lambda_u \sum_{i=1}^{8} c_i(\mu)\,Q_i(\mu) \qquad (\mu < m_c)\,, \tag{1}$$

the non-perturbative contribution, contained in the hadronic matrix elements of the four-quark operators Q_i, can be separated from the perturbative Wilson coefficients $c_i(\mu) = z_i(\mu) + \tau y_i(\mu)$ (with $\tau = -\lambda_t/\lambda_u$ and $\lambda_q = V_{qs}^* V_{qd}$). Introducing $\langle Q_i\rangle_I \equiv \langle(\pi\pi)_I|Q_i|K\rangle$, the CP ratio can be written as

$$\frac{\varepsilon'}{\varepsilon} = \frac{G_F}{2}\frac{\omega}{|\varepsilon|}\frac{\mathrm{Im}\lambda_t}{\mathrm{Re}A_0}\left[\left|\sum_i y_i\,\langle Q_i\rangle_0\right|\left(1-\Omega_{\eta+\eta'}\right) - \frac{1}{\omega}\left|\sum_i y_i\,\langle Q_i\rangle_2\right|\right]. \tag{2}$$

$\omega = \mathrm{Re}A_0/\mathrm{Re}A_2 = 22.2$ is the ratio of the CP conserving $K \to \pi\pi$ isospin amplitudes; $\Omega_{\eta+\eta'}$ encodes the effect of the isospin breaking in the quark masses.[3] ε'/ε is dominated by $\langle Q_6\rangle_0$ and $\langle Q_8\rangle_2$ which cannot be fixed from the CP conserving data.[4,5] Beside the theoretical uncertainties coming from the calculation of the $\langle Q_i\rangle_I$ and of $\Omega_{\eta+\eta'}$, the analysis of the CP ratio suffers from the

[a]Talk presented by P.H. Soldan.

uncertainties on the values of various input parameters, in particular of the CKM phase in $\mathrm{Im}\lambda_t$, of $\Lambda_{\rm QCD} \equiv \Lambda^{(4)}_{\overline{\rm MS}}$, and of the strange quark mass.

To calculate the hadronic matrix elements we start from the effective chiral lagrangian for pseudoscalar mesons which involves an expansion in momenta where terms up to $\mathcal{O}(p^4)$ are included.[6] The method we use is the $1/N_c$ expansion.[7,8] In this approach, we expand the matrix elements in powers of the momenta and of $1/N_c$. For the $1/N_c$ corrections we calculated chiral loops as described in [9,10]. Especially important to this analysis are the non-factorizable corrections, which are UV divergent and must be matched to the short-distance part. They are regularized by a finite cutoff Λ_c which is identified with the short-distance renormalization scale. The definition of the momenta in the loop diagrams, which are not momentum translation invariant, is discussed in detail in [9]. Other recent work on matrix elements in the $1/N_c$ approach can be found in [11,12].

For the Wilson coefficients we use the leading logarithmic and the next-to-leading logarithmic values.[4] The absence of any reference to the renormalization scheme in the low-energy calculation, at this stage, prevents a complete matching at the next-to-leading order.[13] Nevertheless, a comparison of the numerical results obtained from the LO and NLO coefficients is useful as regards estimating the uncertainties and testing the validity of perturbation theory.

2 Analysis of ε'/ε

Analytical formulas for all matrix elements, at next-to-leading order in the twofold expansion in powers of momenta and of $1/N_c$, are given in [9,10]. In the pseudoscalar approximation, the matching has to be done below 1 GeV. Varying Λ_c between 600 and 900 MeV, the bag factors $B_1^{(1/2)}$ and $B_2^{(1/2)}$ take the values $8.2-14.2$ and $2.9-4.6$; quadratic terms in $\langle Q_1\rangle_0$ and $\langle Q_2\rangle_0$ produce a large enhancement which brings the $\Delta I = 1/2$ amplitude in agreement with the data.[10] Corrections beyond the chiral limit were found to be small.

For $\langle Q_6\rangle_0$ and $\langle Q_8\rangle_2$ the leading non-factorizable loop corrections, which are of $\mathcal{O}(p^0/N_c)$, are only logarithmically divergent.[9] Including terms of $\mathcal{O}(p^0)$, $\mathcal{O}(p^2)$, and $\mathcal{O}(p^0/N_c)$, $B_6^{(1/2)}$ and $B_8^{(3/2)}$ take the values $1.10-0.72$ and $0.64-0.42$. As a result the experimental range for ε'/ε can be accommodated in the standard model only if there is a conspiracy of the input parameters.[b] However, since the leading $\mathcal{O}(p^0)$ contribution vanishes for Q_6, corrections from higher order terms beyond the $\mathcal{O}(p^2)$ and $\mathcal{O}(p^0/N_c)$ are expected to be large. In [15] we investigated the $\mathcal{O}(p^2/N_c)$ contribution, i.e., the $1/N_c$ correction at the next order in the chiral expansion, because it brings about, for the

[b] For supersymmetric contributions to ε'/ε see [14] and references therein.

Table 1: Numerical values for ε'/ε (in units of 10^{-4}) as explained in the text.

LO	$14.8 \leq \varepsilon'/\varepsilon \leq 19.4$	$6.1 \leq \varepsilon'/\varepsilon \leq 48.5$
NDR	$12.5 \leq \varepsilon'/\varepsilon \leq 18.3$	$5.2 \leq \varepsilon'/\varepsilon \leq 49.8$
HV	$7.0 \leq \varepsilon'/\varepsilon \leq 14.9$	$2.2 \leq \varepsilon'/\varepsilon \leq 38.5$

first time, quadratic corrections on the cutoff. It is still desirable to calculate the contributions from logarithms and finite terms. Numerically, we observe a large positive correction from the quadratic term in $\langle Q_6 \rangle_0$. This point was already emphasized in [16]. The slope of the correction is qualitatively consistent and welcome since it compensates for the logarithmic decrease at $\mathcal{O}(p^0/N_c)$. Varying Λ_c between 600 and 900 MeV, the $B_6^{(1/2)}$ factor takes the values 1.50–1.62. Q_6 is a $\Delta I = 1/2$ operator, and the enhancement of $\langle Q_6 \rangle_0$ indicates that at the level of the $1/N_c$ corrections the dynamics of the $\Delta I = 1/2$ rule applies to Q_6 as to Q_1 and Q_2.

Using the quoted values for $B_6^{(1/2)}$ together with the full leading plus next-to-leading order B factors for the remaining operators [15] we calculated ε'/ε. The results for the three sets of Wilson coefficients LO, NDR, and HV and for Λ_c between 600 and 900 MeV are given in Tab. 1. The numbers are close to the measured value for central values of the parameters (first column). They are obtained by assuming zero phases from final state interactions.[c] This approximation is very close to the results we would get if we used the small imaginary part obtained at the one-loop level.[15]

Performing a scanning of the parameters [125 MeV $\leq m_s(1\,\text{GeV}) \leq 175$ MeV, $0.15 \leq \Omega_{\eta+\eta'} \leq 0.35$, $1.04 \cdot 10^{-4} \leq \text{Im}\lambda_t \leq 1.63 \cdot 10^{-4}$, and 245 MeV $\leq \Lambda_{\text{QCD}} \leq 405$ MeV] we obtain the numbers in the second column of Tab. 1. They can be compared with the results of [5,19,20,21,22]. The values for $B_6^{(1/2)}$ can also be compared with [11]. The large ranges reported in the table can be traced back to the large ranges of the input parameters. The parameters, to a large extent, act multiplicatively, and the large range for ε'/ε is due to the fact that the central value(s) for the ratio are enhanced roughly by a factor of two compared to the results obtained with B factors for Q_6 and Q_8 close to the VSA. More accurate information on the parameters, from theory and experiment, will restrict the values for ε'/ε. In order to reduce the scheme dependence in the result, appropriate subtractions would be necessary.[11,23]

[c] The effect of the final state interaction phases on the numbers in Tab. 1 has been estimated in [15]. Two recent dispersive calculations of the rescattering effects can be found in [17,18].

Acknowledgments

This work was supported by BMBF, 057D093P(7), Bonn, FRG, and DFG Antrag PA-10-1. T.H. acknowledges support from EEC, TMR-CT980169.

References

1. A. Alavi-Harati *et al.*, Phys. Rev. Lett. **83**, 22 (1999); A. Roodman, these proceedings.
2. V. Fanti *et al.*, Phys. Lett. **B465** (1999) 335; A. Nappi, these proceedings.
3. G. Ecker *et al.*, hep-ph/9912264, and references therein.
4. A.J. Buras *et al.*, Nucl. Phys. **B370**, 69 (1992), ibid. **B400**, 37 (1993), **B400**, 75 (1993), **B408**, 209 (1993). See also: M. Ciuchini *et al.*, Phys. Lett. **B301**, 263 (1993); Nucl. Phys. **B415**, 403 (1994).
5. S. Bosch *et al.*, hep-ph/9904408, A.J. Buras hep-ph/9908395.
6. J. Gasser, H. Leutwyler, Nucl. Phys. **B250**, 465 (1985).
7. D. Tadic, J. Trampetic, Phys. Lett. **B114**, 179 (1982).
8. W.A. Bardeen, A.J. Buras, J.-M. Gérard, Phys. Lett. **B180**, 133 (1986); Nucl. Phys. **B293**, 787 (1987); Phys. Lett. **B192**, 138 (1987).
9. T. Hambye *et al.*, Phys. Rev. **D58**, 014017 (1998).
10. T. Hambye, G. Köhler, P. Soldan, Eur. Phys. J. **C10** (1999) 271.
11. J. Bijnens, J. Prades, JHEP **9901**, 23 (1999), JHEP **0001**, 2 (2000); J. Bijnens, hep-ph/9910415, hep-ph/9907514; J. Prades, hep-ph/9909245.
12. M. Knecht, S. Peris, E. de Rafael, hep-ph/9910396; Phys. Lett. **B457**, 227 (1999), ibid. **B443**, 255 (1998); S. Peris, M. Perrottet, E. de Rafael, JHEP **9805**, 11 (1998).
13. A.J. Buras, Les Houches lectures (1998), hep-ph/9806471.
14. A. Masiero, these proceedings; A. Masiero, H. Murayama, Phys. Rev. Lett. **83**, 907 (1999); A.J. Buras et al., hep-ph/9908371.
15. T. Hambye *et al.*, hep-ph/9906434 (to appear in Nucl. Phys. B); T. Hambye, P. Soldan, hep-ph/9908232.
16. P. Soldan, hep-ph/9608281.
17. E. Pallante, A. Pich, hep-ph/9911233.
18. E.A. Paschos, hep-ph/9912230.
19. S. Bertolini, these proceedings; S. Bertolini, J.O. Eeg, M. Fabbrichesi, hep-ph/9802405; S. Bertolini *et al.*, Nucl. Phys. **B514**, 93 (1998), ibid. **B476**, 225 (1996), **B449**, 197 (1995).
20. A.A. Bel'kov *et al.*, hep-ph/9907335.
21. M. Ciuchini *et al.*, hep-ph/9910237, hep-ph/9910236.
22. T. Blum, A. Soni, hep-lat/9909108; T. Blum *et al.*, hep-lat/9908025.
23. W.A. Bardeen, Nucl. Phys. Proc. Suppl. **7A**, 149 (1989).

FUTURE PROSPECTS OF $K_L \to \pi^0 \nu\bar{\nu}$ EXPERIMENT AT FERMILAB

Y. B. HSIUNG
MS122, Fermilab, P.O.Box 500, Batavia,
IL 60510, USA
E-mail: hsiung@fnal.gov

We reviewed the current status of a proposed KAMI (Kaon at Main Injector) experiment at Fermilab to measure the direct CP-violating $K_L \to \pi^0 \nu\bar{\nu}$ decay. Good progress and encouraging results have been made in the past two years for measuring the required photon veto inefficiencies for both CsI and lead-scintillator detectors in a test beam at INS-KEK Japan. New beam test with 150 GeV Main Injector protons has also been scheduled in January 2000 at Fermilab using the existing KTeV detector with two new beam calorimeters. Prospects of a feasible KAMI experiment in the future is discussed here.

Current status of the CP violation experiments has now just firmly established the effect of "direct" CP violation through the non-zero measurement [1] of ϵ'/ϵ in neutral kaon decays to 2π. This important result makes the test of Standard Model CP hypothesis, that a single phase in the CKM matrix is the sole source of CP-violation, much more interesting. Over-constraining the parameters of CKM Unitarity Triangle through both B and K decays will be the ultimate test to the Standard Model, which will either reveal any new physics or furhter constraint the type of new physics beyond the Standard Model in explaining the origin of CP-violation.

Within the Standard Model the decay $K_L \to \pi^0 \nu\bar{\nu}$ violates CP-invariance in the decay process through the Z-penguin diagram, so called direct CP-violation. The contributions from indirect CP-violating and CP conserving effect are highly suppressed,[2] because the $K_L \to \pi^0 \nu\bar{\nu}$ decay is a flavour changing neutral current process through direct CP-violation only. Furthermore, the decay $K_L \to \pi^0 \nu\bar{\nu}$ is not affected by the long distance effects [3] and the theoretical ambiguity due to the QCD corrections is quite small. [4] Therefore, a discovery of the $K_L \to \pi^0 \nu\bar{\nu}$ decay would be a clean measurement to the CP-violating CKM parameter, η. The current knowledge for the CKM parameters predicts $B(K_L \to \pi^0 \nu\bar{\nu})$ to be $(2.8 \pm 1.7) \times 10^{-11}$. Because of the theoretical cleanness, an observation of the $K_L \to \pi^0 \nu\bar{\nu}$ decay outside the predicted range in branching ratio would indicate exciting new physics.

The current experimental limit[5] from KTeV is $B(K_L \to \pi^0 \nu\bar{\nu}) < 5.9 \times 10^{-7}$, using the π^0 Dalitz decay as the tagging method. There are 3 new initiatives for measuring $K_L \to \pi^0 \nu\bar{\nu}$, KEK-391, BNL-926 (KOPIO) and KAMI. [6]

KAMI DETECTOR LAYOUT

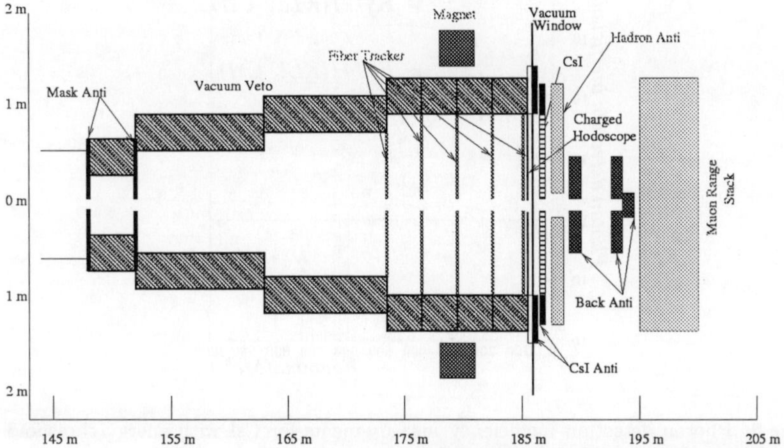

Figure 1: Plan view of the proposed KAMI aparatus for $K_L \to \pi^0\nu\bar{\nu}$ detection.

The goal of KAMI experiment is to make a measurement of $K_L \to \pi^0\nu\bar{\nu}$ using the decay $\pi^0 \to \gamma\gamma$. The signature of the $K_L \to \pi^0\nu\bar{\nu}$ mode is then two γs with large, unbalanced transverse momentum (p_t). Due to the copious π^0s in the K_L decays, the detector has to be hermatic covered for γ detection.

A schematic of the planned apparatus is shown in Figure 1. A few key features and challenges in KAMI detector will be emphasized here:

1) In order to determine the p_t of the π^0, an intense single pencil or flat K_L beam would be produced by the high intensity source of 120 GeV protons (up to 10^{13} Hz) from Fermilab Main Injector. To reach the single event sensitivity well below 10^{-11}, at least 3 MHz K_L decays have to be produced which correspond to at least 30 MHz K_L flux.

2) A new high efficiency hermatic photon veto system and a precision photon detector (CsI) would be used to cover a 34-meter decay region. The KTeV 3100 blocks of pure CsI calorimeter is the ideal photon detector to have better than $1\%/\sqrt{E}$ energy resolution and 1 mm position resolution.

3) An efficient hadronic veto and muon identification with range stacks.

4) An efficient Beam hole photon veto in the presence of high n/K flux.

5) An analyzing magnet and scintillating fiber tracker in the vacuum decay region can provide measurements for charged particles and precision calibration of detector elements. This provides KAMI a diverse program for many other interesting physics, such as $K_L \to \pi^0 e^+ e^-$, $K_L \to \pi^0 \mu^+ \mu^-$, $K_L \to \pi^0 \mu^\pm e^\mp$ and

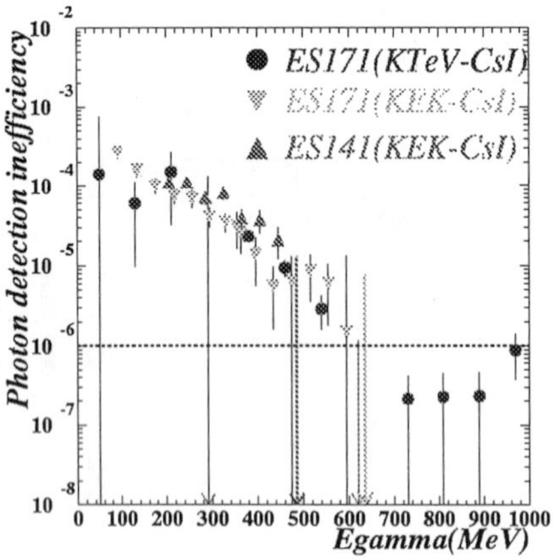

Figure 2: Photon detection inefficiency measurements for CsI with energy threshold of 10 MeV. The inefficiency falls below 10^{-6} (dotted line) for $E_\gamma > 600 MeV$.

$K_L \to \pi^+\pi^-e^+e^-$, in searching for new physics.

Because the $K_L \to 3\pi^0$ decay is so copious, the p_t cut must be above the kinematic cutoff ($\sim$ 150 MeV/c) to avoid being swamped with $\pi^0\pi^0\pi^0$ background. The p_t cut also removes background from Λ and Ξ decays. The longitudinal position of the decay vertex could then be determined by requiring that the invariant mass of the two observed photons equal the π^0 mass. The remaining background mainly comes from $K_L \to \pi^0\pi^0$, especially the asymmetric odd pairing γs from different π^0 decay. Stringent photon veto efficiency over a wide energy range is required to reduce such background. The fully reconstructed $K_L \to \pi^0\pi^0$ also serves as the normalization mode for $K_L \to \pi^0\nu\bar{\nu}$.

There are two options for KAMI experiment. The KAMI-far option will use the existing KTeV beam line geometry. Since target is 145 meter upstream of the decay region, the neutral kaon beam is then quite clean. The KAMI-near option requires new target/dump station very close to the detector to enhance the neutral kaon flux for the improvement of the sensitivity. Debunched proton beam from Main Injector with high duty factor in KAMI-near is required to reduce the accidental background or double decays.

A beam test at INS KEK-Tanashi, ES171, using a tagged photon beam and liquid scintillator neutron tagging detectors, was performed to measure the photon detection inefficiency due to the dominant contribution from photonuclear effect between 100 MeV to 1 GeV for E_γ.[7] Several calorimeter samples were examined with different energy thresholds (1 MeV, 5 MeV and 10 MeV):

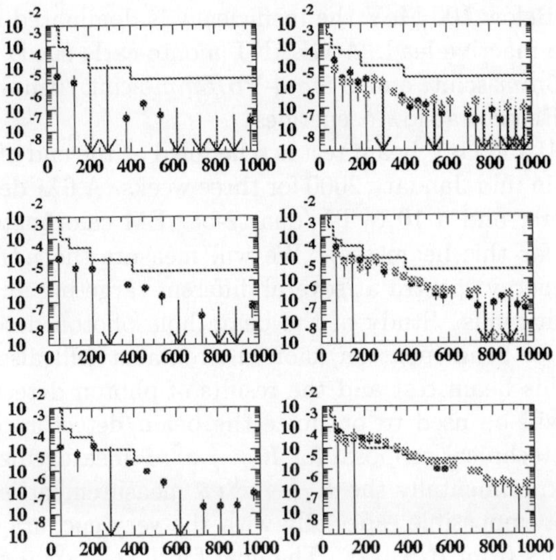

Figure 3: Photon veto inefficiency versus E_γ for CsI (left) and Pb/scintillator (right) calorimeters. The energy thresholds are 1 MeV, 5 MeV and 10 MeV respectively from top to bottom. Also shown for comparison are the KAMI requirements (solid line).

1) CsI crytals, 18 X_0 KEK-CsI (from ES147) and 27 X_0 KTeV-CsI;

2) Pb/scintillator sampling calorimeters, $1mm/3mm$ $(18X_0)$, $1mm/5mm$ $(18X_0)$ and $0.5mm/5mm$ $(21X_0)$ Pb/scintillator sampling prototypes.

Three measurements were made for the two types of CsI crystals and the detection inefficiency for all three measurements with a 10 MeV energy threshold are all consistent, as shown in Fig. 2, confirms the earlier ES147 results. [8] This result shows that with $E_\gamma > 600$ MeV the inefficiency is well below 10^{-6}. This is truly a very encouraging result. The residual of the nuclear resonance absorption above 140 MeV (pion-production threshold) is clearly visible but small.

The inefficiency results for KTeV-CsI, $1mm$-Pb/$5mm$-scint. sampling and $0.5mm$-Pb/$5mm$-scint. sampling prototypes are shown in Fig. 3. The comparison with the KAMI requirement is also shown in this energy region. The plots on left-side are for KTeV-CsI and on the right-side are for the Pb-scintillator sampling calorimeters. The plots from top to bottom have the threshold of 1 MeV, 5 MeV and 10 MeV applied, respectively. From this figure it is quite clear that better than 10^{-6} inefficiency can be achieved with photon energy above 1 GeV with the veto threshold at 10 MeV or lower in both CsI and

Pb-scintillator. Below 100 MeV the inefficiency is dominated by the sampling effect due to the inactive lead. A GEANT monte-carlo study has shown that both $1mm$-Pb/$5mm$-scint. and $0.5mm$-Pb/$5mm$-scint. samplings can satisfy the KAMI requirement at low γ energies.

A new KAMI 150 GeV beam test is scheduled at the end of '99 fixed target run at Fermilab in mid-January 2000 for three weeks. A $6\lambda_I$ depleted Uranium hadron calorimeter and a $10X_0$ Pb/quartz-bar EM calorimeter are currently in construction for this beam test. We will measure the n/K neutral beam flux, ratio and energy spectra at several different targeting angles and several absorber configurations. Study of the beam hole photon detection efficiency in the presence of beam neutron/kaon environment will also be performed. The results of this beam test and the results of photon detection inefficiency from INS test will be used to optimize the beam/detector configuration for preparing a full technical proposal for $K_L \to \pi^0 \nu\bar{\nu}$ measurement at KAMI.

Although experimentally the $K_L \to \pi^0 \nu\bar{\nu}$ measurement is quite challenging, but it is not impossible especially with the very encouraging γ detection inefficiency results from INS test. The goal for $K_L \to \pi^0 \nu\bar{\nu}$ experiment is to collect more than 100 events with low backgrounds to make a 10% measurement in branching ratio (or a 5% measurement in η). Such accuracy together with 10% measurement from $K^+ \to \pi^+ \nu\bar{\nu}$ and $sin(2\beta)$ from $B \to \psi K_S$ as well as another measurement from $B-\overline{B}$mixing can over-constraint the CKM Unitarity Triangle and provide useful information on CP-violation.

References

1. A. Alavi-Harati *et al.*, *Phys. Rev. Lett.* **83**, 22 (1999);
   V. Fanti *et al.*, *Phys. Lett.* B **465**, 335 (1999).
2. L.S. Littenberg, *Phys. Rev.* D **39**, 3322 (1989);
   G. Buchalla and G. Isidori, hep-ph/9806501(1998).
3. D. Rein and L.M. Sehgal, *Phys. Rev.* D **39**, 3325 (1989).
4. G. Buchalla and A.J. Buras, *Nucl. Phys.* B **400**, 225 (1993);
   A.J. Buras, *Phys. Lett.* B **333**, 476 (1994).
5. A. Alavi-Harati *et al.*, hep-ex/9907014, submitted to *Phys. Rev.* D.
6. E. Cheu *et al.*, Fermilab-Pub-97-321-E, KAMI EOI, (1997);
   I.-H. Chiang *et al.*, BNL-926 proposal, (1996);
   T. Inagaki *et al.*, KEK Internal 96-13 (KEK-PS-391 proposal), (1996).
7. T. Tsukamoto, in the Proceedings of the International Workshop on CP Violation in K, edited by S. Sugimoto and T. Yamanaka, KEK Proceedings 99-3, p237-246, (1999).
8. S. Ajimura *et al.*, KEK-Preprint 97-243 (1997).

SEARCH FOR $D^0 - \overline{D}^0$ MIXING

DAVID M. ASNER

Department of Physics, University of California, Santa Barbara, Califo rnia 93106, USA

We report on a search for $D^0 - \overline{D}^0$ mixing made by studying the 'wrong-sign' process $D^0 \to K^+\pi^-$ using an integrated luminosity of 9.0 fb^{-1} of recorded with the CLEO II.V detector.[1] We measure the rate of the 'wrong-sign' process $D^0 \to K^+\pi^-$ relative to that of the Cabibbo-favored process $\overline{D}^0 \to K^+\pi^-$ to be $R = (0.332^{+0.063}_{-0.065} \pm 0.040)\%$. We study $D^0 \to K^+\pi^-$ as a function of decay time to distinguish direct doubly Cabibbo-suppressed decay from $D^0 - \overline{D}^0$ mixing. The amplitudes that describe $D^0 - \overline{D}^0$ mixing, x' and y', are consistent with zero. At the 95% C.L. and without assumptions concerning charge-parity (CP) violating parameters, we find $(1/2)x'^2 < 0.041\%$ and $-5.8\% < y' < 1.0\%$.

Studies of the evolution of a K^0 or B_d^0 into the respective anti-particle, a $\overline{K}^0$ or $\overline{B}_d^0$, have guided the form and content of the Standard Model, and permitted useful estimates of the masses of the charm and top quark prior to their direct observation.[2] In the Standard Model, the evolution of the D^0 into the $\overline{D}^0$ is doubly-Cabibbo and GIM[3] suppressed. Therefore, the principal motivation in our search is to probe new physics.

A D^0 can evolve into a $\overline{D}^0$ through on-shell intermediate states, such as K^+K^- with mass, $m_{K^+K^-} = m_{D^0}$, or through off-shell intermediate states, such as those that might be present due to new physics. We denote the amplitude through the former (latter) states by $-iy$ (x), in units of $\Gamma_{D^0}/2$.[4] Many predictions for x in the $D^0 \to \overline{D}^0$ amplitude have been made.[5] Signatures of new physics include $|x| \gg |y|$, or charge-parity (CP) violating interference between x and y, or between x and a direct decay amplitude. In order to assess the origin of a $D^0 - \overline{D}^0$ mixing signal, the effects described by y must be distinguished from those described by x.

The wrong-sign (WS) process, $D^0 \to K^+\pi^-$, can proceed either through direct doubly Cabibbo-suppressed decay (DCSD), or through state-mixing followed by the Cabibbo-favored decay (CFD), $D^0 \to \overline{D}^0 \to K^+\pi^-$. Both processes could contribute to the time integrated WS rate. To disentangle these processes, we study the distribution of WS final states as a function of the proper decay time t of the D^0. We describe the proper decay time in units of the mean D^0 lifetime, $\tau_{D^0} = 415 \pm 4\,\text{fs}$.[6] The differential WS rate is [7]

$$r(t) \equiv [R_D + \sqrt{R_D}y't + \frac{1}{4}(x'^2 + y'^2)t^2]e^{-t}. \tag{1}$$

The modified mixing amplitudes x' and y' in Eqn. 1 are given by $y' \equiv y\cos\delta - x\sin\delta$, $x' \equiv x\cos\delta + y\sin\delta$ and $R_M \equiv \frac{1}{2}\left(x^2+y^2\right) = \frac{1}{2}\left(x'^2+y'^2\right)$ where δ is a possible strong phase between the DCSD and CFD amplitudes. The symbol R_D (R_M) represents the DCSD (mixing) rate relative to the CFD rate. There are theoretical arguments that δ is small.[8] We modify Eqn. 1 to allow CP asymmetries in state-mixing (A_M) and direct decay (A_D) and we allow a weak phase (ϕ) between those two processes. Specifically, both x' and y' are scaled by $(1 \pm A_M)^{\frac{1}{2}}$, $R_D \to R_D\,(1 \pm A_D)$, $\delta \to \delta \pm \phi$ in Eqn. 1[9]. The plus (minus) sign is used for an initial D^0 ($\overline{D}^0$).

We reconstruct the energy (Q) released in the decay sequence $D^{*+} \to D^0\pi_s^+$. The charge of the slow pion (π_s^+ or π_s^-) identifies the charm state at $t=0$ as either D^0 or $\overline{D}^0$. We reconstruct the mass (M) of the D^0 producing either the $K^+\pi^-$ (WS) or $K^-\pi^+$ (right-sign (RS)) final state. Resolutions of $\sigma_Q = 190 \pm 6$ keV and $\sigma_M = 6.4 \pm 0.1\,\mathrm{MeV}$ are measured with the nearly background free RS sample. Our signal for the RS process contain 13,527 candidates; our signal for the WS process is shown in Fig. 1. We determine the WS background levels by performing a fit to the two-dimensional region of $0 < Q < 10\,\mathrm{MeV}$ versus $1.76 < M < 1.97\,\mathrm{GeV}$.[10] Monte Carlo samples are used to estimate the background shapes in the $Q-M$ plane.

We performed maximum-likelihood fits to the D^0 decay time of WS candidates in the $Q-M$ plane within $4\,\sigma$ [11] of the RS values. The WS signal is described by Eqn. 1, either modified to describe all three forms of CP violation (Fit A), without modification to describe mixing alone (Fit B), or with the mixing parameters constrained to be zero (Fit C). The one-dimensional, 95% confidence intervals for x', y' and R_D are given in Table 1 and the contours, in the two-dimensional plane of y' versus x' that contain the true value of x' and y' with 95% confidence, are shown in Fig. 2.

The dominant systematic errors stem from potential misunderstanding of the shapes, acceptances and decay time of our backgrounds. We assess a systematic error on R_D, x' and y' of $\pm 0.040\%$, $\pm 0.2\%$ and $\pm 0.3\%$, respectively.

If we assume that δ is small, then $x' \approx x$ and we can indicate the impact of our work in limiting predictions of $D^0-\overline{D}^0$ mixing from extensions to the Standard Model. The 95% C.L. interval for x from Fit A has some inconsistency with eighteen of the predictions recently tabulated.[5]

In conclusion, our data are consistent with no D^0–$\overline{D}^0$ mixing. We limit the mixing amplitudes, x' and y', to be $(1/2)x'^2 < 0.041\%$ and $-5.8\% < y' < 1.0\%$ at the 95% C.L., without assumptions concerning CP violating parameters. We observe no evidence for CP violation. These results are a substantial advance in sensitivity to the phenomena that contribute to the wrong-sign process $D^0 \to K^+\pi^-$.

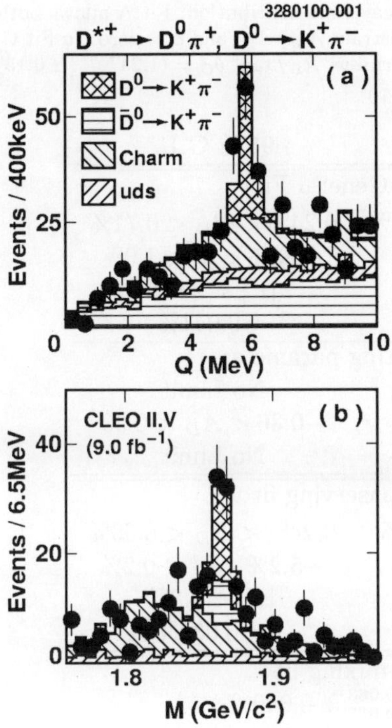

Figure 1. Signal for the WS process $D^0 \to K^+\pi^-$. The data are the full circles with error bars, the projection of the fit for the signal is cross-hatched, and the projections of the fit for the backgrounds from charm and light quark production are single-hatched. For part a), the reconstructed $K\pi$ mass, M, is within 2σ of the CFD value, and for b), the energy released in the $D^{*+} \to D^0\pi_s^+$ decay, Q, is within 2σ [11] of the CFD value.

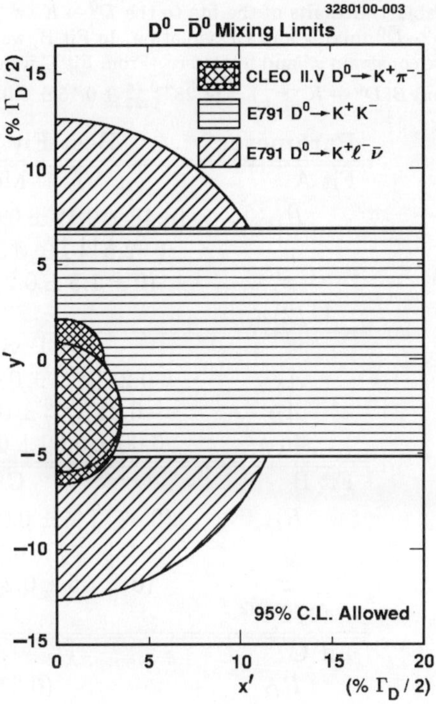

Figure 2. Allowed regions, at 95% C.L., in the y' vs. x' plane. The entire kidney shaped region, filled with tight cross-hatching, is allowed under Fit A of Table 1, while Fit B, in which CP conservation is assumed, allows the smaller region, filled with looser cross-hatching. The allowed regions from studies comparable to Fit A, using $D^0 \to K^+K^-$, [12] for which we assume $\delta = 0$, and $D^0 \to K^+\ell^-\overline{\nu}_\ell$, [13] are shown as single-hatched regions.

References

1. Y. Kubota *et al.*, Nucl. Instrum. Meth. A **320**, 66 (1992); T. S. Hill, Nucl. Instrum. Meth. A **418**, 32 (1998).
2. R.H. Good *et al.*, Phys. Rev. **124**, 1223 (1961); H. Albrecht *et al.*, Phys. Lett. B **192**, 245 (1987).

Table 1. Results of the fits to the $D^0 \to K^+\pi^-$ decay time distribution. Fit A allows both D^0–$\overline{D}^0$ mixing and CP violation. In Fit B, we constrain A_M, A_D, and ϕ to zero. In Fit C, we constrain x' and y' to zero. From Fit C we determine, $R_D/\tan^4\theta_C = (1.24^{+0.23}_{-0.24} \pm 0.15)$ and $\mathcal{B}(D^0 \to K^+\pi^-) = (1.28^{+0.24}_{-0.25} \pm 0.15 \pm 0.03) \times 10^{-4}$.

Parameter	Best Fit	95% C.L.
Fit A	Most General Fit	
R_D	$(0.48 \pm 0.12 \pm 0.04)\%$	$0.24\% < R_D < 0.71\%$
y'	$(-2.5^{+1.4}_{-1.6} \pm 0.3)\%$	$-5.8\% < y' < 1.0\%$
x'	$(0 \pm 1.5 \pm 0.2)\%$	$\lvert x'\rvert < 2.9\%$
$(1/2)x'^2$		$< 0.041\%$
	CP violating parameters	
A_M	$0.23^{+0.63}_{-0.80} \pm 0.01$	No Limit
A_D	$-0.01^{+0.16}_{-0.17} \pm 0.01$	$-0.36 < A_D < 0.30$
$\sin\phi$	$0.00 \pm 0.60 \pm 0.01$	No Limit
Fit B	CP conserving fit	
R_D	$(0.47^{+0.11}_{-0.12} \pm 0.04)\%$	$0.24\% < R_D < 0.69\%$
y'	$(-2.3^{+1.3}_{-1.4} \pm 0.3)\%$	$-5.2\% < y' < 0.2\%$
x'	$(0 \pm 1.5 \pm 0.2)\%$	$\lvert x'\rvert < 2.8\%$
$(1/2)x'^2$		$< 0.038\%$
Fit C	No-mixing fit	
R_D	$(0.332^{+0.063}_{-0.065} \pm 0.040)\%$	
A_D	$-0.02^{+0.19}_{-0.20} \pm 0.01$	$-0.43 < A < 0.34$

3. S. L. Glashow, J. Illiopolous, L. Maiani, Phys. Rev. D **2**, 1285 (1970).
4. T.D. Lee, R. Oehme, and C.N. Yang, Phys. Rev. **106**, 340 (1957); A. Pais and S. B. Treiman, Phys. Rev. D **12**, 2744 (1975).
5. H. N. Nelson, hep-ex/9908021 (unpublished).
6. Particle Data Group, C. Caso *et al.*, Eur. Phys. J. C **3**, 1 (1998).
7. S. B. Treiman and R. G. Sachs, Phys. Rev. **103**, 1545 (1956).
8. T. E. Browder and S. Pakvasa, Phys. Lett. B **383**, 475 (1996).
9. Y. Nir, hep-ph/9911321, (unpublished).
10. R. Godang *et al.*, hep-ex/0001060, submitted to Phys. Rev. Lett.; D. M. Asner, Ph.D. Dissertation, Univ. of Calif. at Santa Barbara, 2000.
11. The standard deviation σ for Q and M is computed from the trajectory reconstruction covariance matrix of the daughters of each D^{*+} candidate.
12. E. M. Aitala *et al.*, Phys. Rev. Lett. **83**, 32 (1999).
13. E. M. Aitala *et al.*, Phys. Rev. Lett. **77**, 2384 (1996).

CP VIOLATION TESTS IN THE CHARM SECTOR

BRIAN O'REILLY

Dept. of Physics, University of Colorado, Campus Box 390, Boulder CO 80309, U.S.A. E-mail: irish@fnal.gov

We present preliminary results on CP violation limits in the decays $D^0 \to K^+K^-$, $D^+ \to K^+K^-\pi^+$ and $D^+ \to K_sK^+$ from the FOCUS[1] experiment at Fermilab.

1 Introduction

Theoretical predictions for CP violating effects in the charm sector are typically on the order of 10^{-3} due to the interference of tree level and penguin processes.[2] An observation, with our current statistics, of a non-zero CP asymmetry in D meson decays could be an indication of physics beyond the Standard Model.

2 Technique

The CP asymmetry A_{CP} for a decay may be written as:

$$A_{\text{CP}} = \frac{\eta^+ - \eta^-}{\eta^+ + \eta^-} \tag{1}$$

$$\eta^\pm = \frac{N(D^0(\bar{D}^0) \to K^\mp K^\pm)}{N(D^0(\bar{D}^0) \to K^\mp \pi^\pm)}. \tag{2}$$

Where $N(D^0(\bar{D}^0) \to K^\mp K^\pm)$ is the efficiency corrected number of decays. Production asymmetries are cancelled by normalising to the Cabibbo favoured (CF) mode. There is the added advantage that many systematic effects will cancel, especially when the singly Cabibbo suppressed decay (SCSD) and the CF decay are very similar.

For D^0 decays the asymmetry may be a combination of direct and indirect CP violating contributions. Indirect CP violation may arise due to $D^0 - \bar{D}^0$ mixing. For a recent review see H. Park's review talk in these proceedings.[3]

$D^+ \to K_sX^+$ decays present a difficult experimental challenge due to the absence of a well defined secondary vertex. Approximately 90% of the K_s decays in FOCUS occur outside of the region which is instrumented with silicon vertex detectors. We use the silicon tracking information from the

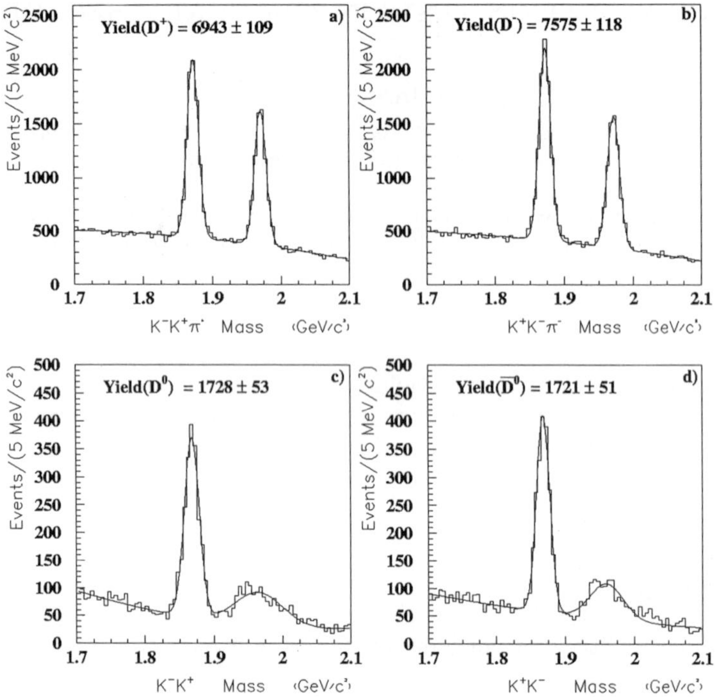

Figure 1. Invariant mass plots for $D^+ \to K^-K^+\pi^+$ and $D^0 \to K^-K^+$.

charged track and the momentum information from the K_s decay in order to find the production and decay vertices.

Measuring the CP asymmetry for decays which contain a K_s is complicated by the presence of CP violation in the decays of the K^0. Buccella *et. al.* [4] observe that a measurement of A_{CP} will be "essentially hopeless" if the value is a lot less than the asymmetry in K_L semileptonic decays, i.e. $\simeq 10^{-3}$.

3 Preliminary Results

In Figure 1 we plot the invariant mass signals for the decays $D^+ \to K^-K^+\pi^+$ and $D^0 \to K^-K^+$. Figure 2 shows the signals for the decays $D^+ \to K_sK^+$ and $D^+ \to K_s\pi^+$. These plots are not efficiency corrected. The efficiency correction is calculated and applied as a function of the D momentum in

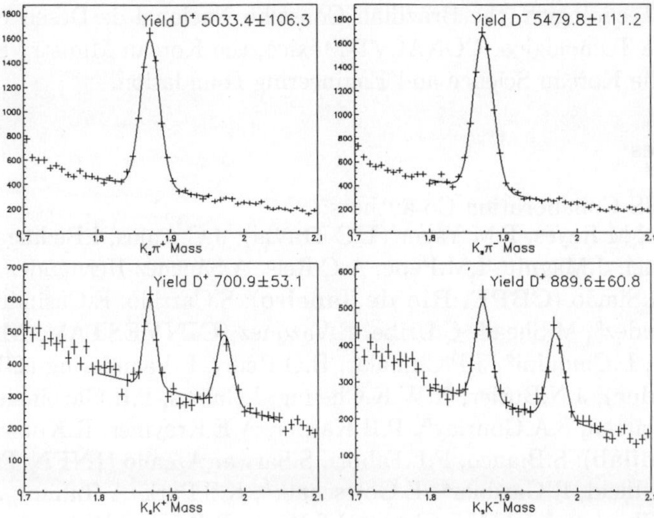

Figure 2. Invariant mass plots for $D^+ \to K_s K^+$ and $D^+ \to K_s \pi^+$.

order to remove any momentum dependent production effects. For the decay $D^+ \to K^- K^+ \pi^+$ we calculate a preliminary CP asymmetry of -0.012 ± 0.011 and for $D^0 \to K^- K^+$ we find $A_{\rm CP} = 0.0002 \pm 0.0219$. Both these values are consistent with zero. The previous best limits from E791 are -0.014 ± 0.029 for $D^+ \to K^- K^+ \pi^+$ and $-0.10 \pm 0.049 \pm 0.012$ for $D^0 \to K^- K^+$. [5,6] Our preliminary results indicate a factor of 3–4 increase in sensitivity for these modes. There are no previous measurements for CP asymmetries in decays involving the K_s. We expect the statistical sensitivity for the $D^+ \to K_s K^+$ decay to be $\simeq 5.3\%$. In the future we plan to measure the CP asymmetry for (among others) the decays $D^+ \to K_s \pi^+$ and $D^0 \to \pi^- \pi^+$. An analysis of the Dalitz plot for the decay $D^+ \to K^- K^+ \pi^+$ is currently in progress.

Acknowledgments

We wish to acknowledge the assistance of the staffs of Fermi National Accelerator Laboratory, the INFN of Italy and the physics departments of the collaborating institutions. This research was supported in part by the National Science Foundation, the U.S. Department of Energy, the Italian Istituto Nazionale di Fisica Nucleare and Minsitero dell'Università e della Ricerca Scientifica e Tecnologica, the Brazilian Conselho Nacional de Desenvolvimento Científico e Tecnológico, CONACyT-México, the Korean Ministry of Education and the Korean Science and Engineering Foundation.

References

1. FOCUS Collaboration Co-authors:
J.Link, M.Reyes, P.M.Yager (**UC Davis**); J.C.Anjos, I.Bediaga, C.Gobel, J.Magnin, I.M.Pepe, A.C.Reis, A.Sánchez-Hernández, F.R.A.Simão (**CBPF, Rio de Janeiro**); S.Carrillo, E.Casimiro[a], H.Mendez[b], M.Sheaff, C.Uribe, F.Vazquez (**CINVESTAV, México City**); L.Cinquini[c], J.P.Cumalat, B.O'Reilly, E.Vaandering (**CU Boulder**); J.N.Butler, H.W.K.Cheung, I.Gaines, P.H.Garbincius, L.A.Garren, S.A.Gourlay[d], P.H.Kasper, A.E.Kreymer, R.Kutschke (**Fermilab**); S.Bianco, F.L.Fabbri, S.Sarwar,A.Zallo (**INFN Frascati**); C.Cawlfield, R.Gardner[e], E.Gottschalk[f], K.S.Park, A.Rahimi, J.Wiss (**UI Champaign**); Y.S.Chung, J.S.Kang, B.R.Ko, J.W.Kwak, K.B.Lee, S.S.Myung, H.Park (**Korea University, Korea**); G.Alimonti, M.Boschini, B.Caccianiga, A.Calandrino, P.D'Angelo, M.DiCorato, P.Dini, M.Giammarchi, P.Inzani, F.Leveraro, S.Malvezzi, D.Menasce, M. Mezzadri, L.Milazzo, L.Moroni, D.Pedrini, F.Prelz, A.Sala, S.Sala (**INFN and Milano**); T.F.Davenport III (**UNC Asheville**); V.Arena, G.Boca, G.Bonomi[g], G.Gianini, G.Liguori, M.Merlo, D.Pantea[h], S.P.Ratti, C.Riccardi, L.Viola, P.Vitulo (**INFN and Pavia**); A.M.Lopez, L.Mendez, A.Mirles, E.Montiel, D.Olaya, J.E.Ramirez[i], C.Rivera, Y.Zhang[j] (**Mayaguez, Puerto Rico**); N.Copty, W.E.Johns[k], M.Purohit, J.R.Wilson (**USC Columbia**); K.Cho, T.Handler (**UT Knoxville**); D.Engh, M.Hosack, M.Nehring[l], M.Sales, P.D.Sheldon, M.Webster (**Vanderbilt**); K.Stenson[k](**Wisconsin, Madison**); Y.Kwon (**Yonsei University, Korea**).
2. Buccella et. al. *Phys. Rev.* D **51**, 3478 (1995).
3. H. Park $D^0 - \bar{D}^0$ *Mixing and Rare Decays*, In these proceedings.
4. Buccella et. al. *Phys. Lett.* B **302**, 319 (1993).
5. Aitala et. al. *Phys. Lett.* B **403**, 377 (1997).
6. Aitala et. al. *Phys. Lett.* B **421**, 405 (1998).

Present Addresses:
[a]University of Milano, Italy [b]University of Puerto Rico, Mayaguez [c]NCAR, Boulder, CO [d]Lawrence Berkeley Laboratory, CA [e]Indiana University, Bloomington, IN [f]Fermilab, IL [g]University of Brescia, Italy [h]Nat. Inst. of Phys. and Nucl. Eng., Bucharest, Romania [i]University of Colorado, Boulder, CO [j]Lucent Technology [k]Vanderbilt University, Nashville, TN [l]Adams State College, Alamosa, CO

D MESON LIFETIMES

Kwei-Chou Yang
Department of Physics, Chung Yuan Christian University, Chung-Li, Taiwan 320, R.O.C.
E-mail: kcyang@cycu.edu.tw

The QCD-based operator-product-expansion technique can be systematically applied to the study of the heavy hadron lifetime problems. Unlike the case of B mesons, we stress that the sizable nonzero momentum of the spectator light quark inside charmed mesons should be taken into account [1]. We have applied the QCD sum rule approach to estimate the hadronic matrix elements of 4-quark operators relevant to the nonspectator effects. The destructive Pauli-interference of the nonspectator quark effect in D^+ decays can explain the experimental observation $\tau(D^+)/\tau(D^0) \approx 2.5$. We obtain the ratio $\tau(D_s)/\tau(D^0) \approx 1.08 \pm 0.04$ which is larger than some earlier theoretical estimates, but still smaller than the recent measurements by CLEO and E791. However, the ratio $\tau(D_s)/\tau(D^0)$ could be enhanced if nonvacuum intermediate states contribute sizably to the 4-quark condensate $\langle \bar{q}\gamma_\mu(1-\gamma_5)q\bar{q}\gamma^\mu(1-\gamma_5)q\rangle$.

It is well known that the inclusive heavy hadron decay widths can be treated systematically in a QCD-based operator-product-expansion (OPE) formulation. Because the final state effects are eliminated, based on the hypothesis of quark-hadron duality, after summing all final states together, the OPE approach thus can be employed for such the smeared physical quantities. To test the validity of quark-hadron duality [2], it is very important to have a reliable estimate of the heavy hadron lifetimes within the OPE framework and compare them with experiment.

The observed lifetime difference between the D^+ and D^0 is attribute to the nonspectator effects: the destructive Pauli interference (PI) in D^+ decays and/or the constructive W-exchange (WE) contribution to D^0 decays [3]. By contrast, the D_s^+ and D^0 lifetimes are theoretically expected to be close to each other [4]. The recent D_s^+ lifetime measurements by Fermilab and CLEO have been much improved. The Fermilab E791 [5] measurement of the D_s^+ lifetime, combining with the world average D^0 lifetime, yields the ratio $\tau(D_s^+)/\tau(D^0) = 1.25 \pm 0.04$. Meanwhile, the CLEO [6] measurement of D_s^+ and D^0 lifetimes shows $\tau(D_s^+)/\tau(D^0) = 1.19 \pm 0.04$.

Up to the order of $1/m_b^2$, all charmed mesons almost have the same lifetimes in the parton picture. The nonspectator effects due to four-quark interactions are of order $1/m_c^3$, but their contributions can be potentially significant due to a phase-space enhancement by a factor of $16\pi^2$. As a result, the lifetime differences of heavy hadrons come mainly from the above-mentioned nonspectator effects [7,8,9].

The total decay width of the charmed meson is given by

$$\Gamma(D) = \Gamma_{\rm NL,spec} + \Gamma_{\rm NL,nspec} + \Gamma_{\rm SL} + \Gamma_{\rm lep}, \tag{1}$$

where $\Gamma_{\rm NL,spec}$ and $\Gamma_{\rm NL,nspec}$ denote the spectator and nonspectator contributions in nonleptonic decay widths, respectively, and $\Gamma_{\rm SL}$ and $\Gamma_{\rm lep}$ are the semileptonic and pure leptonic decay widths, respectively. Note that there is an additional purely leptonic decay contribution to D_s^+, namely $D_s^+ \to \tau\bar{\nu}$. In units of $\Gamma_0 = G_F^2 m_c^5/(192\pi^3)$, we obtain $\Gamma_{\rm NL,spec} = 4.84\,\Gamma_0$, $\Gamma_{\rm SL} = 1.02\,\Gamma_0$ and $\Gamma_{\rm lep}(D_s^+ \to \tau\bar{\nu}_\tau + \mu\bar{\nu}_\mu) = 0.169\,\Gamma_0$ for $m_c = 1.65$ GeV, $m_s = 125$ MeV, $c_1(m_c) = 1.30$ and $c_2(m_c) = -0.57$. It is very important, as stressed by Chernyak [10], for calculations with charmed mesons to account for the momentum of the nonperturbatively initial spectator quarks, which is order of several hundred MeV. The sum rule analysis in [11] suggests that the spectator antiquark carries about $\sim 20\%$ of the parent charmed meson momentum. Therefore, the PI (WE, WA(W-annihilation)) in D decay is suppressed (enhanced) by a factor of $(\langle p_c\rangle \mp \langle p_q\rangle)^2/\langle p_c\rangle^2 \sim 1 \mp 2\langle p_q\rangle/\langle p_c\rangle \sim 0.6$ (1.4), where $\langle p_c\rangle$ and $\langle p_q\rangle$ are the momenta of the c and the light quarks, respectively, in the D meson; that is, the PI is subject to a large suppression and will not overcome the leading c quark decay width. If the momentum of the spectator quark in the D^+ meson is neglected, the destructive PI in D^+ decay is found to be $\Gamma_-^{\rm int}(D^+) = -8.5\,\Gamma_0$, which largely overcomes the c-quark decay rate $\Gamma_{\rm NL,spec}$. Consequently, $\Gamma_{\rm tot}(D^+)$ becomes negative, which is of course of no sense. Since the decay width of D^+ involves a large cancellation between two terms, it is very sensitive to the parameters m_c, f_D and $\langle p_q\rangle$. For $f_D = 190$ MeV and $\langle p_q\rangle = 350$ MeV, we found that the pole mass m_c is preferred at ~ 1.65 GeV.

The predictions of D meson lifetime ratios depend on four unknown hadronic parameters B, δB, ε and $\delta\varepsilon$, where B and δB parametrize the matrix elements of color singlet-singlet four-quark operators and ε and $\delta\varepsilon$ the matrix elements of color octet-octet operators,

$$\begin{aligned} \langle D_q(p^D)|O^q_{\mu\nu}|D_q(p^D)\rangle &= (Bp^D_\mu p^D_\nu + \delta B\, g_{\mu\nu} m^2_{D_q})\frac{f^2_{D_q}}{4}\,, \\ \langle D_q(p^D)|T^q_{\mu\nu}|D_q(p^D)\rangle &= (\varepsilon p^D_\mu p^D_\nu + \delta\varepsilon\, g_{\mu\nu} m^2_{D_q})\frac{f^2_{D_q}}{4}\,, \end{aligned} \tag{2}$$

with

$$\begin{aligned} O^q_{V-A} &= \bar{c}_L\gamma_\mu q_L\,\bar{q}_L\gamma^\mu c_L\,, & O^q_{S-P} &= \bar{c}_R\, q_L\,\bar{q}_L\, c_R\,, \\ T^q_{V-A} &= \bar{c}_L\gamma_\mu t^a q_L\,\bar{q}_L\gamma^\mu t^a c_L\,, & T^q_{S-P} &= \bar{c}_R\, t^a q_L\,\bar{q}_L\, t^a c_R\,, \end{aligned} \tag{3}$$

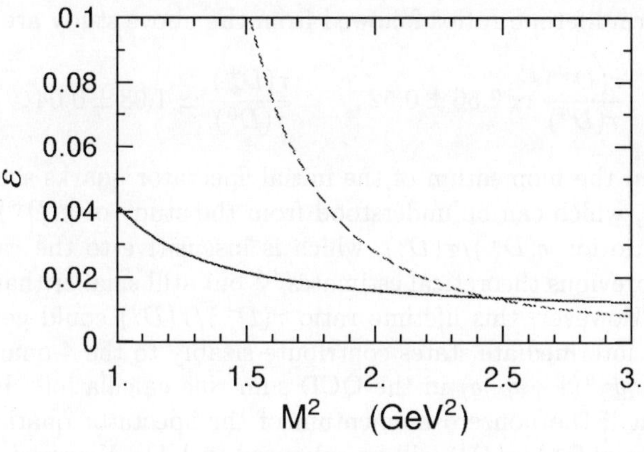

Figure 1: ε $(= -\delta\varepsilon)$ as a function of the Borel mass squared M^2. The solid and dashed curves are for $\varepsilon(D^{0,+})$ and $\varepsilon(D_s^+)$, respectively.

and $q_{R,L} = (1 \pm \gamma_5)q/2$.

The QCD sum rule result [1] of ε $(= -\delta\varepsilon)$ is shown in Fig. 1. Within the Borel window 2.0 GeV^2 $< M^2 <$ 3.0 GeV^2, we obtain $\varepsilon(D^{0,+}) = -\delta\varepsilon(D^{0,+}) = 0.015 \pm 0.010$ and $\varepsilon(D_s^+) = -\delta\varepsilon(D_s^+) = 0.015^{+0.015}_{-0.010}$, where the errors come partially from the uncertainties of input parameters. Also $B \approx 1$ and δB is proportional to $\langle \bar{q}\gamma_\mu(1-\gamma_5)q\bar{q}\gamma^\mu(1-\gamma_5)q\rangle$ which is zero under the vacuum insertion. It should be emphasized that $\tau(D_s^+)/\tau(D^0)$ is quite sensitive to δB. Even if δB deviates from zero by a small amount, say 0.005, the ratio $\tau(D_s^+)/\tau(D^0)$ will be enhanced by $\sim$6%.

Taking the above values and $\delta B = 0$, the resulting non-spectator effects are

$$\begin{aligned} \Gamma^{\rm exc}(D^0) &= (0.46 \pm 0.30)\Gamma_0\,, \\ \Gamma^{\rm int}_-(D^+) &= -(3.29 \pm 0.40)\Gamma_0\,, \\ \Gamma^{\rm ann}(D_s^+) &= (0.19 \pm 0.13)\Gamma_0\,, \\ \Gamma^{\rm int}_-(D_s^+) &= -(0.35 \pm 0.05)\Gamma_0\,. \end{aligned} \tag{4}$$

Collecting all the contributions, we find $\tau(D^0) = 0.38\,ps$, $\tau(D^+) = 0.96\,ps$, $\tau(D_s^+) = 0.41\,ps$. It is clear from our calculations that the lifetime of D_s^+ is longer than that of D^0 because the Cabibbo-allowed WA and leptonic contributions to $\Gamma(D_s^+)$ are compensated by the Cabibbo-suppressed PI.

The charm lifetime ratios followed from the above study are

$$\frac{\tau(D^+)}{\tau(D^0)} \simeq 2.56 \pm 0.52\,, \qquad \frac{\tau(D_s^+)}{\tau(D^0)} \simeq 1.08 \pm 0.04\,. \tag{5}$$

In conclusion, the momentum of the initial spectator quarks should be taken into account, which can be understood from the study of $\tau(D^+)/\tau(D^0)$. The predicted ratio for $\tau(D_s^+)/\tau(D^0)$, which is insensitive to the value of m_c, is larger than previous theoretical estimates [3,4] but still smaller than recent measurements. However, this lifetime ratio $\tau(D_s^+)/\tau(D^0)$ could get enhanced if non-vacuum intermediate states contribute sizably to the 4-quark condensate $\langle\bar{q}\gamma_\mu(1-\gamma_5)q\bar{q}\gamma^\mu(1-\gamma_5)q\rangle$ in the QCD sum rule calculation. It is worth remarking that if the nonzero momentum of the spectator quark is neglected, then the ratio $\tau(D_s^+)/\tau(D^0)$ will be enhanced to 1.11. However, as stressed in passing, it is meaningless to have a negative lifetime for the D^+.

Acknowledgments

It is a pleasure to thank H.Y. Cheng for the collaboration on this subject. This work was supported in part by the National Science Council of R.O.C. under Grant No. NSC88-2112-M-033-010.

References

1. H.Y. Cheng and K.C. Yang, *Phys. Rev.* D **61**, 014008 (2000).
2. I.I. Bigi, arXiv:hep-ph/0001003.
3. G. Bellini, I.I. Bigi, and P.J. Dornan, Phys. Rep. **289**, 1 (1997).
4. I.I. Bigi and N.G. Uraltsev, Z. Phys. **C 62**, 623 (1994).
5. E791 Collaboration, E.M. Aitala *et al., Phys. Lett.* B **445**, 449 (1999).
6. CLEO Collaboration, G. Bonvicini *et al., Phys. Rev. Lett.* **82**, 4586 (1999).
7. H.Y. Cheng, *Phys. Rev.* D **56**, 2783 (1997).
8. M. Neubert and C. T. Sachrajda, *Nucl. Phys.* B **483**, 339 (1997).
9. H.Y. Cheng and K.C. Yang, *Phys. Rev.* D **59**, 014011 (1999).
10. V. Chernyak, *Nucl. Phys.* B **457**, 96 (1995).
11. V.L. Chernyak and A.R. Zhitnitsky, Phys. Rep. 112, 173 (1984).

CURRENT STATUS OF RADIATIVE B DECAYS

JOSIP TRAMPETIĆ

Theoretical Physics Division, Rudjer Bošković Institute, Zagreb, Croatia
E-mail: josip@thphys.irb.hr

A current status report on radiative inclusive and exclusive B decays is presented.

1 Introduction

The experimental challenge of finding new physics in direct search may still take some time if new particles or their effects set in only at several hundred GeV. Complementary to these direct signals at highest available energies are the measurements of the effects of new "heavy" particles in loops, through either precision measurements or detection of processes occurring only at one loop in the standard model (SM). Among these are the transitions induced by flavor-changing neutral currents (FCNC), such as $b \to s\gamma$.

The first observations of the exclusive $B \to K^*\gamma$ and inclusive $B \to X_s\gamma$ decays were reported in 1993/94 by the CLEO collaboration.[1] Skwarnicki from the CLEOII&II.V reported the recent exclusive branching ratios:[1]

$$BR(B^0 \to K^{*0}\gamma) = 4.5 \pm 0.7 \pm 0.3 \times 10^{-5}, \tag{1}$$

$$BR(B^- \to K^{*-}\gamma) = 3.8 \pm 0.9 \pm 0.3 \times 10^{-5}, \tag{2}$$

$$BR(B \to K_2^*(1430)\gamma) = 1.7 \pm 0.5 \pm 0.1 \times 10^{-5}, \tag{3}$$

$$BR(B \to K^*(1410)\gamma) < 12.7 \times 10^{-5}, \tag{4}$$

$$BR(B \to K_2^*(1430)\gamma)/BR(B \to K^*\gamma) = 0.4 \pm 0.1. \tag{5}$$

The measurements envolving higher resonant K^*-states are still preliminary, since the error of $\sim 100MeV$ in the final-state mass spectrum of means that we do not know what resonant state has been observed[1]. This year, CLEOII&II.V collaborations have also reported the inclusive branching ratio

$$BR(B \to X_s\gamma) = 3.15 \pm 0.35 \pm 0.41 \times 10^{-4}. \tag{6}$$

Using the latest results for inclusive and exclusive branching ratios (BR) , we have obtained the following central value for the so-called hadronization rate: $R_{K^*}^{exp} \simeq 13\%$. Note here that Skwarnicki has also reported preliminary results for exclusive modes based on the quark $b \to d\gamma$ electroweak transition:[1]
$BR(B^0 \to \rho^0\gamma) < 4 \times 10^{-5}; BR(B^{0(-)} \to \omega(\rho^-)\gamma) < 1 \times 10^{-5}$.

2 NNL QCD Corrected $b \to s\gamma$ Transition

The $b \to s\gamma$ process is a one-loop electroweak process, where the extra gluon exchange (QCD corrections) changes the nature, i.e. the functional structure of the GIM cancellation:[2] $(m_t^2 - m_c^2)/m_w^2 \to \ln(m_t^2/m_c^2)$.

In the standard model, B decays are described by the effective Hamiltonian obtained by integrating out the top-quark and W-boson fields:

$$H^{\rm eff}_{\Delta B=1}(b \to s\gamma) = \frac{G_F}{\sqrt{2}} V_{tb} V_{ts}^* \sum_{i=1}^{8} c_i(\mu)\, O_i(\mu), \tag{7}$$

where c_i's are the well-known Wilson coefficients. The SM theoretical prediction, up to Next-to-Leading Order (NLO)[3] in $\alpha_s ln(m_w/m_b)$, is considerably larger than the lowest-order result.[4] Buras et al.[5] performed a new analysis by using expansions in powers of α_s and reported a slightly higher, short-distance (SD) result: $BR(B \to X_s\gamma)_{\rm NLO} = (3.60 \pm 0.33) \times 10^{-4}$.

3 Exclusive Radiative B Decays

Exclusive modes are, in principle, affected by large theoretical uncertainties due to the poor knowledge of nonperturbative dynamics and of a correct treatment of large recoil-momenta, determining form factors.

The Lorentz decomposition of the $B - K^*$ matrix element of the operator O_7, taking into account the gauge condition, the current conservation, the spin symmetry and for the on-shell photon, gives the following hadronization rate R_{K^*}:[6]

$$R_{K^*} = \frac{\Gamma(B \to K^*\gamma)}{\Gamma(b \to s\gamma)} = [\frac{m_b(m_B^2 - m_{K^*}^2)}{m_B(m_b^2 - m_s^2)}]^3 (1 + \frac{m_s^2}{m_b^2})^{-1} |T_1^{K^*}(0)|^2. \tag{8}$$

Since the first calculation of the hadronization rate $R_{K^*} \approx 6\%$ by Deshpande at al.[6] a large number of papers have reported R_{K^*} from the range of 5% to (unrealistic) 80%. Different methods have been employed, from simple quark models, QCD sum rules, HQET and chiral symmetry, QCD on the lattice, light cone sum rules, to the perturbative QCD type of evaluations of exclusive modes.[6] In any event, the above form factor will be obtained in the future from first-principle calculations on the lattice. Recently, it looks like that the hadronization rate calculations have stabilized around 10%.

If long-distance (LD) and other nonperturbative effects are neglected, two exclusive modes are connected by a simple relation

$$BR(B \to \rho\gamma) = \xi^2\, |V_{td}/V_{ts}|^2\, BR(B \to K^*\gamma), \tag{9}$$

where ξ measures the SU(3) breaking effects. They are typically of the order of 30%.[6] Misiak has reported the SD BR's:[7]

$$BR(b \to d\gamma) = 1.61 \times 10^{-5}; BR(B^+ \to \rho^+\gamma) = [1,4] \times 10^{-6}, \tag{10}$$

$$BR(B^0 \to \rho^0\gamma) = BR(B^0 \to \omega\gamma) = [0.5, 2] \times 10^{-6}. \tag{11}$$

4 LD Contributions to Inclusive/Exclusive B Decays

First note that LD corrections cannot be computed from first principles; it is possible to estimate them phenomenologically.[8] For instance, the operators $O_{1,2}$ contain the $\bar{c}c$ current. So one could imagine the $\bar{c}c$ pair propagating through a long distance, forming intermediate $\bar{c}c$ states (off-shell J/ψ's), which turn into a photon via the vector meson dominance (VMD).

The total (SD + LD) amplitude for $b \to d(s)\gamma$ is [8]

$$M(b \to d\gamma) = -\frac{eG_F}{2\sqrt{2}}[V_{tb}V_{td}^*(\frac{m_b}{4\pi^2}c_7^{eff}(\mu) - \frac{2}{3}a_2\sum_i \frac{g_{\psi_i}^2(0)}{m_{\psi_i}^2 m_b})$$
$$-\frac{a_2}{m_b}V_{ub}V_{ud}^*(\frac{2}{3}\sum_i \frac{g_{\psi_i}^2(0)}{m_{\psi_i}^2} - \frac{1}{2}\frac{g_\rho^2(0)}{m_\rho^2} - \frac{1}{6}\frac{g_\omega^2(0)}{m_\omega^2})]\bar{d}\sigma^{\mu\nu}(1+\gamma_5)bF_{\mu\nu}. \tag{12}$$

If in the above equation we replace the d by the s quark and forget the last three terms, then we obtain the total amplitude for the $b \to s\gamma$ decay. Deshpande et al[8] have found a strong suppression when extrapolating $g_\psi(m_\psi^2)$ to $g_\psi(0)$: $g_{\psi(1S)}^2(0)/g_{\psi(1S)}^2(m_\psi^2) = 0.13 \pm 0.04$. The LD contributions to an inclusive mode and to its exclusive modes are all found to be small, typically one to two orders of magnitude below the SD's.[8]

5 Discussion and Conclusions

Note that the so-called spectator-quark contributions[9] and the first calculable nonperturbative, essentialy long distance, correction[10] to the inclusive rate are of the order of a few percent. It has also been proved that the fermionic (quarks and leptons) and photonic loop corrections to $b \to s\gamma$ reduce $BR(b \to s\gamma)/BR(b \to ce\bar{\nu})$ by $\sim 8 \pm 2\%$.[11] Consequently, it is more appropriate to use $\alpha_{em} = 1/137$ for the real photon emission.[11,7]

In general, we can conclude that, in theory, more effort is required in calculating quark (inclusive) decays through higher loops. A better understanding of bound states of heavy-light quarks (B-meson ect.) and highly recoiled light-quark bound states (K^*, ρ, ...) is desirable. This can be achived by inventing new, more sophisticated (perturbative) methods[12] and apply them to the calculation of radiative B-meson decays, which incorporate the full spectrum of quark bound states (K^*, ρ, K_1^*, ...). In experiment, with a larger amount of data we might expect a stable but small increase of $BR(b \to s\gamma)$, a decrease of $BR(B \to K^*\gamma)$, determinations of $BR(B \to K_1^*\gamma)$ and $BR(B \to K_2^*\gamma)$ and first measurements of $BR(b \to d\gamma)$ and $BR(B \to \rho\gamma)$.

References

1. R. Ammar *et al*, *Phys. Rev. Lett.* **71**, 674 (1993); M.S. Alam *et al*, *Phys. Rev. Lett.* **74**, 2885 (1995); T. Skwarnicki, this Proceedings.
2. S. Bertolini, F. Borzumati, A. Masiero, *Phys. Rev. Lett.* **59**, 180 (1987); N.G. Deshpande, P. Lo, J. Trampetić, G. Eilam, P. Singer, *Phys. Rev. Lett.* **59**, 183 (1987).
3. K. Chetyrkin, M. Misiak, M. Münz, *Phys. Lett.* B **400**, 206 (1997); C. Greub, T. Hurth, PRD **56**, 2934 (1997); M. Ciuchini, G. Degrassi, P. Gambino, G.F. Giudice, *Nucl. Phys.* B **527**, 21 (1998).
4. B. Grinstein, R. Springer, M.B. Wise, *Nucl. Phys.* B **339**, 269 (1990);
5. A. Buras, A. Kwiatkowski, N. Pott, *Phys. Lett.* B **414**, 157 (1997); *Phys. Lett.* B **434**, 459 (1998)(erratum).
6. N.G. Deshpande *et al*, *Phys. Rev. Lett.* **59**, 183 (1987); *Z. Phys.* **C40**, 369 (1988). P. Colangelo, C.A. Dominguez, G. Nardulli, N. Paver, *Phys. Lett.* B **317**, 183 (1993); R. Casalbuoni, A. Deandrea, N. Di Bartolomeo, R. Gatto, G. Narduli, *Phys. Lett.* B **312**, 315 (1993); K.C. Bowler *et al*, *Phys. Rev. Lett.* **72**, 1398 (1994); L. Del Debbio, J. Flynn, L. Lellouch, J. Nieves, *Phys. Lett.* B **416**, 392 (1998); P. Ball, V.M. Braun, *Phys. Rev.* D **58**, 094016 (1998); H.H. Asatryan, H.M. Asatryan, D. Wyler, *Phys. Lett.* B **470**, 223 (1999).
7. M. Misiak, hep-ph/0002007, this Proceedings.
8. E. Golowich, S. Pakvasa, *Phys. Lett.* B **205**, 393 (1988); J. Trampetić, *Fizika* **B2**, 121 (1993); and a paper in preparation; N.G. Deshpande, X.-G. He, J. Trampetić, *Phys. Lett.* B **367**, 362 (1996).
9. J.F. Donoghue, A.A. Petrov *Phys. Rev.* D **53**, 3664 (1996).
10. M.B. Voloshin, *Phys. Lett.* B **397**, 275 (1997).
11. A. Czarnecki, W. Marciano, *Phys. Rev. Lett.* **81**, 277 (1998);
12. H. Simma, D. Wyler, *Phys. Lett.* B **272**, 395 (1991); B. Melić, *Phys. Rev.* D **59**, 074005 (1999); M. Benecke, G. Buchalla, M. Neubert, C.T. Sachrajda, *Phys. Rev. Lett.* **83**, 1914 (1999).

ATLAS AND CMS B PHYSICS PROSPECTS

P. VIKAS
University of Minnesota
EP Division, CERN, CH-1211 Geneva 23, Switzerland
E-mail: Pratibha.Vikas@cern.ch

Although, the main focus of the ATLAS and CMS physics programmes is the search for and study of physics beyond the Standard Model, an important range of B-physics studies is also planned.The copious production of B hadrons at the LHC opens the way for significant measurements of CP violation effects in the B system, precise measurements of the periods of flavour oscillations in B_s^0 mesons and searches for and measurements of very rare decays which are strongly suppressed in the Standard Model. The expected sensitivity of the two experiments for these measurements is presented.

1 Introduction

ATLAS and CMS have primarily been conceived and optimised for the high luminosity LHC environment, to search for physics beyond the Standard Model. However, the first three years of low luminosity operation will allow the investigation of B physics issues, thereby influencing the design of the two detectors. The expected LHC luminosity during this phase is , $\mathcal{L} \sim 10^{33} cm^{-2}s^{-1}$ giving $10^4 pb^{-1}$/year. Thus, 10^{12} $b\bar{b}$ events/year, with millions fully reconstructed, are expected by each of the two experiments. The enormous statistics and full spectrum, open up rich possibilities for B physics measurements.

All the studies presented here have been performed with events generated by Pythia [1], assuming $\sigma_{b\bar{b}} = 500\mu b$, followed by GEANT-based detector simulations.

1.1 ATLAS and CMS as B Physics Detectors

The CMS and ATLAS detectors and triggers are both extensively described in [3] and [2], respectively. Both the detectors have foreseen high resolution vertex detectors for B physics.

The ATLAS level 1 trigger requires a muon with $p_t^\mu > 6$ GeV in $|\eta| < 2.4$. The K^0 reconstruction efficiency is ~75% ($R < 25$ cm). The proper time resolution for $B_s^0 \to D_s^- \pi^+$ decays [a] is 0.073 ps and the mass resolution for $B_d^0 \to J/\psi K_s^0, J/\psi \to \mu^+\mu^-$ is 19 MeV. The CMS Level 1 trigger uses both muons and calorimetric information with η dependent thresholds. The overall

[a] Charge conjugation is assumed throughout this paper.

Table 1. List of channels studied by ATLAS and CMS and their physics interests.

Channel	Interest
$B_d \to J/\psi K_s^0$	β
$B_d \to \pi^+\pi^-$	α
$B_s \to K^+K^-$	γ
$B_s^0 \to D_s^-\pi^+$	Δm_s
$B_s^0 \to J/\psi\phi$	$\delta\gamma$
$B_{s,d} \to \mu^+\mu^-, X\mu^+\mu^-$	New Physics

K^0 reconstruction efficiency is ~35%. The proper time resolution for $B_s^0 \to D_s^-\pi^+$ is 0.073 ps and the mass resolution for $B_d^0 \to J/\psi K_s^0, J/\psi \to \mu^+\mu^-$ is 20 MeV.

Both the detectors suffer from the drawback of possessing little or no capability for hadron identification. However, this can be overcome by their good mass resolution in some cases, as has already been shown by CDF. Table 1 lists the channels that have been studied by the two experiments and the physics interests behind them. In addition, it will also be possible to make precision measurements of masses, lifetimes and polarisations of different species of B hadrons.

1.2 *Flavour Tagging*

The knowledge of the hadron flavour at birth is necessary for most CP violation measurements. The feasibility of using flavour tagging methods devised and used by the LEP experiments and CDF have been studied by both experiments, yielding results comparable to those from CDF. The opposite side lepton tag has the highest dilution but its efficiency is limited by the B semileptonic branching ratio. The same-side pion (either from B^{**} or b fragmentation and the same or opposite jet charge have low dilution but high efficiency. Typically, all the analyses use the combination of all the tags available in an event.

2 CP Violation Measurements

2.1 $B_d^0 \to J/\psi K_s^0$

Excellent measurements of $\sin 2\beta$ from this channel are expected from $\Upsilon(4S)$ and Tevatron by the time LHC turns on. However, the LHC statistics will allow for a true precision measurement. With one year's data ATLAS and

CMS the expected sensitivity of ATLAS [4] and CMS [2] for this measurement is 0.017 and 0.018, respectively.

2.2 $B_d^0 \to \pi^+\pi^-$

Traditionally, the CP asymmetry in this channel is seen as the primary method to measure $\sin 2\alpha$. The B factories will be hampered by the low branching ratio ($\sim 10^{-5}$) and will have < 1000 events in total by 2005. The prospects at the LHC are much better and both ATLAS and CMS expect ~ 2000 reconstructed events/year. The two body background to this channel with its own CP asymmetry is also a problem. ATLAS have devised a method [4] to simultaneously fit the CP asymmetries for the signal and background and expect to measure α with an accuracy of $\sim 2°$ with three years of data. The CMS [2] sensitivity to this measurement has been estimated to be $\sim 3°$.

3 B_s^0 Physics

B_s^0 physics is the "El-Dorado" of the LHC era since, the B factories will only measure the B_d and B_u systems. Mixing in the $B_s^0 - \bar{B}_s^0$ system is studied through the time-dependent asymmetry between B_s^0 at time $t = 0$ that have oscillated or not to $\bar{B}_s^0$ at time t. ATLAS has used B_s^0 decays to $D_s^-\pi^+$ and $D_s^- a^+$ and $D_s^- \to \phi\pi^-$. CMS studies have used $B_s^0 \to D_s^-\pi^+$ followed by $D_s^-\pi^+ \to \phi\pi^-, K^{*0}K^-$. Table 2 summarises the ATLAS [4] and CMS [7] expectations with one year's data. Adding more channels will further improve the sensitivity.

Table 2. ATLAS and CMS potential for Δm_s measurement.

	ATLAS	CMS
Proper time resolution	50 fs(60%), 93 fs(40%)	65 fs
5σ measurement upto	30 ps^{-1}	22 ps^{-1}
95% C.L. exclusion		31 ps^{-1}

In addition, $\sim 10^5$ reconstructed $B_s^0 \to J/\psi\phi$ events are expected by each of the experiments. These can be used for a variety of measurements [4,6].

4 Rare Decays $B_{d,s}^0 \to \mu\mu(X)$

Rare semileptonic and leptonic decays are sensitive to new physics and may give the indications of its existence in the first year(s) of LHC running. ATLAS and CMS have the possibilities to measure branching ratios, $|V_{td}|/|V_{ts}|$, influence of form factors on branching ratios and the dynamics of these decays. Table 3 summarises the SM branching ratios, current experimental status and the ATLAS [4] and CMS [5] expectations with three years data for some rare channels.

Table 3. The current experimental status and the ATLAS and CMS expectations for signal(background) with three years data.

	SM BR	Current Limit(CDF)	ATLAS	CMS
$B_d^0 \to \mu^+\mu^- K^{*0}$	$\sim 10^{-6}$	$< 2.5 \times 10^{-5}$	2000(290)	4200(435)
$B_d^0 \to \mu^+\mu^- \rho^0$	$\sim 10^{-7}$		220(950)	350(340)
$B_d^0 \to \mu^+\mu^- \phi$	$\sim 10^{-6}$	$< 2.5 \times 10^{-5}$	410(140)	1200(70)
$B_s^0 \to \mu^+\mu^-$	$\sim 10^{-9}$	$< 2.0 \times 10^{-6}$	27(93)	31($\leq$3)
$B_d^0 \to \mu^+\mu^-$	$\sim 10^{-10}$	$< 6.8 \times 10^{-7}$	4(93)	1.1(0.7)

Searches for purely muonic decays will also be possible at high luminosity. At high luminosity, both the experiments expect to be able to clearly observe $B_s \to \mu^+\mu^-$ decay with one year's data.

5 Conclusions

The full spectrum of B hadrons with high statistics will be available to ATLAS and CMS on the LHC. Thanks to their good mass and proper time resolutions, both the experiments will be able to make precision measurements in the CKM sector and perform searches for very rare B decays.

Acknowledgements

The author wishes to thank the members of the ATLAS and CMS B physics groups, especially thr participants of the LHC SM workshop, for all their help.

References

1. H.-U. Bengtsson and T. S:jostrand, Comp.Phys.Comm. **46**, 43 (1987)
2. CMS Technical Proposal, CERN/LHCC/94-38.
3. ATLAS Technical Proposal, CERN/LHCC/94-43.
4. ATLAS Detector and Physics Performance Technical Design Report, CERN/LHCC/99-15.
5. A. Nikitenko *et al.*, CMS NOTE 1999/039.
6. F. Charles, CMS NOTE 1999/025.
7. A. Starodumov, CMS NOTE 1997/072.

$B \to X_s\tau^+\tau^-$ in a CP spontaneously broken two Higgs doublet model

Chao-Shang Huang [a] and Shou Hua Zhu
Institute of Theoretical Physics, Academia Sinica, P.O.Box 2735, Beijing 100080, P.R.China

The differential branching ratio, CP asymmetry and lepton polarization P_N for $B \to X_s\tau^+\tau^-$ in a CP spotaneously broken two Higgs doublet model are computed. It is shown that contributions of neutral Higgs bosons to the decay are quite significant when $\tan\beta$ is large and P_N can reach five persent.

The origin of the CP violation has been one of main issues in high energy physics. The measurements of electric dipole moments of the neutron and electron and the matter-antimatter asymmetry in the universe indicate that one needs new sources of CP violation in addition to the CP violation come from CKM matrix, which has been one of motivations to search new theoretical models beyond the standard model (SM).

The minimal extension of the SM is to enlarge the Higgs sectors of the SM. It has been shown that if one adheres to the natural flavor conservation (NFC) in the Higgs sector, then a minimum of three Higgs doublets are necessary in order to have spontaneous CP violations [1]. However, the constraint can be evaded if one allows the real and image parts of $\phi_1^+\phi_2$ have different self-couplings and adds a linear term of $\mathrm{Re}(\phi_1^+\phi_2)$ in the Higgs potential (see below Eq. (1)). Then, one can construct a CP spontaneously broken (and Z_2-symmetry softly broken) two Higgs doublet (2HDM), which is the minimal among the extensions of the SM that provide new source of CP violation.

Rare decays $B \to X_s l^+l^- (l = e, \mu)$ have been extensively investigated in both SM and the beyond [2,3]. The inclusive decay $B \to X_s\tau^+\tau^-$ has also been investigated in the SM, the model II 2HDM and SUSY models with and without including the contributions of NHB [4,5]. In this note we investigate the inclusive decay $B \to X_s\tau^+\tau^-$ with emphasis on CP violation effect in a CP spontaniously broken 2HDM in which the up-type quarks get masses from Yukawa couplings to the one Higgs doublet H_2 and down-type quarks and leptons get masses from Yukawa couplings to the another Higgs doublet H_1. The contributions from exchanging neutral Higgs bosons now is enhanced roughly by a factor of $tg^2\beta$ and can compete with those from exchanging γ, Z when $tg\beta$ is large enough. We shall be interested in the large $\tan\beta$ limit in this note.

[a] Presenting author

Consider two complex $Y = 1$ $SU(2)_w$ doublet scalar fields, ϕ_1 and ϕ_2. The Higgs potential which spontaneously breaks $SU(2) \times U(1)$ down to $U(1)_{EM}$ can be written in the following form [5,6]:

$$V(\phi_1, \phi_2) = \sum_{i=1,2} [m_i^2 \phi_i^+ \phi_i + \lambda_i (\phi_i^+ \phi_i)^2] + m_3^2 Re(\phi_1^+ \phi_2)$$
$$+\lambda_3[(\phi_1^+ \phi_1)(\phi_2^+ \phi_2)] + \lambda_4 [\mathrm{Re}(\phi_1^+ \phi_2)]^2 + \lambda_5 [\mathrm{Im}(\phi_1^+ \phi_2)]^2 \quad (1)$$

Hermiticity requires that all parameters are real so that the potential is CP conservative. It is easy to see that the minimum of the potential is at $<\phi_1>= v_1$ and $<\phi_2>= v_2 e^{i\xi}$, thus breaking $SU(2) \times U(1)$ down to $U(1)_{EM}$ and simutaneously breaking CP, as desired.

In the case of large $\tan\beta$, if we neglect all terms proptional to c_β and taken $s_\beta = 1$ in the mass-squared matrix of neutral Higgs, one would get that one of the Higgs boson mass is 0, obviously which is conflict with current experiments. So instead, below we will discuss a special case in which a analytical solution of the mass matrix can be obtained.

Assuming $4\lambda_1 c_\beta^2 = \lambda_4 - \lambda_5$ and neglecting other terms proptional to c_β in the mass-squared matrix, one obtains the masses of neutral Higgs bosons

$$m_{H_1^0}^2 = 2(\lambda_4 - \lambda_5) v^2 s_\psi^2, \quad m_{H_2^0}^2 = 2(\lambda_4 - \lambda_5) v^2 c_\psi^2, \quad m_{H_3^0}^2 = 4(\lambda_2 + \lambda_3) v^2 \quad (2)$$

with the mixing angle $\psi = \frac{\xi}{2}$.

Then it is straightforward to obtain the couplings of neutral Higgs to fermions

$$H_1^0 \bar{f} f : -\frac{igm_f}{2m_w c_\beta}(-s_{\xi/2} + ic_{\xi/2}\gamma_5); \quad H_2^0 \bar{f} f : -\frac{igm_f}{2m_w c_\beta}(c_{\xi/2} + is_{\xi/2}\gamma_5), \quad (3)$$

where f represents down-type quarks and leptons. The coupling of H_3^0 to f is not enhenced by $\tan\beta$ and will not be given explicitly. The couplings of the charged Higgs bosons to fermions are the same as those in the model II 2HDM. This is in contrary with the model III [7] in which the couplings of the charged Higgs to fermions are quite different from model II. It is easy to see from Eq. (3) that the contributions come from exchanging NHB is proportional to $\sqrt{2} G_F s_{\xi/2} c_{\xi/2} m_f^2 / \cos^2\beta$, so that we have the constraint $\sqrt{|\sin\xi|}\tan\beta < 50$ from the neutron EDM, and the constraint from the electron EDM is not stronger this. The constraints on $\tan\beta$ due to effects arising from the charged Higgs are the same as those in the model II and can be found in ref. [8,4].

The transition rate for $b \to s\tau^+\tau^-$ can be computed in the framework of the QCD corrected effective weak Hamiltonian

$$H_{eff} = \frac{4G_F}{\sqrt{2}} V_{tb} V_{ts}^* (\sum_{i=1}^{10} C_i(\mu) O_i(\mu) + \sum_{i=1}^{10} C_{Q_i}(\mu) Q_i(\mu)) \tag{4}$$

where $O_i (i = 1, \cdots, 10)$ is the same as that given in the ref.[2], Q_i's come from exchanging the neutral Higgs bosons and are defined in Ref. [4].

The coefficients C_i's at $\mu = m_W$ have been given in the ref.[2] and C_{Q_i}'s are (neglecting the $O(tg\beta)$ term)

$$
\begin{aligned}
C_{Q_1}(m_W) &= \frac{m_b m_\tau tg^2\beta x_t}{2sin^2\theta_W} \{ \sum_{i=H_1,H_2} \frac{A_i}{m_i^2} (f_1 B_i + f_2 E_i) \}, \\
C_{Q_2}(m_W) &= \frac{m_b m_\tau tg^2\beta x_t}{2sin^2\theta_W} \{ \sum_{i=H_1,H_2} \frac{D_i}{m_i^2} (f_1 B_i + f_2 E_i) \}, \\
C_{Q_3}(m_W) &= \frac{m_b e^2}{m_\tau g_s^2} (C_{Q_1}(m_W) + C_{Q_2}(m_W)), \\
C_{Q_4}(m_W) &= \frac{m_b e^2}{m_\tau g_s^2} (C_{Q_1}(m_W) - C_{Q_2}(m_W)), \\
C_{Q_i}(m_W) &= 0, \quad i = 5, \cdots, 10
\end{aligned} \tag{5}
$$

where

$$
\begin{aligned}
A_{H_1} &= -s_{\xi/2}, \quad D_{H_1} = ic_{\xi/2}, \quad A_{H_2} = c_{\xi/2}, \quad D_{H_2} = is_{\xi/2}, \\
B_{H_1} &= \frac{ic_{\xi/2} - s_{\xi/2}}{2}, \qquad B_{H_2} = \frac{c_{\xi/2} + is_{\xi/2}}{2} \\
E_{H_1} &= \frac{1}{2}(-s_{\xi/2} c_1 + c_{\xi/2} c_2), \quad E_{H_2} = \frac{1}{2}(c_{\xi/2} c_1 + s_{\xi/2} c_2) \\
c_1 &= -x_{H^\pm} + \frac{c_\xi}{2s_{\xi/2}^2} x_{H_1}(c_\xi + is_\xi) + \frac{1}{2s_{\xi/2}^2} x_{H_1} \\
c_2 &= i[-x_{H^\pm} + \frac{c_{\xi/2} x_{H_1}}{s_{\xi/2}} (s_\xi - ic_\xi)].
\end{aligned} \tag{6}
$$

The QCD corrections to coefficients C_i and C_{Q_i} can be incooperated in the standard way by using the renormalization group equations [4]. The explict expressions of the invariant dilepton mass distribution, CP asymmetries A_{CP}^i (i=1,2) in branching ratio and in forward-backward (F-B) asymmetry and normal polarization of lepton P_N for $B \to X_s \tau^+ \tau^-$ can be found in ref. [5].

The following parameters have been used in the numerical calculations: $m_t = 175Gev$, $m_b = 5.0Gev$, $m_c = 1.6Gev$, $m_\tau = 1.77Gev$, $\eta = 1.724$, $m_{H_1} = 100Gev$, $m_{H^\pm} = 200Gev$.

From the numerical results, the following conclusions can be drawn. A. The contributions of NHB to the differential branching ratio $d\Gamma/ds$ are significant when $\tan\beta$ is not smaller than 30 and the masses of NHB are in the reasonable region. For $\tan\beta$=30, $d\Gamma/ds$ is enhenced by a factor of 6 compared to SM. B. The direct CP violation in branching ratio A^1_{CP} is of order 10^{-4} which is hard to be measured. C. The direct CP violation in F-B A^2_{CP} can reach about a few percents and is strongly dependent of the CP violation phase ξ and comes mainly from exchanging NHBs, as expected. D. The CP-violating polarization P_N is also strongly dependent of the CP violation phase ξ and can be as large as 5% for some values of ξ, which should be within the luminosity reach of coming B factord comes mainly from NHB contributions in the most of range of ξ. So it is possible to discriminate the model from the other 2HDMs by measuring the CP-violated observables such as A^2_{CP}, P_N if the nature chooses large $\tan\beta$. [b]

1. T. D. Lee, Phys. Rev. **D8** (1973) 1226; Phys. Rep. **9c** (1974) 143; P. Sikivie, Phys. Lett. **B65** (1976) 141; S. Weinberg, Phys. Rev. Lett. **37** (1976) 657; G. C. Branco, Phys. Rev. **D22** (1980) 2901; K. Shizuya and S.-H. H. Tye, Phys. Rev. **D23** (1981) 1613.
2. B.Grinstein, M.J.Savage and M.B.Wise, *Nucl.Phys.***B319** (1989)271.
3. C.S. Huang, W. Liao and Q.S. Yan, Phys. Rev. **D59** (1999) 011701; T. Goto et.al., hep-ph/9812369; S. Baek and P. Ko, hep-ph/9904283; Y.G. Kim, P. Ko and J.S. Lee, Nucl. Phys. **B544** (1999) 64, and references therein. For the earlier references, see, the references in ref. [4].
4. Y.B. Dai, C.S. Huang and H.W. Huang, Phys. Lett. **B390** (1997) 257; C.S. Huang and Q.S. Yan, Phys. Lett. **B442** (1998) 209.
5. Chao-Shang Huang and Shou-Hua Zhu, hep-ph/9905463, Phys. Rev. **D61** (2000) 015011, references therein.
6. H. Georgi, Hadronic Jour. 1 (1978) 155; I. Vendramin, hep-ph/9909291.
7. see, for example, D. Bowser-Chao, K. Cheung and W. Y. Keung, hep-ph/9811235, and references therein.
8. ALEPH Collaboration (D. Buskulic et al.), Phys. Lett. B343 (1995) 444; J.Kalinowski, *Phys.Lett.* **B245** (1990) 201; A.K.Grant, *Phys.Rev.* **D51** (1995) 207.

[b]This research was supported in part by the National Nature Science Foundation of China and the post doctoral foundation of China. S.H. Zhu gratefully acknowledges the support of K.C. Wong Education Foundation, Hong Kong.

EFFECTS OF FLAVOR CONSERVING CP VIOLATING PHASES IN SUSY MODELS

P. KO

Department of Physics, KAIST, Taejon 305-701, KOREA
E-mail: [a] pko@muon.kaist.ac.kr

I summarize our recent works on the effects of flavor conserving CP violating phases in SUSY models on B and K phenomenology.

1 Introduction

The minimal supersymmetric standard model (MSSM) has many CP violating (CPV) phases beyond the KM phase in the standard model (SM). These SUSY CPV phases, depending on their flavor structures, are strongly constrained by ϵ_K or electron/neutron electric dipole moment (EDM), and have been considered very small ($\delta \leq 10^{-2}$ for $M_{\rm SUSY} \sim O(100)$ GeV).[1] Another way to solve these problems is to consider effective SUSY models, where decouplings of the 1st/2nd generation sfermions are invoked to evade the EDM constraints and also SUSY FCNC/CP problems.[2] In such cases, these new SUSY phases may affect B and K physics. One strong motivation for new CP violating phases beyond the KM phase is related with the baryon number asymmetry of the universe. Electroweak baryogenesis is possible in a certain region of the MSSM parameter space, especially for light stop (120 GeV $\leq m_{\tilde{t}_1} \leq 175$ GeV) with CP violating phases in μ and A_t parameters.[3] This light stop and new CP violating phases in μ and A_t parameters can affect B and K physics, although these phases are flavor conserving. In this talk, we report our three recent works related with this subject. [4,5,6] The topics covered here are the following : the effects of ϕ_μ and ϕ_{A_t} on B physics in the MMSSM, and fully supersymmetric CP violations in the kaon system.

2 Effects of μ and A_t phases on B physics in the more minimal supersymmetric standard model (MMSSM)

In the MMSSM we consider in this section, only the third family squarks and charginos can be light enough to affect $B \to X_s \gamma$ and $B^0 - \overline{B^0}$ mixing. We also ignore possible flavor changing squark mass matrix elements that could generate gluino-mediated flavor changing neutral current (FCNC) process, dicussions of which can be found in the literatures,[7,8], for example.

Ignoring such contributions, the only source of the FCNC in our model may be attributed to the CKM matrix, whereas there are new CPV phases coming from the phases of μ and A_t parameters in the flavor preserving sector in addition to the KM phase δ_{KM} in the flavor changing sector.

Even if the 1st/2nd generation squarks are very heavy and degenerate, there is another important edm constraints considered by Chang, Keung and Pilaftsis (CKP) for large $\tan\beta$.[9] This constraint comes from the two loop diagrams involving stop/sbottom loops, and is independent of the masses of the 1st/2nd generation squarks. Therefore, this CKP edm constraints can not be simply evaded by making the 1st/2nd generation squarks very heavy, and it turns out that this puts a strong constraint on the possible new phase shift in the $B^0-\overline{B^0}$ mixing. We scanned over the broad parameter space and imposed the various experimental constraints including $BR(B\to X_s\gamma)$.It has to be emphasized that this parameter space is larger than that in the constrained MSSM (CMSSM) where the universality of soft terms at the GUT scale is assumed.

The $B^0-\overline{B^0}$ mixing is generated by the box diagrams with $u_i-W^{\pm}(H^{\pm})$ and $\tilde{u}_i-\chi^{\pm}$ running around the loops in addition to the SM contribution. The gluino and neutralino contributions are negligible in our model. The chargino exchange contributions to $B^0-\overline{B^0}$ mixing is generically complex relative to the SM contributions, and this effect can be in fact significant for large $\tan\beta(\simeq 1/\cos\beta)$, since the chargino contribution is proportional to $(m_b/M_W\cos\beta)^2$. However, the CKP edm constraint puts a strong constraint for large $\tan\beta$ case. The result is that the CKP edm constraint on $2\theta_d$ is in fact very important for large $\tan\beta$, and we have $|2\theta_d|\leq 1^\circ$. This observation is important for the CKM phenomenology, since time-dependent CP asymmetries in neutral B decays into $J/\psi K_S,\pi\pi$ etc. would still measure directly three angles of the unitarity triangle even if ϕ_{A_t} and ϕ_μ are nonzero. We also find that the dilepton asymmetry (proportional to $\mathrm{Re}(\epsilon_B)$) is very small as in the SM, but Δm_B can be enhanced as much as 60%. See Ref. [4] for more details.

The radiative decay of B mesons, $B\to X_s\gamma$, is described by the effective Hamiltonian including (chromo)magnetic dipole operators. Interference between $b\to s\gamma$ and $b\to sg$ (where the strong phase is generated by the charm loop via $b\to c\bar{c}s$ vertex) can induce direct CP violation in $B\to X_s\gamma$.[10] The SM predicts a very small asymmetry smaller than 0.5%, so the larger asymmetry will be a clean signal for new source of CP violating phases. In our model, we find that $A_{\rm CP}^{b\to s\gamma}$ can be as large as $\simeq\pm16\%$ if chargino is light enough, even if we impose the edm constraints. So this mode may be one of the good place for probing new CPV phases.

Next let us next consider R_{ll}, the ratio of the branching ratio for $B\to$

$X_s l^+ l^-$ in our model to that in the SM. In the presence of the new phases ϕ_μ and ϕ_{A_t}, $R_{\mu\mu}$ can be as large as 1.85, and the deviations from the SM prediction can be large, if $\tan\beta > 8$. As noticed in Ref. [8], the correlation between the Br$(B \to X_s\gamma)$ and R_{ll} is distinctly different from that in the minimal supergravisty case.[11]

3 Fully SUSY CP violation in the kaon system

In the MSSM with many new CPV phases, there is an intriguing possibility that the observed CP violation in $K_L \to \pi\pi$ is fully supersymmetric due to the complex parameters μ and A_t in the soft SUSY breaking terms which also break CP softly, or CP violating $\tilde{g} - q_i - \tilde{q}_j$. Our study on the first possibility in the MMSSM [5] indicates that the supersymmetric ϵ_K (namely, for $\delta_{KM} = 0$) is less than $\sim 2 \times 10^{-5}$, which is too small compared to the observed value : $|\epsilon_K| = (2.280 \pm 0.019) \times 10^{-3}$. (See also Ref. [12].)

Although one cannot generate enough CP violations in the kaon system through flavor preserving μ and A_t phases in the MSSM, it is possible if one considers the flavor changing SUSY CPV phases. In the mass insertion approximation (MIA), the folklore was that if one saturates the ϵ_K with $(\delta^d_{12})_{LL}$, the corresponding $\epsilon^{'}/\epsilon_K$ is far less than the observed value. On the other hand, if one saturates $\epsilon^{'}/\epsilon_K$ with $(\delta^d_{12})_{LR}$, the resulting ϵ_K is again too small compared to the data. Therefore one would need two independent parameters $|(\delta^d_{12})_{LL}| \sim O(10^{-3})$ and $|(\delta^d_{12})_{LR}| \sim O(10^{-5})$, each of which has a $\sim O(1)$ phase. Recently, Masiero and Murayama argued that such a large value of $(\delta^d_{12})_{LR}$ is not implausible in general MSSM, e.g., if the fundamental theory is a string theory.[13] In their model, the large $(\delta^d_{12})_{LR}$ is intimately related with the large $(\delta^d_{11})_{LR}$, so that their prediction on the neutron edm is very close to the current experimental limit.

In recent work, we pointed out it is possible in fact to generate both ϵ_K and $\mathrm{Re}(\epsilon^{'}/\epsilon_K)$ with a single complex number $(\delta^d_{12})_{LL} \sim O(10^{-2} - 10^{-3})$ with an order $\sim O(1)$ phase, if one goes beyond the single mass insertion approximation.[6] Namely, the ϵ_K is generated by $(\delta^d_{12})_{LL}$, whereas $\epsilon^{'}/\epsilon_K$ is generated by a flavor preserving $\tilde{s}_R - \tilde{s}_L$ transition followed by flavor changing $\tilde{s}_L - \tilde{d}_L$ transition. The former is proportinal to $m_s(A_s - \mu\tan\beta)/\tilde{m}^2$ where $\tilde{m}$ is the common squark mass in the MIA. *This induced LR mixing is present generically in any SUSY models, if* $|\mu\tan\beta| \sim 10 - 20$ *TeV*. The only relevant question would be how one can have an $\sim O(1)$ phase in the $(\delta^d_{12})_{LL}$. For example, the gluino mass can have a CPV phase ϕ_3 which is flavor preserving. After we redefine the gluino field so that the the gluino mass parameter becomes real, the phase ϕ_3 will be tranferred to the $\tilde{g} - q_i - \tilde{q}_j$

vertex, thereby generating CP violations in both flavor preserving and flavor changing gluino mediated strong interactions.[14]

If the KM phase were not zero in this scenario, we cannot use the constraints coming from ϵ_K or ΔM_B, since new physics would contribute to both $\Delta S = 2$ and $\Delta B = 2$ amplitudes. In particular, even the third or fourth quadrant in the (ρ, η) plane should be possible, in principle. More detailed discussions on these points will be presented elsewhere. Finally let us note that the recent observation on CP asymmetry in $B^0 \to J/\psi K_S$ depends on different CP violating parameter $(\delta^d_{i3})_{AB}$ where $i = 1$ or 2, and $A, B = L$ or R, and is independent of the kaon sector we considered here in the mass insertion approximation.

Acknowledgments

I am grateful to S. Baek, J.-H. Jang, Y.G. Kim, J.S. Lee and J.H. Park for enjoyable collaborations on the works presented in this talk, and also to W.S. Hou and H.Y. Cheng for their nice organization of this conference. This work is supported in part by grant No. 1999-2-111-002-5 from the interdisciplinary research program of the KOSEF and BK21 project of Ministry of Education.

References

1. W. Bernreuther and M. Suzuki, Rev. Mod. Phys. **63**, 313 (1991).
2. A.G. Cohen, D.B. Kaplan, A.E. Nelson, Phys. Lett. **B388**, 588 (1996).
3. J.M. Cline and G.D. Moore, Phys. Rev.Lett. **81**, 3315 (1998).
4. Seungwon Baek and P. Ko, Phys. Rev. Lett. **83**, 488 (1999).
5. Seungwon Baek and P. Ko, Phys. Lett **B462**, 95 (1999).
6. S. Baek, J.-H. Jang, P. Ko and J.H. Park, KAIST-TH 99/04, hep-ph/9907572.
7. A. G. Cohen, David B. Kaplan, F. Lepeintre, Ann E. Nelson, Phys. Rev. Lett.**78**, 2300 (1997).
8. Y.G. Kim, P. Ko and J.S. Lee, Nucl. Phys. **B 544**, 64 (1999).
9. D. Chang, W.-Y. Keung and A. Pilaftsis, Phys. Rev. Lett. **82**, 900 (1999).
10. A. Kagan and M. Neubert, Phys. Rev. **D 58**, 094012 (1998) ; Eur. Phys. J. **C 7**, 5 (1999).
11. T. Goto, Y. Okada and Y. Shimizu, Phys. Rev. **D 58**, 094006 (1998).
12. S.A. Abel and J.-M. Frere, Phys. Rev. **D 55**, 1623 (1997).
13. A. Masiero and H. Murayama, Phys. Rev. Lett. **83**, 907 (1999).
14. M. Brhlik, G.J. Good and G.L. Kane, hep-ph/9810457.

DIRECT CP VIOLATION IN B DECAYS IN SUPERSYMMETRY WITHOUT R PARITY

GAUTAM BHATTACHARYYA

Saha Institute of Nuclear Physics, 1/AF Bidhan Nagar, Calcutta 700064, India

E-mail: gb@tnp.saha.ernet.in

Supersymmetric models without R-parity predict CP asymmetry in the $B^+ \to \pi^+ K$ channel which is significantly larger than predicted in the standard model. Upcoming B factories will test this scenario.

1 Introduction to direct CP violation

Measurements of CP-violating asymmetries in different B decay modes in the upcoming B factories might reveal new physics with new phases. The purpose of this talk [1] is to demonstrate that if the measured value of *direct* CP-violating asymmetry in the channel $B^+ \to \pi^+ K^0$ (at the quark level $\bar{b} \to \bar{s} d \bar{d}$) turns out to be large, it would constitute an unambiguous signal of new physics.

For the purpose of illustration, we assume that a generic B^+ decay amplitude receives contributions from two diagrams. The amplitude can be written as $A(B^+ \to f) = |A_1| e^{i\phi_1^W} e^{i\phi_1^S} + |A_2| e^{i\phi_2^W} e^{i\phi_2^S}$. Here ϕ_i^W and ϕ_i^S are the weak and strong phases, respectively, for the ith diagram. The CP asymmetry, defined as $a_{CP} \equiv [\mathcal{B}(B^+ \to f) - \mathcal{B}(B^- \to \bar{f})]/[\mathcal{B}(B^+ \to f) + \mathcal{B}(B^- \to \bar{f})]$, can be expressed as

$$a_{CP} = \frac{2|A_1||A_2| \, \sin(\phi_1^W - \phi_2^W) \, \sin(\phi_1^S - \phi_2^S)}{|A_1|^2 + |A_2|^2 + 2|A_1||A_2| \, \cos(\phi_1^W - \phi_2^W) \, \cos(\phi_1^S - \phi_2^S)}. \tag{1}$$

Clearly, in order to produce a large a_{CP} the following conditions will have to be satisfied: (i) $|A_1| \approx |A_2|$, (ii) $\sin(\phi_1^W - \phi_2^W) \approx 1$ and (iii) $\sin(\phi_1^S - \phi_2^S) \approx 1$.

2 The standard model prediction

In the standard model (SM), only penguins contribute to $\bar{b} \to \bar{s} d \bar{d}$. The decay amplitude can be written as

$$A^{\rm SM}(B^+ \to \pi^+ K^0) = -A\lambda^2(1 - \lambda^2/2) \left[1 + \rho e^{i\theta} e^{i\gamma}\right] |P_{tc}| e^{i\delta_{tc}}, \tag{2}$$

where $\lambda = 0.22$ is the Wolfenstein parameter; $A \equiv |V_{cb}|/\lambda^2 = 0.81 \pm 0.06$; $\gamma \equiv -{\rm Arg}(V_{ub}^* V_{ud}/V_{cb}^* V_{cd})$ is the CKM weak phase; θ and δ_{tc} are CP-conserving strong phases; $P_{tc} \equiv P_t^S - P_c^S + P_t^W - P_c^W$ (the difference between top- and charm-mediated strong and electroweak penguins); and finally, ρ depends on the dynamics of the up- and charm-penguins. Using factorization technique, one can express $|P_{tc}| \approx G_F f(\bar{C}_i)\mathcal{F}/\sqrt{2}$, where $f(\bar{C}_i) \approx 0.09$ is an analytic function of the Wilson coefficients, and $\mathcal{F} = (m_{B_d}^2 - m_\pi^2) f_K F_{B\pi}$ with $F_{B\pi} = 0.3$. The NLO estimate of $\mathcal{B}(B^\pm \to \pi^\pm K) \equiv 0.5[\mathcal{B}(B^+ \to \pi^+ K^0) + \mathcal{B}(B^- \to \pi^- \bar{K^0})]$ lies in the range $(1.0 - 1.8) \times 10^{-5}$ for $\rho = 0$.

Using Eqs. (1) and (2), one obtains

$$a_{CP}^{\rm SM} = -2\rho \sin\theta \sin\gamma / (1 + \rho^2 + 2\rho\cos\theta\cos\gamma). \quad (3)$$

In the perturbative limit, $\rho = \mathcal{O}(\lambda^2 R_b) \sim 1.7\%$, where $R_b = |V_{ub}|/\lambda|V_{cb}| = 0.36 \pm 0.08$. On the other hand, rescattering effects, such as, $B^+ \to \pi^0 K^+ \to \pi^+ K^0$, can enhance ρ to as large as $\sim 10\%$, which means a_{CP} can go up to $\sim 20\%$. Therefore, if experiments measure a significantly larger a_{CP}, it would establish an undisputed evidence of new physics.

3 Effects of R-parity violation

3.1 New diagram

In the MSSM, due to the presence of additional penguins, a_{CP} can go up to $\sim 30\%$. But admitting R-parity-violating ($\not{R}$) $\lambda'_{ijk} L_i Q_j D_k^c$ couplings generates new *tree level* diagrams contributing to this process. The key point is that the interference between this lepton-number-violating tree interactions and the SM penguins may trigger large a_{CP}.

3.2 New weak phase

A non-leptonic B decay amplitude involves the product of the type $\lambda'_{ij3}\lambda'^{*}_{ilm}$ which are in general complex. Thus we have a new weak phase whose value is arbitrary.

3.3 New strong phase

The QCD penguin (dominant) operator does not carry any isospin, while the $\not{R}$ tree operator carries an isospin (0 or 1). As a result, the QCD penguin always leads to a isospin 1/2 final state, while the $\not{R}$ tree diagram may yield both isospin 1/2 and 3/2 final states. The final state interaction between states with different isospins may generate a relative strong phase.

3.4 Computation of the new diagram

Consider that λ'_{i13} and λ'_{i12} are the only non-vanishing couplings. This constitutes a sneutrino mediated decay. The new amplitude can be written as

(the negative sign in front is just our convention)

$$A^{\not R}(B^+ \to \pi^+ K^0) = -(|\lambda'_{i13}\lambda'^*_{i12}|/8\tilde{m}^2)\mathcal{F}e^{i\gamma_{\not R}} \equiv -|\Lambda_{\not R}|e^{i\gamma_{\not R}}, \tag{4}$$

where $\gamma_{\not R}$ denotes the weak phase associated with the product of λ's, and $\tilde{m}$ is the mass of the sneutrino of the ith generation, leading to the total amplitude

$$A(B^+ \to \pi^+ K^0) = -A\lambda^2(1-\lambda^2/2)|P_{tc}|e^{i\delta_{tc}}\left(1 + \rho e^{i\gamma}e^{i\theta} + \rho_{\not R}e^{i\gamma_{\not R}}e^{i\theta_{\not R}}\right), \tag{5}$$

where

$$\rho_{\not R} \equiv |\Lambda_{\not R}|/A\lambda^2(1-\lambda^2/2)|P_{tc}|. \tag{6}$$

It is important to note that $\rho_{\not R}$ is free from uncertainties due to factorization. Numerically $\rho_{\not R}$ could be order one.

3.5 The CP *asymmetry*

Assuming $\rho_{\not R} \gg \rho$, one obtains

$$a_{CP} \approx -\frac{2\rho_{\not R}\sin\theta_{\not R}\sin\gamma_{\not R}}{1 + \rho_{\not R}^2 + 2\rho_{\not R}\cos\theta_{\not R}\cos\gamma_{\not R}}. \tag{7}$$

We first choose $\tilde{m} = 100$ GeV throughout our analysis. Employing the current upper limits [2] on $\lambda'_{i13}\lambda'_{i12}$, we obtain, for $i = 1$, 2, and 3, $\rho_{\not R} \lesssim 0.17$, 3.45, and 4.13, respectively. Therefore, it is possible to arrange $\rho_{\not R} = 1$ (for $i = 2, 3$). This implies that a 100% CP asymmetry is very much attainable, requiring $\gamma_{\not R} = \theta_{\not R} = \pi/2$. *We assert that such a drastic hike of* CP *asymmetry is hard to obtain from any other sources of new physics.*

The minimum $\rho_{\not R}$ required to generate a given a_{CP} is given by (for $\rho = 0$)

$$\rho_{\not R} \gtrsim \left(1 - \sqrt{1 - a_{CP}^2}\right)/|a_{CP}|. \tag{8}$$

Eq. (8) has been derived by minimizing $\rho_{\not R}$ with respect to $\gamma_{\not R}$ and $\theta_{\not R}$ for a given a_{CP}. Numerically,

$$\rho_{\not R} \gtrsim 1.0\ (1.0), 0.50\ (0.8), 0.33\ (0.6), 0.21\ (0.4), 0.10\ (0.2); \tag{9}$$

where the numbers within brackets refer to the corresponding a_{CP}'s.

3.6 New bounds on λ' product couplings

The experimental upper limit on $\mathcal{B}(B^\pm \to \pi^\pm K)$ constrains the size of $\rho_{\not R}$. According to the latest CLEO measurement, $\mathcal{B}^{\rm exp} \leq 1.9 \times 10^{-5}$ (1σ) [3]. On the other hand, $\mathcal{B}^{\rm SM} \sim (1.0 - 1.8) \times 10^{-5}$ in view of the present uncertainty of the SM. Hence one can accommodate a multiplicative new physics effect by at most a factor 1.9 (at 1σ). One can check from the denominator of Eq. (7) that $\not{R}$ effects modify the SM prediction of the branching ratio by a multiplicative factor $(1 + \rho_{\not{R}}^2 + 2\rho_{\not{R}} \cos\theta_{\not{R}} \cos\gamma_{\not{R}})$. To evaluate the maximum value of $\rho_{\not{R}}$, we set one of the two angles appearing in this factor to zero and the other to π. This leads to the conservative upper bound

$$\rho_{\not{R}} \lesssim 2.4; \quad \text{which implies} \quad |\lambda'_{i13}\lambda'^{*}_{i12}| \lesssim 5.7 \times 10^{-3} \quad (1\sigma). \tag{10}$$

For $i = 2, 3$, these limits are already stronger than the existing ones.

4 Extraction of the CKM parameter γ

$B \to \pi K$ decays are expected to provide useful information on γ. But we have seen that some $\not{R}$ couplings change the theoretical prediction for this branching ratio. So unless the new physics effects are properly taken into account, this might lead to a wrong estimate of γ. The best way to tackle this problem is to consider different channels for measuring γ and compare their consistency.

5 Conclusions

We identified one distinctive feature of R-parity violation that it can enhance both the CP asymmetry and the branching ratio simultaneously. As a result it would be easier to capture these effects experimentally. Our study is a prototype analysis in a particular channel, which can be extended to a multi-channel analysis combining all kinds of $B \to \pi\pi, \pi K, DK$ modes. Who knows whether CP violation in B physics might turn out to be a Trojan horse?!

Acknowledgements

I thank Amitava Datta for a very fruitful collaboration. I also thank the organizers of BCP3 for a stimulating meeting they organized at Taipei.

References

1. This talk is based on G. Bhattacharyya and A. Datta, *Phys. Rev. Lett.* **83**, 2300 (1999). See references therein.
2. G. Bhattacharyya, hep-ph/9709395, Invited talk at 'BEYOND 97', Ringberg Castle, Tegernsee, Germany, June 1997.
3. CLEO Collaboration, M. Artuso *et al.*, CLEO-CONF 98-20 (presented at ICHEP'98, Vancouver, 1998).

CP VIOLATION INDUCED BY THE R-L DOWN SQUARK MIXING

CHUN-KHIANG CHUA
Physics Department, National Taiwan University, Taipei, Taiwan 10764, R.O.C.

We show that the squark mixings in the SUSY model with an abelain horizontal symmetry and non-standard soft breaking terms has interesting CP asymmetry in $B \to \rho\gamma$, $\omega\gamma$ decay. The mixing induced radiative B decay asymmetry is maximal and the decay rate can also be enhanced up to a factor of 2. Different sensitivities of LR(RL) sector in the decay amplitude and in the box diagram make it possible to satisfy existing $\Delta m(B_d)$ data while interesting results in B decay still survive.

It is exciting that the CP asymmetry in $B^0 \to J/\psi K_S^0$ has been observed in several exprimental groups [1]. It is well known that the asymmetry $a_{J/\psi K_S^0} = \sin 2\beta \sin \Delta m_d t$ contains the fundamental CP violation parameter $\sin 2\beta$. The CDF collaboration gives $\sin 2\beta = 0.79^{+0.41}_{-0.44}$, while the ALEPH collaboration gives, $\sin 2\beta = 0.93^{+0.64+0.36}_{-0.08-0.24}$. The mixing induced CP violation in B decay is observed. It is a good time to study other related CP violation in B decay systems, especially those are standard model suppressed, to look for a sign of new physics. In models beyond the SM interesting CP violating asymmetry such as $a_{\rho\gamma}$ in mixing induced radiactive B decay can occur [2]. It has some similar feature related to the $J/\psi K_s^0$ golden mode. Firstly, the hadronic uncertainties factor out due to the CP self-conjugate hadronic final eigenstate(s). Secondly, they need B and $\bar{B}$ to decay to the same final state where the inteference effect can take place. Therefore the formular of the asymmetry $a_{M^0\gamma}$, where M^0 is a CP self-conjugate hadron such as ρ^0, w, ϕ, K^{*0}, resamble that of $a_{J/\psi K_s^0}$. One has $a_{M^0\gamma} = \xi \sin 2\vartheta \sin[2\beta - \phi(C_{7\gamma}) - \phi(C'_{7\gamma})] \sin \Delta m\, t$ and $\sin 2\vartheta \equiv 2|C_{7\gamma}C'_{7\gamma}| \,/\, (|C_{7\gamma}|^2 + |C'_{7\gamma}|^2)$, where ξ is the CP eigenvalue of M^0, $C^{(\prime)}_{7\gamma}$ is the coefficient of magnetic dipole interaction in $b \to d_{L(R)}\gamma$ and $\phi(C^{(\prime)}_{7\gamma})$ is the phase of $C^{(\prime)}_{7\gamma}$. Note that one needs both $C_{7\gamma}$ and $C'_{7\gamma}$ for the interference to occur, therefore the asymmetry is suppressed in the standard model and hence sensitive to new physics. Furthermore, the asymmetry in $B \to \rho^0\gamma$ is more accessible then that in $B \to K^{*0}\gamma$, due to the vertex tagging for the time dependence information [2].

In supersymmetry (SUSY), the gluino-squark interaction contributes to neutral current flavor changing process due to squark mixings [3]. For almost degenerate squarks the contributions to $b \to d\gamma$ can be parametrize by the squark mixing angle δ in the quark mass eigenstate basis, $\delta^{ij}_{AB} \equiv$

$(V_A^\dagger\, \widetilde{m}^2_{AB} V_B)^{ij}/\widetilde{m}^2$, where $A, B = L, R$, $\widetilde{m}$ is the squark average mass, i, j are generation indices and $V_{A,B}$ are quark mixing matrices. The squark contributions on $C^{(\prime)}_{7\gamma}$ can be found in Ref. [3,4]. There is a enhancement factor $M_{\tilde{g}}/m_b$ comes with $\delta_{LR,RL}$. The factor m_b shows the normalization with respect to the SM result. Similar machanism has been discussed in Ref. [5].

To obtain large asymmetry $a_{\rho\gamma}$ we require a comparable size of $C_{7\gamma}$ and $C'_{7\gamma}$. We need large δ^{13}_{RR} and/or $\delta^{13}_{RL}\, M_{\tilde{g}}/m_b$ and small δ_{LL} and $\delta^{13}_{LR}\, M_{\tilde{g}}/m_b$ to enhance $C'^{\,\mathrm{New}}_{7\gamma}$ without much enhancement on $C_{7\gamma}$. This is indeed the case in supersymmetric models with abelian horizontal symmetry. Abelian horizontal symmetries are used to give hierarchy patterns in the quark masses and mixing angles [6,7]. In these models the mixing angles in $V_{u,dR}$ are in general greater than those in $V_{u,dL}$. The mixing angles in $V_{u,dL}$ are retricted by the CKM matrix and cannot be too large when 3rd generation quarks are involved. The hierachy pattern in squark mass matrices are related to quark mass matrices, $\widetilde{M}^{2ij}_{u(d)LR} \sim \widetilde{m}\, M^{ij}_{u(d)}$. One should note that the values of these matrix elements are order of magnitude estimations.

We get order of magnitude in δ s of down type squark: $\delta^{13}_{LL} \sim \lambda^3$, $\delta^{13}_{RR} \sim \lambda$, while $\delta^{13}_{LR}\,\frac{M_{\tilde{g}}}{m_b} \sim \lambda^3$, $\delta^{13}_{RL}\,\frac{M_{\tilde{g}}}{m_b} \sim \lambda$. Therefore $\delta^{13}_{RR} \sim \delta^{13}_{RL}\, M_{\tilde{g}}/m_b$ is larger then $\delta_{LL} \sim \delta^{13}_{LR}\, M_{\tilde{g}}/m_b$ from the abelian symmetry. By taking the average squark mass $\widetilde{m}$ be 300 GeV and the gluino mass $M_{\tilde{g}}$ be 1 TeV, this model indeed gives maximal $\sin 2\vartheta$ and the decay rate can be enhanced by a factor of 2. There is however some servere low energy constrains, especially due to kaon mixing [8]. The SUSY contribution to the $\Delta S = 2$ effective Hamiltonian (and similar expressions for $D - \bar{D}$ and $B - \bar{B}$ mixing) and LO QCD corrected one are given by Ref. [4] and Ref. [9], respectively. There is no $m_{\tilde{g}}/m_b$ enhacement factor in box diagram contributions. This is due to the fact that one does not need massive external quark in the SM to give a non-vanishing box diagram. These different sensivities of $\delta_{LR,RL}$ in $b \to d\gamma$ and in $B_d - \bar{B}_d$ mixing make it possible for SUSY to have significant contribution in the $b \to d\gamma$ while not violating the almost SM saturated $\Delta M(B_d)$ data.

In the SUSY model with the abelian horizonal symmetry, the $\delta^{12}_{LL,RR} \sim \lambda$ give too large contributions to Δm_K and ϵ. The box diagram is more sensitive to $\delta_{LL}\delta_{RR}$, less sensitive to $\delta^2_{LR}(\delta_{LR}\delta_{RL})$ and much less sensitive to $\delta^2_{LL,RR}$. The main point to suppress the SUSY contribution is to have zero $\delta_{LL}\delta_{RR}$. This can be done by using the quark-squark aligment (QSA) method which make use of the holomorphy in SUSY models [7]. One then has suppressed V^{12}_{dL} and δ^{12}_{LL}, which give small contribution to kaon mixing. On the other hand the suppression of V^{12}_{dL} implies $V^{12}_{uL} \sim V^{12}_{CKM}$. This will generate $D - \bar{D}$ mixing due to up squark mixing, which is a common feature to QSA models [7]. A recent search for $D^0 - \bar{D}^0$ mixing suggest that $x \equiv \Delta m_D/\Gamma$ is less then 3.2% at 95% C.L. [10]. In Fig. 1(a) we show X_D with gluino mass at 3 TeV, with few TeV squark average mass ($\widetilde{m}$), it is possible to have x of a few percent.

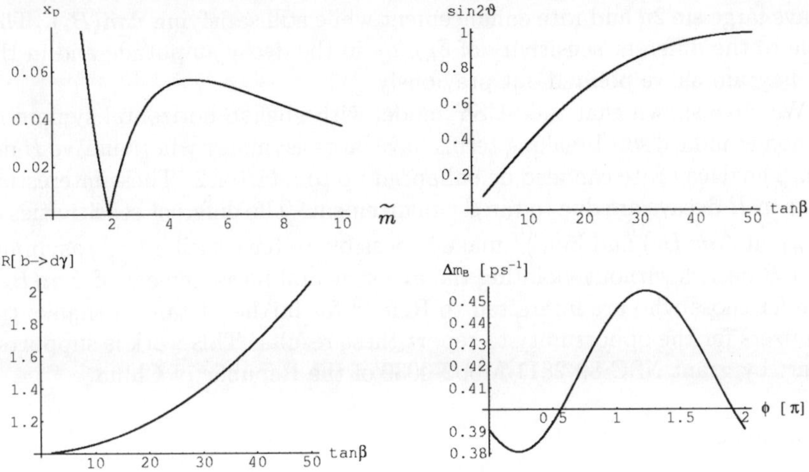

Figure 1. The figure on upper left, upper right, lower left and lower right corners are denoted as 1(a), 1(b), 1(c) and 1(d), respectively. They are explained in the text.

Note that a zero can be acheived when $\widetilde{m} = 1925$ GeV. By choicing mass parameters arround this point one also has neglecgible contributions on kaon mixing. We find that there is a $\tan\beta$ enhancement on $C_{7\gamma}^{(\prime)\mathrm{New}}$ due to non-standard soft breaking terms to give interesting CP asymmetry even with these heavy masses. Non-standard soft breaking terms can survive without inducing quadratic divergent if there is no gauge single particle in the spectrum, as those in the minimal supersymmetry standard model (MSSM) [12]. In particular, the non-standard soft breaking terms includes $\phi_i^*\phi_j\phi_k$ which is called the C term. The $\widetilde{M}^2_{dLR}$ now contains a (standard) soft breaking A term $A_d < H_d > Y'Q\bar{d}$ and also a non-standard term, the C term, $C < H_u^* > Y'Q\bar{d}$. The natural value of A_d as well as C is $\widetilde{m}$. Therefore by adding the C term, one has a $\tan\beta$ enhancement on $\widetilde{M}^2_{dLR,RL}$. In Fig. 1(b) and 1(c), we show the $\sin 2\vartheta$ and $Br(B \to X_d\gamma)/Br(B \to X_d\gamma)_{\mathrm{SM}}$ with respect to $\tan\beta$, respectively. We see that the maximal value of $\sin 2\vartheta$ can be reached and the rate can be enhanced up to a factor 2 for large $\tan\beta$. These make the asymmetry $a_{\rho\gamma}$ even more interesting. We note that the constraint from $Br(B_s \to X_s\gamma)$ can also be evaded by using the quark-squark aligment mechanism. In Fig. 1(d) we show the $\Delta m(B_s)$ including the SM and the SUSY contributions with respect to a phase (ϕ) assigned in δ_{LL}^{13}. For the SM part, we use $|V_{td}| = 8 \times 10^{-3}$ and $\gamma = \deg 65$. We see that the $\Delta m(B_d)$ has a lot of parameter space in ϕ to lie in the eperimental value $\Delta m_{expt} = 0.470 \pm 0.019\,\mathrm{ps}^{-1}$ [11]. It is quit amazing to have large $\sin 2\vartheta$ and rate enhancement while still satisfying $\Delta m(B_s)$. This is due to the different sensitivity of $\delta_{LR,RL}$ in the decay amplitude and in the box diagram as we pointed out previously.

We have shown that the SUSY model with abelian horizontal symmetry and non-standard soft breaking terms have large asymmetry in radiative B decays. The decay rate can also be enhanced up to a factor 2. These interesting effects in B decays are due to $\tan\beta$ enhancement. The different sensitivities of $\delta_{LR,RL}$ in $\Delta m(B_d)$ and in $C_{7\gamma}^{(\prime)}$ make it possible to have such a large asymmetry in B decays without violating the experimental measurement of $\Delta m(B_d)$. We refer those who are interested to Ref. [13] for further details. I thanks the organizers for the opportunity to report these results. This work is supported in part by grant NSC 89-2811-M002-0039 of the Republic of China.

References

1. T. Affolder *et al.*, CDF collaboration, hep-ex/9909003; R. Forty, ALEPH collaboration, talk presented on BCP3 (Taipei, 1999).
2. D. Atwood, M. Gronau and A. Soni, *Phys. Rev. Lett.* **79**, 185 (1997).
3. S. Bertolini, F. Borzumati and A. Masiero, *Nucl. Phys.* B **284**, 321 (1987); S. Bertolini *et al.*, *ibid.* B **353** 591 (1991).
4. F. Gabbiani et al., *Nucl. Phys.* B **477**, 321 (1996).
5. K. Fujikawa and A. Yamada, *Phys. Rev.* D **49**, 5890 (1994);
6. C.D. Froggatt and H.B. Nielsen, *Nucl. Phys.* B **147**, 277 (1979); M. Leurer, Y. Nir and N. Seiberg, *Nucl. Phys.* B **398**, 319 (1993).
7. Y. Nir and N. Seiberg, *Phys. Lett.* B **309**, 337 (1993); M. Leurer, Y. Nir and N. Seiberg, *Nucl. Phys.* B **420**, 468 (1994).
8. F. Gabbiani and A. Masiero, *Nucl. Phys.* B **322**, 235 (1989); J.S. Hagelin, S. Kelley and T. Tanaka,*Nucl. Phys.* B **415**, 293 (1994); E. Gabrielli, A. Masiero and L. Silvestrini,*Phys. Lett.* B **374**, 80 (1996).
9. J.A. Bagger, K.T. Matchev, R.-J. Zhang, *Phys. Lett.* B **412**, 77 (1997).
10. H.N. Nelson (CLEO Collaboration), hep-ex/9909028, presented at 1999 Chicago Conference on Kaon Physics (K 99), Chicago, IL, Jun 1999.
11. Particle Data Group, C. Caso *et al.*, *Eur. Phys. J.* C **3**, 1 (1998).
12. L.J. Hall and L. Randall, *Phys. Rev. Lett.* **65**, 2939 (1990).
13. C.-K. Chua and W.-S. Hou, in preparation.

T VIOLATION AND OTHER CP/CPT STUDIES

T. YAMANAKA
Physics Department, Osaka University, Toyonaka, Osaka 560-0043, Japan
E-mail: taku@hep.sci.osaka-u.ac.jp

Recent results from kaon experiments to study T violation, CP violation, and CPT violation are described.

Kaons are useful tools to study various symmetry violations. Even now, the kaon system is the only system where CP violation has been observed definitively. This talk will describe recent results from T violation, CPT violation, and CP violation studies using kaons.

1 T Violation in $K^+ \to \pi^0 \mu^+ \nu$

The transverse polarization of muons in $K^+ \to \pi^0 \mu^+ \nu$ decay is a sensitive probe to search for T violation caused by new physics beyond the Standard Model. The transverse polarization is expressed as:

$$P_T = \vec{\sigma}_\mu \cdot (\vec{p}_{\pi^0} \times \vec{p}_\mu)/|(\vec{p}_{\pi^0} \times \vec{p}_\mu)|, \quad (1)$$

where σ_μ is the muon spin, $\vec{p}_{\pi^0}$ and $\vec{p}_\mu$ are the momentum of π^0 and μ^+, respectively. Experiment E246 at KEK has recently measured P_T. The 660MeV/c K^+'s were stopped in a target surrounded by a cylindrical detector. The π^0 from the decay was detected by the CsI photon detector. The muon from the decay passed through a hole of the CsI array, and after its momentum was analyzed by a toroidal magnet, it stopped in a stack of pure Al plates. The transverse component of the muon spin was held by a fringing field of the magnet, and the number of e^+'s from the muon decay was measured by scintillation counters located on the both sides of the Al plates. Most of the systematic errors canceled due to the cylindrical geometry and by comparing events with π^0 in the forward versus backward directions.

The transverse polarization was determined to be [1]: $P_T = (-4.2 \pm 4.9(stat) \pm 0.9(syst)) \times 10^{-3}$, where the major systematic errors were due to the understanding of the magnetic field, analysis cuts, and chamber alignments. The upper limit on the $|P_T|$ was set to 0.011 with the 90% confidence level.

2 T Violation in $K^0\ \overline{K^0}$ Mixing

Most of the CP violating effects which have been observed so far are due to an unequal mixture of K^0 and $\overline{K^0}$ amplitudes in the K_L state. This stable unbalance is caused by a slight difference in the transition rates between $K^0 \to \overline{K^0}$ and $\overline{K^0} \to K^0$.

The CPLEAR experiment at CERN observed this difference directly for the first time. The 200MeV/c antiprotons were stopped and annihilated inside a gaseous hydrogen target surrounded by tracking devices, Cherenkov/scintillation counters for particle identification, and a lead/gas sampling electromagnetic calorimeter.

The initial state of the kaon was tagged in $\overline{p}+p \to \pi^+K^-K^0$ and $\overline{p}+p \to \pi^-K^+\overline{K^0}$ interactions. The state of the neutral kaon at the time of the decay was tagged by semileptonic decays, $K^0 \to \pi^-e^+\nu_e$ and $\overline{K^0} \to \pi^+e^-\overline{\nu_e}$. The measured asymmetry between the transition rates was [2]:

$$\begin{aligned} A_T &= \frac{R(\overline{K^0}_{(t=0)} \to e^+\pi^-\nu_{(t=\tau)}) - R(K^0_{(t=0)} \to e^-\pi^+\overline{\nu}_{(t=\tau)})}{R(\overline{K^0}_{(t=0)} \to e^+\pi^-\nu_{(t=\tau)}) + R(K^0_{(t=0)} \to e^-\pi^+\overline{\nu}_{(t=\tau)})} \\ &= (6.6 \pm 1.3(stat) \pm 1.0(syst)) \times 10^{-3}, \end{aligned}$$

in the region $\tau_S < \tau < 20\tau_S$, where τ_s is the lifetime of K_S, and $R(\overline{K^0}_{(t=0)} \to e^+\pi^-\nu_{(t=\tau)})$ is the rate for $\overline{K^0}$ production at t=0 and decay to $e^+\pi^-\nu$ at t=τ, and vice versa. Since this means that the transition rates are different if the initial state and the final state are exchanged, this is the first direct observation of T-violation. A later analysis using Bell-Steinberger relations gave [3]

$$A_T = (6.60 \pm 0.10) \times 10^{-3} - (0.8 \pm 1.2) \times 10^{-3},$$

where the first term is a measurement of 4Re(ϵ), and the second term is the size of T violation beyond the contribution from CP violation.

3 CPT Violation

The violation of CPT was also searched for by the CPLEAR experiment. The whole analysis is rather complicated, so I shall just describe the basic concept here. If the CP and CPT are conserved, then $R(K^0 \to K^0; t)$ which is the rate of kaon produced as K^0 at t=0 decayed as K^0 at t=τ, is exactly the same as its counterpart for $\overline{K^0}$, $R(\overline{K^0} \to \overline{K^0}; t)$. However, if CP is conserved but CPT is violated, then $R(K^0 \to K^0; t)$ and $R(\overline{K^0} \to \overline{K^0}; t)$ can be different. In reality since CP is violated ($\epsilon \neq 0$), to search for CPT violation, one should look for the difference between $R(K^0 \to K^0; t)$ and $R(\overline{K^0} \to \overline{K^0}; t)$ beyond the contribution by the CP violation. The CPT asymmetry, A_δ, was measured to be [3]: $A_\delta = 8 \times (2.4 \pm 2.8) \times 10^{-4}$ based on a global analysis using the Bell-Steinberger relationship.

4 CP Violation in $K_L \to \pi^+\pi^-e^+e^-$

A new CP violating phenomena in the decay $K_L \to \pi^+\pi^-e^+e^-$ was found by the Fermilab KTeV experiment. Let ϕ be the angle between the e^+e^- plane and $\pi^+\pi^-$ planes in the kaon's center of mass frame. The normal vectors of these planes are defined as:

$$\vec{n}_e = \frac{\vec{e}^+ \times \vec{e}^-}{|\vec{e}^+ \times \vec{e}^-|}, \vec{n}_\pi = \frac{\vec{\pi}^+ \times \vec{\pi}^-}{|\vec{\pi}^+ \times \vec{\pi}^-|}, \tag{2}$$

where $\vec{e}$ and $\vec{\pi}$ are the momentum vectors for e and π, respectively. The product,

$$\sin\phi\cos\phi = (\vec{n}_e \times \vec{n}_\pi) \cdot \vec{p}_{\pi\pi}/|\vec{p}_{\pi\pi}| \cdot (\vec{n}_e \cdot \vec{n}_\pi), \tag{3}$$

where $\vec{p}_{\pi\pi}$ is the momentum of the $\pi^+\pi^-$, changes its sign under a CP transformation.

At KTeV, an 800GeV proton beam struck a target to produce neutral kaons with a momentum around 70GeV decaying in the decay region. The momentum of charged particles from the decay were analyzed by four sets of drift chambers and an analysis magnet. The energy and hit position of electrons (and photons) were measured by an electromagnetic calorimeter made of 3100 pure CsI crystals.

KTeV has collected 1811 ± 43 $K_L \to \pi^+\pi^-e^+e^-$ events after subtracting 45 ± 11 background events for this analysis. The asymmetry, $A = (N_+ - N_-)/(N_+ + N_-)$, where $N_+(N_-)$ is the number of events in $\sin\phi\cos\phi > 0 (< 0)$ region, is $23.3 \pm 2.3\%$ for the observed events. Since the experiment is not sensitive to the low m_{ee} region where the asymmetry is small, the asymmetry after correcting for the acceptance is [4] $13.6 \pm 2.5 \pm 1.2\%$.

This is consistent with the 14% asymmetry predicted by a theoretical calculation using the CP violation parameter, ϵ, assuming CPT conservation[5]. This is the first CP violating phenomena observed in a kinematical distribution.

5 CP Violation in $K_L \to \pi^0 e^+e^-, K_L \to \pi^0 e^+e^-, K_L \to \pi^0 \nu\bar{\nu}$

The rare decays, $K_L \to \pi^0 \ell\bar{\ell}, (\ell = e, \mu, \nu)$, are good probes for studying CP violation. One of the major diagrams contributing to these decays is a penguin diagram. In this diagram, the decay amplitude for $K^0 \to \pi^0 \ell\bar{\ell}$ is mostly proportional to V_{td}, and since $|K_L >\simeq 1/\sqrt{2}(|K^0 > -|\overline{K^0} >)$, the amplitude of $K_L \to \pi^0 \ell\bar{\ell}$ is proportional to $Im(V_{td})$ which is the origin of CP violation according to the Standard Model. In the case of $K_L \to \pi^0 e^+e^-$ and $K_L \to \pi^0 \mu^+\mu^-$, there are other contributions involving one or two intermediate photons, in addition to the penguin diagram.

5.1 $K_L \to \pi^0 e^+e^-$

Fermilab KTeV E799-II has searched for $K_L \to \pi^0 e^+e^-$ decays. Events were required to have two electrons and two photons consistent with a π^0, and $m_{ee} > 140 MeV/c^2$ to remove $K_L \to \pi^0\pi^0$ decays with $\pi^0 \to e^+e^-\gamma$. In order to reduce $K_L \to e^+e^-\gamma\gamma$ background which has the same final state particles as the signal, further cuts were made on the minimum angle between a photon and an electron, and the angle between a photon and the π^0 in the $\gamma\gamma$ center of mass frame. After all the cuts were defined, 2 events were found in the signal region where 1.51 background events were expected. Since this is consistent with the background, a preliminary upper limit $BR(K_L \to \pi^0 e^+e^-) < 5.64 \times 10^{-10}$ at the 90% confidence level [6] was set. This is a factor 8 improvement over the previous limit [7].

KTeV has also collected $K_L \to e^+e^-\gamma\gamma$ events, and measured a preliminary branching ratio $BR(K_L \to e^+e^-\gamma\gamma; E_\gamma^* > 5MeV) = (6.31 \pm 0.14(stat) \pm 0.43(syst)) \times 10^{-7}$ based on 1911.4 events after subtracting 76.6 background events [6]. This is by far the best measurement, and the result is consistent with the QED prediction [8] of $BR = 5.8 \times 10^{-7}$.

5.2 $K_L \to \pi^0 \mu^+\mu^-$

Fermilab KTeV E799-II also searched for $K_L \to \pi^0 \mu^+\mu^-$ decays. Two muons were identified by a scintillator hodoscope placed downstream of steel wall totaling 5m in thickness. After cutting backgrounds from $K_L \to \mu^+\mu^-\gamma\gamma$ events, and $K_L \to \pi\mu\nu$ or $K_L \to \pi^+\pi^-\pi^0$ decays where charged pions either decayed or punched through the steel, 2 events were found in the signal region where 0.87 ± 0.15 events were expected. This is still consistent with the background, and a preliminary upper limit on the branching ratio was set to be[9] $BR(K_L \to \pi^0 \mu^+\mu^-) < 3.4 \times 10^{-10}$ at the 90% confidence level. This is a factor 15 improvement over the previous limit[10].

KTeV E799-II has also looked for the $K_L \to \mu^+\mu^-\gamma\gamma$ decay. This is a radiative decay of $K_L \to \mu^+\mu^-\gamma$, and is a major background for the $K_L \to \pi^0\mu^+\mu^-$ decay. They have observed 4 events for the first time, where 0.16±0.08 background events were expected. Based on this observation, a preliminary branching ratio was determined to be[11] $BR(K_L \to \pi^0\mu^+\mu^-; m_{\gamma\gamma} > 1MeV/c^2) = (10.4^{+7.5}_{-5.9} \pm 0.7) \times 10^{-9}$, which is consistent with the QED prediction of $(9.1 \pm 0.8) \times 10^{-9}$.

5.3 $K_L \to \pi^0\nu\overline{\nu}$

The decay $K_L \to \pi^0\nu\overline{\nu}$ is completely dominated by the CP violating amplitude, because unlike $K_L \to \pi^0 e^+e^-$ and $K_L \to \pi^0\mu^+\mu^-$, other contributions involving one or two virtual photons do not exist. Theoretical uncertainty in calculating the branching ratio for a given η is on the order of 1%. Therefore, by measuring the branching ratio of the decay $K_L \to \pi^0\nu\overline{\nu}$, one can directly measure the imaginary part of the CKM matrix, η, in Wolfenstein's parametrization. For example, Buchalla and Buras have pointed out[12] that kaon experiments measuring $K_L \to \pi^0\nu\overline{\nu}$ and $K^+ \to \pi^+\nu\overline{\nu}$ branching ratios to 10% of themselves have comparable sensitivities to $\sin 2\beta$ and η, as B factory experiments.

However, from an experimental point of view, this is a challenging decay mode, since one has to identify the decay by two photons emerging from vacuum, and the major background comes from $K_L \to \pi^0\pi^0$ decay where two photons escaped detection. The expected branching ratio is around 3×10^{-11}.

For now, KTeV has searched for the decay using the existing detector and beamline. They looked for $K_L \to \pi^0\nu\overline{\nu}$ where π^0 decayed to $e^+e^-\gamma$. This way, they could reconstruct the decay vertex, and the mass of the π^0 to identify the decay. A further cut on the transverse momentum of the π^0 removed backgrounds from $\Lambda \to p\pi^0$. They found no events in the signal region where $0.12^{+0.05}_{-0.04}$ background events were expected, and determined the upper limit on the branching ratio to be $BR(K_L \to \pi^0\nu\overline{\nu}) < 5.9 \times 10^{-7}$ at the 90% confidence level [13].

There are experiments at BNL, FNAL, and KEK currently being planned to observe and measure the branching ratio of this decay in the future.

6 Summary

Kaon physics has been pushing the frontier of CP violation, T violation, and CPT violation studies, and many new results have come out recently. The kaon decays will continue to be sensitive probes for searching for T and CPT violations, and to measure the CP violating parameter, η, in the CKM matrix.

References

1. M.Abe *et al.*, *Phys. Rev. Lett.* **83**, 4253 (1999).
2. Angelopoulos *et al.*, *Phys. Lett.* B **444**, 43 (1998).
3. Apostelakis *et al.*, *Phys. Lett.* B **456**, 297 (1999).
4. A. Alavi-Harati *et al.*, *Phys. Rev. Lett.* **84**, 408 (2000).
5. L.M. Sehgal and M. Wanninger, *Phys. Rev.* D **46**, 1035 (1992); *ibid.* **D46**, 5209(E)(1992).
6. T.Yamanaka, presented at XXXIVth Rencontres de Moriond, 1999.
7. D.A. Harris, *et al.*, *Phys. Rev. Lett.* **71**, 3918 (1993).
8. H.B.Greenlee, *Phys. Rev.* D **42**, 3724 (1990).
9. A.Alavi-Harati *et al.*, submitted to Phys. Rev. Lett., hep-ex/0001006.
10. D.A. Harris, *et al.*, *Phys. Rev. Lett.* **71**, 3914 (1993).
11. A.Alavi-Harati *et al.*, submitted to Phys. Rev. Lett., hep-ex/0001005.
12. G.Buchalla and A.J.Buras, *Phys. Rev.* D **54**, 6782 (1996).
13. A.Alavi-Harati *et al.*, To be published in Phys. Rev. D..

CP VIOLATION IN HYPERON DECAYS

S. PAKVASA
Department of Physics
University of Hawaii at Manoa
Honolulu, HI 96822
Email: pakvasa@phys.hawaii.edu

The theory and phenomenology of CP violation in hyperon decays is summarized.

1 Introduction

The CPT theorem was proved in 1955[1] and soon thereafter Lüders and Zumino[2] deduced from it the equality of masses and lifetimes between particles and anti-particles. In 1958 Okubo[3] observed that CP violation allows hyperons and antihyperons to have different branching ratios into conjugate channels even though their total rates must be equal by CPT. Somewhat later, this paper inspired Sakharov[4] to his famous work on cosmological baryon-antibaryon asymmetry. In fact, he called this the "Okubo effect", perhaps a better phrase than the current dull use of "direct CP violation." Pais[5] extended Okubo's proposal to asymmetry parameters in Λ and $\bar{\Lambda}$ decays. The subject was revived in the '80s and a number of calculations were made[6,7]. Only now, over 40 years after Okubo's paper, are these proposals about to be tested in the laboratory.

The reason for the current interest is the need to find CP violation in places other than just $K_L - K_S$ complex. Only a number of different observations of CP violation in different channels will help us pin down the source and nature of CP violation in or beyond the standard model (SM). From this point of view, hyperon decay is one more weapon in our arsenal in addition to the K system, the B system, the D system, etc.

2 Phenomenology of Hyperon Decays

I summarize here the salient features of the phenomenology of non-leptonic hyperon decays[8]. Leaving out Ω^- decays, there are seven decay modes $\Lambda \to N\pi$, $\Sigma^\pm \to N\pi$ and $\Xi \to \Lambda\pi$. The effective matrix element can be written as

$$i\ \bar{u}_{\bar{p}}(a + b\gamma_5)u_\Lambda\ \phi \tag{1}$$

for the mode $\Lambda \to p + \pi^-$, where a and b are complex in general. The corresponding element for $\bar{\Lambda} \to \bar{p} + \pi^+$ is then:

$$i\ \bar{v}_{\bar{p}}(-a^* + b^*\gamma_5)v_{\bar{\Lambda}}\phi^+ \tag{2}$$

It is convenient to express the observables in terms of S and P and write the matrix element as

$$S + P\ \sigma.\hat{\mathbf{q}} \tag{3}$$

where $\mathbf{q}$ is the proton momentum in the Λ rest frame and S and P are:

$$S = a\sqrt{\frac{\{(m_\Lambda + m_p)^2 - m_\pi^2\}}{16\pi\ m_\Lambda^2}}$$

$$P = b\sqrt{\frac{\{(m_\Lambda - m_p)^2 - m_\pi^2\}}{16\pi\ m_\Lambda^2}} \tag{4}$$

In the Λ rest-frame, the decay distribution is given by:

$$\begin{aligned}\frac{d\Gamma}{d\Omega} = \frac{\Gamma}{8\pi}\{[1 + \alpha < \sigma_\Lambda > .\hat{\sigma}] \\ + < \sigma_p > .[(\alpha + < \sigma_\Lambda > .\hat{\mathbf{q}})\hat{\mathbf{q}} + \beta < \sigma_\Lambda > \times\hat{\mathbf{q}} \\ + \gamma(\hat{\mathbf{q}} \times (< \sigma_\Lambda > \times\hat{\mathbf{q}})]\}\end{aligned} \tag{5}$$

where Γ is the decay rate and is given by:

$$\Gamma = 2 \mid \mathbf{q} \mid \{\mid S \mid^2 + \mid P \mid^2\} \tag{6}$$

α, β and γ are given by

$$\alpha = \frac{2Re(S^*P)}{\{\mid S \mid^2 + \mid P \mid^2\}},$$

$$\beta = \frac{2Im(SP^*)}{\{\mid S \mid^2 + \mid P \mid^2\}}$$

$$\gamma = \frac{\{\mid S \mid^2 - \mid P \mid^2\}}{\{\mid S \mid^2 + \mid P \mid^2\}} \tag{7}$$

For a polarized Λ, the up-down asymmetry of the final proton is given by α(α is also the longitudinal polarization of the proton for an unpolarized Λ). β and γ are components of the transverse polarization of proton[9].

The observed properties of the hyperon decays can be summarised as: (i) the $\Delta I = 1/2$ dominance i.e. the $\Delta I = 3/2$ amplitudes are about 5% of the

$\Delta I = 1/2$ amplitudes; (ii) the asymmetry parameter α is large for Λ and Ξ decays, Ξ decays and $\Sigma^+ \to p\pi^0$ and is near zero for $\Sigma^\pm \to n\pi^\pm$; and (iii) the Sugawara-Lee triangle sum rule $\sqrt{3}A(\Sigma^+ \to p\pi^0) - A(\Lambda \to p\pi^-) = 2A(\Xi \to \Lambda\pi^-)$ is satisfied to a level of 5% in both s and p wave amplitudes.

3 CP Violating Observables

Let a particle P decay into several final states f_1, f_2 etc. The amplitude for P $\to f_1$ is in general:

$$A = A_1 e^{i\delta 1} \quad + A_2\ e^{i\delta 2} \tag{8}$$

where 1 and 2 are strong interaction eigenstates and δ_i are corresponding final state phases. Then the amplitude for $\bar{P} \to \bar{f}_1$ is

$$\bar{A} = A_1^* e^{i\delta 1} \quad + A_2^*\ e^{i\delta 2} \tag{9}$$

If $| A_1 | >> | A_2 |$, then the rate asymmetry $\Delta (= (\Gamma - \bar{\Gamma})/(\Gamma + \bar{\Gamma}))$ is given by:

$$\Delta \approx -2 \mid A_2/A_1 \mid sin(\phi_1 - \phi_2) sin(\delta_1 - \delta_2) \tag{10}$$

where $A_i = \mid A_i \mid e^{i\phi_i}$. Hence, to get a non-zero rate asymmetry, one must have (i) at least two channels in the final state, (ii) CPV weak phases must be different in the two channels, and (iii) <u>unequal</u> final state scattering phase shifts in the two channels[6]. A similar calculation of the asymmetry of α[10] shows that for a single isospin channel dominance,

$$A = (\alpha + \bar{\alpha})/(\alpha - \bar{\alpha}) = 2tan(\delta_s - \delta_p)\ tan(\phi_s - \phi_p) \tag{11}$$

In this case the two channels are orbital angular momenta 0 and 1; and even a single isospin mode such as $\Xi^- \to \Lambda\pi^-$ can exhibit a non-zero A. In B decays an example of a single isospin mode exhibiting CP violating rate asymmetry is $B \to \pi\pi$, i.e. In this case the two eigen-channels with different weak CP phases and different final state phases are $B \to D\overline{D}\pi \to \pi\pi$ and $B \to \pi\pi \to \pi\pi$[11].

To define the complete set of CP violating observables, consider the example of the decay modes $\Lambda \to p\pi^-$ and $\bar{\Lambda} \to \bar{p}\pi^+$. The amplitudes are:

$$S = -\sqrt{\frac{2}{3}} S_1 e^{i(\delta_1 + \phi_1^s)} + \frac{1}{\sqrt{3}} S_3 e^{i(\delta_3 + \phi_3^s)}$$
$$P = -\sqrt{\frac{2}{3}} P_1 e^{i(\delta_{11} + \phi_1^p)} + \frac{1}{\sqrt{3}} P_3 e^{i(\delta_3 + \phi_3^p)} \tag{12}$$

where S_i, P_i are real, i refers to the final state isospin (i=2I) and ϕ_i are the CPV phases. With the knowledge that S_3/S_1, $P_3/P_1 << 1$; one can write[12,13]

$$\begin{aligned}\Delta_\Lambda &= \frac{(\Gamma-\overline{\Gamma})}{(\Gamma+\overline{\Gamma})} \cong \sqrt{2}\ (S_3/S_1) sin(\delta_3-\delta_1) sin(\phi_3^s-\phi_1^s) \\ A_\Lambda &= \frac{(\alpha+\overline{\alpha})}{(\alpha-\overline{\alpha})} \cong -tan(\delta_{11}-\delta_1) tan(\phi_1^p-\phi_1^s) \\ B_\Lambda &= \frac{(\beta+\overline{\beta})}{(\beta-\overline{\beta})} \cong cot(\delta_{11}-\delta_1) tan(\phi_1^p-\phi_1^s) \end{aligned} \tag{13}$$

The last one B_Λ has the peculiar feature that it blows up as the phase shift difference vanishes. The reason is that in the limit of CP conservation $\beta+\overline{\beta}=0$ but in the limit of no final state phase difference $\beta-\bar{\beta}=0$. For πN final states, the phase shifts at $E_{c.m.}=m_\Lambda$ are known and are: $\delta_1=6^0$, $\delta_3=-3.8^0$, $\delta_{11}=1.1^0$ and $\delta_{31}=-0.7^0$ from the 1965 analysis of Roper et al.[14] with errors estimated at 10%. The CPV phases ϕ_i have to be provided by theory.

Similar expressions can be written for other hyperon decays. For example, for $\Lambda \rightarrow n\pi^0$, Δ is $-2\Delta_\Lambda$ and A and B are identical to A_Λ and B_Λ. For $\Xi^- \rightarrow \Lambda\pi^-$ (and $\Xi^0 \rightarrow \Lambda\pi^0$) the asymmetries are[13]:

$$\begin{aligned}\Delta_\Xi &= 0 \\ A_\Xi &= -tan(\delta_{21}-\delta_2) tan(\phi^p-\phi^s) \\ B_\Xi &= cot(\delta_{21}-\delta_2) tan(\phi^p-\phi^s)\end{aligned} \tag{14}$$

where δ_{21} and δ_2 are the p and s-wave $\Lambda\pi$ phase shifts at m_Ξ respectively. Somewhat more complicated expressions can be and have been written for Σ decays[13].

4 Calculating CP Phases

In standard model description of the non-leptonic hyperon decays, the effective $\Delta S=1$ Hamiltonian is

$$H_{eff} = \frac{G_F}{\sqrt{2}}\, U_{ud}^*\, U_{us} \sum_{i=1}^{12} c_i(\mu)\; O_i(\mu) \tag{15}$$

after the short distance QCD corrections (LLO + NLLO) where $c_i = z_i + y_i\tau(\tau = -U_{td}\, U_{ts}^*/U_{ud}\, U_{us})$, and $\mu \sim 0(1 \text{ GeV})^{15}$. For CP violation, the most important operator is:

$$O_6 = \bar{d}\ \lambda_i \gamma_\mu (1+\gamma_5) s \bar{q} \lambda_i \gamma_\mu (1-\gamma_5) q \tag{16}$$

and $y_6 \approx -0.1$ at $\mu \sim 1\ GeV$. To estimate the CP phases in Eq. (12), one adopts the following procedure. The real parts(in the approximation that the imaginary parts are very small) are known from the data on rates and asymmetries. The real parts of the amplitudes have also been evaluated in SM with reasonable success with some use of chiral perturbation theory(current algebra and soft pion theorems) and a variety of choices for the baryonic wave functions. The MIT bag model wave function is one such choice which gives conservative results. The same procedure is adopted for calculating the imaginary parts using O_6. The major uncertainty is in the hadronic matrix elements and the fact that the simultaneous fit of s and p waves leaves a factor of 2 ambiguity [16]. In the SM, with the Kobayashi-Maskawa phase convention[17] there is no CPV in $\Delta I = 3/2$ amplitudes; and for Λ and Ξ decays $\phi_3 = 0$. There is a small electroweak penguin contribution to ϕ_3 which can be safely neglected. The rate asymmetry is dominated by the s wave amplitudes and the asymmetry A_Λ is dominated by the $\Delta I = 1/2$ amplitudes. Evaluating the matrix elements in the standard way and with the current knowledge of the K-M matrix elements one finds for the decays[13,18] $\Lambda \to p\pi^-$ and $\Xi^- \to \Lambda\pi^-$:

$$\begin{aligned} \phi^s_\Lambda - \phi^p_\Lambda &\cong 3.5.10^{-4} \\ \phi^s_\Xi - \phi^p_\Xi &\cong -1.4.10^{-4} \end{aligned} \tag{17}$$

With the $N\pi$ phase shifts known to be

$$\delta_s - \delta_p \cong 7^0 \tag{18}$$

one finds for the asymmetry A_Λ in the standard model a value of about -4.10^{-5}. For the $\Xi \to \Lambda\pi^-$ decay mode the phase shifts are not known experimentally and have to be determined theoretically. There are calculations from 1965 [19] which gave large values for $\delta_s - \delta_p \sim -20^0$; however, all recent calculations based on chiral perturbation theory, heavy baryon approximation etc. agree that $\delta_s - \delta_p$ lies between 1^0 and 3^0 [20]. These techniques have been tested in π-N scattering where they reproduce the known phase shifts within a factor of two[21]. In this case the asymmetry A_Ξ is expected to be $\sim -(0.2$ to $0.7)10^{-5}$. In the Table 1, the SM results for the expected asymmetries in SM are given. Using very crude back of the envelope estimates, similar results are obtained. What is needed is some attention to these matrix elements from the Lattice community.

An experimental measurement of the phase shifts $\delta_s - \delta_p$ in $\Lambda\pi$ system will put the predictions for A_Ξ on a firmer basis. There is an old proposal due to Pais and Treiman [22] to measure $\Lambda\pi$ phase shifts in $\Xi \to \Lambda\pi e\nu$, but this does not seem practical in the near future. Another technique, more feasible, it to measure β and α to high precision in Ξ and $\bar{\Xi}$ decays. Then the combination.

$$(\beta - \bar{\beta})/(\alpha - \bar{\alpha}) = tan\ (\delta_s - \delta_p) \tag{19}$$

can be used to extract $\delta_s - \delta_p$. To the extend CP phases are negligible one can also use the approximate relation:

$$\beta/\alpha \approx tan(\delta_s - \delta_p) \tag{20}$$

In Σ decays, some asymmetries are quite large[13] but in difficult to measure channels e.g. B_Σ. In $\Omega^- \to \Xi\pi$ decays the rate asymmetry is larger due to the larger $\Delta I = 3/2$ amplitudes[23]. There are no experimental proposals to measure CP asymmetries in Σ or Ω^- decays at this time.

5 Beyond Standard Model

Can new physics scenarios in which the source of CP violation is not K-M matrix yield large enhancements of these asymmetries? We consider some classes of models where these asymmetries can be estimated more or less reliably[12,13]. It should be kept in mind that any estimates with new physics are at least as uncertain as SM and usually much more prone to uncertainty for obvious reasons.

First there is the class of models which are effectively super-weak[24]. Examples are models in which the K-M matrix is real and the observed CP violation is due to exchange of heavier particles; heavy scalars with FCNC, heavy quarks etc. In all such models direct CP violation is negligible and unobservable and so all asymmetries in hyperon decays are essentially zero. Furthermore, they need to be modified to accommodate the fact that direct CP violation ("Okubo effect") has now been seen in the kaon decays(the fact that ϵ'/ϵ is not zero). In the three Higgs doublet model with flavor conservation imposed, the charged Higgs exchange tends to give large effects in direct CP violation as well as large neutron electric dipole moment[25].

There are two generic classes of left-right symmetric models: (i) Manifest Left- Right symmetric model without $W_L - W_R$ mixing[26] and (ii) with $W_L - W_R$ mixing[27]. In (i) U^L_{KM} = real and U^R_{KM} complex with arbitrary phases but angles given by U^L_{KM}. Then one gets the "isoconjugate" version in which

$$H_{eff} = \frac{G_F\ U_{us}}{\sqrt{2}} \left[J^\dagger_{\mu L}\ J_{\mu L} + \ \eta e^{i\beta} J^\dagger_{\mu R}\ J_{\mu R} \right] \tag{21}$$

where $\eta = m^2_{WL}/m^2_{WR}$ and β is the relevant CPV phase. Then $H_{p.c.}$ and $H_{p.v.}$ have overall phases $(1 + i\eta\beta)$ and $(1 - i\eta\beta)$ respectively. To account for the observed CPV in K-decay, $\eta\beta$ has to be of order $4.5.10^{-5}$. In this model, $\epsilon'/\epsilon = 0$ and there are no rate asymmetries in hyperon decays but the asymmetries A and B are not zero and e.g. A goes as $2\eta\beta\sin(\delta_s - \delta_p)$. In the class of models where $W_L - W_R$ mixing is allowed, the hyperon asymmetries can be enhanced, and also ϵ'/ϵ is not zero in general[27] (see Table 1).

In MSSM (Minimum Supersymmetric Standard Model) there are new CP violating phases and potentially new contributions to many observables. Until recently the conventional thinking was that the most relevant phase was the one in the squark LL mass terms:

$$m_{12}^2\ \bar{\tilde{d}}_L\ \tilde{s}_L \tag{22}$$

and well constrained by ϵ so that the contribution to ϵ'/ϵ would be less than 2.10^{-4} (similarly for hyperon decays). The new wisdom, painfully learnt after the new results on ϵ'/ϵ, is that this is not the whole story. There are several ways in which supersymmetric contributions can arise for K and hyperon decays.

One example is the lack of degeneracy of $\tilde{d}_R$ and $\tilde{u}_R$ masses[28]. This gives rise to I-spin breaking and in turn can enhance ImA_2 and contribute to ϵ'/ϵ at a level of 10^{-3}. For hyperon decays this would lead to a mild enhancement of rate asymmetries but would have no effect on the asymmetries A being probed by E871.

Another possibility is the existence of phases in the L-R squark mass terms[29]. The effect of these on the s-d gluon dipole operator can be parameterised as:

$$\left\{a_{LR}\ \bar{d}_L\ \lambda^a\ \sigma_{\mu\nu}\ s_R\ + a_{RL}\ \bar{d}_R\ \lambda^a\ \sigma_{\mu\nu}\ s_L\right\} G^a_{\mu\nu} + h.c. \tag{23}$$

In terms of a_{LR} and a_{RL}, ϵ'/ϵ and $A(\Lambda)$ can be written as[30]:

$$\begin{aligned} \epsilon'/\epsilon\ \ &\alpha\ \ \ \ \ Im(a_{LR} - a_{RL}) \\ A(\Lambda)\ \ &\alpha\ Im(0.2\ a_{LR} + 2.6 a_{RL}). \end{aligned} \tag{24}$$

The figure shows the range of $A(\Lambda)$ for various allowed values of a_{LR} and a_{RL}. Note that a_{RL} can yield values for $A(\Lambda)$ as large as 10^{-3} easily probed by E871. This operator is also enhanced in models where CP violation arises thru the exchange of charged scalars such as the Weinberg model[25].

6 Experiments

There have been several proposals to measure hyperon decay asymmetries in $\bar{p}p \to \bar{\Lambda}\Lambda$, $\bar{p}p \to \bar{\Xi}\Xi$ and in $e^+e^- \to J/\psi \to \Lambda\bar{\Lambda}$ but none of these were approved[31]. The only approved and on-going experiment is E871 at Fermilab.

Table 1: Expectations for Hyperon CPV Asymmetries.

	SM	2-Higgs	FCNC Superweak	L-R-S (1)	L-R-S (2)
Δ_Λ	10^{-6}	10^{-5}	0	0	0
A_Λ	-4.10^{-5}	-2.10^{-5}	0	-10^5	6.10^{-4}
B_Λ	10^{-4}	2.10^{-3}	0	7.10^{-4}	-
Δ_Ξ	0	0	0	0	0
A_Ξ	-4.10^{-6}	-3.10^{-4}	0	2.10^{-5}	10^{-4}
B_Ξ	10^{-3}	4.10^{-3}	0	3.10^{-4}	-

In this experiment Ξ^- and $\overline{\Xi}^+$ are produced and the angular distribution of $\Xi^- \to \Lambda\pi^- \to p\pi^-\pi^-$ and $\overline{\Xi}^+$ compared. This experiment effectively measures $A_\Lambda + A_\Xi$ and will be described in detail by Kam-Biu Luk [32]. To summarize the implications for the measurement of $A_\Lambda + A_\Xi$ by E871: the SM expectation is about -4.10^{-5} with a factor of two uncertainty; if new physics should contribute it could be as large as 10^{-3}. A measurement by E871 at the 10^{-4} level, therefore, will already be a strong discriminant. Eventually, it will be important to know A_Λ and A_Ξ separately and the old proposals[31] should be revived.

7 ϵ'/ϵ and Hyperon Decay Asymmetries

It might seem that now that ϵ'/ϵ has been measured and direct CP violation in $\Delta S = 1$ channel been observed, a study of CP violation in hyperon decays is unnecessary and no new information will be obtained. Why is it worthwhile measuring another $\Delta S = 1$ process like hyperon decay? The point is that there are important differences and the two are not at all identical. First, there are important differences in the matrix elements. Hyperon matrix elements do not have the kind of large cancellations that plague the kaon matrix elements. The hadronic uncertainties are present for both, but are different. Next, a very important difference is the fact that the $K \to \pi\pi$ decay (and hence ϵ') is only sensitive to CP violation in the parity violating amplitude and cannot yield any information on parity conserving amplitudes. Hyperon decays, by contrast, are sensitive to both. Thus, ϵ'/ϵ and hyperon decay CP asymmetries are different and complimentary. The hyperon decay measurements are as important and significant as ϵ'/ϵ.

Conclusion

The searches for direct CPV are being pursued in many channels: $\Lambda \to N\pi$, B decays and D decays. Any observation of a signal would be the first outside of $K^0 - \overline{K}^0$ system and would be complimentary to the measurement of ϵ'/ϵ. This will constitute one more step in our bid to confirm or demolish the Standard Kobayashi-Maskawa description of CP violation.

Hyperon decays offer a rich variety of CP violating observables, each with different sensitivity to various sources of CP violation. For example, Δ_Λ is mostly sensitive to parity violating amplitudes, $\Delta_{\Sigma+}$ is sensitive only to parity conserving amplitudes, A is sensitive to both etc. The size of expected signals vary inversely with the ease of making measurements, i.e. $\Delta < A < B$. Probably because of that, the number of experimental proposals is rather small so far. The one on-going experiment Fermilab E871 can probe A to a level of 10^{-4} which is already in an interesting range. In addition to more experiments, this subject sorely needs more attention devoted to calculating the matrix elements more reliably.

Acknowledgment

I am grateful to my collaborators Alakabha Datta, Xiao-Gang He, Hitoshi Murayama, Pat O'Donnell, German Valencia and John Donoghue and to members of the E871 collaboration for many discussions. The hospitality of Hai-Yang Cheng, George Hou and their colleagues and staff was memorable and the atmosphere of the conference was most stimulating. This work is supported in part by USDOE under Grant #DE- FG-03-94ER40833.

References

1. G. Lüders, *Ann. Phys.* 2, 1 (1957); G. Lüders, *Dan. Mat. Fys. Medd.* 28, No. 5 (1954); J.S. Bell, *Proc. Roy. Soc.* A231, 79 (1955); J. Schwinger, Phys. Rev. 91, 713 (1953); W. Pauli, "*Niels Bohr and the Development of Physics,*" McGraw-Hill, New York, and Pergammon Press, London 1955.
2. G. Lüders and B. Zumino, *Phys. Rev.* 106, 385 (1957).
3. S. Okubo, *Phys. Rev.* 109, 984 (1958).
4. A.D. Sakharov, Zh. EK. Teor. Fiz. 5, 32 (1967) (English translation, *JETP Letters* 5, 24 (1967)).
5. A. Pais, *Phys. Rev. Lett.* 3, 242 (1959).
6. T. Brown, S. F. Tuan and S. Pakvasa, *Phys. Rev. Lett.* 51, 1238 (1983).

7. L-L. Chau and H. Y. Cheng, Phys. Lett. B131, 202 (1983); D. Chang and L. Wolfenstein, *Intense Medium Energy Sources of Strangeness*, edited by T. Goldman, H.E. Haber, and H.F.-W. Sadrozinski, AIP Conf. Proc. No. 102 (AIP, New York, 1983), p. 73; C. Kounnas, A. B. Lahanas and P. Pavlopoulos, *Phys. Lett.* B127, 381 (1983).
8. S. P. Rosen and S. Pakvasa, *Advances in Particle Physics*, Ed. R.E. Marshak and R.L. Cool, Wiley-Interscience, NY 1968, p. 473; S. Pakvasa and S.P. Rosen, "*The Past Decade in Particle Theory*", Ed. E.C.G. Sudarshan and Y. Ne'eman, Gordon and Breach , 1973, p. 437.
9. T. D. Lee and C. N. Yang, *Phys. Rev. 108*, 1645 (1957); R. Gatto, *Nucl. Phys. 5*, 183 (1958); T. D. Lee, J. Steinberger, G. Feinberg, P.K. Kabir and C.N. Yang, *Phys. Rev.* 106, 1367 (1957).
10. O. E. Overseth and S. Pakvasa, *Phys. Rev.* 184, 1663 (1969).
11. L. Wolfenstein, *Phys. Rev.* D43, 151 (1990); N. G. Deshpande, X-G. He and S. Pakvasa, hep-ph/9606259.
12. J. F. Donoghue and S. Pakvasa, *Phys. Rev. Lett.* 55, 162 (1985).
13. J. F. Donoghue, X-G. He and S. Pakvasa, *Phys. Rev.* D39, 833 (1986).
14. L.D. Roper, R. M. Wright and B. Feld, *Phys. Rev.* 138, 190 (1965).
15. A. J. Buras, P. Gambino and U. Haisch, hep-ph/9911250.
16. J. F. Donoghue, E. Golowich, and B. Holstein, *Phys. Rep.* 131, 319 (1986); *Dynamics of the Standard Model*, Cambridge Univ. Press, 1992.
17. M. Kobayashi and T. Maskawa, Progr. Theoret. Phys. (Kyoto), 49, 652 (1973).
18. M. J. Iqbal and G. Miller, *Phys. Rev.* D41, 2817 (1990); X-G. He, H. Steger and G. Valencia, *Phys. Lett.* B272, 411 (1991); N.G. Deshpande, X-G. He and S. Pakvasa, *Phys. Lett.* B326, 307 (1994); X-G. He and G. Valencia, *Phys. Rev.* D52, 5257 (1995).
19. B. Martin, *Phys. Rev.* 138, 1136 (1965); R. Nath and A. Kumar, *Nuov. Cim.* 36, 669 (1965).
20. M. Lu, M. Wise and M. Savage, *Phys. Lett.* B337, 133 (1994); A. Datta and S. Pakvasa, *Phys. Lett.* B344, 340 (1995); A. Kamal, *Phys. Rev.* D58, 07750 (1998); A. Datta, P.J. O'Donnell and S. Pakvasa, hep-ph/9806374.
21. A. Datta and S. Pakvasa, *Phys. Rev.* D56, 4322 (1997).
22. A. Pais and S.B. Treiman, *Phys. Rev.* 178, 2365 (1969).
23. J. Tandean and G. Valencia, *Phys. Lett.* B451, 382 (1999).
24. Some examples can be found in the following references: R.N. Mohapatra, J. C. Pati and L. Wolfenstein, *Phys. Rev.* D11, 3319 (1975); T. Brown, N. Despande, S. Pakvasa and H. Sugawara, *Phys. Lett.* B141, 95 (1984); J. M. Soares and L. Wolfenstein, *Phys. Rev.* D46, 256 (1992);

P. Frampton and S.L. Glashow, *Phys. Rev.* D55, 1691 (1997); H. Georgi and S.L. Glashow, hep-ph/9807399.
25. S. Weinberg, *Phys. Rev. Lett.* 37, 657 (1976).
26. R. N. Mohapatra and J. C. Pati, *Phys. Rev.* D11, 566 (1975).
27. D. Chang, X-G. He and S. Pakvasa, *Phys. Rev. Lett.* 74, 3927 (1995).
28. A. Kagan and M. Neubert, hep-ph/9908404.
29. A. Masiero and H. Murayama, *Phys. Rev. Lett.* 83, 907 (1999); K. S. Babu, B. Dutta and R. Mohapatra, hep-ph/9905464.
30. X-G He, H. Murayama, S. Pakvasa and G. Valencia (hep-ph/9909562), Phys. Rev. D., Rapid Comm. (in press).
31. N. Hamman et al. CERN/SPSPLC 92-19, SPSLC/M49; S. Y. Hsueh and P. Rapidis, FNAL Proposal 859 (1992), D. Kaplan, hep-ex/9911035.
32. K-B. Luk, these proceedings.

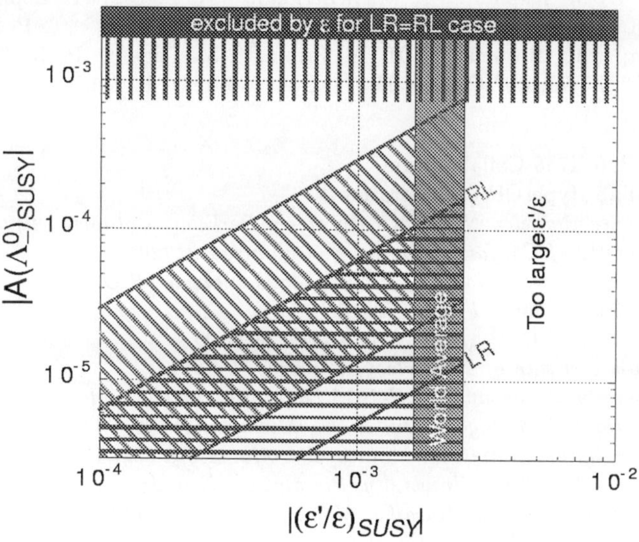

Figure 1: The allowed regions on $(|(\epsilon'/\epsilon)_{SUSY}|, |A(\Lambda^0_-)_{SUSY}|)$ parameter space for three cases: a) only $Im(a_{LR})$ contribution, which is the conservative case (hatched horizontally), b) only $\mathrm{Im}(a_{RL})$ contribution (hatched diagonally), and c) $Im(a_{LR}) = Im(a_{RL})$ case which does not contribute to ϵ' and can give a large $|A(\Lambda^0_-)|$ below the shaded region (or vertically hatched region for the central values of the matrix elements). The last case is motivated by the relation $\lambda = \sqrt{m_d/m_s}$. The vertical shaded band is the world average of ϵ'/ϵ. The region to the right of the band is therefore not allowed.

EXAMINING CP SYMMETRY IN STRANGE BARYON DECAYS

K. B. LUK

Department of Physics, University of California and Physics Division, Lawrence Berkeley National Laboratory, Berkeley CA 94720, USA
E-mail: k_luk@lbl.gov

representing the Fermilab E756 and HyperCP Collaborations

R. A. Burnstein,[6,b] A. Chakravorty,[6,b] A. Chan,[3,b] Y. C. Chen,[3,b] W. S. Choong,[1,2,b] K. Clark,[11,b] T. Diehl,[4,a] E. C. Dukes,[12,b] C. Durandet,[12,b] J. Duryea,[9,a] J. Felix,[5,b] G. Gidal,[1,b] G. Guglielmo,[4,a] H. R. Gustafson,[8,b] K. Heller,[9,a] C. Ho,[3,b] P. M. Ho,[1,a] T. Holmstrom,[12,b] M. Huang,[12,b] C. James,[4,a,b] M. Jenkins,[11,b] K. Johns,[9,a] T. Jones,[1,b] D. M. Kaplan,[6,b] L. M. Lederman,[6,b] N. Leros,[7,b] F. Lopez,[8,b] M. J. Longo,[8,a,b] L. Lu,[12,b] W. Luebke,[6,b] K. Nelson,[12,b] H. K. Park,[8,b] J. P. Perroud,[7,b] D. Rajaram,[6,b] R. Rameika,[4,a] H. A. Rubin,[6,b] S. Teige,[6,a] P. K. Teng,[3,b] G. Thomson,[10,a] J. Volks,[4,b] C. G. White,[6,b] S. L. White,[6,b] P. Zyla[1,b]

[a] Fermilab E756 Collaboration
[b] Fermilab HyperCP Collaboration
[1] *Lawrence Berkeley National Laboratory, Berkeley, California 94720*
[2] *University of California, Berkeley, California 94720*
[3] *Academia Sinica, Nankang, Taipei 11529, Taiwan, R.O.C.*
[4] *Fermilab, Batavia, Illinois 60510*
[5] *University of Guanajuato, 37000 Leon, Mexico*
[6] *Illinois Institute of Technology, Chicago, IL 60616*
[7] *University of Lausanne, CH-1015 Lausanne, Switzerland*
[8] *University of Michigan, Ann Arbor, Michigan 48109*
[9] *University of Minnesota, Minneapolis, Minnesota 55455*
[10] *Rutgers–The State University, Piscataway, New Jersey 08854*
[11] *University of South Alabama, Mobile, Alabama 36688*
[12] *University of Virginia, Charlottesville, Virginia 22901*

Non-conservation of CP symmetry can manifest itself in non-leptonic hyperon decays as a difference in the decay parameter between the strange-baryon decay and its charge conjugate. By comparing the decay distribution in the Λ helicity frame for the decay sequence $\Xi^- \to \Lambda\pi^-$, $\Lambda \to p\pi^-$ with that of $\overline{\Xi}^+$ decay, E756 at Fermilab did not observe any CP-odd effect at the 10^{-2} level. The status of a follow-up experiment, HyperCP (FNAL E871), to search for CP violation in charged $\Xi - \Lambda$ decay with a sensitivity of 10^{-4} is also presented.

1 Introduction

The Standard Model, as well as many other models, predicts the existence of CP violation in non-leptonic strange-baryon decays.[1] One approach to observe this CP-odd effect is to compare the decay distribution of a strange baryon with that of its charge conjugate. In the non-leptonic weak decay of hyperon, parity violation leads to a forward-backward asymmetric distribution of the daughter particles with respect to the spin of the hyperon:

$$\frac{dn}{dcos\theta} = \frac{1}{2}\left(1 + \alpha \mathbf{P} \cdot \hat{\mathbf{p}}\right) = \frac{1}{2}\left(1 + \alpha P cos\theta\right) \tag{1}$$

where α is a parameter quantifying the degree of parity violation in the decay, $\mathbf{P}$ is the polarization of the hyperon, and $\hat{\mathbf{p}}$ is the momentum unit vector of the daughter baryon in the rest frame of the hyperon.

Under CP transformation, the strange-baryon decay is related to the corresponding decay of its anti-particle with the condition that

$$\overline{\alpha} = -\alpha \ . \tag{2}$$

Thus, any violation of equation (2) would signal the breaking of CP symmetry in the decay. Alternatively we can define a parameter

$$A = \frac{\alpha + \overline{\alpha}}{\alpha - \overline{\alpha}} \tag{3}$$

for measuring the amount of CP violation in the decay. The latest predictions of A for hyperon decays are summarized in reference 1.

To date all searches of CP nonconservation in strange-baryon sector focussed on the $\Lambda \rightarrow p\pi$ decay, with Λ's and $\overline{\Lambda}$'s produced in $p\overline{p}$ collisions [2,3] or from the J/Ψ decays.[4] The best result of 0.013 ± 0.022 for A_Λ came from LEAR PS185.[3]

In this talk, a new approach for studying CP symmetry in strange-baryon decay is presented. According to equation (1) Λ hyperons with precisely known polarization are needed for determining α_Λ. Significantly polarized Λ's can be created from charged $\Xi \rightarrow \Lambda\pi$ decays. In this case, the polarization of the Λ is given by

$$\mathbf{P}_\Lambda = \frac{(\alpha_\Xi + \mathbf{P}_\Xi \cdot \hat{\mathbf{\Lambda}})\hat{\mathbf{\Lambda}} + \beta_\Xi \mathbf{P}_\Xi \times \hat{\mathbf{\Lambda}} + \gamma_\Xi \hat{\mathbf{\Lambda}} \times (\mathbf{P}_\Xi \times \hat{\mathbf{\Lambda}})}{\left(1 + \alpha_\Xi \mathbf{P}_\Xi \cdot \hat{\mathbf{\Lambda}}\right)} \tag{4}$$

where $\hat{\mathbf{\Lambda}}$ is the unit vector along the Λ momentum in the Ξ rest frame, β_Ξ and γ_Ξ are the other two decay parameters of the $\Xi \rightarrow \Lambda\pi$ decay, and $\mathbf{P}_\Xi$ is the polarization of the Ξ. In the helicity frame of the Λ the angular distribution of the proton is

$$\frac{dn}{dcos\theta_{p\Lambda}} = \frac{1}{2}\left(1 + \alpha_\Lambda\alpha_\Xi \; cos\theta_{p\Lambda}\right) \; , \tag{5}$$

independent of the polarization of the Ξ. A new parameter, $A_{\Xi\Lambda}$, defined as

$$A_{\Xi\Lambda} = \frac{\alpha_\Xi\alpha_\Lambda - \alpha_{\overline{\Xi}}\alpha_{\overline{\Lambda}}}{\alpha_\Xi\alpha_\Lambda + \alpha_{\overline{\Xi}}\alpha_{\overline{\Lambda}}} \approx A_\Xi + A_\Lambda \tag{6}$$

is used for searching for CP asymmetry in the $\Xi - \Lambda$ decay sequence. This new scheme was first demonstrated by E756 at Fermilab with polarized Ξ. A value of 0.012 ± 0.014 for $A_{\Xi\Lambda}$ was obtained.[5] Motivated by this successful study, a dedicated experiment, HyperCP (E871), was proposed and is currently taking data at Fermilab to reach a projected sensitivity of 10^{-4}. The remainder of this talk will highlight the analysis of E756 and present the status of HyperCP.

2 FNAL E756

The primary goals of FNAL E756 were to measure the magnetic moment of the Ω^- [6] and to study the production polarization of hyperons.[7,8,9] Along with the Ω^-'s a large sample of Ξ^- decays was obtained.

This experiment was performed in the FNAL Proton-Center beam line during the 1987 fixed-target run. The plan view of the E756 spectrometer is shown in Fig.1. Polarized Ξ^- hyperons were produced with an 800-GeV proton beam striking a 2 mm $\times$ 2 mm $\times$ 9.2 cm-long beryllium target at an angle of 2.4 mrad in the vertical plane. After the strange baryons emerged from a curved collimator in a 7.3 m-long hyperon magnet, with its field in the vertical direction, they were allowed to decay in a 26 m-long region. The charged particles in the decay sequence $\Xi^- \to \Lambda\pi^-$ and $\Lambda \to p\pi^-$ were momentum analyzed with a simple spectrometer consisting of scintillation counters, silicon strip detectors, multiwire proportional chambers, and two dipole magnets. The trigger was designed to look for events with at least two oppositely charged tracks downstream of the analysis magnets and with three or four tracks upstream. The targetting angle was cycled between 2.4 mrad and -2.4 mrad every few hours to minimize temporal variations between runs. For each production angle, roughly equal number of events were also collected with the polarity of the analysis magnets reversed and the left-right requirements of C8 and C9 interchanged.

Ten days were spent to collect $\overline{\Xi}^+$ decays, the incident proton intensity was reduced from 3×10^{10} per 20-s spill to 1×10^{10} per spill, to maintain a similar singles rate in the spectrometer as the Ξ^- run. The polarity of the hyperon magnet was also reversed but the trigger requirements and the production angles were unchanged, thus reducing potential biases between the Ξ^- and $\overline{\Xi}^+$ runs.

For the CP-symmetry study, the hyperon magnetic field was set for selecting Ξ's with monenta between 240 GeV/c and 450 GeV/c.

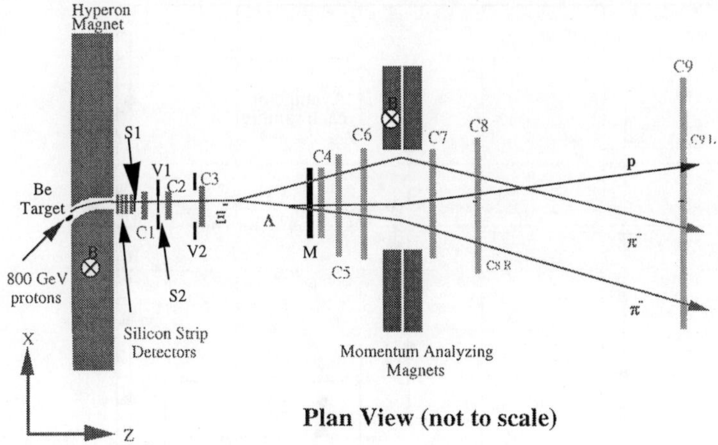

Figure 1. Plan view of E756 spectrometer (not to scale).

Both the Ξ^- and $\overline{\Xi}^+$ data were processed with the same reconstruction program and were subjected to identical event-selection requirements. Further details of the experiment and the event selection can be found in reference 8.

Two different techniques were employed to extract $A_{\Xi\Lambda}$. In the first approach, the value of $\alpha_\Xi\alpha_\Lambda$ was determined with the Hybrid Monte Carlo method[8] for the Ξ^- and the $\overline{\Xi}^+$ samples separately. Based on about 70,000 $\overline{\Xi}^+$ decays, $\alpha_{\overline{\Xi}}\alpha_{\overline{\Lambda}}$ was found to be -0.2894 ± 0.0073. From three independent Ξ^- samples, each with approximately 70,000 events, $\alpha_\Xi\alpha_\Lambda$ was -0.2955 ± 0.0073, -0.3041 ± 0.0073, and -0.2894 ± 0.0073, giving a mean of -0.2963 ± 0.0042. As shown in Fig. 2 these results were in good agreement with the world average[10] and were stable with respect to the momentum of Ξ. This method gave a value of 0.012 ± 0.014 for $A_{\Xi\Lambda}$, with insignificant systematic error that was estimated by varying event-selection requirements.

In the second approach, the difference in $\alpha_\Xi\alpha_\Lambda$ between Ξ^- and $\overline{\Xi}^+$ was determined directly without unfolding the acceptance in $\cos\theta_{p\Lambda}$. For two data samples, a comparison of the $\cos\theta_{p\Lambda}$ distributions can be defined by

$$R(cos\theta_{p\Lambda}) = \frac{\epsilon_1(cos\theta_{p\Lambda})}{\epsilon_2(cos\theta_{p\Lambda})}\frac{[1+(\alpha_\Lambda\alpha_\Xi)_1 cos\theta_{p\Lambda}]}{[1+(\alpha_\Lambda\alpha_\Xi)_2 cos\theta_{p\Lambda}]} \tag{7}$$

where $R(cos\theta_{p\Lambda})$ is the ratio of the probabilities of getting $\cos\theta_{p\Lambda}$ in the two samples, and the ϵ's are the acceptance functions of the $\cos\theta_{p\Lambda}$ distributions.

When two sets of Ξ^- events are compared, R is a measure of how well the acceptances agree. Without any corrections, detailed studies showed that the ac-

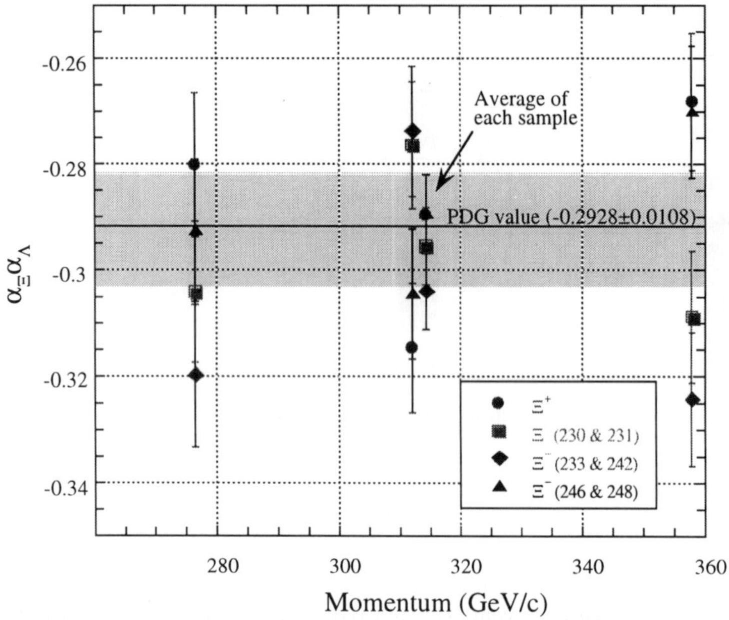

Figure 2. E756 results on $\alpha_\Xi\alpha_\Lambda$ as a function of the momentum of the Ξ. The shaded area is an one-standard-deviation band centered at the world average.

ceptance in $\cos\theta_{p\Lambda}$ was strongly dependent on the momentum of the Ξ^-, but was insensitive to the polarization of the Ξ^- or other variations in the experiment down to a few $\times 10^{-3}$ level.[11] This unique feature is due to the fact that the $\hat{\mathbf{\Lambda}}$ defining the helicity frame changes from event to event over the entire phase space in the Ξ rest frame. Any systematic bias due to imperfection of the experiment in the laboratory is mapped into a wide range of $\cos\theta_{p\Lambda}$, thus highly diluted.

In the study of CP symmetry, a sample of Ξ^- events was selected in such a way that the resulting Ξ^- momentum spectrum was identical to that of the $\overline{\Xi}^+$ sample. This removed the difference due to different mechanism for producing particles and anti-particles by protons, and ensured that $\epsilon(cos\theta_{p\Lambda})$ was identical for both data sets. With $\alpha_\Xi\alpha_\Lambda$ taken to be -0.2928[10] the difference in $\alpha_\Xi\alpha_\Lambda$ between the Ξ^- and $\overline{\Xi}^+$ samples was determined by fitting R as a function of $\cos\theta_{p\Lambda}$ according to equation (7). With approximately 70,000 Ξ^- events along with equal number of $\overline{\Xi}^+$ decays, the difference was -0.011 ± 0.009. This implied that $A_{\Xi\Lambda}$ was 0.019 ± 0.015, which was consistent with the result obtained with the Hybrid Monte Carlo method. As a check, another sample of Ξ^- events was picked to repeat the measurement which yielded a result of 0.008 ± 0.015 for $A_{\Xi\Lambda}$.

3 HyperCP (FNAL E871)

HyperCP is a dedicated experiment designed to search for CP violation in charged-Ξ-Λ decay with about 100 times better sensitivity than E756. This requires a high-rate spectrometer which is sufficiently simple for controlling systematic effects.

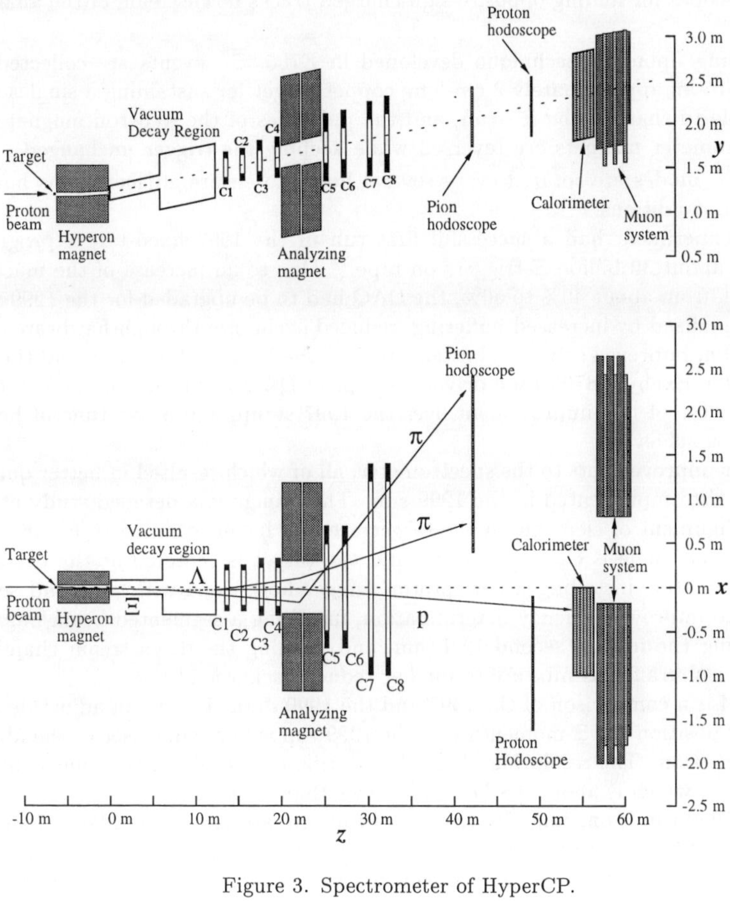

Figure 3. Spectrometer of HyperCP.

As shown in Fig. 3, an 800-GeV proton beam with a typical intensity of 7.5×10^9 per sec incidents on a 2 mm $\times$ 2 mm $\times$ 3 cm-long copper target at 0 mrad. The secondary beam is bent through a 6.1 m-long shaped channel in the vertical plane by a dipole magnet with a field of about 1.667 T. The momentum range of the selected Ξ^-'s is between 120 GeV/c and 220 GeV/c. The acceptance of this channel is about a factor of two larger than the one in E756. Downstream of a 13 m-long evacuated decay region is a spectrometer consisting of eight multiwire proportional chambers and two analysis magnets. There are about two times more anode planes and finer wire pitch than the E756 spectrometer, offering better reconstruction efficiency and better momentum resolution.[a] The trigger elements are a hadron calorimeter for detecting particles with energy greater than 70 GeV at close to 100% efficiency, and two hodoscopes for finding opposite-sign charged tracks downstream of the analysis magnets.

Building upon the technique developed in E756, $\overline{\Xi}^+$ events are collected by switching to an approximately 2 cm-long copper target for sustaining a singles rate comparable to that for the Ξ^- run, and the polarities of the hyperon magnet and the spectrometer magnets are reversed while keeping the trigger unchanged. The Ξ^- and $\overline{\Xi}^+$ modes are rotated every several hours to ensure adjacent runs having similar run conditions.

The experiment had a successful first run in the 1997 fixed-target program, collecting about 30 billion Ξ triggers on tape.[12] Due to an increase of the machine duty cycle, from about 40% to 50%, the DAQ had to be upgraded for the 1999 run. This was realized by increased buffering, reduced event size through hardware data compression, improved software, better layout of the DAQ architecture, and the use of the faster Exabyte 8705 tape drives. A typical DAQ bandwidth of 23 Mbyte/s, about a factor of two improvement over the 1997 setup, and a live time of better than 75% is achieved.

Other improvements to the spectrometer, all of which resulted in better quality data, are also implemented in the 1999 run. These include a detailed study of the relative alignment of elements in the target area by beam scanning to ensure that the targetting angle is close to 0 mrad and the target is centered at the entrance of the collimator, installing a new proton trigger hodoscope with a second set of counters to improve efficiency determination, installing a segmented beam monitor for handling the intense secondary beam, and running the downstream chambers with CF_4-isobutane gas mixture to further reduce background hits.

Fig. 4 is a comparison of the 1997 and the 1999 data. Due to an adjustment in the target position the Ξ momentum in the 1999 run is lowered, closer to the Monte Carlo prediction. The resolution of the $\Lambda\pi$ invariant mass of the two runs is about 1.6 MeV/c^2, which is about 0.8 MeV/c^2 better than in E756.

The 1999 run is ongoing. By the end of the run, around the middle of January, 2000, about a factor of two more data than the 1997 run is expected to be logged. The combination of the two runs is expected to yield about 3×10^9 reconstructed Ξ^- decays and 0.75×10^9 $\overline{\Xi}^+$ events. This corresponds to a projected statistical sensitivity for the strange-baryon CP asymmetry of about 1.3×10^{-4}, which is comparable to the proposed goal.

[a] The muon system at the end of the spectrometer is used for studying rare decays.

4 Conclusion

Based on approximately 70,000 $\overline{\Xi}^{+}$ and 210,000 Ξ^{-} decays the Fermilab E756 Collaboration has searched for CP violation in hyperon decay by measuring $A_{\Xi\Lambda}$. A result of 0.012 ± 0.014 was obtained. Systematic biases to the measurement, even without making any correction, have been shown to be insignificant down to the 10^{-3} level.

HyperCP has two successful runs, amassing probably the largest sample of strange-baryon decays in the world. By the end of the 1999 run, the projected

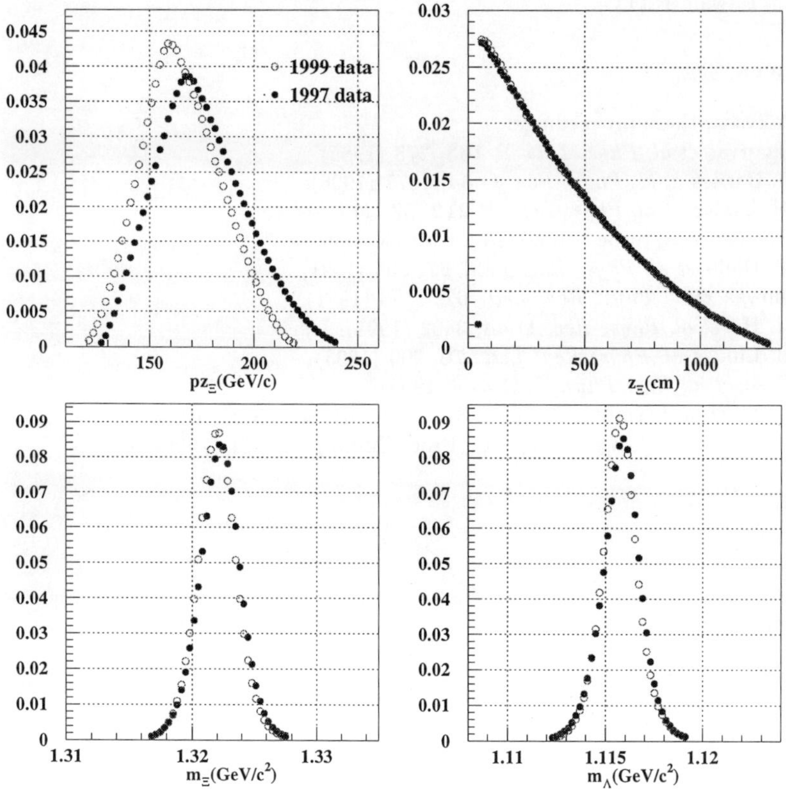

Figure 4. Comparison of momentum, decay position, $\Lambda\pi$ invariant mass, and $p\pi$ invariant mass between the 1997 and the 1999 run in HyperCP.

numbers of fully reconstructed Ξ^- and $\overline{\Xi}^+$ events are approximately 3×10^9 and 0.75×10^9 respectively. This should yield a statistical precision of 1.3×10^{-4} for $A_{\Xi\Lambda}$.

Acknowledgments

K. B. Luk would like to thank the organizers, Professor George Hou and Dr. Hai-Yang Cheng, for the invitation and their hospitality. This work was supported by the Director, Office of Energy Research, Office of High Energy and Nuclear Physics, Division of High Energy Physics of the U.S. Department of Energy under Contract DE-AC03-76SF00098, the National Science Foundation, and the National Science Council of Taiwan, R.O.C..

References

1. S. Pakvasa, these proceedings.
2. P. Chauvat *et al*, *Phys. Lett.* B **163**, 273 (1985).
3. P. D. Barnes *el al*, *Phys. Rev.* C **54**, 1877 (1996).
4. M. H. Tixier *et al*, *Phys. Lett.* B **212**, 523 (1988).
5. K. B. Luk *et al*, to be submitted for publication.
6. H. T. Diehl *et al*, *Phys. Rev. Lett.* **67**, 804 (1991).
7. J. Duryea *et al*, *Phys. Rev. Lett.* **67**, 1193 (1991).
8. P. M. Ho *et al*, *Phys. Rev.* D **44**, 3402 (1991).
9. K. B. Luk *et al*, *Phys. Rev. Lett.* **70**, 900 (1993).
10. C. Caso *et al*, *Eur. Phys. J.* C **3**, 1 (1998).
11. J. Antos *et al*, Fermilab Proposal 871 (revised), March 26, 1994.
12. C. G. White *et al*, *Nucl. Phys.* B (Proc. Suppl.) **71**, 451 (1999).

The CDF Run II Upgrade

R. E. Hughes
Department of Physics,
The Ohio State University,
Columbus, OH 43210 USA
E-mail: hughes@mps.ohio-state.edu

Representing the CDFII Collaboration

The CDF detector is in the final stages of an upgrade in preparation for Run II at the Tevatron. The upgrades include a complete replacement of the central tracking system, including 3 separate silicon strip detector systems and a new drift chamber. A new time-of-flight system is being constructed, to enhance particle identification for B decays. A new calorimeter covering the region $1.1 < |\eta| < 3.6$ is being installed, which uses scintillating tile-fiber technology. The muon detection system will be extended to provide additional acceptance both in the central and forward directions. Finally, new front-end electronics and a new data aquisition and trigger system are being constructed to handle the 132 ns bunch spacing of an upgraded Tevatron. These upgrades will enhance CDF's ability in the low P_T physics regime of B physics, including the search for and study of CP violation, as well as the high P_T physics regime of top physics, the search for the Higgs boson, and the search for extensions beyond the standard model. The status of these upgrades will be reviewed and the future schedule summarized.

1 Introduction

The Collider Detector at Fermilab (CDF) was first commissioned in 1985. In 1987, the first engineering run accumulated 30 nb^{-1}, followed by the first run - Run 0 - from 1988-89, which accumulated 4.7 pb^{-1}. CDF's Run I occurred in 1992-96, accumulating 110 pb^{-1}, and highlighted by the discovery[1,2,3] of the top quark (an honored shared by CDF with D0), and the first measurement[4] of sin 2β . During the past 3 years, the CDF collaboration has focused its efforts on a massive upgrade[5] of the CDF detector, in preparation for Run II at the Tevatron, scheduled to begin in March 2001.

Run II at the Tevatron will see the commissioning of the Main Injector Facility[6]. The Main Injector is a new 150 GeV accelerator half the circumference of the Tevatron, which will increase the anti-proton intensity into the Tevatron. This will allow the instantaneous luminosity in the Tevatron to reach $2 \times 10^{32} cm^{-1} s^{-1}$, a factor of 10 higher than Run I. The initial integrated luminosity goal is 2 fb^{-1} over the years 2001-2002. Beyond that, a further goal is the accumulation of much larger data samples, possibly as high as 15 fb^{-1} before 2007.

At the high luminosity required to achieve such goals, it is important to keep the number of multiple interactions as low as possible. This will be achieved by reducing the spacing between proton and antiproton bunches from the Run I value of 3.5 μs to 396 ns and eventually to 132 ns. The shortening of the bunch spacing and the increase in the luminosity requires the complete replacement and redesign of the trigger and data aquisition systems and the front-end electronics, as well as replacement of a number of detector subsystems.

However, it is clear that with such large data samples there are also enormous physics opportunities. These range from the investigation of CP violation in the B system to the continued study of the top quark, to the precision measurement of the W mass, and the search for new phenomena. The upgraded CDF detector will be capable of excellence in all of these diverse areas: the expected uncertainty on sin2β in Run II is ≈ 0.07, the uncertainty on the top quark mass is anticipated[5,7] to be as low as 1-2 GeV/c^2, and the precision on the W mass is expected[5,7] to reach ≈ 40 MeV/c^2. In the sections below, I give a short review of the major upgrades to the CDF detector. Those interested in more detail are encouraged to consult the references or visit the public CDF Web page: http://www-cdf.fnal.gov/upgrades/upgrades.html.

2 Tracking System

CDf has always placed a high priority on excellent charged particle tracking. An entirely new integrated tracking system is being developed to meet the challenges of the Run II environment. In order of their distance from the beam line, the system is composed of 3 silicon microvertex detectors: the single layer Layer 00, the 5 layer silicon microstrip vertex detector SVXII, and the additional layers of Intermediate Silicon Layers. These are surrounded by a drift chamber called the Central Outer Tracker. For tracks passing completely through all 4 tracking systems, the momentum resolution is expected to be better than $\Delta P_T/P_T^2 < 0.1\%$ for $P_T > 4$ GeV/c. The entire system is placed in magnetic field generated by a superconducting solenoid at $r \approx 1.5$ m.

2.1 Silicon Layer 00 (L00)

Layer 00 is a low-mass radiation-hard single-sided axial-strip silicon detector, located just outside the beam pipe, at a radius of approximately 1.6cm. This detector uses sensors which support high bias voltages which allow good signal-to-noise even after the high radiation doses expected at such a small radius. There will be 6 readout groups along the z-direction, and alternating wide (6 total) and narrow (6 total) sensors in phi. The total channel count is approximately 14,000, which is less that 2% of the total combined silicon system.

Layer 00 is expected to improve the impact parameter resolution for 1 GeV/c tracks from ~ 50 microns to ~ 25 microns. As a result, we expect a large improvement in the vertex-finding resolution contribution to the determination of the proper time for b hadron decays.

2.2 Silicon Vertex Detector (SVXII)

This 5 layer silicon-microstrip vertex detector[8] covering the radial region from 2.4cm to 10.6cm builds upon the experience gained with the SVX and SVX′ systems during Run I. Besides one additional layer, a number of important improvements have been made. First, the electronics for the front end of SVXII will be more radiation hard than previous versions. It is expected that the readout chips (SVX3 Chip) will withstand 2-4 Mrad of dose. This corresponds to about 3-5 fb^{-1} of integrated luminosity for the inner most layer. Second, SVXII is nearly twice as long as the SVX and SVX′ detectors. SVXII will cover ≈87% of the luminous region. Third, the silicon layers are double sided, providing both r-ϕ and r-z information for the tracks. The r-z information is obtained from 2 layers with strips at 1.2° stereo angle and 3 layers with 90° stereo angle. Three dimensional standalone tracking will be possible with SVXII.

An important feature of the SVX3 readout chip is its ability to acquire data while digitizing and reading out the results of a previous beam crossing. This dramatically reduces the deadtime caused by this system. In addition, the SVXII system has two high speed readout paths. One supplies data to the normal data aquisition path for mass storage. The other path supplies the hit information to the Level 2 trigger. Having this information available at the trigger level allows for a trigger requirement on track impact parameter (see Section 6). Therefore, it provides the opportunity for CDF to trigger on hadronic b-quark decays and extend its reach in areas such as the study of CP violation in the B system.

2.3 Intermediate Silicon Layers (ISL)

The Intermediate Silicon Layers[9] provide an important link between the SVXII track hits and the tracks reconstructed by the Central Outer Tracker (COT). The system consists of a complete layer of silicon at $r \approx 20$ cm and a partial layer at $r \approx 28$ cm and $1 < |\eta| < 2$. Many components of the ISL are similar to the components of the SVXII system. The parameters of the ISL were optimized to (i) best bridge the gap between the SVXII and COT and (ii) extend precision tracking into the plug calorimeter to assist with lepton identification and b-quark tagging at low angles ($1 < |\eta| < 2$). The SVXII and the ISL provide tracking in the region $1 < |\eta| < 2$ with $\Delta P_T/P_T^2 = 0.4\%$ ($P_T > 5$ GeV/c).

2.4 Central Outer Tracker (COT)

The Central Outer Tracker[10] is based on the Central Tracking Chamber (CTC) from Run I. However, a number of changes have been made to correct short comings of the CTC, to respond to changing accelerator conditions, and to simplify the construction. Like the CTC, the COT consists of 8 "superlayers". Four of these layers have axial sense wires to provide $r - \phi$ measurements. Four of the superlayers have wires tilted at a ~3° stereo angle to provide z positions of charged particles. Each superlayer is divided into many identical cells. The total number of cells is 2520, four times more cells than the CTC. A cell has an average maximum drift distance of 0.88 cm with a maximum drift time which is less than the beam crossing time of 132 ns. Therefore, there is no "pile-up" of events in the chamber. A cell has 12 sense wires and 17 potential wire which are 40 micron diameter gold plated tungsten. The total number of sense wires in the COT is 30,240. The cathodes for the cells are sheets of gold plated mylar. The cells are tilted at 35° for the Lorentz angle.

3 Time of Flight (TOF)

The Time-of-Flight system lies in the space between the outer radius of the COT (138 cm) and the inner radius of the cryostat for the solenoid. The TOF system uses 216 2.8 meter long scintillator bars, with an approximate $4cm \times 4cm$ trapezoidal cross section. Fine-mesh phototubes are used for readout on each end, and precision ($\sim$ 25 ps) timing electronics should achieve a resolution on individual particles of $\sim$ 100 ps. This leads to a $K - \pi$ separation at the 2σ level for particle momenta $p <$ 2 GeV/c.

The primary purpose of the TOF is to improve B flavor tagging in the measurement of CP asymmetries[11]. The TOF will be used to increase the flavor tagging efficiency of B^0 decays, by selecting pions over kaons for a same-side tag, and by selecting kaons for an opposite-side tag. In B_s^0 decays, the TOF will be used to select kaons over pions for the same-side tag. The combination of the TOF system, the Layer 00 system, and the remaining detector lead to an expected reach in x_s of up to 63.

4 Plug Upgrade

For Run II, the calorimeters in the region $3.6 < |\eta| < 1.1$ are being replaced with a single fast, hermetic, scintillating tile-fiber calorimeter[5]. The electromagnetic calorimeter is 21 radiation length deep and uses 4 mm lead plates. Located at a depth of approximately 6 radiation lengths is a shower maximum detector to measure the precise position of electron and photon showers and provide rejection against π^o's. The hadronic calorimeter uses steel absorber and is 6.6 interaction lengths deep.

Based on test beam results, the resolution of the electromagnetic calorimeter is expected to be $\sigma(E)/E = 16\%/\sqrt{E}$, while the resolution of the the hadronic calorimeter is expected to be $\sigma(E)/E = 75\%/\sqrt{E} \bigoplus 5.3\%$, both substantially better than the previous plug calorimeter. The improved resolution along with the improved tracking into the plug region should dramatically improve lepton identification and b-quark tagging in this region.

5 Muon System Upgrade

The muon detection system for CDF has been improved several times over the last few years and improvements are continuing for Run II. The existing system consists of the CMU ($|\eta| < 0.6$), CMP ($|\eta| < 0.6$ behind more shielding), and CMX ($0.6 < |\eta| < 1.0$). The primary improvement for Run II in these systems lies in filling gaps in fiducial coverage. The coverage of the CMP system is being increased by 17% while the coverage for the CMX is being increased by 45%. In addition, shielding for the CMX system is being added to reduce accidental triggers from the scatter of low energy particles near the beam line. Finally, an additional muon detector subsystem called the Intermediate Muon system (IMU) is being added[5]. It covers the region of forward-backward going muons ($1.0 < |\eta| < 2.0$) and covers 270° in azimuthal angle. Including all the enhancements to the muon system, the total coverage for muons increases by a factor of 2 when compared to Run I.

6 Data Aquisition and Trigger Systems

Just as in Run I, CDF will employ a 3 Level trigger system[5] in selecting events to write out to mass storage. The Level 1 decision is based upon primitives from various detector subsystems, such as the calorimetry, the muon systems, and the tracking system. With 132 nsec between bunches, there is insufficient time to make a Level 1 decision based on the current crossing, before the next crossing arrives. To accommodate this, the Level 1 trigger electronics is pipelined, with on-board buffering for 42 beam crossings, or 5.5 μsec. The maximum accept rate for Level 1 is 50 kHz. Upon a Level 1 trigger accept, the data on each trigger subsystem are transferred to one of four local Level 2 buffers. Level 2 consists of Alpha processor nodes that examine the events in more detail. The Level 2 trigger is an asynchronous system with an average decision time of 20 μsec. The maximum Level 2 accept rate is 300-1000 Hz. The system is designed to be 90% live at 90% of the maximum bandwidth. Upon a Level 2 accept, the data is collected in DAQ buffers. Event building is performed via a network switch between these buffers and the Level 3 trigger, which is a farm of PC's running Linux. Level 3 uses an optimzed version of full data reconstruction to make a final decision whether or not to write the event to mass storage for offline analysis, at an accept rate of less than 30 Hz.

In Run I, the trigger had to reduce the raw collision rate by a factor of 10^5 to reach the Level 3 output rate of < 10 Hz. In Run II, the rejection factor from collision rate to output rate needs to increase by another factor of 10. The improved background rejection will be accomplished by providing additional detector information to the Level 1 and Level 2 systems. The most significant change for Level 1 is the addition of the trigger track processor, the eXtremely Fast Tracker (XFT)[12]. In Run I, this capability was available only at Level 2. The XFT uses information from the COT to determine the P_T and ϕ of charged tracks with $P_T \geq 1.5$ GeV/c. This allows a track to be matched to an electromagnetic-calorimeter cluster for electron identification, or to be matched to a stub in the muon system for muon identification. The most important addition to the Level 2 trigger is the presence of hit data from the SVXII detector. A Silicon Vertex Trigger (SVT)[13] will use Level 1 XFT tracks as seeds for linking hits in the silicon system. The resulting track will have an improved momentum resolution and an impact parameter measurement. The performance for this device is expected to be $\Delta P_T/P_T^2 = 0.3\%$ with a 35 micron resolution on the impact parameter. This feature will allow trigging on displaced tracks and is a tremendous advantage for the study of the b quark.

7 Summary

Although characterized as an upgrade to an existing detector, the CDF detector for Run II is really a new and powerful tool for studying $p\bar{p}$ collisions. All aspects of the upgrade are well underway, with most detector components well into the production and construction stages. Initial commissioning with cosmic rays of some subsystems and electronics has already started, and will continue through the beginning of summer 2000. Most of the electronics and the detector systems - with the exception of the silicon systems - will be ready by mid summer 2000. A $p\bar{p}$ commissioning run of this part of the CDF detector will take place from August to November of 2000. The silicon systems will be complete by September 2000 and installed by January 2001. The complete detector will be ready to observe collisions in March 2001, and will help to begin a exciting and promising new era in high energy physics.

Acknowledgements

The upgrade to CDF is supported by the U.S. Department of Energy; the National Science Foundation; the Italian Istituto Nazionale di Fisica Nucleare; the Ministry of Education, Science and Culture of Japan; the Natural Sciences and Engineering research Council of Canada; the National Science Council of the Republic of China; the Swiss National Science Foundation; and the A.P. Sloan Foundation.

References

1. F. Abe *et al.* (CDF Collaboration), Phys. Rev. D **50**, 2966 (1994); F. Abe *et. al.* (CDF Collaboration), Phys. Rev. Lett. **73**, 225 (1994).
2. F. Abe *et al.* (CDF Collaboration), Phys. Rev. Lett. **74**, 2626 (1995).
3. S. Abachi *et al.* (D0 Collaboration), Phys. Rev. Lett. **74**, 2632 (1995).
4. T. Affolder *et al.* (CDF Collaboration), Fermilab-Pub-99-225-E, Sep 1999.
5. The CDF II Collaboration, *The CDF II Detector: Technical Design Report*, Fermilab-Pub-96-390-E, 1996.
6. S. D. Holmes *et al.*, *Status of the Main Injector and Recycler*, Proceedings of the 1997 Particle Accelerator Conference.
7. Editors D Amidei, R. Brock *Future ElectroWeak Physics at the Fermilab Tevatron: Report of the TeV2000 Study Group*, Fermilab-Pub-96/082, 1996.
8. J. A. Valls *et al.*, *The SVX II Silicon Vertex Detector at CDF*, Nucl. Phys. Proc. Suppl. 78:311-314, 1999.
9. P. Azzi-Bacchetta *et al.*, *The CDF Intermediate Silicon Layers Detector*, Nucl. Phys. Proc. Suppl. 78:307-310, 1999.
10. K. T. Pitts *et al.*, *The CDF Central Outer Tracker*, Presented at the 5th International Conference on Advanced Technology and Particle Physics (COMO 96), Como, Italy, Fermilab-COnf-96-443-E, 1997.
11. T. Miao *et al.*, *CP Violation and Mixing of B Mesons at CDF*, Fermilab-Conf-99-295-E, 1999.
12. S. Holm *et al.*, *System Architecture and Hardware Design of the CDF XFT Online Track Processor*, Fermilab-Conf-99-317, 1999.
13. W. Ashmanskas *et al.*, *Silicon Vertex Tracker: A Fast Precise Tracking Trigger for CDF*, Fermilab-Conf-99-219-E, 1999.

FUTURE B EXPERIMENTS FROM THE BTEV/LHC-B PERSPECTIVE

SHELDON STONE

Physics Department, Syracuse University, Syracuse N. Y., USA, 13244-1130
E-mail: stone@phy.syr.edu

Many measurements are necessary in the program of studying mixing, CP violation and rare decays of b and c quarks. These measurements require large numbers of B^o, B_s, B^- and D^{*+} hadrons. Fortunately, copius production of particles containing b and c quarks will occur at Tevatron and the LHC. The crucial measurements are described here, as well as the design of the two experiments, LHC-b and BTeV, that can exploit the $4-20\ \times 10^{11}$ b hadrons produced every 10^7 seconds.

1 Introduction

The basic experimental goals are to make exhaustive search for physics beyond the Standard Model and to precisely measure Standard Model parameters. I first discuss what studies need to be done, not just what studies can be done in the near future. Measurements are necessary on CP violation in B^o and B_s mesons, B_s mixing, rare b decay rates, and mixing, CP violation and rare decays in the charm sector.

2 The CKM Matrix and CP Violation

2.1 The 6 Unitarity Triangles

The base states of quarks, the mass eigenstates, are mixed to form the weak eigenstates as described by the 3×3 complex Cabibbo-Kobayashi-Maskawa matrix,[1] that can be expressed in terms of 4 fundamental constants of nature, that need to be determined experimentally. In the Wolfenstein approximation[2] the matrix is written as:

$$\begin{matrix} 1-\lambda^2/2 & \lambda & A\lambda^3(\rho-i\eta(1-\lambda^2/2)) \\ -\lambda & 1-\lambda^2/2-i\eta A^2\lambda^4 & A\lambda^2(1+i\eta\lambda^2) \\ A\lambda^3(1-\rho-i\eta) & -A\lambda^2 & 1 \end{matrix} . \quad (1)$$

This expression is accurate to order λ^3 in the real part and λ^5 in the imaginary part. It is necessary to express the matrix to this order to have a complete formulation of the physics we wish to pursue.

Non-zero η allows for CP violation. CP violation thus far has only been seen in the neutral kaon system. By exploring CP violation in the b and c

systems we can see if the CKM model works or perhaps discover new physics that goes beyond the model, if it does not.

The unitarity of the CKM matrix allows us to construct six relationships. These equations may be thought of triangles in the complex plane. They are shown in Fig. 1.

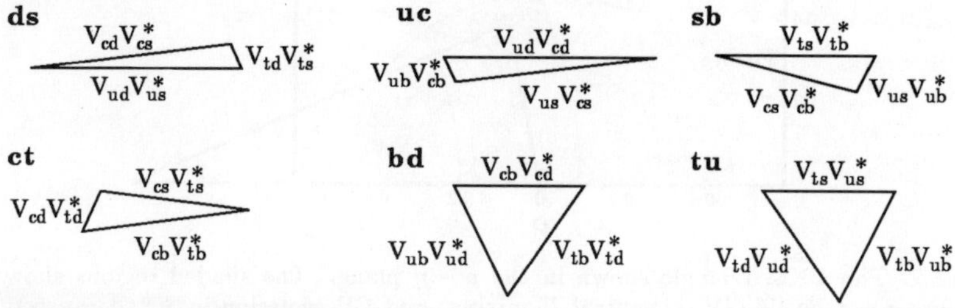

Figure 1. The six CKM triangles. The bold labels, e.g. **ds** refer to the rows or columns used in the unitarity relationship.

All six of these triangles can be constructed knowing four and only four independent angles.[3,4,5] These are defined as:

$$\beta = arg\left(-\frac{V_{tb}V_{td}^*}{V_{cb}V_{cd}^*}\right), \qquad \gamma = arg\left(-\frac{V_{ub}^*V_{ud}}{V_{cb}^*V_{cd}}\right),$$
$$\chi = arg\left(-\frac{V_{cs}^*V_{cb}}{V_{ts}^*V_{tb}}\right), \qquad \chi' = arg\left(-\frac{V_{ud}^*V_{us}}{V_{cd}^*V_{cs}}\right). \tag{2}$$

Two of the phases β and γ are probably large while χ is estimated to be small $\approx$0.02, but measurable, while χ' is likely to be much smaller.

In the **bd** triangle, the one usually considered, the angles are all thought to be relatively large. Since $V_{cd}^* = \lambda$, this triangle has sides 1, $\frac{1}{\lambda}\left|\frac{V_{td}}{V_{ts}}\right|$, $\frac{1}{\lambda}\left|\frac{V_{ub}}{V_{cb}}\right|$. This CKM triangle is depicted in Fig. 2, with constraints from other measurements.[6] Also shown are the angles α, β, and γ. Since they form a triangle the "real" α, β and γ must sum to 180°; therefore measuring any two of these determines the third.

It has been pointed out by Silva and Wolfenstein[3] that measuring these angles may not be sufficient to detect new physics. For example, suppose there is new physics that arises in $B^o - \overline{B}^o$ mixing. Let us assign a phase θ to this new physics. If we then measure CP violation in $B^o \rightarrow J/\psi K_S$ and eliminate any Penguin pollution problems in using $B^o \rightarrow \pi^+\pi^-$, then

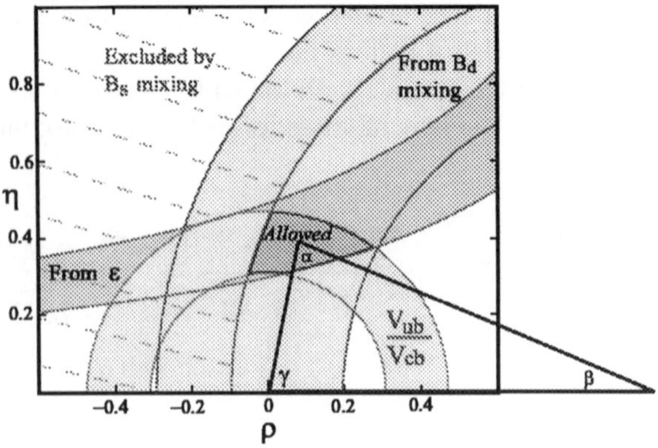

Figure 2. The CKM triangle shown in the $\rho - \eta$ plane. The shaded regions show $\pm 1\sigma$ contours given by $|V_{ub}/V_{cb}|$, neutral B mixing, and CP violation in K_L^o decay (ϵ). The dashed region is excluded by B_s mixing limits. The allowed region is defined by the overlap of the 3 permitted areas, and is where the apex of the CKM triangle sits.

we actually measure $2\beta' = 2\beta + \theta$ and $2\alpha' = 2\alpha - \theta$. So while there is new physics, we miss it, because $2\beta' + 2\alpha' = 2\alpha + 2\beta$ and $\alpha' + \beta' + \gamma = 180°$.

2.2 Ambiguities

In measuring CP phases there are always ambiguities. For example, any determination of $\sin(2\phi)$, has a four-fold ambiguity; ϕ, $\pi/2 - \phi$, $\pi + \phi$, $3\pi/2 - \phi$ are all allowed solutions. Often the point of view taken is that we know η is a positive quantity and thus we can eliminate two of the four possibilities. However, this is dangerous as it could lead to our missing new physics. Evidence that η is positive is derived from the measurements of ϵ and ϵ' using theoretical models. Even accepting that K_L decays give $\eta > 0$, it would be foolhardy to miss new physics just because we now assume that η must be positive rather than insisting on a clean measurement of the angles that could show a contradiction.

2.3 Technique for Measuring α

It is well known that $\sin(2\beta)$ can be measured without problems caused by Penguin processes using the reaction $B^o \to J/\psi K_S$. The simplest reaction that can be used to measure $\sin(2\alpha)$ is $B^o \to \pi^+\pi^-$. This reaction can proceed via both the Tree and Penguin diagrams shown in Fig. 3.

Current CLEO results[8] are $\mathcal{B}(B^o \to K^{\mp}\pi^{\pm}) = (1.88^{+0.28}_{-0.26} \pm 0.13) \times 10^{-5}$

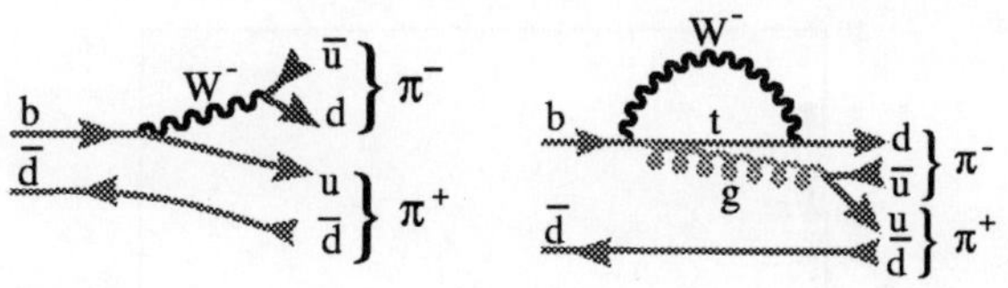

Figure 3. Processes for $B^o \to \pi^+\pi^-$: Tree (left) and Penguin (right).

and $\mathcal{B}(B^o \to \pi^+\pi^-) = (0.47^{+0.18}_{-0.15} \pm 0.06) \times 10^{-5}$, showing a relatively large Penguin amplitude that cannot be ignored. The Penguin contribution to $\pi^+\pi^-$ is roughly half the Tree *amplitude.* Thus the effect of the Penguin must be determined in order to extract α. The only model independent way of doing this was suggested by Gronau and London, but requires the measurement of $B^\mp \to \pi^\mp\pi^o$ and $B^o \to \pi^o\pi^o$, the latter being rather daunting.

There is however, a theoretically clean method to determine α. The interference between Tree and Penguin diagrams can be exploited by measuring the time dependent CP violating effects in the decays $B^o \to \rho\pi \to \pi^+\pi^-\pi^o$ as shown by Snyder and Quinn.[9]

The $\rho\pi$ final state has many advantages. First of all, it has been seen with a relatively large rate. The branching ratio for the $\rho^o\pi^+$ final state as measured by CLEO[10] is $(1.5\pm0.5\pm0.4)\times10^{-5}$, and the rate for the neutral B final state $\rho^\pm\pi^\mp$ is $(3.5^{+1.1}_{-1.0}\pm0.5)\times10^{-5}$, while the $\rho^o\pi^o$ final state is limited at 90% confidence level to $< 5.1\times10^{-6}$. These measurements are consistent with some theoretical expectations.[11] Secondly, the associated vector-pseudoscalar Penguin decay modes have conquerable or smaller branching ratios. Furthermore, since the ρ is spin-1, the π spin-0 and the initial B also spinless, the ρ is fully polarized in the (1,0) configuration, so it decays as $cos^2\theta$, where θ is the angle of one of the ρ decay products with the other π in the ρ rest frame. This causes the periphery of the Dalitz plot to be heavily populated, especially the corners. A sample Dalitz plot is shown in Fig. 4. This kind of distribution is good for maximizing the interferences, which helps minimize the error. Furthermore, little information is lost by excluding the Dalitz plot interior, a good way to reduce backgrounds.

To estimate the required number of events Snyder and Quinn preformed an idealized analysis that showed that a background-free, flavor-tagged sample of 1000 to 2000 events was sufficient. The 1000 event sample usually yields good results for α, but sometimes does not resolve the ambiguity. With the 2000 event sample, however, they always succeeded.

This technique not only finds $\sin(2\alpha)$, it also determines $\cos(2\alpha)$, thereby

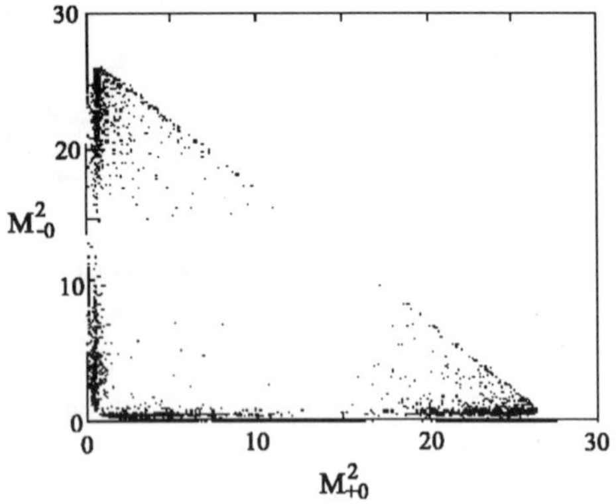

Figure 4. The Dalitz plot for $B^o \to \rho\pi \to \pi^+\pi^-\pi^o$ from Snyder and Quinn.

removing two of the remaining ambiguities. The final ambiguity can be removed using the CP asymmetry in $B^o \to \pi^+\pi^-$ and a theoretical assumption.[12]

2.4 Techniques for Measuring γ

In fact, it may be easier to measure γ than α. There have been at least four methods suggested.

(1) Time dependent flavor tagged analysis of $B_s \to D_s^{\pm}K^{\mp}$. This is a direct model independent measurement.[13]

(2) Measure the rate differences between $B^- \to \overline{D}^o K^-$ and $B^+ \to D^o K^+$ in two different D^o decay modes such as $K^-\pi^+$ and K^+K^-. This method makes use of the interference between the tree and doubly-Cabibbo suppressed decays of the D^o, and does not depend on any theoretical modeling.[14,15]

(3) Rate measurements in two-body $B \to K\pi$ decays. A cottage industry has developed. However, all methods are model dependent.[16]

(4) Use U-spin symmetry to relate $B^o \to \pi^+\pi^-$ and $B_s \to K^+K^-$.[17]

2.5 Required Measurements Involving β

The phase of $B^o - \overline{B^o}$ mixing will soon be measured by e^+e^- b-factories using the $J/\psi K_S$ final state. New physics could be revealed by measuring other final states such as ϕK_S, $\eta' K_S$ or $J/\psi\pi^o$.

It is also important to resolve the ambiguities. There are two suggestions on how this may be accomplished. Kayser[18] shows that time dependent measurements of the final state $J/\psi K^o$, where $K^o \to \pi\ell\nu$, give a direct measurement of $\cos(2\beta)$ and can also be used for CPT tests. Another suggestion is to use the final state $J/\psi K^{*o}$, $K^{*o} \to K_S\pi^o$, and to compare with $B_s \to J/\psi\phi$ to extract the sign of the strong interaction phase shift assuming SU(3) symmetry, and thus determine $\cos(2\beta)$.[19]

2.6 A Critical Check Using χ

The angle χ, defined in equation 2, can be extracted by measuring the time dependent CP violating asymmetry in the reaction $B_s \to J/\psi\eta^{(\prime)}$, or if one's detector is incapable of quality photon detection, the $J/\psi\phi$ final state can be used. However, there are two vector particles in the final state, making this a state of mixed CP a requiring time-dependent angular analysis to find χ that requires large statistics.

Measurements of the magnitudes of CKM matrix elements all come with theoretical errors. Some of these are hard to estimate; we now try and view realistically how to combine CP violating phase measurements with the magnitude measurements to best test the Standard Model.

The best measured magnitude is that of $\lambda = |V_{us}/V_{ud}| = 0.2205 \pm 0.0018$. Silva and Wolfenstein[3,4] show that the Standard Model can be checked in a profound manner by seeing if:

$$\sin\chi = \left|\frac{V_{us}}{V_{ud}}\right|^2 \frac{\sin\beta\ \sin\gamma}{\sin(\beta+\gamma)} \quad . \tag{3}$$

Here the precision of the check will be limited initially by the measurement of $\sin\chi$, not of λ. This check can reveal new physics, even if other measurements have not shown any anomalies. There are other checks using $\left|\frac{V_{ub}}{V_{cb}}\right|$ or $\left|\frac{V_{td}}{V_{ts}}\right|$.[6]

2.7 Other Critical CKM Measurements and Summary

Magnitudes of the CKM elements are important to measure as precisely as possible. Current measurements of $|V_{cb}|$ and $|V_{ub}|$ are discussed elsewhere.[20]

It has been predicted that $\Delta\Gamma/\Gamma$ for the B_s system is of the order of 10%. This can be determined by measuring the lifetimes in different final states such as $D_s^-\pi^+$ (mixed CP), $J/\psi\eta'$ (CP $-$) and K^+K^- (CP $+$). A finite $\Delta\Gamma$ would allow many other interesting measurements of CP violation.[21]

Table 1 lists the most important physics quantities and the suggested decay modes. The necessary detector capabilities include the ability to collect purely hadronic final states, the ability to identify charged hadrons, the ability to detect photons with good efficiency and resolution and excellent time resolution required to analyze rapid B_s oscillations.

Table 1. Required CKM Measurements for b's

Physics Quantity	Decay Mode	Hadron Trigger	$K\pi$ sep	γ det	Decay time σ
$\sin(2\alpha)$	$B^o \to \rho\pi \to \pi^+\pi^-\pi^o$	√	√	√	
$\cos(2\alpha)$	$B^o \to \rho\pi \to \pi^+\pi^-\pi^o$	√	√	√	
$\text{sign}(\sin(2\alpha))$	$B^o \to \rho\pi$ & $B^o \to \pi^+\pi^-$	√	√	√	
$\sin(\gamma)$	$B_s \to D_s^\pm K^\mp$	√	√		√
$\sin(\gamma)$	$B^- \to \overline{D}^0 K^-$	√	√		
$\sin(\gamma)$	$B^o \to \pi^+\pi^-$ & $B_s \to K^+K^-$	√	√		√
$\sin(2\chi)$	$B_s \to J/\psi\eta'$, $J/\psi\eta$			√	√
$\sin(2\beta)$	$B^o \to J/\psi K_s$				
$\cos(2\beta)$	$B^o \to J/\psi K^o$, $K^o \to \pi\ell\nu$				
$\cos(2\beta)$	$B^o \to J/\psi K^{*o}$ & $B_s \to J/\psi\phi$				√
x_s	$B_s \to D_s^+\pi^-$	√			√
$\Delta\Gamma$ for B_s	$B_s \to J/\psi\eta'$, $D_s^+\pi^-$, K^+K^-	√	√	√	√

3 Searches for New Physics

Because new physics at much larger mass scales can appear in loops, rare process such as $b \to s\gamma$, $d\gamma$, $s\ell^+\ell^-$ and $d\ell^+\ell^-$ have the promise to reveal new physics. Searches in both exclusive and inclusive final states are important.

Charm decays also offer the possibility of finding new physics in the study of either mixing or CP violation as the Standard Model prediction is small. The current experimental measurement of mixing is $r_D < 5 \times 10^{-3}$, while the SM expectation[22] is $10^{-7} - 10^{-6}$. For CP violation the current limits are about 10%, while the expectation[23] is 10^{-3}.

4 The Next Generation of Experiments: LHC-b and BTeV

4.1 Rationale

To over constrain the CKM matrix and look for new physics, all the quantities listed in Table 1 must be measured. This requires large samples of b-flavored hadrons, and detectors capable of tolerating large interaction rates and having excellent lifetime resolution, particle identification and γ/π^o detection capabilities.

Fortunately, large samples of b quarks are available. With the Fermilab Main Injector, the Tevatron collider will produce $\approx 4 \times 10^{11}$ b hadrons/10^7 s at a luminosity of $2 \times 10^{32}\text{cm}^{-2}\text{s}^{-1}$. Rates are ≈ 5 times larger at the LHC. These compare very favorably to e^+e^- machines operating at the $\Upsilon(4S)$. At a luminosity of 3×10^{33} they produce 6×10^7 B's/10^7 s. Furthermore B_s, Λ_b and other b-flavored hadrons are accessible for study at hadron colliders. Also important are the large charm rates, ~10 times larger than the b rate.

4.2 b Production Characteristics

Let us explore the reasons for the choice of the forward direction. It is often customary to characterize heavy quark production in hadron collisions with the two variables p_t and η, where $\eta = -ln(\tan(\theta/2))$, and θ is the angle of the particle with respect to the beam direction. According to QCD based calculations of b quark production, the B's are produced "uniformly" in η and have a truncated transverse momentum, p_t, spectrum characterized by a mean value approximately equal to the B mass.[24] The distribution in η is shown in Fig. 5. Note that at larger values of $|\eta|$, the B boost, $\beta\gamma$, increases rapidly.

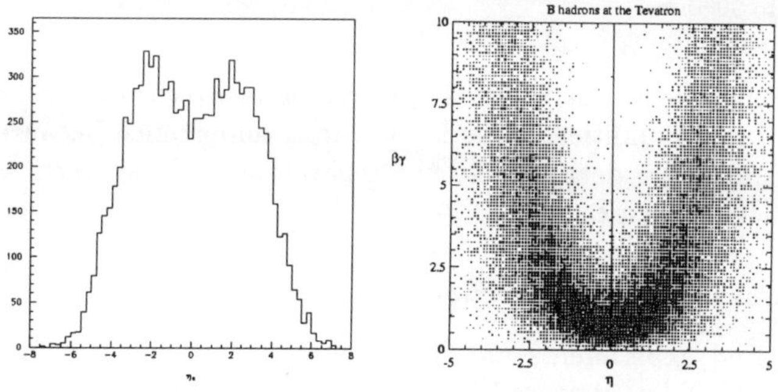

Figure 5. The B yield versus η (left). $\beta\gamma$ of the B versus η (right). Both plots are for the Tevatron.

The "flat" η distribution hides an important correlation of $b\bar{b}$ production at hadronic colliders. In Fig. 6 the production angles of the hadron containing the b quark is plotted versus the production angle of the hadron containing the $\bar{b}$ quark according to the Pythia generator. Many important measurements require the reconstruction of a b decay and the determination of the flavor of the other $\bar{b}$, thus requiring both b's to be observed in the detector. There is a very strong correlation in the forward (and backward) direction: when the B is forward the $\overline{B}$ is also forward. This correlation is not present in the central region (near zero degrees). By instrumenting a relative small region of angular phase space, a large number of $b\bar{b}$ pairs can be detected. Furthermore the B's populating the forward and backward regions have large values of $\beta\gamma$.

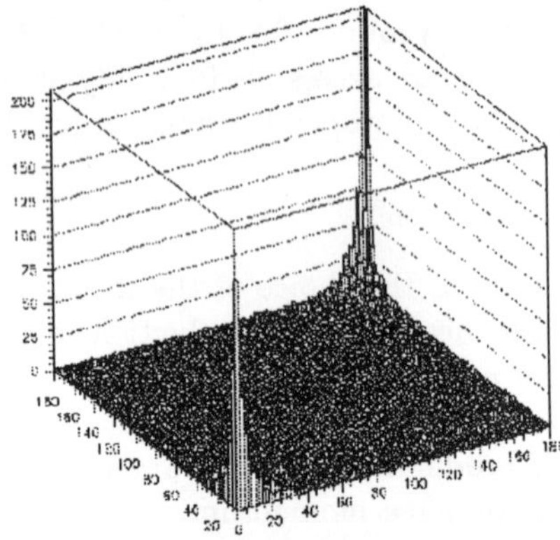

Figure 6. The production angle (in degrees) for the hadron containing a b quark plotted versus the production angle for a hadron containing $\bar{b}$ quark. (For the Tevatron.) η of zero corresponds to 90° here.

Both experiments use forward spectrometers which utilizes the boost of the B's at large rapidities. This is of crucial importance because the main way to distinguish b decays is by the separation of decay vertices from the main interaction vertex.

4.3 General Design Considerations

Two dedicated experiments are contemplated, LHC-b which has been approved and BTeV, an approved R&D program at Fermilab that was requested to prepare a proposal for submission in summer of 2000.[25] They both look for b decays in the "forward" direction close to the beam line to exploit the large b momenta and the correlated $b\bar{b}$ production. As a result, they both have long and narrow interaction regions. The C0 interaction region at Fermilab was constructed to allow the BTeV experiment to fit.

There are problems that heavy quark experiments at hadron colliders must overcome. First of all, the hugh b rate is accompanied by an even larger rate of uninteresting interactions. At the Tevatron the b-fraction is only 1/500, and is only 5 times larger at the LHC. In searching for rare processes, at the level of parts per million, the background from b events is dominant. (Of course all b experiments have this problem.) The large data rate of b's must be handled. For example, BTeV, has 1 kHz into the detector, and these events must be selected and written out. The electromagnetic calorimeter must robust enough to deal with the particles from the underlying event and still have useful efficiency. Furthermore, radiation damage can destroy detector elements.

The strategy BTeV employes to overcome these obstacles includes triggering on events with detached vertices in the first trigger level. Both BTeV and LHC-b use the excellent detached vertex resolution, $\sim$ 40 fs, to reject backgrounds in their analyses. Both experiments incorporate deadtimeless trigger and data acquisition systems, and have excellent Cherenkov Ring Imaging detectors to select charged hadrons. Furthermore BTeV has an excellent Electromagnetic calorimeter made from $PbWO_4$ crystals, based on the design of CMS. Both experiments can measure B_s mixing up to x_s values of 60-80.

LHC-b employs a somewhat different trigger strategy. They have a first level trigger (which they somewhat annoyingly call level-0) that selects events on the basis of having hadronic tracks or leptons with transverse momenta in the range of a few GeV/c. They then ask for detached verticies in the next trigger level.

4.4 Short Detector Descriptions

The LHC-b detector is shown in Fig. 7. The vertex detector is made from silicon strips. It is located in a magnetic field free region, and placed as close as 10 mm from the beam line. Two Ring Imaging Cherenkov detectors are used with different gases to cover the full momentum range for K/π separation. In addition, to separate kaons from protons, an aerogel radiator will likely be attached to the entrance window on the first RICH. There is a "Shaslik" style electromagnetic calorimeter made from scintillating fibers embedded in lead and a hadron calorimeter whose primary use is to veto crossing with more than one interactions, since these can confuse the trigger.

The BTeV Detector is shown in Fig. 8. It has two "forward" arms. This is done to increase the accepted b rate. BTeV must make up a factor of $\approx$5 to compete with LHC-b. BTeV has the vertex detector in the magnetic field. It is a pixel detector and is placed 6 mm from the beam line. BTeV believes that the detached vertex triggering is enhanced by eliminating low momentum, large multiple scattering tracks, from consideration. The single RICH, also with aerogel, is followed by a $PbWO_4$ electromagnetic calorimeter based on the design of CMS for radiation hard crystals. The crystals are

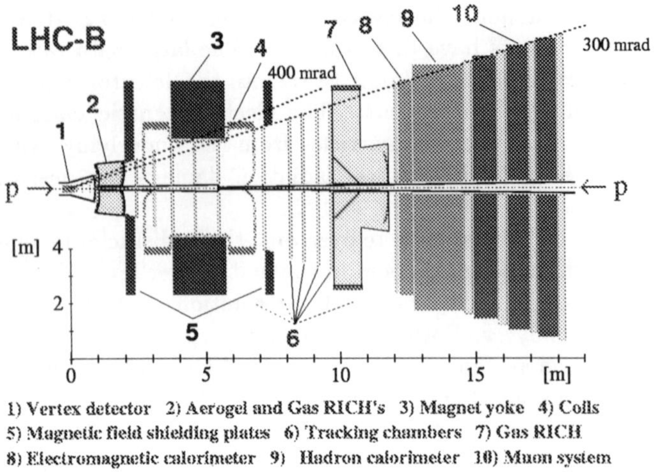

Figure 7. Schematic of the LHC-b detector.

approximately 26 mm x 26 mm x 220 mm. This gives good segmentation. Note that the lower momenta and multiplicities allow each arm of BTeV to be about half the length of LHC-b.

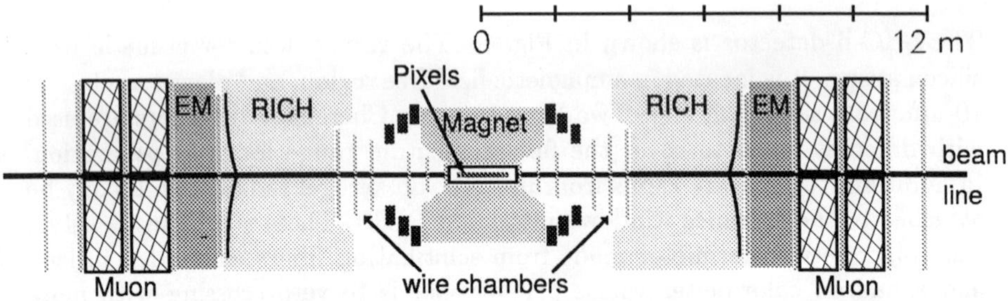

Figure 8. Schematic of the BTeV detector.

4.5 Some Sensitivity Projections

Space precludes a through discussion of sensitivities. I will discuss one illustrative example, the CP asymmetry in $B^o \to \pi^+\pi^-$. I use a BTeV simulation. The B^o momentum distribution for events in the detector is shown in Fig. 9.

Also shown is the error in decay distance. The peak of the B^o momentum

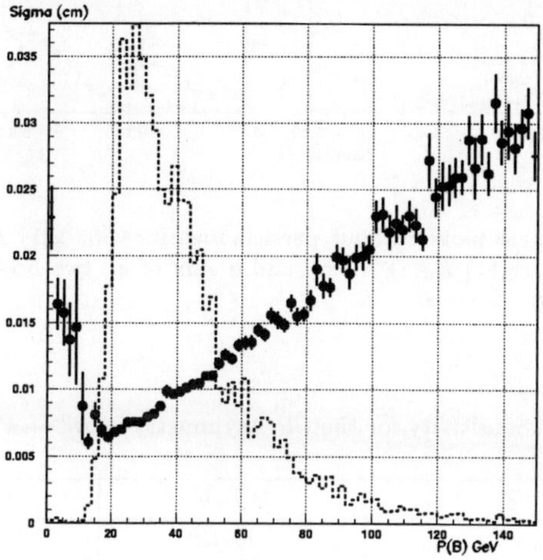

Figure 9. The momentum distribution (dashed) of B's accepted by BTeV in the $\pi^+\pi^-$ decay mode along with the error on the vertex position measurment (solid points).

distribution is about 30 GeV/c. Since the average decay length goes as 480 $\mu\text{m}\times p_B/m_B$, 30 GeV/c B's go about 3 mm. Below about 20 GeV/c the error on measuring the decay distance grows, while above the decay distance error grows linearly with B momentum. The error growth at low momentum is due to multiple scattering. Since the key variable is decay distance divided by the error on decay distance, L/σ, it is best to have B's above 20 GeV/c. L/σ is key in triggering as well as rejecting background. However, another significant source of background could be the other two-body decays into two pseudoscalars. These include $B^o \to K^\pm\pi^\mp$ and K^+K^-, and $B_s \to K^+K^-$ and $K^\pm\pi^\mp$. Fig. 10 shows the reconstructed mass spectra using the current CLEO measured branching ratios[8] for $K^-\pi^+$ and $\pi^+\pi^-$, and assuming the corresponding modes in B_s decay have the same branching fractions. The $\pi^+\pi^-$ signal is completely swamped. However, the RICH particle identification should remove almost all the background. The relevant sensitivity calculations are shown for LHC-b and BTeV in Table 2

Comparing with e^+e^- machines at a luminosity of 3×10^{33}, the sensitivity is vastly greater; there are only 13 flavor tagged $\pi^+\pi^-$ events in e^+e^-.

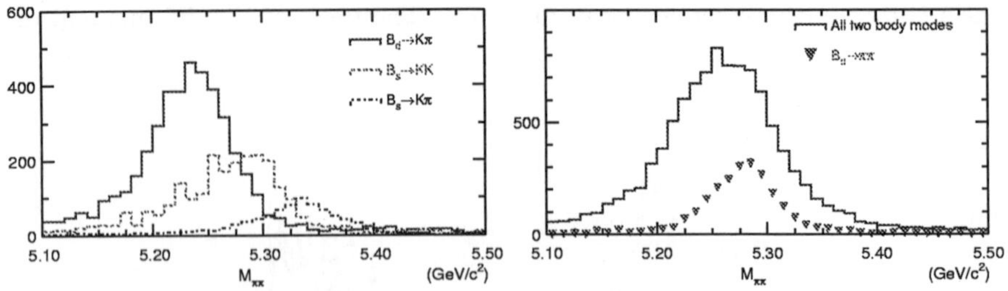

Figure 10. Two-body mass plots without particle identification left) $B_d \to K^+\pi^-$, $B_s \to \pi^+K^-$, $B_s \to K^+K^-$, right) $B_d \to \pi^+\pi^-$ and a sum of all two body decay modes. All particles are assumed to be pions.

Table 2. Sensitivity for the CP Asymmetry in $B^o \to \pi^+\pi^-$

Quantity	BTeV	LHC-b
b cross-section	100 μb	500 μb
Luminosity	$2 \times 10^{32}\text{cm}^{-2}\text{s}^{-1}$	$2 \times 10^{32}\text{cm}^{-2}\text{s}^{-1}$
# B^o/Year (10^7 s)	1.4×10^{11}	7.0×10^{11}
$\mathcal{B}(B^o \to \pi^+\pi^-)$	0.47×10^{-6}	0.47×10^{-6}
Reconstruction efficiency	0.06	0.03
Triggering efficiency	0.50	0.17
# of $(\pi^+\pi^-)$	20,000	18,700
ϵD^2 for flavor tags	0.1	0.1
Signal/Background	0.4	0.67
Error in Asymmetry	±0.042	±0.037

5 Conclusions

Both LHC-B and BTeV will be able to make what we now consider to be the quintessential measurements in b and charm physics. The sensitivities on the angle γ are estimated to be better than 10°. It is likely that at least two of the ambiguities in $\sin(2\beta)$ will be resolved. To measure α the reaction $B^o \to \rho\pi$ looks most promising. Here preliminary estimates give BTeV a factor of 50 larger efficiency. However, neither experiment has estimated the backgrounds. Both experiments aim toward measuring χ. BTeV, with its excellent electromagnetic calorimeter will use primarily the $B_s \to J/\psi\eta'$ mode, while both experiments will also use the $J/\psi\phi$ final state. Besides the precise measurement of CKM angles and resolving ambiguities, the search for new physics will be of paramount importance.

6 Acknowledgments

Support for this work was provided by the National Science Foundation. I thank George Hou and Hai-Yang Cheng for organizing an excellent and fruitful meeting. I thank Marina Artuso, Boris Kayser and Tomasz Skwarnicki for many useful discussions.

References

1. N. Cabibbo, *Phys. Rev. Lett.* **10**, 531 (1963); M. Kobayashi and K. Maskawa, *Prog. Theor. Phys.* **49**, 652 (1973).
2. L. Wolfenstein, *Phys. Rev. Lett.* **51**, 1945 (1983).
3. J. P. Silva, L. Wolfenstein, *Phys. Rev.* D **55**, 5331 (1997) [hep-ph/9610208].
4. R. Aleksan, B. Kayser and D. London, *Phys. Rev. Lett.* **73**, 18 (1994) [hep-ph/9403341].
5. I. I. Bigi and A. I. Sanda, "On the Other Five KM Triangles," [hep-ph/9909479].
6. S. Stone, "Future of Heavy Flavour Physics: Experimental Perspective," presented at Heavy Flavours 8, Southapton, UK, 1999, [hep/ph xxxx].
7. T. Blum *et al.*(RIKEN-BNL-Columbia), hep-lat/9908025.
8. Y. Kwon *et al.* (CLEO), "Study of Charmless Hadronic B Decays into the Final States $K\pi$, $\pi\pi$ and KK, with the First Observations of $B^o \to \pi^+\pi^-$and $B^o \to K^o\pi^o$," Conf 99-14, [hep-ex/9908039].
9. A. E. Snyder and H. R. Quinn, Phys. Rev. **D 48**, 2139 (1993).
10. Y. Gao and F. Würthwein, CLEO preprint [hep-ex/9904008].
11. A. Ali, G. Kramer, and C.D. Lu, *Phys. Rev.* **D 59**, 014005 (1999) [hep-ph/9805403].
12. Y. Grossman and H. R. Quinn, *Phys. Rev.* D **56**, 7259 (1997) [hep-ph/9705356].
13. D. Du, I. Dunietz and Dan-di Wu, *Phys. Rev* **D 34**, 3414 (1986); R. Aleksan, I Dunietz, and B. Kayser, *Z. Phys.* **C 54**, 653 (1992); R. Aleksan, A. Le Yaouanc, L. Oliver, O. Pène and J.-C. Raynal, Z. Phys. **C67** (1995) 251 [hep-ph/9407406].
14. D. Atwood, I. Dunietz and A. Soni, *Phys. Rev. Lett.* **78**, 3257 (1997).
15. M. Gronau and D. Wyler, *Phys. Lett.* B **265**, 172 (1991).

16. R. Fleischer, and T. Mannel, *Phys. Rev.* **D 57**, 2752 (1998) [hep-ph/9704423]; M. Neubert and J. L. Rosner, *Phys. Rev. Lett.* **81**, 5076 (1998) [hep-ph/9809311]; M. Gronau, and J. L. Rosner, *Phys. Rev.* **D 57**,6843 (1998) [hep-ph/9711246]; M. Gronau, and D. Pirjol, [hep-ph/9902482]; M. Gronau and J. L. Rosner, *Phys. Rev.* D **59**, 113002 (1999) [hep-ph/9809384]; J.-M. Gerard and J. Weyers, *Eur. Phys. J* **C7**, 1 (1999) [hep-ph/9711469].
17. R. Fleischer, *Phys. Lett.* B **459**, 306 (1999) [hep-ph/9903456].
18. B. Kayser, "Cascade Mixing and the CP-Violating Angle Beta," [hep-ph/9709382]. Previous work in this area was done by Y. Azimov, *Phys. Rev.* **D 42**, 3705 (1990).
19. A. Dighe, I. Dunietz and R. Fleischer, *Phys. Lett.* B **433**, 147 (1998) [hep-ph/9804254].
20. Marina's EPS
21. For a recent calculation of $\Delta\Gamma_s$, see M. Beneke, G. Buchalla, C. Greub, A. Lenz and U. Nierste, *Phys. Lett.* B **459**, 631 (1999) [hep-ph/9808385]. See also I. Dunietz, *Phys. Rev.* **D 52**, 3048 (1995).
22. H. Georgi, *Phys. Lett.* B **297**, 353 (1992); T. Ohl *et al.*, *Nucl. Phys.* B **403**, 603 (1993).
23. F. Buccella *et al.*, *Phys. Lett.* B **302**, 319 (1993); F. Buccella *et al.*, *Phys. Lett.* B **379**, 249 (1996).
24. M. Artuso,"Experimental Facilities for b-Quark Physics," in *B Decays* revised 2nd Edition, Ed. S. Stone, World Scientific, Sinagapore (1994).
25. Harnew has recently discussed the prospects for these experiments and CMS and ATLAS as well. See N. Harnew, "Prospects for LHC-b, BTeV, ATLAS, CMS," in proceedings of HF8.

MANIFESTATION OF SUSY IN B DECAYS

YASUHIRO OKADA
Institute for Particle and Nuclear Studies, KEK,
Oho 1-1, Tsukuba 305-0801, Japan
E-mail: yasuhiro.okada@kek.jp

SUSY effects on various flavor changing neutral current processes are discussed in the minimal supergravity model and the SU(5) grand unified theory with right-handed neutrino supermultiplets. In particular, in the latter case the neutrino Yukawa coupling constants can be a source of the flavor mixing in the right-handed-down-type-squark sector. It is shown that due to this mixing the time-dependent CP asymmetry of radiative B decay can be as large as 30% and the ratio of B_s–$\bar{B}_s$ mixing and B_d–$\bar{B}_d$ mixing deviates from the prediction in the standard model and the minimal supergravity model without the neutrino interaction.

1 Introduction

In order to explore supersymetry (SUSY) indirect search experiments can play a complementary role to direct search for SUSY particles at collider experiments. Since SUSY particles may affect flavor changing neutral current (FCNC) processes and CP violation in B and K meson decays, it is possible that new experiments in B decay at both e^+e^- colliders and hadron machines reveal new physics signals which can be interpreted as indirect evidence of SUSY.

In the context of SUSY models flavor physics has important implications. Because the squark and the slepton mass matrices become new sources of flavor mixings generic mass matrices would induce too large FCNC and lepton flavor violation (LFV) effects if the superpartners' masses are in a few-hundred-GeV region. For example, if we assume that the SUSY contribution to the $K^0 - \bar{K}^0$ mixing is suppressed because of the cancellation among the squark contributions of different generations, the squarks with the same gauge quantum numbers must be highly degenerate in masses at least for the first two generations.

There are several scenarios to solve this flavor problem. In the minimal supergravity model flavor problem are avoided by taking SUSY soft-breaking terms as flavor-blind structure. The scalar mass terms are assumed to be common for all scalar fields at the Planck scale and therefore there are no FCNC effects nor LFV from the squark and slepton sectors at this scale. The physical squark and slepton masses are determined taking account of renormalization effects from the Planck to the weak scale. This induces sizable

SUSY contributions to various FCNC and LFV processes.

In this talk we consider two types of SUSY models and discuss FCNC processes. The first one is the minimal supersymmetric standard model (MSSM) with a universal SUSY breaking terms at the Planck scale which is realized in the minimal supergravity model. The other is the SU(5) grand unified theory with right-handed neutrino supermultiplets. This model incorporates the see-saw mechanism for neutrino mass generation. In the latter case the neutrino Yukawa coupling constants can be a source of the flavor mixing in the right-handed-down-type-squark sector and due to this mixing the time-dependent CP asymmetry of radiative B decay can be as large as 30% and the ratio of B_s–$\bar{B}_s$ mixing and B_d–$\bar{B}_d$ mixing deviates from the prediction for the standard model (SM) and the MSSM without the neutrino interaction.

2 Update of FCNC Processes in the Supergravity Model

In the minimal SM various FCNC processes and CP violation in B and K decays are determined by the Cabibbo-Kobayashi-Maskawa (CKM) matrix. Constraints on the parameters of the CKM matrix can be conveniently expressed in terms of the unitarity triangle. With CP violation at B factory as well as rare K decay experiments we will be able to check consistency of the unitarity triangle and at the same time search for effects of physics beyond the SM. In order to distinguish possible new physics effects it is important to identify how various models can modify the SM predictions.

Although general SUSY models can change the lengths and the angles of the unitarity triangle in variety ways, the supergravity model predicts a specific pattern of the deviation from the SM.[1] Namely, we can show that the SUSY loop contributions to FCNC amplitudes approximately have the same dependence on the CKM elements as the SM contributions. In particular, if we assume that there are no CP violating phases from SUSY breaking sectors, the complex phase of the $B^0 - \bar{B}^0$ mixing amplitude does not change even if we take into account the SUSY contributions. The case with supersymmetric CP phases was also studied within the minimal supergravity model and it was shown that effects of new CP phases on the $B^0 - \bar{B}^0$ mixing amplitude and the direct CP asymmetry in the $b \to s\gamma$ process are small once constraints from neutron and electron EDMs are included.[2]

We calculate various FCNC processes in the supergravity model with universal soft breaking terms at a high energy scale. The results are summarized as follows.

1. The amplitude for $b \to s\gamma$ can receive a large contribution from the

SUSY and the charged-Higgs-top-quark loop diagrams. The experimental branching ratio puts a strong constraint on SUSY parameter space. Since the SUSY contribution can interfere with other contributions either constructively or destructively we cannot exclude the light charged Higgs boson region unlike the non-SUSY type II two Higgs doublet model.[3]

2. When the sign of the $b \to s\gamma$ amplitude is opposite to that of the SM, $B(b \to sl^+l^-)$ can be twice larger than the SM prediction. This can occur for a large $\tan\beta$ region where $\tan\beta$ is the ratio of two vacuum expectation values of Higgs fields. The deviation is also evident in the differential branching ratio and the lepton forward-backward asymmetry.[4]

3. In terms of consistency check of the unitarity triangle the supergravity model has the following features.[5] (i)ΔM_{B_d} and ϵ_K are enhanced by the SUSY and charged-Higgs loop effects. When these quantities are normalized by the corresponding quantities in the SM they are almost independent of the CKM matrix element, and the enhancement factors for ΔM_{B_d} and ϵ_K are almost equal. (ii) The branching ratios for $K^+ \to \pi^+ \nu \overline{\nu}$ and $K_L \to \pi^0 \nu \overline{\nu}$ processes are suppressed compared to the SM prediction. Again the suppression factor are almost the same for two branching ratios and does not depend strongly on the CKM matrix element. (iii) CP asymmetries in various B decay modes such as $B \to J/\psi K_S$ and the ratio of ΔM_{B_s} and ΔM_{B_d} are the same as the SM prediction.

In Fig. 1 we present the correlation between ΔM_{B_d} and $B(K_L \to \pi^0 \nu \overline{\nu})$ normalized by the corresponding quantities in the SM for $\tan\beta = 3$. The constraint on the SUSY parameter space from the recent improved SUSY Higgs search is implemented.[5] We have calculated the SUSY particle spectrum based on two different assumptions on the initial conditions of renormalization group equations. The minimal case corresponds to the minimal supergravity where all scalar fields have a common SUSY breaking mass at the GUT scale. For "nonminimal" we enlarge the SUSY parameter space by relaxing the initial conditions for the SUSY breaking parameters, namely all squarks and sleptons have a common SUSY breaking mass whereas an independent SUSY breaking parameter is assigned for Higgs fields. The square(dot) points correspond to the minimal (enlarged) parameter space of the supergravity model. We can see that the ΔM_{B_d} (and ϵ_K) can deviated from the SM by 20% whereas the deviation in $K_L \to \pi^0 \nu \overline{\nu}$ and $K^+ \to \pi^+ \nu \overline{\nu}$ processes are small. These deviations may be evident in future when B factory experiments provide additional information on the CKM parameters.

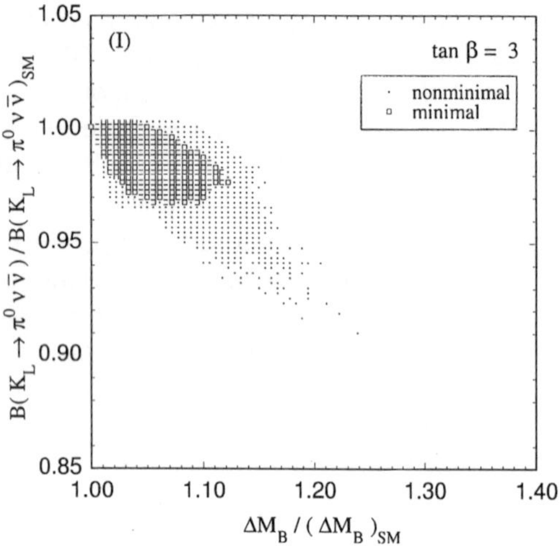

Figure 1. Correlation between ΔM_{B_d} and ϵ_K normalized by the SM value for $\tan\beta = 3$. The square(dot) points correspond to the minimal (enlarged) parameter space of the supergravity model.

3 FCNC in SUSY GUT with Right-handed Neutrino

In this section we consider FCNC and LFV of charged lepton decays in the model of a SU(5) SUSY GUT which incorporates the see-saw mechanism for the neutrino mass generation.[6] In this model sources of the flavor mixing are Yukawa coupling constant matrices for quarks and leptons as well as that for the right-handed neutrinos. Because the quark and lepton sectors are related by GUT interactions, the flavor mixing relevant to the CKM matrix can generate LFV such as $\mu \to e\gamma$ and $\tau \to \mu\gamma$ processes [7] in addition to FCNC in hadronic observables.[8] In the SUSY model with right-handed neutrinos, branching ratios of the LFV processes can become large enough to be measured in near-future experiments. [9] When we consider the right-handed neutrinos in the context of GUT, the flavor mixing related to the neutrino oscillation can be a source of the flavor mixing in the squark sector. We show that due to the large mixing of the second and third generations suggested by the atmospheric neutrino anomaly, B_s–$\bar{B}_s$mixing, the time-dependent CP asymmetry of the $B \to M_s\,\gamma$ process, where M_s is a CP eigenstate including the strange quark, can have a large deviation from the SM prediction.[6]

The Yukawa coupling part and the Majorana mass term of the superpotential for the SU(5) SUSY GUT with right-handed neutrino supermultiplets is given by $W = \frac{1}{8} f_U^{ij} \Psi_i \Psi_j H_5 + f_D^{ij} \Psi_i \Phi_j H_{\bar{5}} + f_N^{ij} N_i \Phi_j H_5 + \frac{1}{2} M_\nu^{ij} N_i N_j$,where Ψ_i, Φ_i and N_i are **10**, $\bar{\mathbf{5}}$ and **1** representations of SU(5) gauge group. $i, j = 1, 2, 3$ are the generation indices. H_5 and $H_{\bar{5}}$ are Higgs superfields with **5** and $\bar{\mathbf{5}}$ representations.

The renormalization effects due to the Yukawa coupling constants induce various FCNC and LFV effects from the mismatch between the quark/lepton and squark/slepton diagonalization matrices. In particular the large top Yukawa coupling constant is responsible for the renormalization of the $\tilde{q}_L$ and $\tilde{u}_R$ mass matrices. At the same time the $\tilde{e}_R$ mass matrix receives sizable corrections between the Planck and the GUT scales and various LFV processes are induced. In a similar way, if the neutrino Yukawa coupling constant f_N^{ij} is large enough, the $\tilde{l}_L$ mass matrix and the $\tilde{d}_R$ mass matrix receive sizable flavor changing effects due to renormalization between the Planck and the right-handed neutrino mass scales and the Planck and the GUT scales, respectively. These are sources of extra contributions to LFV processes and various FCNC processes such as $b \to s\gamma$, the B^0–$\overline{B}^0$ mixing and the K^0–$\overline{K}^0$ mixing.

The flavor mixing in the $\tilde{d}_R$ sector can induce large time-dependent CP asymmetry in the $B \to M_s\gamma$ process. Using the Wilson coefficients c_7 and c_7' in the effective Lagrangian for the $b \to s\gamma$ decay $\mathcal{L} = c_7 \bar{s}\sigma^{\mu\nu} b_R F_{\mu\nu} + c_7' \bar{s}\sigma^{\mu\nu} b_L F_{\mu\nu} + H.c.$, the asymmetry is written as $\frac{\Gamma(t)-\bar{\Gamma}(t)}{\Gamma(t)+\bar{\Gamma}(t)} = \xi A_t \sin \Delta m_d t$, $A_t = \frac{2\mathrm{Im}(e^{-i\theta_B} c_7 c_7')}{|c_7|^2+|c_7'|^2}$, where $\Gamma(t)$ ($\bar{\Gamma}(t)$) is the decay width of $B^0(t) \to M_s\gamma$ ($\bar{B}^0(t) \to M_s\gamma$) and M_s is some CP eigenstate ($\xi = +1(-1)$ for a CP even (odd) state) such as $K_S\pi^0$.[10] $\Delta m_d = 2|M_{12}(B_d)|$ and $\theta_B = \arg M_{12}(B_d)$ where $M_{12}(B_d)$ is the B_d–$\bar{B}_d$ mixing amplitude. Because the asymmetry can be only a few percent in the SM, a sizable asymmetry is a clear signal of new physics beyond the SM.

We calculated various FCNC and LFV observables in this model under the assumption of the universal soft breaking terms at the Planck scale. As typical examples of the neutrino parameters, we consider the following parameter set corresponding to the Mikheyev-Smirnov-Wolfenstein (MSW) small mixing case. $m_\nu = \left(2.236 \times 10^{-3}, 3.16 \times 10^{-3}, 5.92 \times 10^{-2}\right)$ eV and the Maki-Nakagawa-Sakata (MNS) matrix is given by

$$V_{\mathrm{MNS}} = \begin{pmatrix} 0.999 & 0.0371 & 0 \\ -0.0262 & 0.707 & 0.707 \\ 0.0262 & -0.707 & 0.707 \end{pmatrix}.$$

We also take M_ν to be proportional to a unit matrix with a diagonal element of

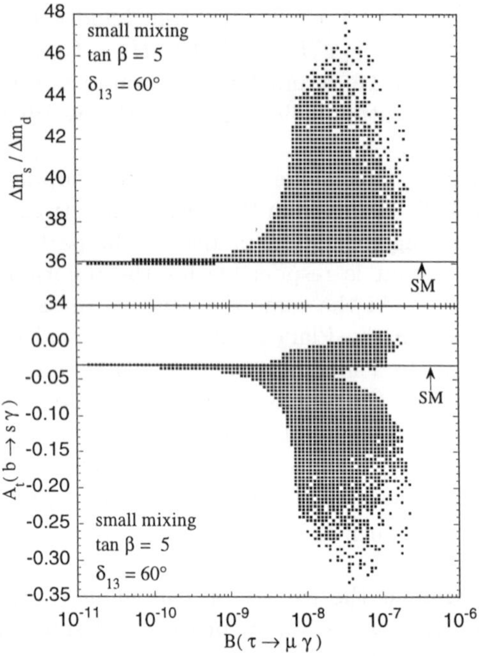

Figure 2. The ratio of B_s–$\bar{B}_s$ and B_d–$\bar{B}_d$ mass splittings $\Delta m_s/\Delta m_d$ and the magnitude factor A_t of the time-dependent CP asymmetry in the $B \to M_s\,\gamma$ process as a function of $\mathrm{B}(\tau \to \mu\,\gamma)$ for the small mixing.

$M_R = 4 \times 10^{14}$ GeV. We fix CKM parameters as $V_{cb} = 0.04$, $|V_{ub}/V_{cb}| = 0.08$ and $\delta_{13} = 60°$. We take $\tan\beta = 5$ and vary other SUSY parameters. We take account of various phenomenological constraints on SUSY parameters including $\mathrm{B}(b \to s\,\gamma)$. We also calculated $\mathrm{B}(\mu \to e\,\gamma)$ and ϵ_K and imposed constraints from these quantities. The upper part of Fig.2 shows a correlation between $\Delta m_s/\Delta m_d$ (ratio of B_s–$\bar{B}_s$ and B_d–$\bar{B}_d$ mass splittings) and $\mathrm{B}(\tau \to \mu\,\gamma)$. We can see that $\Delta m_s/\Delta m_d$ can be enhanced up to 30% compared to the SM prediction. This feature is quite different from the minimal supergravity model without the GUT and right-handed neutrino interactions [5] where $\Delta m_s/\Delta m_d$ is almost the same as the SM value. A_t for the same parameter set is shown as a function of $\mathrm{B}(\tau \to \mu\,\gamma)$ in the lower part of Fig. 2. We can see that $|A_t|$ can be close to 30% when $\mathrm{B}(\tau \to \mu\,\gamma)$is larger than 10^{-8}. The large asymmetry arises because the renormalization effect due to f_N induces sizable contribution to c_7' through gluino–$\tilde{d}_R$ loop diagrams. In this example possible new physics signals in $\mathrm{B}(\tau \to \mu\gamma)$, B_s–$\bar{B}_s$ mixing and A_t all come from the renormalization effect on squark and slepton mass matrices from the large neutrino Yukawa coupling constant. Because these signals provide quite different signatures compared to the SM and the minimal supergravity model without GUT and right-handed neutrino interactions, future experiments in B physics and LFV can provide us important clues on the interactions at very high energy scale.

The work was supported in part by the Grant-in-Aid of the Ministry of Education, Science, Sports and Culture, Government of Japan (No.09640381), Priority area "Supersymmetry and Unified Theory of Elementary Particles" (No. 707), and "Physics of CP Violation" (No.09246105).

References

1. S. Bertolini, *et al. Nucl. Phys.* B **353**, 591 (1991).
2. T. Nihei,*Prog. Theor. Phys.* **98**, 1157 (1997); T. Goto, Y.Y. Keum, T. Nihei, Y. Okada and Y. Shimizu, *Phys. Lett.* B **460**, 333 (1999).
3. T. Goto and Y. Okada, *Prog. Theor. Phys.* **94**, 407 (1995) and references therein.
4. A. Ali, G. Giudice and T. Mannel, *Z. Phys.* C **67**, 417 (1995); P. Cho, M. Misiak and D. Wyler, *Phys. Rev.* D **54**, 3329 (1996); T. Goto, Y. Okada, Y. Shimizu and M. Tanaka, *Phys. Rev.* D **55**, 4273 (1997); J. Hewett, J.D. Wells, *Phys. Rev.* D **55**, 5549 (1997).
5. T. Goto, T. Nihei and Y. Okada, *Phys. Rev.* D **53**, 5233 (1996); Erratum-*ibid.* **D54**, 5904 (1996); T. Goto, Y. Okada and Y. Shimizu, *Phys. Rev.* D **58**, 094006 (1998); KEK Preprint 99-72, KEK-TH-611, hep-ph/9908499.
6. S. Baek, T. Goto, Y. Okada and K. Okumura, KEK Preprint 99-176, KEK-TH-677, hep-ph/0002141.
7. R. Barbieri and L. J. Hall,*Phys. Lett.* B **338**, 212 (1994); J. Hisano *et al.*,*Phys. Lett.* B **391**, 341 (1997).
8. R. Barbieri, L. Hall and A. Strumia,*Nucl. Phys.* B **445**, 219 (1995); N. G. Deshpande, B. Dutta and S. Oh, *Phys. Rev. Lett.* **77**, 4499 (1996).
9. F. Borzumati and A. Masiero,*Phys. Rev. Lett.* **57**, 961 (1986); J. Hisano *et al.*,*Phys. Lett.* B **357**, 579 (1995); J. Hisano, *et al.*,*Phys. Rev.* D **53**, 2442 (1996); J. Hisano, D. Nomura and T. Yanagida,*Phys. Lett.* B **437**, 351 (1998); J. Hisano and D. Nomura, *Phys. Rev.* D **59**, 116005 (1999);

J. Ellis, *et al.*, hep-ph/9911459.
10. D. Atwood, M. Gronau and A. Soni, *Phys. Rev. Lett.* **79**, 185 (1997); C. Chua, X. He and W. Hou, *Phys. Rev.* D **60**, 014003 (1999).

CP Violation Beyond the Standard Model in Hadronic B Decays

Xiao-Gang He [a]
Department of Physics, National Taiwan University, Taipei, Taiwan 10764, R.O.C.

Abstract

In this talk I discuss three different methods, using $B_d \to J/\psi K_S$, $J/\psi K_S \pi^0$, $B_d \to K^-\pi^+, \pi^+\pi^-$ and $B_u \to K^-\pi^0, \bar{K}^0\pi^-, \pi^-\pi^0$, to extract hadronic model independent information about new physics.

1. Introduction

The Gold-plated mode for the study of CP violation in the Standard Model (SM) is $B \to J/\psi K_S$. This decay mode can be used to measure $\sin 2\beta$ in the SM without hadronic uncertainties. CDF and ALEPH data obtain [1] $\sin 2\beta = 0.91 \pm 0.35$ which is consistent with $\sin 2\beta = 0.75^{+0.055}_{-0.065}$ from fitting other experimental data[2]. The best fit value for γ from the same fitting gives[2] $\gamma_{best} = 59.9°$. Rare hadronic B decays also provide important information [3]. Using factorization calculations for rare hadronic B decays measured by CLEO [4], γ is determined to be [5] 114^{+25}_{-21} degrees. It seems that there is a conflict between γ obtained from rare B decay data and from other fits. At present it is not possible to conclude whether this possible inconsistence is due to experimental and theoretical uncertainties or due to new physics. It is important to find ways to test the SM and to study new physics without large hadronic uncertainties. In this talk I discuss three methods, using $B_d \to J/\psi K_S$, $J/\psi K_S \pi^0$ [6], $B_d \to K^-\pi^+, \pi^+\pi^-$ [7] and $B_u \to K^-\pi^0, \bar{K}^0\pi^-, \pi^-\pi^0$ [8,9], to extract hadronic model independent information about the SM and models beyond.

In the SM the Hamiltonian for hadronic B decays is given by, $H_{eff} = (4G_F/\sqrt{2})[V_{fb}V^*_{fq}(c_1 O_1^{f(q)}(L) + c_2 O_2^{f(q)}(L)) - \sum_{i=(3-12),j=(u,c,t)} V^*_{jb}V_{jq}c_i^j O_i]$. $O_{11,12}$ are the gluonic and electromagnetic dipole operators and O_{1-10} are defined in Ref. [10]. I use the values of c_i obtained for the SM in Ref. [10] with $(c_1, c_2,\ c_3^t, c_4^t,\ c_5^t, c_6^t,\ c_8^t, c_9^t, c_{10^t}, c_{11}^t, c_{12}^t)$ =(-0.313,1.150,0.017,-0.037, 0.010, -0.046, -0.001α_{em}, 0.049α_{em}, -1.321α_{em}, 0.267 α_{em}, -0.3, -0.6). $c_i^{c,u}$ are due to c and u quarks in the loop and can be found in Ref. [10]. When going beyond the SM, there are new contributions. I take three models for illustrations.

Model i): R-parity violation model

In R-parity violating supersymmetric (SUSY) models, there are new CP violating phases. Here I consider the effects due to, $L = (\lambda''_{ijk}/2)U_R^{ci}D_R^{cj}D_R^{ck}$, R-

[a] e-mail: hexg@phys.ntu.edu.tw

parity violating interaction. Exchange of $\tilde{d}_i$ squark can generate the following effective Hamiltonian at tree level, $H_{eff} = (4G_F/\sqrt{2})\, V^*_{fb}V_{fq}\, c^{f(q)}[O_1^{f(q)}(R) - O_2^{f(q)}(R)]$. Here $O_1^{f(q)}(R) = \bar{f}\gamma_\mu R f \bar{q}\gamma^\mu R b$ and $O_2^{f(q)}(R) = \bar{f}_\alpha\gamma_\mu R f_\beta \bar{q}_\beta\gamma^\mu R b_\alpha$. The operators $O_{1,2}(R)$ have the opposite chirality as those of the tree operators $O_{1,2}(L)$ in the SM. The coefficients $c^{f(q)}$ with QCD corrections are given by $(c_1 - c_2)(\sqrt{2}/(4G_F V_{fb}V^*_{fq}))(-\lambda''_{fqi}\lambda''^*_{f3i}/2m^2_{\tilde{d}^i})$. Here $m_{\tilde{d}^i}$ is the squark mass. $B_u \to \pi^-\bar{K}^0$ and $B_u \to K^-\bar{K}^0$ data constrain $|c^{c(q)}|$ to be less than $O(1)$ [11]. The new contributions can be larger than the SM ones. I will take the values to be 10% of the corresponding values for the SM with arbitrary phases $\delta^{f(q)}$ for later discussions.

Model ii): SUSY with large gluonic dipole interaction

In SUSY models with R-parity conservation, potential large contributions to B decays may come from gluonic dipole interaction by exchanging gluino at loop level with left- and right-handed squark mixing. In the mass insertion approximation [12], $c_{11}^{new} = -(\sqrt{2}\pi\alpha_s/(G_F V_{tb}V_{ts}m_{\tilde{g}}))(m_{\tilde{q}}/m_b)\delta^{LR}_{qb}G(m^2_{\tilde{g}}/m^2_{\tilde{q}})$, where $m_{\tilde{g}}$ is the gluino mass, and δ^{LR}_{qb} is the squark mixing parameter. $G(x)$ is from loop integral. c_{11}^{new} is constrained by experimental data from $b \to s\gamma$ which, however, still allows c_{11}^{new} to be as large as three times of the SM contribution in magnitude with an arbitrary CP violating phase δ_{dipole} [12]. I will take c_{11}^{new} to be 3 times of the SM value with an arbitrary δ_{dipole}.

Model iii): Anomalous gauge boson couplings

Anomalous gauge boson couplings can modify c_i (mainly on c^t_{7-10}) with the same CP violating source as that for the SM [9,13]. The largest contribution may come from the WWZ anomalous coupling Δg_1^Z. To the leading order in QCD corrections, the new contributions to c^{AC}_{7-10} due to Δg_1^Z are given by [9,13], $(c^t_{7AC}, c^t_{8AC}, c^t_{9AC}, c^t_{10AC}) = \alpha_{em}\Delta g_1^Z(1.397, 0.391, -5.651, 1.143)$. LEP data constrain [14] Δg_1^Z to be within $-0.113 < \Delta g_1^Z < 0.126$ at the 95% c.l.. c_{iAC} can be very different from those in the SM.

2. Test new physics using $\sin 2\beta$ from $B \to J/\psi K_S, J/\psi K_S\pi^0$

The usual CP violation measure for B decays to CP eigenstates is $\mathrm{Im}\,\xi = \mathrm{Im}\{(q/p)(A^*\bar{A}/|A|^2)\}$, where $q/p = e^{-2i\phi_B}$ is from B^0–$\bar{B}^0$ mixing, while A, $\bar{A}$ are B, $\bar{B}$ decay amplitudes. For $B \to J/\psi K_S$, the final state is P-wave hence CP odd. Setting the weak phase in the decay amplitude to be $2\phi_0 = Arg(A/\bar{A})$, one has , $\mathrm{Im}\,\xi(B \to J/\psi K_S) = -\sin(2\phi_B + 2\phi_0) \equiv -\sin 2\beta_{J/\psi K_S}$.

For $B \to J/\psi K^* \to J/\psi K_S\pi^0$, the final state has both P-wave (CP odd) and S- and D-wave (CP even) components. If S- and D-wave have a common weak phase $\tilde{\phi}_1$ and P-wave has a weak phase ϕ_1 [6],

$$\begin{aligned}\mathrm{Im}\,\xi(B \to J/\psi K_S \pi^0) &= \mathrm{Im}\{e^{-2i\phi_B}[e^{-2i\phi_1}|P|^2 - e^{-2i\tilde{\phi}_1}(1-|P|^2)]\} \\ &\equiv (1-2|P|^2)\sin 2\beta_{J/\psi K_S\pi^0}, \end{aligned} \quad (1)$$

where $|P|^2$ is the fraction of P-wave component. In the SM one has $\phi_B = \beta$ and $2\phi_0 = 2\phi_1(2\tilde{\phi}_1) = Arg[\{V_{cb}V_{cs}^*\{c_1 + a(a')c_2\}\}/\{V_{cb}^*V_{cs}\{c_1 + a(a')c_2\}\}] = 0$, in the Wolfenstein phase convention. Here a and a' are parameters which indicate the relative contribution from $O_2^{c(s)}(L)$ compared with $O_1^{c(s)}(L)$ for the P-wave and (S-, D-) wave. In the factorization approximation $1/a = 1/a' = N_c$ (the number of colors). Therefore $\sin 2\beta_{J/\psi K_S} = \sin 2\beta_{J/\psi K_S\pi^0} = \sin 2\beta$. $|P|^2$ has been measured with a small value[15] $0.16 \pm 0.08 \pm 0.04$ by CLEO which implies that the measurement of $\sin 2\beta$ using $B \to J/\psi K_S \pi^0$ is practical although there is a dilution factor of 30%.

When one goes beyond the SM, $\sin 2\beta_{J/\psi K_S} = \sin 2\beta_{J/\psi K_S\pi^0}$ is not necessarily true. The difference $\Delta \sin 2\beta \equiv \sin 2\beta_{J/\psi K_S} - \sin 2\beta_{J/\psi K_S\pi^0}$ is a true measure of CP violation beyond the SM without hadronic uncertainties.

Let us now analyze the possible values for $\Delta \sin 2\beta$ for the three models described in the previous section. Because $B \to J/\psi K_S, J/\psi K^*$ are tree dominated processes, Models ii) and iii) would not change the SM predictions significantly. $\Delta \sin 2\beta$ is not sensitive to new physics in Models ii) and iii). However, for Model i), the contributions can be large. The weak phases are given by $2\phi_0 = 2\phi_1(2\tilde{\phi}_1) = Arg[\{V_{cb}V_{cs}^*\{c_1 + ac_2 + (-)c^{c(s)}\{1 - a(a')\}\}\}/\{V_{cb}^*V_{cs}\{c_1 + ac_2 + (-)c^{c(s)*}\{1 - a(a')\}\}\}]$. Taking the new contributions to be 10% of the SM ones, one obtains $\phi_0 = \phi_1 \approx -\tilde{\phi}_1 \approx 0.1 \sin\delta^{c(s)}$. From this, $\Delta \sin 2\beta \approx 4((1-|P|^2)/(1-2|P|^2))\cos 2\phi_B(0.1\sin\delta^{c(s)}) \approx 0.5\cos 2\phi_B \sin\delta^{c(s)}$. ϕ_B may be different from the SM one due to new contributions. Using the central value $\sin 2\beta_{J/\psi K_S} = 0.91$ measured from CDF and ALEPH, $\Delta \sin 2\beta \approx 0.2 \sin\delta^{c(s)}$ which can be as large as 0.2. Such a large difference can be measured at B factories. Information about new CP violating phase $\delta^{c(s)}$ can be obtained.

3. Test new physics using rate differences between $B_d \to \pi^+ K^-, \pi^+\pi^-$

Theoretical calculations for rare hadronic B decays have large uncertainties. Information extracted using model calculations is unreliable. I now show that hadronic model independent information about CP violation can be obtained using SU(3) analysis for rare hadronic B decays.

The operators $O_{1,2}$, $O_{3-6,11,12}$, and O_{7-10} transform under SU(3) symmetry as $\bar{3}_a + \bar{3}_b + 6 + \overline{15}$, $\bar{3}$, and $\bar{3}_a + \bar{3}_b + 6 + \overline{15}$, respectively. These properties enable one to write the decay amplitudes for $B \to PP$ in only a few SU(3) invariant amplitudes. There are relations between different decays. When small annihilation contributions are neglected, one has

$$A(B_d \to \pi^+\pi^-) = V_{ub}V_{ud}^*T + V_{tb}V_{td}^*P,$$
$$A(B_d \to \pi^+K^-) = V_{ub}V_{us}^*T + V_{tb}V_{ts}^*P.$$

Due to different KM matrix elements involved in these decays, although the amplitudes have similarities, the branching ratios are not simply related. However, when considering the rate difference, $\Delta(PP) = \Gamma(B_d \to PP) - \Gamma(\bar{B}_d \to \bar{P}\bar{P})$, the situation is dramatically different because the simple relation $Im(V_{ub}V_{ud}^*V_{tb}^*V_{td}) = -Im(V_{ub}V_{us}^*V_{tb}^*V_{ts})$. In the SU(3) limit,

$$\Delta(\pi^+\pi^-) = -\Delta(\pi^+K^-). \tag{2}$$

This non-trivial equality dose not depend on detailed models for the hadronic physics and provides test for the SM [7]. Including SU(3) breaking effect from factorization calculation, one has, $\Delta(\pi^+\pi^-) \approx -\frac{f_\pi^2}{f_K^2}\Delta(\pi^+K^-)$. Although there is correction, the relative sign is not changed.

When going beyond the SM, there are new CP violating phases leading to violation of the equality above. Among the three models described in the first section, Models i) and ii) can alter the equality significantly, while Model iii) can not because the CP violating source is the same as that in the SM. To illustrate how the situation is changed in Models i) and ii), I calculate the normalized asymmetry $A_{norm}(PP) = \Delta(PP)/\Gamma(\pi^+K^-)$ using factorization approximation following Ref. [16]. The new effects may come in such a way that only $B_d \to \pi^+K^-$ is changed but not $B_d \to \pi^+\pi^-$. This scenario leads to maximal violation of the equality discussed here. I will take this iscenario for illustration.

The results are shown in Figure 1. The solid curve is the SM prediction for $A_{norm}(\pi^+K^-)$ as a function of γ. For γ_{best} $A_{norm}(\pi^+K^0) \approx 10\%$. It is clear from Figure 1 that within the allowed range of the parameters, new physics effects can dramatically violate the equality discussed above. Experimental study of the rate differences $\Delta(\pi^+\pi^-, \pi^+K^-)$ can provide model independent information about CP violation beyond the SM in the near future.

4. Test new physics using SU(3) relation for $B_u \to \pi^-\bar{K}^0, \pi^0K^-, \pi^0\pi^-$

I now discuss another method which provides important information about new physics using $B_u \to \pi^-\bar{K}^0, \pi^0K^-, \pi^0\pi^-$. Using SU(3) relation and factorization estimate for the breaking effects, one obtains [8,9]

$$A(B_u \to \pi^-\bar{K}^0) + \sqrt{2}A(B_u \to \pi^0K^-) \quad = \quad \epsilon A(B_u \to \pi^-\bar{K}^0)e^{i\Delta\phi}(e^{-i\gamma} - \delta_{EW}),$$

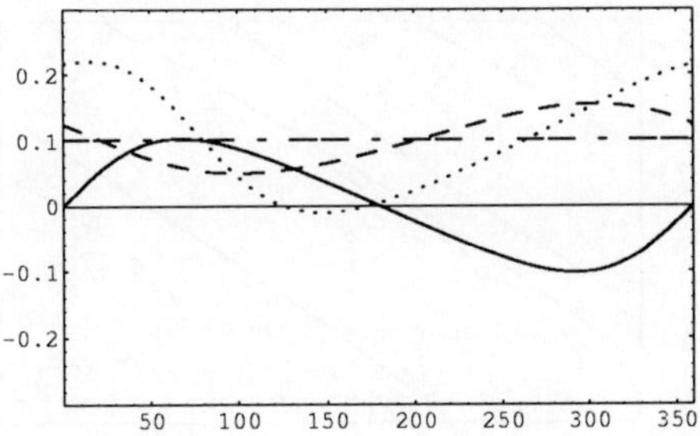

Figure 1: $A_{norm}(\pi^+K^-)$ vs. phases γ, $\delta^{u(s)}$ and δ_{dipole} for the SM (solid), Models i) (dashed) and ii) (dotted). For Models i) and ii), $\gamma = 59.9^o$ is used. $A_{norm}(\pi^+K^-) = -(f_\pi/f_k)^2 A_{norm}(\pi^+\pi^-)$ is satisfied in the SM. For Models i) and ii) $-(f_\pi/f_k)^2 A_{norm}(\pi^+\pi^-)$ is approximately the same as that in the SM (dot-dashed curve).

$$\delta_{EW} = -\frac{3}{2}\frac{|V_{cb}||V_{cs}|}{|V_{ub}||V_{us}|}\frac{c_9 + c_{10}}{c_1 + c_2}, \qquad \epsilon = \sqrt{2}\frac{|V_{us}|}{|V_{ud}|}\frac{f_K}{f_\pi}\frac{|A(\pi^+\pi^0)|}{|A(\pi^+K^0)|},$$

where $\Delta\phi$ is the difference of the final state rescattering phases for $I = 3/2, 1/2$ amplitudes. For $f_K/f_\pi = 1.22$ and $Br(B^\pm \to \pi^\pm\pi^0) = (0.54^{+0.21}_{-0.20} \pm 0.15) \times 10^{-5}$ [4], one obtains $\epsilon = 0.21 \pm 0.06$.

Neglecting small tree contribution to $B_u \to \pi^-\bar{K}^0$, one obtains

$$\cos\gamma = \delta_{EW} - \frac{(r_+^2 + r_-^2)/2 - 1 - \epsilon^2(1 - \delta_{EW}^2)}{2\epsilon(\cos\Delta\phi + \epsilon\delta_{EW})}, \quad r_+^2 - r_-^2 = 4\epsilon\sin\Delta\phi\sin\gamma,$$

where $r_\pm^2 = 4Br(\pi^0K^\pm)/[Br(\pi^+K^0) + Br(\pi^-\bar{K}^0)] = 1.33 \pm 0.45$ [4].

The relation between γ and δ_{EW} is complicated. However it is interesting to note that even in the most general case, bound on $\cos\gamma$ can be obtained [9]. For $\Delta = (r_+^2 + r_-^2)/2 - 1 - \epsilon^2(1 - \delta_{EW}^2) \geq (\leq) > 0$, we have

$$\cos\gamma \leq (\geq)\delta_{EW} - \frac{\Delta}{2\epsilon(1 + \epsilon\delta_{EW})}, \quad \text{or} \quad \cos\gamma \geq (\leq)\delta_{EW} - \frac{\Delta}{2\epsilon(-1 + \epsilon\delta_{EW})}. \tag{3}$$

The bounds on $\cos\gamma$ as a function of δ_{EW} are shown in Fig. 2 by the solid curves for three representative cases: a) Central values for ϵ and $r_\pm^2$; b) Central values for ϵ and 1σ upper bound $r_\pm^2 = 1.78$; and c) Central value for ϵ and 1σ lower bound $r_\pm^2 = 0.88$. The bounds with $|\cos\gamma| \leq 1$ for a), b) and c) are indicated by the curves (a1, a2), (b) and (c1, c2), respectively. For cases a) and c) there are two allowed regions, the regions below (a1, c1) and the regions above (a2, c2). For case b) the allowed range is below (b).

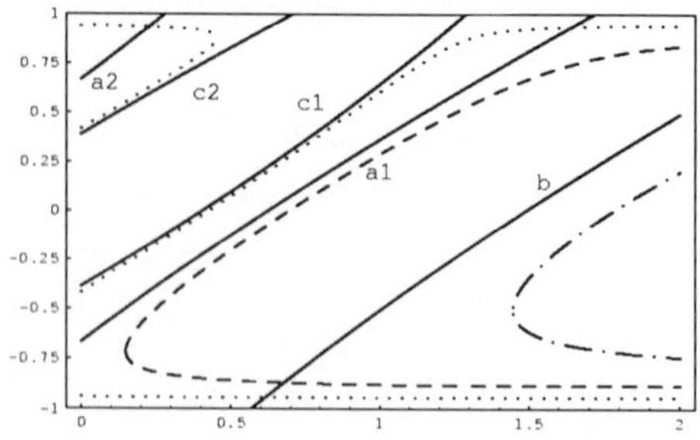

Figure 2: $\cos\gamma$ vs. δ_{EW}. The solutions for the cases a), b) and c) are indicated by the dashed, dot-dashed and dotted curves respectively.

When the decay amplitudes for $B^\pm \to K\pi$, $B^\pm \to \pi^\pm\pi^0$ and the rate asymmetries for these decays are determined to a good accuracy, γ can be determined, one obtains,

$$(1-\cos^2\gamma)[1-(\frac{\Delta}{2\epsilon(\delta_{EW}-\cos\gamma)}-\epsilon\delta_{EW})^2]-\frac{(r_+^2-r_-^2)^2}{16\epsilon^2}=0. \qquad (4)$$

To have some idea about the details, I analyze the solutions of $\cos\gamma$ as a function of δ_{EW} for the three cases discussed earlier with a given value for the asymmetry $A_{asy} = (r_+^2 - r_-^2)/(r_+^2 + r_-^2) = 15\%$.

In the SM for $r_v = |V_{ub}|/|V_{cb}| = 0.08$ and $|V_{us}| = 0.2196$, $\delta_{EW} = 0.81$. The central values for $r_\pm$ and ϵ prefers $\cos\gamma < 0$ which is different from the result obtained in Ref. [2] by fitting other data. The parameter δ_{EW} is sensitive to new physics in the tree and electroweak sectors. Model i) has large corrections to the tree level contributions. However, the contribution is proportional to the sum of the coefficients of operators $O_{1,2}^{u(s)}(R)$ which is zero in Model i). The above method does not provide information about new physics due to Model i). This method would not provide information about new physics due to Model ii) neither because the gluonic dipole interaction transforms as $\bar{3}$ which does not affect δ_{EW}. Model iii) can have large effect on δ_{EW}. In this model $\delta_{EW} = 0.81(1+4.33\Delta g_1^Z)$ which can vary in the range $0.40 \sim 1.25$. Constraint on $\cos\gamma$ can be very different from the SM prediction.

For case a), in the SM $\cos\gamma < 0.18$ which is inconsistent with $\cos\gamma_{best} \approx 0.5$. In Model iii) $\cos\gamma$ can be consistent with $\cos\gamma_{best}$. For case b), $\cos\gamma$ is less than zero in both the SM and Model iii). If this is indeed the case, other types of new physics is needed. For case c) $\cos\gamma$ can be close to $\cos\gamma_{best}$ for both the SM and Model iii). Improved measurements of ϵ and $r_\pm$ can provide important information about γ and new physics.

5. Conclusion

From discussions in previous sections, it is clear that using $B_d \to J/\psi K_S$, $J/\psi K_S \pi^0$, $B_u \to \pi^- K^+, \pi^+\pi^-$ and $B^- \to \pi^0 K^-, \pi^- \bar{K}^0, \pi^0\pi^-$ important information free from uncertainties in hadronic physics about the Standard Model and models beyond can be obtained. These analyses should be carried out at B factories.

Acknowledgements

This work was partially supported by National Science Council of R.O.C. under grant number NSC 89-2112-M-002-016. I thank Deshpande, Hou, Hsueh and Shi for collaborations on materials presented in this talk.

1. R. Forty et al., ALEPH 99-099/CONF 99054; G. Bauer, hep-ex/9908055.
2. F. Parodi, P. Roudeau and A. Stocchi, e-print hep-ex/9903063.
3. N. Deshpande et al, Phys. Rev. Lett. **82**,2240(1999); X.-G. He, W.-S. Hou and K.-C. Yang, Phys. Rev. Lett. **83**, 1100(1999); W.-S. Hou and K.-C. Yang, hep-ph/9908202; M. Gronau and J. Rosner, hep-ph/9909478.
4. Y. Kwon et al., CLEO CONF 99-14.
5. W.-S. Hou, J. Smith and F. Wurthwein, hep-ex/9910014.
6. X.-G. He and W.-S. Hou, Phys. Lett. **B445**, 344(1999).
7. N. Deshpande, and X.-G. He, Phys. Rev. Lett. **75**, 3064(1995); X.-G. He, Eur. Phys. J. **C9**, 443(1999); M. Suzuki, hep-ph/9908420.
8. M. Neubert and J. Rosner, Phys. Lett. **B441**, 403(1998); M. Neubert and J. Rosner, Phys. Rev. Lett. **81**, 5074(1998); M. Neubert, JHEP 9902:014 (1999); Y. Grosssman, M. Neubert and A. Kagan, hep-hep/9909297
9. X.-G. He, C.-L. Hsueh and J.-Q. Shi, Phys. Rev. Lett. **84**, 18(2000).
10. N. Deshpande and X.-G. He, Phys. Lett. **B336**, 471(1994).
11. C. Carlson and M. Sher, Phys. Lett. **B357**, 99(1995).
12. M. Ciuchini, E. Gabrielli and G. Giudice, Phys. Lett. B**388**, 353 (1996).
13. X.-G. He and B. McKellar, Phys. Rev. **D51**, 6484(1995); X-G. He, Phys. Lett. **B460**, 405(1999).
14. ALEPH Collaboration, ALEPH 99/072, CONF 99/046.
15. C.P. Jessop et al., Phys. Rev. Lett. **79**, 4533 (1997).
16. X.-G. He, W.-S. Hou and K.-C. Yang, Phys. Rev. Lett. **81**, 5738(1999).

THE B SYSTEM AS A WINDOW TO NEW PHYSICS

J.L. HEWETT
Stanford Linear Accelerator Center, Stanford CA 94309, USA

The opportunities to explore physics beyond the Standard Model in the B meson system are described. After denoting some overall features which are generic consequences of the search for new physics, we concentrate on the effects of new interactions in flavor changing neutral current transitions, focusing on supersymmetry and the left-right symmetric model as specific examples of new physics scenarios.

1 Overview

The B-meson system promises to yield a fertile testing ground of the Standard Model (SM). With the impressive start-up of the SLAC and KEK B-Factories, as well as the large data samples which will be acquired over the next decade at CESR, the Tevatron, and the LHC, the SM willed be probed at an unprecedented level of precision. It is well-known that precision measurements of low-energy processes can provide insight to very high energy scales via the indirect effects of new interactions. As such, the B sector offers a complementary probe of new physics, and in some cases may yield constraints which surpass those from direct collider searches or exclude entire classes of models.

The existence of new physics may modify the low-energy effective Hamiltonian governing B decays by (i) giving new contributions to the SM operators, (ii) generating new operators, or (iii) with the presence of new CP violating phases. In fact, calculations [1] of electroweak baryogenesis within the SM show that the produced baryon asymmetry is too small by many orders of magnitude and hence suggests that new CP violating phases must exist in nature. However, new physics cannot change properties that we know to be true, such as $a_{CP}(\psi K_s) = -a_{CP}(\psi K_L)$, and is unlikely to compete in processes which receive large SM contributions, such as $B(b \to c\bar{u}d)$.

The best way to probe for new physics is to employ observables which are easy to measure, have small SM uncertainties, and are likely to be sensitive to new interactions. New physics may manifest itself in the B system in several ways, for example, inconsistencies with the SM may be found between precision measurements of the sides and angles of the unitarity triangle, in rare decays, in B meson mixing, or in processes which are predicted to vanish in the SM. In particular, one would expect large effects from new interactions in 1-loop processes or in observables which are suppressed in the SM. Processes which provide good examples of the latter scenario are the CP asymmetries in $B_s \to \psi\phi$ and $B_s \to D_s^+ D_s^-$, the direct CP asymmetries in $B^\pm \to \psi K^\pm$ and $b \to$

$s\gamma$, the semi-leptonic asymmetry in $B_{d,s}$ mixing, and $D - \bar{D}$ mixing. It is anticipated that additional tree-level contributions to B decay are suppressed, since the scale of new physics is expected to be large compared to M_W,

Here, we concentrate on the loop effects of new interactions in flavor changing neutral current (FCNC) transitions. We note that most classes of models which induce large effects in the FCNC decays also affect $B_d^0 - \bar{B}_d^0$ mixing, and that measurements of several different loop-level processes may elucidate the origin of new interactions. In addition, long distance effects are expected to play less of a role due to the heavy B mass, and hence rare processes are essentially short distance dominated.

2 New Physics Constraints from $B_d - \bar{B}_d$ Mixing

Constraints on the magnitude and phase of potential new physics contributions to $B_d - \bar{B}_d$ mixing can be obtained[2] from the recent CDF measurement[3] of the CP asymmetry in $B \to \psi K_S$ decays. This measurement yields a value for the unitarity triangle $\sin 2\beta$ from an interference of the phases in the amplitudes for B_d mixing and tree-level decay. Assuming that the 3×3 CKM matrix is unitary and that the tree-level decay, in particular its phase, is dominated by SM contributions, the new physics effects in B_d mixing can then be isolated. The modification to the magnitude and phase of the B_d mixing amplitude, M_{12}, can be expressed in a model independent fashion by

$$M_{12} = r_d^2 e^{2i\theta_d} M_{12}^{SM} . \tag{1}$$

The direct measurement of Δm_B provides bounds on the magnitude of new physics contributions, r_d, while the CDF results constrains new phases, θ_d. Taking into account the uncertainties on the values of the relevant CKM angles and hadronic matrix elements, Barenboim *et al.*[2] find

$$0.3 \lesssim r_d^2 \lesssim 5 , \tag{2}$$

and

$$\sin 2\theta_d \gtrsim -0.6\,(-0.87) , \tag{3}$$

at 1σ (95% C.L.). It is clear that large contributions to this process from new interactions are still allowed, and may hence admit for an exciting discovery as future measurements improve.

3 Formalism for $b \to s$ Transitions

The observation[4] of radiative penguin mediated processes, in both the exclusive $B \to K^*\gamma$ and inclusive $B \to X_s\gamma$ channels, has placed the study of rare B

decays on a new footing and has provided powerful constraints on classes of models.

The effective field theory for $b \to s$ transitions is summarized at this meeting by Misiak and Morozumi,[5] however, we briefly review the features which are essential to the remainder of this talk. Incorporating QCD corrections, these transitions are governed by the Hamiltonian

$$\mathcal{H}_{eff} = \frac{-4G_F}{\sqrt{2}} V_{tb} V_{ts}^* \sum_{i=1}^{10} C_i(\mu) \mathcal{O}_i(\mu) , \tag{4}$$

where the $\mathcal{O}_i$ are a complete set of renormalized operators of dimension six or less which mediate these transitions. The C_i represent the corresponding Wilson coefficients which are evaluated perturbatively at the electroweak scale, where the matching conditions are imposed, and then evolved down to the renormalization scale $\mu \approx m_b$. The expressions for $C_i(M_W)$ at the electroweak scale in the SM are given by the Inami-Lim functions.[6]

3.1 $B \to X_s \ell^+ \ell^-$ in the Standard Model

For $B \to X_s \ell^+ \ell^-$ this formalism leads to the physical decay amplitude (neglecting m_s and m_ℓ)

$$\begin{aligned} \mathcal{M} &= \frac{\sqrt{2} G_F \alpha}{\pi} V_{tb} V_{ts}^* \Big[C_9^{eff} \bar{s}_L \gamma_\mu b_L \bar{\ell} \gamma^\mu \ell + C_{10} \bar{s}_L \gamma_\mu b_L \bar{\ell} \gamma^\mu \gamma_5 \ell \\ & \quad -2 C_7^{eff} m_b \bar{s}_L i \sigma_{\mu\nu} \frac{q^\nu}{q^2} b_R \bar{\ell} \gamma^\mu \ell \Big] , \end{aligned} \tag{5}$$

where q^2 represents the momentum transferred to the lepton pair. The next-to-leading order (NLO) analysis for this decay has been performed in Buras *et al.*[7] The $1/m_b^2$ and $1/m_c^2$ heavy quark corrections have been computed and are found to be small.[8] The resulting inclusive branching fractions are found to be $(6.25^{+1.04}_{-0.93}) \times 10^{-6}$, $(5.73^{+0.75}_{-0.78}) \times 10^{-6}$, and $(3.24^{+0.44}_{-0.54}) \times 10^{-7}$ for $\ell = e, \mu$, and τ, respectively.

3.2 $B \to X_s \gamma$ in the Standard Model

The basis for the decay $B \to X_s \gamma$ contains the first eight operators in the effective Hamiltonian of Eq. (4). The next-to-leading order logarithmic QCD corrections have been computed, leading to a much reduced renormalization scale dependence in the branching fraction. The higher-order QCD calculation involves several steps, requiring corrections to both C_7 and the matrix element of $\mathcal{O}_7$. In addition, the 2-loop electroweak corrections have also been computed [9] and are found to slightly reduce the branching fraction. The numerical value of the branching fraction is then found to be (again, scaling to semi-leptonic decay)

$$B(B \to X_s\gamma) = (3.28 \pm 0.30) \times 10^{-4}\,. \tag{6}$$

This is well within the range observed by CLEO and ALEPH[4] which is $B = (3.15 \pm 0.35 \pm 0.41) \times 10^{-4}$ and $B = (3.38 \pm 0.74 \pm 0.85) \times 10^{-4}$, respectively, with the 95% C.L. bounds of $2 \times 10^{-4} < B(B \to X_s\gamma) < 4.5 \times 10^{-4}$. The inclusive decays are measured by analyzing the photon energy spectrum, which has been found to retain its SM predicted shape, but not necessarily its normalization, in the presence of new physics.[10]

4 Model Independent Tests for New Physics in $b \to s$ Transitions

Measurements of $B(B \to X_s\gamma)$ constrain the magnitude, but not the sign, of $C_7(\mu)$. The coefficients at the matching scale ($\mu = M_W$) can be written in the form $C_i(M_W) = C_i^{SM}(M_W) + C_i^{new}(M_W)$, where $C_i^{new}(M_W)$ represents the contributions from new interactions. Due to operator mixing, $B \to X_s\gamma$ then limits the possible values for $C_i^{new}(M_W)$ for $i = 7, 8$. These bounds are summarized in Fig. 1. Here, the solid bands correspond to the constraints obtained from the current CLEO measurement, taking into account the variation of the renormalization scale $m_b/2 \leq \mu \leq 2m_b$, as well as the allowed ranges of the other input parameters. The dashed bands represent the constraints when the scale is fixed to $\mu = m_b$. We note that large values of $C_8^{new}(M_W)$ (which would yield an anomalous rate for $b \to sg$) are allowed even in the region where $C_7^{new}(M_W) \simeq 0$.

Measurement of the kinematic distributions[11,12] associated with the final state lepton pair in $B \to X_s\ell^+\ell^-$ as well as the rate for $B \to X_s\gamma$ allows the determination of the sign and magnitude of all the Wilson coefficients for the contributing operators in a model independent fashion. We have performed a Monte Carlo analysis in order to ascertain how much quantitative information will be obtainable at future B-Factories and follow the procedure outlined in Ref. [13]. For the process $B \to X_s\ell^+\ell^-$, we consider the lepton pair invariant mass distribution and forward-backward asymmetry[11] for $\ell = e, \mu, \tau$, and the tau polarization asymmetry[12] for $B \to X_s\tau^+\tau^-$. A three dimensional χ^2 fit to the coefficients $C_{7,9,10}(\mu)$ is performed for three values of integrated luminosity, 3×10^7, 10^8, and 5×10^8 $B\bar{B}$ pairs, corresponding to the expected e^+e^- B-Factory luminosities of one year at design, one year at an upgraded accelerator, and the total accumulated luminosity at the end of these programs. The 95% C.L. allowed regions (including statistical errors only for $B \to X_s\ell^+\ell^-$ and a

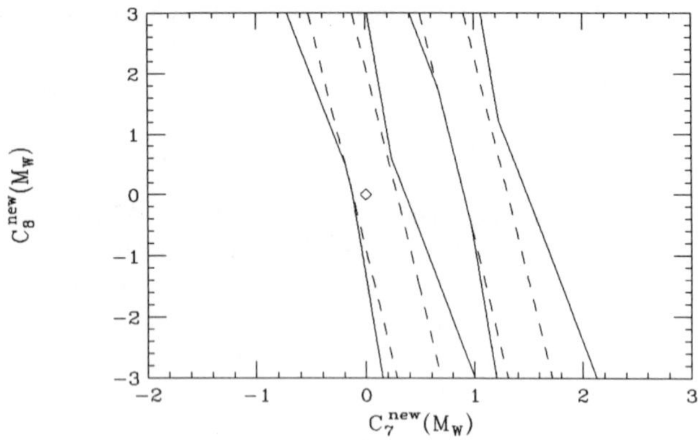

Figure 1: Bounds on the contributions from new physics to $C_{7,8}$. The region allowed by the CLEO data corresponds to the area inside the solid diagonal bands. The dashed bands represent the constraints when the renormalization scale is set to $\mu = m_b$. The diamond at the position (0,0) represents the Standard Model.

flat 10% error on $B \to X_s\gamma$) as projected onto the $C_9(\mu) - C_{10}(\mu)$ and $C_7(\mu) - C_{10}(\mu)$ planes are depicted in Figs. 2(a-b), where the diamond represents the central value for the expectations in the SM. We see that the determinations are relatively poor for 3×10^7 $B\bar{B}$ pairs and that higher statistics are required in order to focus on regions centered around the SM.

The presence of new physics may be probed via this global fit by discovering that either the values of the Wilson coefficients are altered from their SM predictions, or by obtaining a poor χ^2 for the fit, indicating that the operator basis must be extended.

5 Supersymmetric Effects in $b \to s$ Transitions

These model independent bounds can be compared with model dependent predictions for the Wilson coefficients in order to ascertain at what level specific new interactions can be probed. First, we consider supersymmetric extensions to the SM. Supersymmetry (SUSY) contains many potential sources for flavor violation. The spectroscopy of the supersymmetric states is quite model dependent and we analyze two possibilities. The first is the familiar minimal supergravity model; in this instance all the supersymmetric states follow from a common scalar mass and a common gaugino mass at the high scale. The second case is where the condition of common scalar masses is relaxed and they are allowed to take on uncorrelated values at the low scale while still preserving gauge invariance.

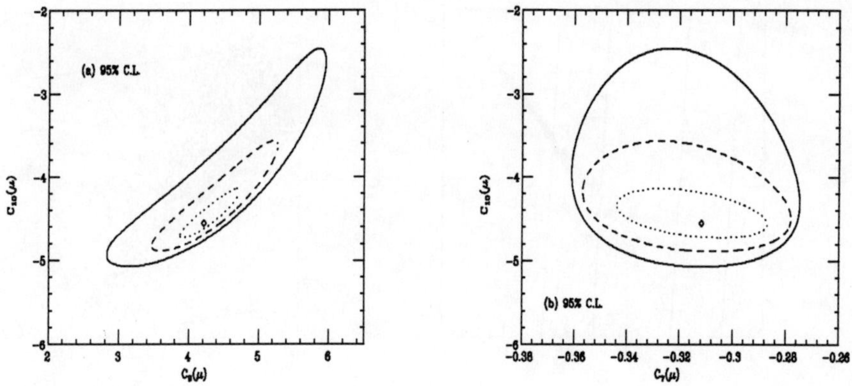

Figure 2: The 95% C.L. projections in the (a) $C_9 - C_{10}$ and (b) $C_7 - C_{10}$ planes, where the allowed regions lie inside of the contours. The solid, dashed, and dotted contours correspond to 3×10^7, 10^8, and 5×10^8 $B\bar{B}$ pairs. The central value of the SM prediction is labeled by the diamond.

We analyze the supersymmetric contributions to the Wilson coefficients [13] in terms of the quantities

$$R_i \equiv \frac{C_i^{susy}(M_W)}{C_i^{SM}(M_W)} - 1 \equiv \frac{C_i^{new}(M_W)}{C_i^{SM}(M_W)}, \tag{7}$$

where $C_i^{susy}(M_W)$ includes the full SM plus superpartner contributions. R_i is meant to indicate the fractional deviation from the SM value. We explore the full parameter space of the minimal supergravity model, calculate the R_i for each generated point in the supersymmetric parameter space, and then compare with the expected ability of B Factories to measure the R_i as determined by our global fit to the Wilson coefficients. Each supersymmetric solution is kept only if it is not in violation of present constraints from LEP II and Tevatron direct sparticle production limits. Our results are shown in the scatter plots of Fig. 3 in the (a) $R_7 - R_8$ and (b) $R_9 - R_{10}$ planes. The diagonal bands represent the bounds on the Wilson coefficients as previously determined from our global fit. We see from Fig. 3(a) that the current CLEO data on $B \to X_s\gamma$ already place signigicant restrictions on the supersymmetric parameter space, whereas the minimal supergravity contributions to $R_{9,10}$ are predicted to be essentially unobservable.

A second, more phenomenological approach is now adopted. The maximal

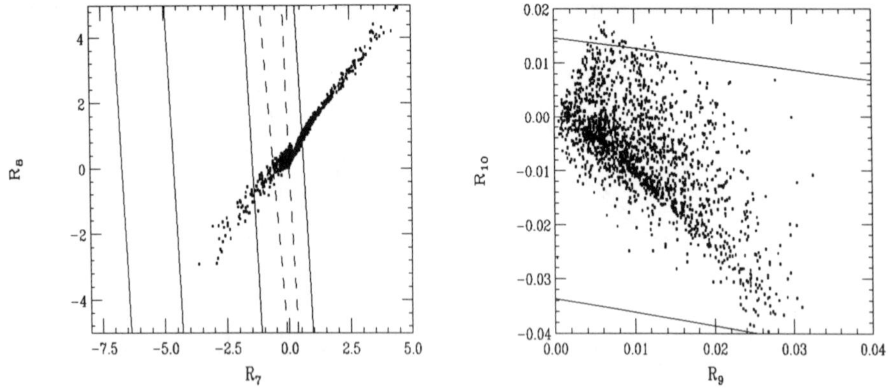

Figure 3: (a) Parameter space scatter plot of R_7 vs. R_8 in the minimal supergravity model. The allowed region from CLEO data lies inside the solid diagonal bands. The dashed band represents the potential future 10% measurement of $B \to X_s\gamma$ as described in the text. (b) Parameter space scatter plot of R_9 vs. R_{10}. The global fit to the coefficients obtained with 5×10^8 $B\bar{B}$ pairs corresponds to the region inside the diagonal bands.

effects for the parameters R_i can be estimated for a superparticle spectrum, independent of the high scale assumptions. The most important features which result in large contributions are a light $\tilde{t}_1$ state present in the SUSY spectrum and at least one light chargino state. For the dipole moment operators a light Higgsino state is also important. A pure higgsino and/or pure gaugino state have less of an effect than two mixed states when searching for maximal effects in R_9 and R_{10}. Fig. 4 displays the maximum contribution to $R_{9,10}$ versus an applicable SUSY mass scale. The other sparticle masses which are not shown ($\tilde{t}_i$, $\tilde{l}_L$, etc.) are chosen to be just above the reach of LEP II or the Tevatron, whichever yields the best bound. We see that the maximum size of $R_{9,10}$ is somewhat larger than what was allowed in the minimal supergravity model, due to the lifted restriction on mass correlations.

Given the sensitivity of the observables it is instructive to narrow our focus to the coefficient of the magnetic dipole operator. The first case we examine is that where the lightest chargino is a pure Higgsino and the lightest stop is purely right-handed: $\chi_1^\pm \sim \tilde{H}^\pm$, $\tilde{t}_1 \sim \tilde{t}_R$. The resulting contribution to R_7 is shown as a function of the $\tilde{t}_R$ mass in Fig. 5 (dashed line) for the case of $m_{\chi_1^\pm} \gtrsim M_W$. Note that the SUSY contribution to $C_7(M_W)$ in this limit always adds constructively to that of the SM. Next we examine the limit where the light chargino is a pure Wino, this contribution is shown in Fig. 5 (dotted line). Our third limiting case is that of a highly mixed $\tilde{t}_1$ state. We find

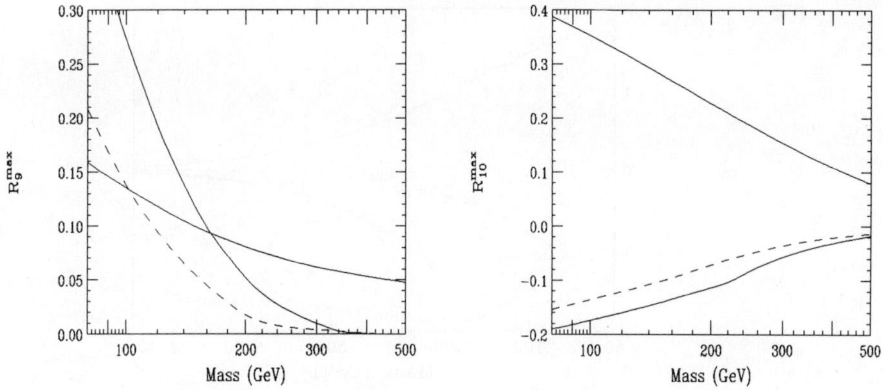

Figure 4: The maximum value of (a) R_9 and (b) R_{10} achievable for general supersymmetric models. The top solid line comes from the $t - H^\pm$ contribution and is displayed versus the $H^\pm$ mass. The bottom solid line is from the $\tilde{t}_i - \chi_j^\pm$ contribution with $\tan\beta = 1$ and is shown versus the $\chi_i^\pm$ mass. The dashed line is the $\tilde{t}_i - \chi_j^\pm$ contribution with $\tan\beta = 2$. The other mass parameters which are not plotted are chosen to be just above the reach of LEP II and the Tevatron.

that in this case large $\tan\beta$ solutions ($\tan\beta \gtrsim 40$) can yield greater than $\mathcal{O}(1)$ contributions to R_7 even for SUSY scales of 1 TeV! Low values of $\tan\beta$ can also exhibit significant enhancements; this is demonstrated for $\tan\beta = 2$ in Fig. 5 (solid line).

Lastly, we compare the reach of rare B decays in probing SUSY parameter space with that of high energy colliders. We examine a set of five points in the minimal supergravity (SUGRA) parameter space that were chosen at Snowmass 1996[14] for the study of supersymmetry at the NLC. Point #3 is the so-called "common" point used for a comparison of SUSY studies at the NLC, LHC, and Tev33. Once these points are chosen the sparticle mass spectrum is obtained, as usual, via the SUGRA relations and their contributions to $B \to X_s\gamma$ can be readily computed. The results are displayed in the $R_7 - R_8$ plane in Fig. 6 (labeled $1-5$ for each SUGRA point), along with the bounds previously obtained from our fits to the present CLEO data and to anticipated future data assuming the SM is realized. We see that four of the points should be discernable from the SM in future measurements, and that one of the points is already excluded by CLEO! We thus conclude that rare B decays are indeed complementary to high energy colliders in searching for supersymmetry.

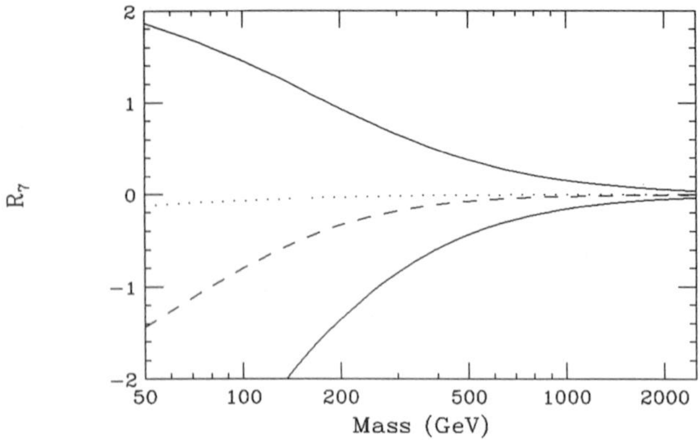

Figure 5: Contributions to R_7 in the different limits described in the text. The top solid line is the charged $H^{\pm}/t$ contribution versus $m_{H^{\pm}}$. The bottom solid line is the $\tilde{\chi}_1^{\pm}/\tilde{t}_1$ contribution versus $m_{\tilde{\chi}^{\pm}}$ where both the chargino and stop are maximally mixed states with $\mu < 0$. The dashed line is the $\tilde{H}^{\pm}/\tilde{t}_R$ contribution, and the dotted line represents the $\tilde{W}^{\pm}/\tilde{t}_1$ contribution. These two lines are both shown as a function of $\tilde{\chi}_1^{\pm}$ mass. All curves are for $\tan\beta = 2$ and $m_t = 175\,\text{GeV}$.

6 Left-Right Symmetric Model in $b \to s$ Transitions

The Left-Right Symmetric Model (LRM), which is based on the extended electroweak gauge group $SU(2)_L \times SU(2)_R \times U(1)$ can lead to interesting new effects in the B system.[15,16] Due to the extended gauge structure there are both new neutral and charged gauge bosons, Z_R and W_R, as well as a right-handed gauge coupling, g_R, which is subject to the constraint $0.55 < \kappa \equiv g_R/g_L < 2.0$ from naturalness and GUT embedding conditions. After complete symmetry breaking the charged W_R mixes with the SM W_L to form the mass eigenstates $W_{1,2}$ (where W_1 is the state which is directly produced at the Tevatron and LEP II). This mixing is described by two parameters: a real mixing angle ϕ and a phase α. In most models $\tan\phi$ is naturally of order of the ratio of masses $r = M_1^2/M_2^2$, or less, in the limit of large M_2. The charged current interactions of the right-handed quarks are governed by a right-handed CKM matrix, V_R, which, in principle, need not be related to its left-handed counterpart V_L. V_R will then involve 3 new angles as well as 6 additional phases, all of which are *a priori* unknown parameters. Phenomenological constraints on the LRM are quite sensitive to variations of V_R. If one assumes manifest left-right symmetry, that is $V_R = V_L$ and $\kappa = 1$, then the $K_L - K_S$ mass difference implies that $M_R > 1.6$ TeV. However, if that assumption is relaxed and V_R (as well as κ) is allowed to vary then W_R masses as low as 500 GeV can be accomodated by

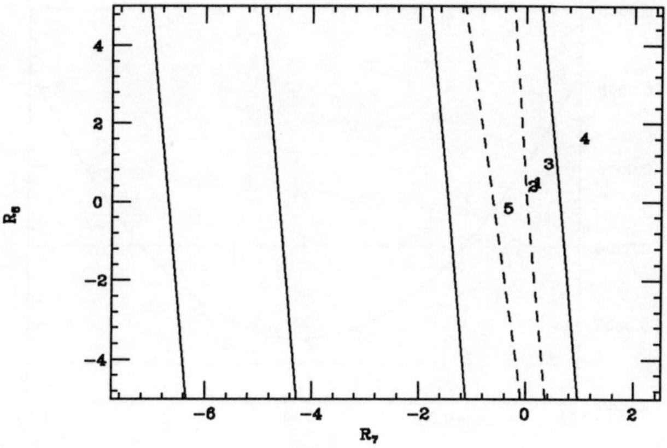

Figure 6: Values in the $R_7 - R_8$ plane for the five Snowmass NLC SUGRA points. The solid and dashed bands represent the present bounds from CLEO and those anticipated from future experiment, respectively, as described in Figure 3.

present data. This implies that the magnitude of $\tan\phi$ is $\leq few \times 10^{-2}$.

Clearly, this model contains many additional sources of CP violation. In addition, the influence of the LRM may be felt in both tree and loop-level B decays. In particular, the possibility of a large right-handed component in the hadronic current describing $b \to c$ transitions has long been a subject of discussion.[15] Here we examine the possibility of using the rare decays $B \to X_s\gamma$ and $B \to X_s\ell^+\ell^-$ as a new tool in exploring the parameter space of the LRM. The exchange of a W_R within a penguin or box diagram, in analogy with the SM W_L exchange, can lead to significant deviations from the SM predictions for the rates and kinematic distributions in these decays.

In the LRM the complete operator basis governing $b \to s$ transitions in Eq. (4) must be expanded to

$$\mathcal{H}_{eff} = \frac{-4G_F}{\sqrt{2}} \sum_{i=1}^{12} C_{iL}(\mu)\mathcal{O}_{iL}(\mu) + L \to R. \tag{8}$$

This includes the right-handed counterparts to the usual 10 purely left-handed operators, as well as two pairs of additional four-quark operators of mixed chirality. The 2 subsets of left- and right-handed operators, $\mathcal{O}_{1-12L,R}$ are decoupled and do not mix under REG evolution. The decay $B \to X_s\gamma$, where

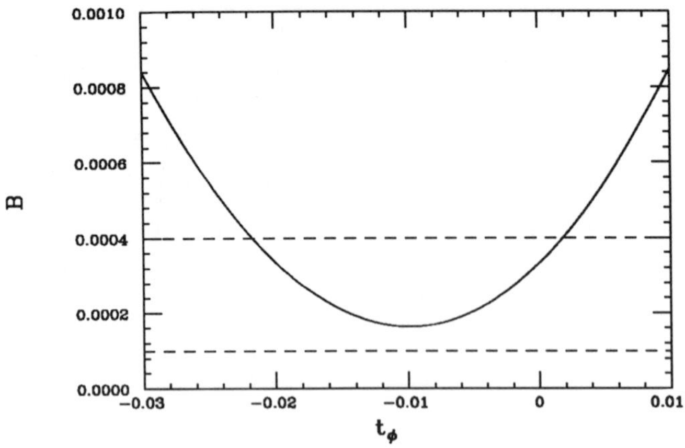

Figure 7: The $B \to X_s\gamma$ branching fraction in the LRM for $m_t(m_t) = 170$ GeV as a function of $\tan\phi$, assuming $\kappa = 1$, $V_R = V_L$, and $M_2 = 1.6$ TeV. The 95% C.L. CLEO results lie inside the dashed lines.

the operators $\mathcal{O}_{1-8,11,12(L,R)}$ contribute, has been studied in some detail.[16] In particular it was shown that the left-right mixing terms associated with $\tan\phi \neq 0$ can be enhanced by a helicity flip factor of $\sim m_t/m_b$ and can lead to significantly different predictions from the SM even when $V_L = V_R$ and W_2 is heavy. This is depicted in Fig. 7 from Rizzo.[16] It is also clear from the figure that not only is the SM result is essentially obtained when $\tan\phi = 0$, but also that a conspiratorial solution occurs when $\tan\phi \simeq -0.02$. From the $B \to X_s\gamma$ perspective, these two cases are indistinguishable, independent of any further improvements in the measurement of the branching fraction.

LRM effects in $B \to X_s\ell^+\ell^-$ have recently been examined in Ref. [17] where it is found that the observables associated with this decay can distinguish the LRM from the SM. Here, all 24 operators in Eq. (8) participate in the renormalization. The determination of the matching conditions at the electroweak scale for these 24 operators is tedious due to the large number of parameters, and in addition to new tree graphs, 116 one-loop diagrams must be evaluated. It is found that this decay can easily distinguish the LRM from the SM, even for the parameter space region which mimics the SM rate for $B \to X_s\gamma$. The extension of the operator basis in Eq. (8) implies that the conventional model independent determination of the Wilson coefficients discussed above will not apply in this case. In fact, this global fit technique has recently been shown to fail[17] for the LRM, resulting in enormous values of $\chi^2/d.o.f.$, and hence provides a powerful probe for the existence of new operators.

7 Conclusions

This talk focused on supersymmetric and left-right symmetric model effects, as well as model independent tests for new physics, in rare B decays. Of course, there are numerous other candidates for physics models beyond the SM, as well as many other reactions where they can be tested. A brief compendium of these is given in Table 1. Here, we display the effects of (*i*) Multi-Higgs-Doublet Models (MHDM), with and without Natural Flavor Conservation (NFC), (*ii*) the Minimal Supersymmetric Standard Model (MSSM), and the supersymmetric models with squark alignment, effective SUSY scales, and R-parity violation, (*iii*) the LRM, with and without manifest left-right symmetry in the quark mixing matrices, (*iv*) a fourth generation, and (*v*) models with Z-boson mediated FCNC. We describe whether these models have the potential to cause large deviations from SM predictions in rare decays and $D^0 - \bar{D}^0$ mixing, whether new phases exist which contribute to $B_d^0 - \bar{B}_d^0$ mixing, and whether the new physics effects cancel in the ratio of mass differences in the B_s to B_d systems. This table is only intended to give a quick indication of potential effects.

In conclusion, we see that the B sector can provide a powerful probe, not only for the existence, but also for the structure of physics beyond the SM.

1. See, for example, P. Huet and E. Sather, *Phys. Rev.* **D51**, 379 (1995).
2. G. Barenboim, G. Eyal, and Y. Nir *Phys. Rev. Lett.* **83**, 4486 (1999).
3. CDF Collaboration, T. Affolder *et al.*, hep-ex/9909003.
4. CLEO Collaboration, S. Ahmed *et al.*, hep-ex/9908022; CLEO Collaboration, M.S. Alam *et al.*, *Phys. Rev. Lett.* **74**, 2885 (1995); CLEO Collaboration, R. Ammar *et al.*, *Phys. Rev. Lett.* **71**, 674 (1993); ALEPH Collaboration, R. Barate *et al.*, *Phys. Lett.* **B429**, 169 (1998).
5. M. Misiak, these proceedings; T. Morozumi, these proceedings.
6. T. Inami and C.S. Lim, Prog. Theor. Phys. **65**, 297 (1981).
7. A.J. Buras and M. Münz, *Phys. Rev.* **D52**, 186 (1995).
8. A. Ali *et al.*, *Phys. Rev.* **D61**, 074024 (2000); G. Buchalla and G. Isidori, *Nucl. Phys.* **B525**, 42 (1998).
9. A. Czarnecki and W. Marciano, *Phys. Rev. Lett.* **81**, 277 (1998).
10. A. Kagan and M. Neubert, Eur. Phys. J **C7**,5 (1999).
11. A. Ali, T. Mannel, and T. Morozumi, *Phys. Lett.* **B273**, 505 (1991); A. Ali, G. F. Giudice, and T. Mannel, *Z. Phys.* **C67**, 417 (1995).
12. J.L. Hewett, *Phys. Rev.* **D53**, 4964 (1996).

Model	Rare Decays	$\frac{\Delta M_s}{\Delta M_d}$	New Phase B_d Mixing	$D^0 - \bar{D}^0$ Mixing
MHDM: with NFC	$B \to X_s\gamma$	$= \text{SM}$	No	Small
: no NFC	Not Really	$\neq \text{SM}$	Yes	Big
MSSM	$B \to X_s\gamma$	$= \text{SM}$	No	Small
Squark Alignment	Small	$\sim \text{SM}$	No	Huge
Effective SUSY	Small	$\sim \text{SM}$	Yes	Small
R-Parity Violation	Big	$\neq \text{SM}$	Yes	Big
LRM: $V_L = V_R$	$B \to X_s\gamma$ and	$= \text{SM}$	No	Small
: $V_L \neq V_R$	$B \to X_s\ell^+\ell^-$	$\neq \text{SM}$	Yes	Big
4th Generation	Big	$\neq \text{SM}$	Yes	Huge
Z-mediated FCNC	Big	$\neq \text{SM}$	Yes	Big

Table 1: Model dependent effects of new physics in various processes.

13. J.L. Hewett and J.D. Wells, *Phys. Rev.* **D55**, 5549 (1997), and references therein.
14. M.N. Danielson *et al.*, Proceedings of *New Directions for High-Energy Physics*, Snowmass, CO, 1996, ed. D.G. Cassel *et al.*.
15. M.B. Voloshin, hep-ph/9704278; M. Gronau and S. Wakaizumi, *Phys. Rev. Lett.* **68**, 1814 (1992); T.G. Rizzo, *Phys. Rev.* **D58**, 055009 (1998); W. Goldberger, hep-ph/9902311.
16. D. Cocolicchio *et al.*, *Phys. Rev.* **D40**, 1477 (1989); G.M. Asatryan and A.N. Ioannisyan, Yad. Fiz. **51**, 1350 (1990); K.S. Babu, K. Fujikawa, and A. Yamada, *Phys. Lett.* **B333**, 196 (1994); P. Cho and M. Misiak, *Phys. Rev.* **D49**, 5894 (1994); T.G. Rizzo, *Phys. Rev.* **D50**, 3303 (1994); G. Bhattacharyya and A. Raychaudhuri, *Phys. Lett.* **B357**, 119 (1995); T.G. Rizzo, hep-ph/9705209.
17. T.G. Rizzo, *Phys. Rev.* **D58**, 114014 (1998); S. Fukae, C.S. Kim, T. Morozumi, and T. Yoshikawa, *Phys. Rev.* **D59**, 074013 (1999); S. Fukae, C.S. Kim, and T. Yoshikawa, hep-ph/0003053.

EXPERIMENTAL SUMMARY

T. NAKADA
CERN, CH-1211 Geneva 23, Switzerland
University of Lausanne, Switzerland
E-mail: tatsuya.nakada@cern.ch

As seen from this conference, B physics is approaching its turning point. While LEP experiments CLEO, and CDF have almost completed their physics analysis with 10^6 to 10^7 B's, we anticipate results from the next generation of B experiments with $> 10^8$ B's, which include the first observation of CP violation outside of the neutral kaon system. Furthermore, B physics with $\sim 10^{10}$ B's will emerge around the year 2005. In this article, we reflect on the past and imagine how B physics will develop in the future.

1 Overview

Physics results in the b quark sector presented in this conference came from CLEO-2, CLEO-2.5, four LEP experiments and CDF (Run-I). They all have completed the data taking and are making all the effort to extract physics. The two HERA experiments gave their results on the b production in the e-p interactions, which are larger then the QCD calculations.

At the same time, the two B factories, KEK-B at KEK and PEP-II at SLAC are rapidly increasing their luminosities. In particular, it is quite impressive how quickly PEP-II archived a luminosity beyond 10^{33} cm^{-2} s^{-1}. The performance of the two detectors, BABAR at PEP-II and BELLE at KEK-B is equally impressive. At the time of the conference, the BELLE detector appears to be better understood. Figure 1 illustrates progress made by PEP-II (D. Boutigny) and BELLE (K. Abe).

The HERA-B collaboration has almost completed the detector installation and commissioning of the trigger has started. Many important results in charm physics were obtained with fixed target experiments working at hadron machines. Very little contribution in the b quark sector has been made so far by the fixed target experiments. These experiments were extremely challenging due to the small fraction of events with b-quarks, expected to be $\sim 10^{-6}$ at those energies. HERA-B tries to overcome this problem with a sophisticated track trigger currently being commissioned.

B physics at hadron colliders operated at TeV energies have been shown to be very competitive by the CDF experiment. The fraction of b-quark events at those energies is 10^{-3} to 10^{-2}, which is similar to that for the events with charm quark at fixed target energies. In order to fully exploit the large number

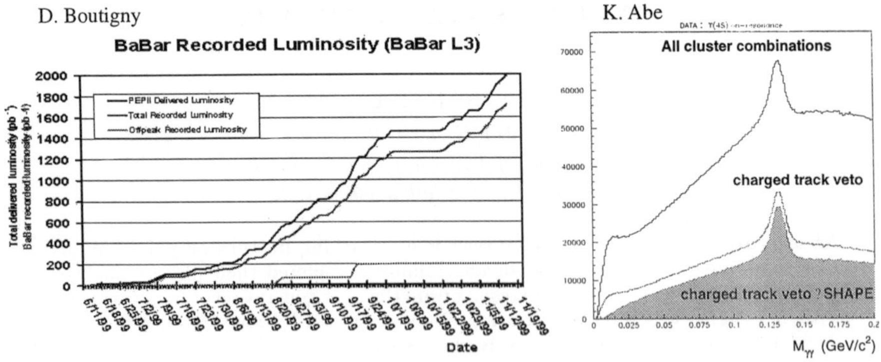

Figure 1. The left plot shows the integrated luminosity achieved by the PEP-II. The upper curve is the PEP-II delivered. The middle curve is recorded by BABAR and the lower curve is the luminosity recorded at center of mass energy off the resonance. The right plot is the reconstructed π^0 signal by BELLE.

of b quarks expected at hadron colliders, a new generation of experiments are being planned.

CP violation in kaon decays is still an active field. Recent results from KTeV and NA48 seem to establish a non-zero value of ε' indicating that CP symmetry is not conserved in the decay amplitude as well as in the K^0-$\overline{K}^0$ oscillation. This excludes the scenario where the Superweak model [1] is solely responsible for CP violation in the neutral kaon system. Future kaon experiments which can measure the branching fractions for $K^+ \to \pi^+ \nu\overline{\nu}$ and $K_L \to \pi^0 \nu\overline{\nu}$ would provide physics competitive to the future B experiments. However this is clearly a very challenging task.

To complete the overview of this conference, I list the subjects of all the experimental talks together with the name of the speakers:

- Physics of b hadrons
  - b quark production (C. Grab)
  - b hadron decays (F. Ukegawa, J. Thaler, A. Wolf, T. Skwarnicki, J. Smith, F. Würthwein, S. L. Wu)
  - B^0-$\overline{B}^0$ and B^0_s-$\overline{B}^0_s$ oscillations (S. Willocq, S. L. Wu)
  - CP violation in B^0 and B^0_s decays (R. Forty, J. Alexander, T. Trischuk)

- CP violation and flavour oscillations in other systems (A. Roodman, A. Nappi, M. Incagli, S. Kettell, H. Park, Y.B. Hsiung, D. Asner, B. O'Reilly, K. B. Luk)
- Extraction of CKM parameters (A. Stocchi, G. Eigen)
- New detectors for b physics:
  - data taking just started: BABAR (D. Boutigny, M. Zito, D. Kirkby), BELLE (K. Abe, T. Iijima, H. Aihara), CLEO-III (T. Skwarnicki)
  - being commissioned: HERA-B (O. Steinkamp, R. Chistov)
  - start in 2001: CDF and D0 Run-II (R. Hughes)
  - in preparation: ATLAS and CMS (P. Vikas), LHCb (S. Stone)
  - under consideration: BTeV (S. Stone)
- e^+e^- B factories (S. Kurokawa)
- CPT symmetry (T. Yamanaka)

As a summary of the conference, we now make a quick tour of b physics and CP violation in the past, present and future.

2 Past Surprises

2.1 Discovery of CP Violation

While the theory of weak interactions with $V - A$ structure naturally explains the observed violation of parity (P) and charge conjugation (C) symmetries, the combined CP symmetry is still conserved in the theory. Therefore, CP violation discovered in the neutral kaon decays [2] in 1964 was completely unexpected. It was understood that CP and T symmetries are violated in the K^0-$\overline{K}^0$ oscillation. There was no obvious way to explain this in the framework of the weak interaction theory and this led to an introduction of hypothetical new interaction, e.g. the Superweak Model [1].

2.2 Discovery of b Quark

In 1973 [3], Kobayashi and Maskawa realised that CP violation can be naturally introduced in the renormalisable gauge theory of the electroweak interactions by having three quark doublets (families). In this theory, CP violation is generated by the complex three by three unitary mass matrix, V given in Eq. (1), which describe the mass eigenstates as linear combinations of the flavour eigenstates. The matrix is often refereed as Cabibbo-Kobayashi-Maskawa (CKM) matrix and an element V_{ij} gives the coupling for the transition of a down-type quark j to an up-type quark i with an emission of W^-. Since the theory remained relatively unknown, the discovery of the b quark through the observation of Υ families [4] came as a big surprise.

$$V = \begin{pmatrix} V_{\rm ud} & V_{\rm us} & V_{\rm ub} \\ V_{\rm cd} & V_{\rm cs} & V_{\rm cb} \\ V_{\rm td} & V_{\rm ts} & V_{\rm tb} \end{pmatrix} \tag{1}$$

2.3 *Discovery of Long b Quark Lifetime*

The next surprise was the discovery of the long lifetime of the b quark. At the beginning of 1980, it was generally believed that the coupling constant for $\rm b \to c + W^-$ described by one of the CKM elements, $|V_{\rm cb}|$, would be similar to that for $\rm s \to u + W^-$, given by $|V_{\rm us}| \approx |V_{\rm cd}|$, i.e. ~ 0.22. Since the b quark is about 3.3 times heavier than the charm quark, the b quark lifetime would be ~ 100 times shorter than that of the charm quark, i.e. of the order of 10^{-14} s. The MAC [5] and Mark-II [6] experiments at SLAC however measured the b quark lifetime to be $\sim 10^{-12}$ s, i.e. as long as the charm quark lifetime. This indicated that $|V_{\rm cb}| \approx |V_{\rm us}|^2$. Inspired by this observation, Wolfenstein proposed an approximation of the CKM matrix [7] based on an expansion in $\lambda = |V_{\rm us}|$ as

$$V \approx \begin{pmatrix} 1-\lambda^2/2 & \lambda & A\lambda^3(\rho - i\eta) \\ -\lambda & 1-\lambda^2/2 & A\lambda^2 \\ A\lambda^3(1-\rho-i\eta) & -A\lambda^2 & 1 \end{pmatrix} \tag{2}$$

where A, ρ and η are real parameters. CP violation is introduced by the nonzero value of η.

2.4 *Discovery of Large* $\rm B^0$-$\rm \overline{B}^0$ *Oscillation Frequency*

The latest surprise was the discovery of the large $\rm B^0$-$\rm \overline{B}^0$ oscillation frequencies. In the Standard Model, $\rm B^0$-$\rm \overline{B}^0$ oscillations are generated by the box diagrams shown in Figure 2 where the top quark is expected to play an important role. The oscillation frequency is given by the twice the absolute value of the $\rm B^0$-$\rm \overline{B}^0$ transition amplitude which is approximately proportional to $m_{\rm t}^2$ where $m_{\rm t}$ is the top quark mass. Toponium had been anticipated in the energy region accessible by PETRA and PEP. Also evidence was announced by UA1 that $m_{\rm t} \approx 40$ GeV/c^2 [9]. This led to the expectation of an oscillation signal too

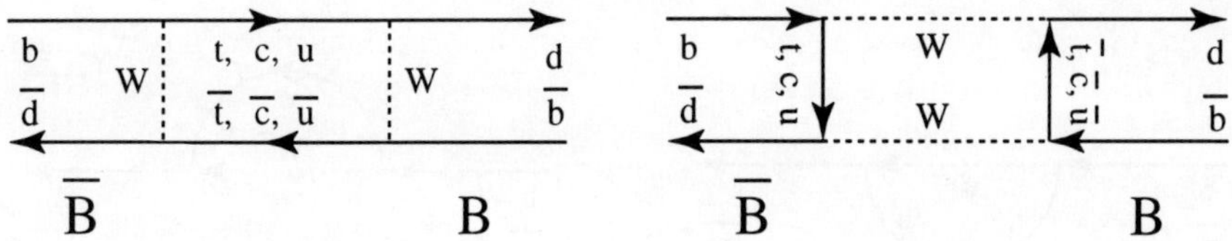

Figure 2. Box diagrams describing B^0-$\overline{B}^0$ oscillations.

small to be detected by ARGUS and CLEO. When ARGUS [8] discovered the B^0-$\overline{B}^0$ oscillation and measured the oscillation frequency to be an order of magnitude larger than expected, it became obvious that the discovery of the top quark would be completely out of reach of the existing e^+e^- colliders. And the B^0-$\overline{B}^0$ transition amplitude is totally dominated by the top quark due to its large mass.

3 Present Situation

3.1 Extraction of CKM Elements

One of the main goals being the accurate extraction of the CKM matrix elements, current heavy flavour physics requires a high precision. The Wolfenstein approximation given in Eq. (2) has to be extended as

$$V \approx \begin{pmatrix} 1-\lambda^2/2 & \lambda & A\lambda^3(\rho - i\eta) \\ -\lambda - iA^2\lambda^5\eta & 1-\lambda^2/2 & A\lambda^2 \\ A\lambda^3(1-\hat{\rho}-i\hat{\eta}) & -A\lambda^2 - iA\lambda^4\eta & 1 \end{pmatrix} \tag{3}$$

where $\hat{\rho}$ and $\hat{\eta}$ are given by $\hat{\rho} = \rho(1-\lambda^2/2)$ and $\hat{\eta} = \eta(1-\lambda^2/2)$ respectively.

There are two motivations behind this goal:

1. In the framework of the Standard Model, elements of the CKM matrix is parameterized with only four parameters. Any deviation from this indicates existence of physics beyond the Standard Model.

2. As for the weak mixing angle, the Standard Model itself does not explain the origin of the CKM matrix. It is hoped that physics at higher energies, which introduces symmetry among different families, may derive it. Values of the matrix elements could give us a clue on the nature of such symmetry.

Absolute values of the six elements, V_{ud}, V_{us}, V_{ub}, V_{cd}, V_{cs} and V_{cb}, are measured from decays generated by the tree diagrams. Two of them, $|V_{cb}|$

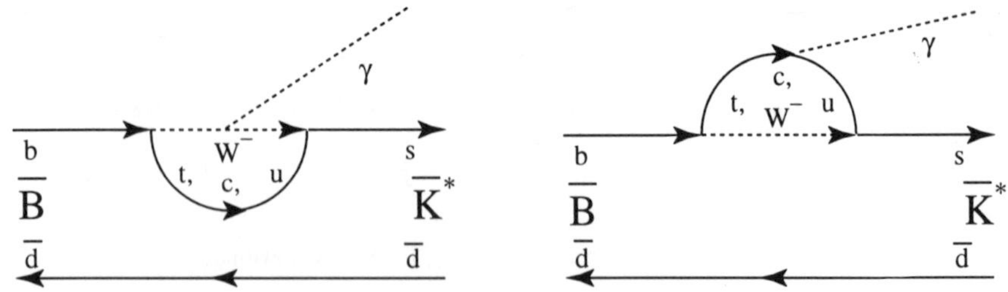

Figure 3. Penguin diagrams describing the $\overline{B}^0_s \to K^{*0}\gamma$ decays.

and $|V_{ub}|$ are obtained from hadrons containing b quarks. Compared with the other four elements they are less well known, in particular the latter. While there exists theoretically well understood procedure for calculating the short range interactions including the higher order effects, the treatment of the long rage hadronic interactions is still not fully under control. Lattice theory could in principe derive solutions from first principles. However, this may still take quite sometime.

In the mean time, many other theoretical ideas have been developed in order to describe the long range hadronic interactions. Here, close collaborative work between theory and experiments are essential. Measured branching fractions in many decay modes can constrain parameters of the models and improve the models. A better model can be developed from more involved experimental knowledge such as differential decay width distributions and helicity polarisation. Many examples of such collaborative work are discussed in the proceedings of this conference.

Assuming that there exits only three quark families, the absolute values of the already known six elements are sufficient to show that $|V_{tb}| \approx 1$ to a very good accuracy. Meaningful measurements on $|V_{td}|$, $|V_{ts}|$ and $|V_{tb}|$ from the tree diagrams generated top quark decays are not realistic in the near future. However, they can be accessed by the loop processes. At the moment, $|V_{td}| \times |V_{tb}|$ is extracted from the frequency, Δm_d, of the B^0-$\overline{B}^0$ oscillation generated by the box diagram. The error on the extracted value is limited by the uncertainties in evaluating the hadronic matrix element in the B^0-$\overline{B}^0$ transition amplitude. Once the oscillation frequency for the B^0_s-$\overline{B}^0_s$ oscillation, Δm_s, is measured, $|V_{td}|/|V_{ts}|$ will be extracted with a much smaller hadronic uncertainty of less than 5% from the frequency ratio, $\Delta m_d/\Delta m_s$. The oscillation frequency of the B_s meson system is expected to be $\sim 1/\lambda^2 =$ 20 times larger than that for the B_d meson system. At the moment, only the

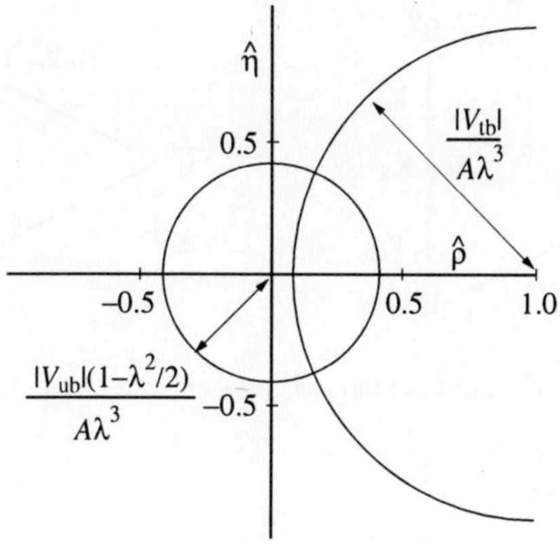

Figure 4. Illustration of the two circles in the $\hat{\rho}$-$\hat{\eta}$ plane obtained from $|V_{ub}|$ and $V_{td}|$.

lower limit on Δm_s is known from the LEP experiments.

Already measured penguin induced decays such as B_s^0 and $\overline{B}_s^0 \to K^{*0}\gamma$, shown in Figure 3, measure $|V_{ts}| \times |V_{tb}|$. However, the theoretical uncertainties in the hadronic interactions are too large to use these measurements for extracting accurately $|V_{ts}| \times |V_{tb}|$. It has to be noted that there are several penguin diagrams contributing to this decay modes. Although the top contribution is dominant, contributions from the other quarks cannot be ignored completely.

3.2 Analysis of the CKM Parameters

A most comprehensive way to analyze the parameters of the CKM matrix is to apply the Wolfenstein parameterization given in Eq. (3). As already discussed, the parameter λ can be extracted using only the first two quark families. Then A can be determined from $|V_{cb}|$. Once λ and A are determined, $|V_{ub}|(1-\lambda^2/2)/A\lambda^3$ gives the radius for a circle centred at (0, 0) in the $\hat{\rho}$-$\hat{\eta}$ plane. Similarly, $|V_{td}|/A\lambda^3$ (or $|V_{td}|/|V_{ts}|\lambda$ obtained from $\Delta m_d/\Delta m_s$) gives the radius for a circle centred at (1, 0) in the $\hat{\rho}$-$\hat{\eta}$ plane. This is illustrated in Figure 4. Errors on $|V_{td}|$ and $|V_{ub}|$ presented in these proceedings are $\sim 20\%$ for both and totally dominated by the theoretical uncertainties.

It is interesting to note that the two circles indeed cross each other and give two possible solutions,

$$(\hat{\rho},\ \hat{\eta}) \approx (0.24,\ 0.33) \tag{4}$$

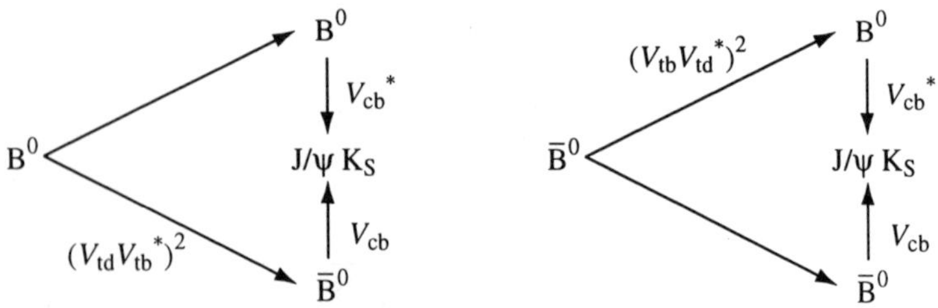

Figure 5. Illustration of relevant CKM matrix elements for CP violation in the $\mathrm{B_d} \to \mathrm{J}/\psi\mathrm{K_S^0}$ decays.

and

$$(\hat{\rho},\ \hat{\eta}) \approx (0.24,\ -0.33) \tag{5}$$

and $\eta = 0$ is disfavoured with a rather high probability.

This analysis made within the framework of the Standard Model uses only the absolute values of the elements of the CKM matrix without including information on CP violation. It leads however to the conclusion that the CKM matrix is complex. The Standard Model, therefore, must generate CP violation.

We now look at the measurements of CP violation. As well known, the CP asymmetry in the initial $\mathrm{B^0}$ and $\overline{\mathrm{B}}^0$ decaying into $\mathrm{J}/\psi\mathrm{K_S^0}$ measures the difference between the phase of the $\mathrm{B^0}$-$\overline{\mathrm{B}}^0$ transition amplitude and twice the phase of the $\mathrm{B^0} \to \mathrm{J}/\psi\mathrm{K_S^0}$ decay amplitude. Since the top quarks completely dominate in the box diagrams, the phase of the transition amplitude is given by the phase of $(V_{\rm tb}V_{\rm tb}^*)^2$ in the Standard Model (Figure 5). In the decay, both the $\overline{\mathrm{b}} \to \overline{\mathrm{c}} + \mathrm{W}^-$ tree diagram and $\overline{\mathrm{b}} \to \overline{\mathrm{s}} + \mathrm{c}\overline{\mathrm{c}}$ penguin diagrams contribute. However, the dominant penguin diagram has the same phase as the tree diagram and the decay amplitude is real to a good approximation. Therefore, the asymmetry must give $\sin 2\phi_1$ where $\phi_1 \equiv \beta$ is the phase of $V_{\rm td}^*$ given by

$$\phi_1 = \tan^{-1}\frac{\hat{\eta}}{\hat{\eta} + (1-\hat{\rho})}.$$

As reported in these proceedings, $\sin 2\phi_1$ is measured to be 0.91 ± 0.35 by combining the results from CDF and LEP experiments shown in Figure 6 (R. Forty). This is consistent with the solution given in Eq. (4).

CP violation parameter in the $\mathrm{K^0}$-$\overline{\mathrm{K}}^0$ oscillation is well measured. The Standard Model analysis based on the box diagrams gives an allowed region of $\hat{\rho}$ and $\hat{\eta}$ entirely consistent with the solution given in Eq. (4). Recent

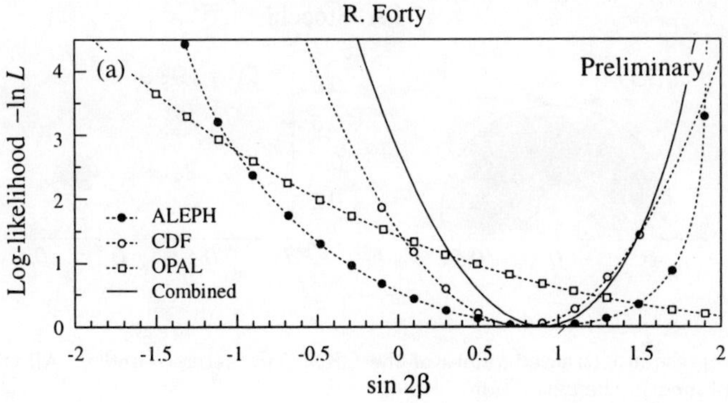

Figure 6. Likelihood distributions versus $\sin 2\beta$ measured with B_d decaying into $J/\psi K_S^0$ measured by various experiments.

measurements of ϵ' strongly indicates that $\Re\epsilon' > 0$. This implies that CP violation exists in $K \to 2\pi$ decay amplitudes. The measured value of $\Re\epsilon'$, i.e. the order 10^{-6}, is consistent with the Standard Model calculation using the solution given in Eq. (4).

In conclusion, all the data analysed in the framework of the Standard Model make a consistent picture that the Standard Model generates CP violation through the complex CKM matrix and no other source of CP violation is needed at the moment. Figure 7 shown by A. Stocchi illustrates the progress in extracting the CKM parameters by combining all the data.

This apparent agreement may be due to the lack of precision. While CP violation in the $B_d \to J/\psi K_S^0$ decays measure $\sin 2\phi_1$ with very little theoretical uncertainty within the framework of the Standard Model, the current experimental error is very large. Analysis of CP violation in the K^0-$\overline{K}^0$ and ϵ' is totally dominated by the large theoretical uncertainties due to the long range hadronic interactions. Recent theoretical development in ϵ' suggests that the theoretical uncertainty is indeed very large. There is also not enough redundant measurements, in particular for CP violation in the B meson systems.

While CP violation measures the sine of the phase of the CKM matrix, the cosine of the phase can be extracted by comparing the branching fractions of many decays. Some decay modes are due to the tree diagram and others are due to the penguin diagrams. There are also decay modes where both the tree and penguins contribute. A study of many branching fractions would

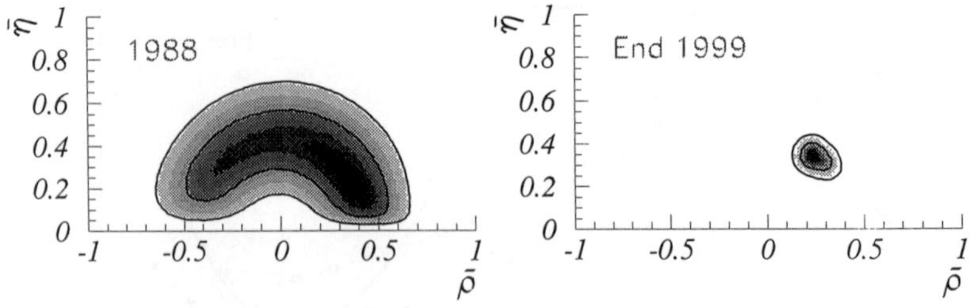

Figure 7. Progress in extracted values of the CKM parameters $\hat{\rho}$ and $\hat{\eta}$. All the available data are combined in the extraction.

allow to isolate the term due to the interference between the tree and penguin processes which contains the cosine of the phase. Although this gives an orthogonal insight to the phase of the CKM matrix, hadronic uncertainties appear to be too large to perform a real precision study.

4 Future Prospect by 2005

For the next five years, we will be improving the precision of the Standard Model analysis by:

1. improving the accuracy of $|V_{\rm cb}|$ and $|V_{\rm ub}|$. Although theoretical uncertainties dominate the errors on $|V_{\rm ub}|$, more data will allow collaborative work between theory and experiment as done for $|V_{\rm cb}|$. They will reduce the error on $|V_{\rm ub}|$ significantly, hopefully below 10%.
2. Obtain $\sqrt{\hat{\eta}^2 + (1-\hat{\rho})^2}$ from $\Delta m_{\rm d}/\Delta m_{\rm s}$ with much reduced error of below 10%.
3. Measure CP asymmetry in $\rm B^0$ and $\rm \overline{B}^0 \to J/\psi K^0_S$ decays with an error on $\sin 2\phi_1$ of < 0.1.
4. Measure other CP asymmetries in the $\rm B_d$ meson decays. such as $\rm D^+D^-$, $\rm D^{*+}D^{*-}$ and $\chi_c \rm K^0_S$.

Combination of all will allow us to make a consistency test of the Kobayashi-Maskawa model with high precision using only the B meson system for the first time.

The first point will be achieved by BABAR, BELLE and CLEO-3 detectors working at $\rm e^+e^-$ machines. The second measurement will be made by

CDF, D0 and possibly by HERA-B at hadron machines. The CP violation measurement will be done by BABAR, BELLE, CDF, D0 and HERA-B.

There are many other CP asymmetries which will be studied by BABAR and BELLE, and also by CLEO-3 if no time development study is required, in particular for $B_d \to \pi^+\pi^-$ and $\pi^+\pi^-\pi^0$ decay modes. Since this decay is generated by $b \to u + W^-$ tree diagram, one can measure $\phi_1 + \phi_3$ where $\phi_3 \equiv \gamma$ is the angle of V_{ub}^* given by

$$\phi_3 = \tan^{-1}\frac{\hat{\eta}}{\hat{\rho}}.$$

Unfortunately, a large contribution from the $b \to d + u\bar{u}$ penguin diagrams are expected in the both decay amplitudes. The penguin diagram producing the largest contribution with the top quark has a different phase to that of the tree diagram. Therefore, quantitative understanding of the penguin contribution is required, introducing larger theoretical uncertainties in calculating the phase of the decay amplitudes. By analysing time dependent Dalitz plot, the penguin contribution can be evaluated in the $\pi^+\pi^-\pi^0$ decay mode. However, some theoretical assumptions on the π-π resonance is still required. Furthermore, a large number of events are required for the analysis since the fit involves many parameters. Another interesting decay mode is $B_d \to K\pi$ which can access ϕ_3. Again theoretical uncertainties will prevent a true precision analysis. All those decay modes will provide important consistency tests. However, none is as clean as the $B_d \to J/\psi K_S^0$ decay.

Measurement of branching fractions for the rare decay modes with expected branching fractions of much less that 10^{-5} and search for forbidden decays may also reveal results inconsistent with the Standard Model prediction.

In five years, the Standard Model analysis using those data may
1) indicate some inconsistency
or
2) still be consistent with the Kobayashi-Maskawa model.
In both cases, further investigation is necessary.

5 Future Prospect beyond 2005

5.1 Challenges

If there is indeed some inconsistency that emerges from the Standard Model analysis based on BABAR, BELLE, CLEO-3, CDF, D0 and HERA-B data, we need to extend the analysis in a way which allows the presence of the contribution from physics beyond the Standard Model. This should make a separate extraction of the Standard Model parameters, i.e. λ, A, ρ and η and parameters for new physics possible.

If there is no sign of new physics, this could be due some numerical cancellation between the Standard Model and new physics contributions hiding a visible effect. The "extended" analysis mentioned above should reveal the contribution from new physics. Finally, insufficient accuracy of the measurements may be the cause for the lack of sign.

New generation of experiments using hadron machines at high energies are designed to undertake these tasks. ATLAS, CMS and LHCb are approved experiments at LHC. At the TEVATRON, the BTeV experiment is under consideration. While ATLAS and CMS are general purpose experiments whose primary goal is to discover Higgs and SUSY particles, LHCb and BTeV are designed to study CP violation using abundantly produced B mesons at high energies. With the new generation of the experiments we can study CP violation
- with B_d mesons with ~ 100 times higher statistics
and
- with B_s mesons,
which opens ways to measure ϕ_3 very cleanly as explained below.

Discussions are started for e^+e^- B factories with a luminosity well above 10^{34} $cm^{-2}s^{-1}$. Although their statistical power would be less than hadron machines, the clean environment at e^+e^- interactions could allow studies of rare decays and CP violation with final states such as $\pi^0\pi^0$, $K_S\pi^0$ and $K^\pm\pi^0$. Their access to the B_s sector would be very limited.

5.2 *Improved Standard Model Analyses*

The decay final state $D^{*-}\pi^+$ can be generated by both B^0 and $\overline{B}^0$ decays through $\overline{b} \to \overline{c} + W^+$ and $b \to u + W^-$ processes respectively. Note that they are tree processes. Then, by studying four time dependent decay rates for the initial B^0 and $\overline{B}^0$ decaying into $D^{*-}\pi^+$ and $D^{*+}\pi^-$, the difference between the phase of the B^0-$\overline{B}^0$ transition amplitude and that of V_{ub} can be extracted. The phase of the transition amplitude is given by V_{td}^2 in the Standard Model and we obtain $2\phi_1 + \phi_3$ (Figure 8). Since only one tree diagram contributes to the decay, no theoretical uncertainty is present in the analysis: it is even cleaner than the case for the $B_d \to J/\psi K_S^0$ decays. A drawback is that the CP violation signal is very small requiring a very large number of B_d mesons.

CP asymmetry obtained from the initial B_s^0 and $\overline{B}_s^0$ decaying into $J/\psi\phi$ can be used to extract the phase of V_{ts}, which is very small as seen from

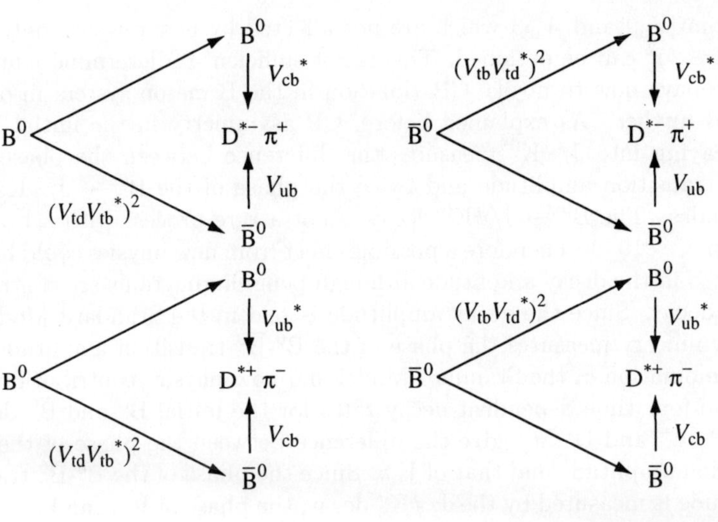

Figure 8. Illustration of relevant CKM matrix elements for CP violation in the $B_d \to D^{*\mp}\pi^{\pm}$ decays.

Eq. (3). Similarly, the four time dependent decay rates for the initial B_s^0 and $\overline{B}_s^0$ decaying into $\overline{D}_sK^+$ and D_sK^- can be used to extract the phase ϕ_3 by applying the measured V_{ts} phase. Unlike the case for the $B_d \to D^*\pi$ decays, here the expected signal is very large. However, the experiment must measure the amplitudes of the B_s^0-$\overline{B}_s^0$ oscillations.

5.3 Analysis Including New Physics

The decay modes described in the previous section are essential to analyse the data without imposing the framework of the Standard Model. If there exist physics beyond the Standard Model introducing new particles, they will contribute very little to processes which are dominated by the tree diagram. Therefore $|V_{cb}|$ and $|V_{ub}|$ determination are still valid even if there exists physics beyond the Standard Model.

On the other hand, the effect of new physics can be sizable in the processes involving loops. In the loop, new particles could be produced as a virtual state. New particles would then generate more box and penguin diagrams in addition to those in the Standard Model. Therefore, the B^0-$\overline{B}^0$ and B_s^0-$\overline{B}_s^0$ oscillations could be affected and the determinations of $|V_{td}|$ and $|V_{ts}|$ are not possible from the measured oscillation frequencies alone.

From $|V_{\rm cb}|$ and $|V_{\rm ub}|$ which are not affected by new physics, only A and $\hat{\eta}^2 + (1-\hat{\rho})^2$ can be obtained. This is not sufficient to determine ρ and η.

We have now to invoke CP violation in the B meson system in order to proceed further. As explained before, CP asymmetry in the initial $\rm B^0$ and $\rm \overline{B}^0$ decaying into $\rm J/\psi K^0_S$ measures the difference between the phase of the $\rm B^0$-$\rm \overline{B}^0$ transition amplitude and twice the phase of the $\rm B^0 \to J/\psi K^0_S$ decay amplitudes. The $\rm B^0 \to J/\psi K^0$ decay is not a rare process with a branching fraction of $\sim 10^{-3}$. Therefore a possible effect from new physics could be safely neglected in the decay amplitude although penguin diagrams are contributing to the decay. Since the decay amplitude is real in the Standard Model, this CP asymmetry measures the phase of the $\rm B^0$-$\rm \overline{B}^0$ transition amplitude which is a combination of the Standard Model and new physics contributions.

The four time dependent decay rates for the initial $\rm B^0$ and $\rm \overline{B}^0$ decaying into $\rm D^{*-}\pi^+$ and $\rm D^{*+}\pi^-$ give the difference between the phase of the $\rm B^0$-$\rm \overline{B}^0$ transition amplitude and that of $V_{\rm ub}$. Since the phase of the $\rm B^0$-$\rm \overline{B}^0$ transition amplitude is measured by the $\rm J/\psi K^0_S$ decay, the phase of $V_{\rm ub}$ can be obtained.

By knowing the phase and absolute value of $V_{\rm ub}$, we can extract ρ and η, i.e. all the CKM parameters are determined and the Standard Model contribution to the $\rm B^0$-$\rm \overline{B}^0$ transition amplitude can be calculated. By comparing this Standard Model contribution with the measured $\rm B^0$-$\rm \overline{B}^0$ oscillation frequency and the phase of the $\rm B^0$-$\rm \overline{B}^0$ transition amplitude, the contribution from new physics to the $\rm B^0$-$\rm \overline{B}^0$ transition amplitude can be extracted.

In a similar way, the phase of $V_{\rm ub}$ can be extracted also by combining the CP asymmetry in the initial $\rm B^0_s$ and $\rm \overline{B}^0_s$ decaying into $\rm J/\psi\phi$ and the study of the four time dependent decay rates for the initial $\rm B^0_s$ and $\rm \overline{B}^0_s$ decaying into $\rm \overline{D}_s K^+$ and $\rm D_s K^-$. Comparison of the phase of $V_{\rm ub}$ measure in the $\rm B_s$ and $\rm B_d$ meson systems provides a consistency test of the method.

In order to make those measurements, experiments must have a trigger which is efficient for the pure hadronic decay modes for collecting enough events. Particle identification, in particular $\rm K/\pi$ separation, is essential to reduce background in the hadronic final states. Furthermore, the CP violation study with rapidly oscillating $\rm B_s$ mesons requires an excellent decay time resolution. BTeV and LHCb are designed to fulfil such requirements.

5.4 Kaons

Within the framework of the Standard Model, $\rm K \to \pi\nu\overline{\nu}$ decays are generated by the Z^0 penguins. Since there is only one hadron in the final state, no final state interaction is present. The hadronic matrix elements can be extracted from the measured semileptonic decays. Therefore, ρ and η can be extracted cleanly from those decay modes. However, detecting $K_L \to \pi^0 \nu\overline{\nu}$ and $K^+ \to \pi^+ \nu\overline{\nu}$ decays are experimentally very difficult. Nevertheless, their branching fractions could be measured with an accuracy of 10 to 20% by $\sim$ 2005 at BNL or FNAL. Since the effect of new physics in the kaon system would be different from that in the B system, kaon experiments could give an interesting complementary information.

6 Epilogue

Physics of CP violation and b quark has been very rewarding. There were surprises which opened new perspectives in understanding of the Standard Model. It appears that we are in a phase of consolidation where many basic facts, such as theoretical uncertainties, need to be put in their proper places. This will be soon followed by quantum leap generated by the data produced by the next generation of experiments. Another quantum leap is expected in $\sim$ 2005 when the new generation of experiments start taking data at high energy hadron machines. So we have all good reason to look forward to the future International Conferences on B physics and CP Violation.

Acknowledgments

The author thanks the local organizer and secretaries of the conference for providing us with a pleasant and scientifically stimulating atmosphere. Roger Forty is very much acknowledged for kindly reading this manuscript with lots of constructive suggestions.

References

1. L. Wolfenstein, *Phys. Rev. Lett.* **13**, 562 (1964).
2. J.H. Christenson *et al*, *Phys. Rev. Lett.* **13**, 138 (1964).
3. M. Kobayashi and T. Maskawa, *Prog. Theor. Phys.* **49**, 652 (1973).
4. S.W, Herb *et al*, PRL **39**, 1977 (.)
5. E. Fernandez *et al*, PRL **51**, 1022 (1983).
6. N.S. Lockyer *et al*, PRL **51**, 1316 (1983).
7. L. Wolfenstein, *Phys. Rev. Lett.* **51**, 1945 (1983).
8. H. Albrecht *et al*, PLB **192**, 245 (1987).
9. G. Arnison *et al*, PLB **147**, 493 (1987).

BCP3: SUMMARY OF THEORY

R. D. PECCEI

Department of Physics and Astronomy, University of California
Los Angeles, California 90095

I discuss a number of the highlights in theory presented at the BCP3 conference. These included new, and more stringent, CKM fits; a critical overview of heavy hadron lifetimes; progress in computing rates and CP-asymmetries in charmless B-decays; a thorough discussion of the implications of the new results on ϵ'/ϵ; and, finally, a peek at the future, trying to estimate how well one is going to be able to measure the unitarity triangle angles.

It is, as usual, very difficult to summarize the myriad of talks in a conference dedicated to a forefront topic in particle physics. BCP3 was no exception. Thus here, asking the indulgence of the speakers whose work I will not mention, I will only touch on a few of the highlights in theory which I thought were particularly interesting.

1 CKM Fits: Where Do We Stand?

During BCP3 there was much discussion of how well we can determine the parameters in the CKM matrix. In practice, this means finding the allowed region, at a certain confidence level, in the $\rho-\eta$ plane - with ρ and η being the usual Wolfenstein [1] parameters. The ingredients of the global fits discussed here particularly by Stocchi [2], as well as by Kim [3] and Eigen [4], are crucially dependent on the errors one assigns to the values one infers experimentally for $|V_{ub}|$ and $|V_{cb}|$. [5 6] Although the data on $\sin 2\beta$ coming from the Tevatron [7], and now also from ALEPH, [8 9] is not accurate enough to have much impact on the allowed region in the $\rho-\eta$ plane, the sharpened LEP bound on Δm_s, presented here by Willocq, [10] provides a significant restriction.

When considering the result of the CKM fits, it is important to distinguish three different ingredients that enter into these fits

i) There are a number of experimental inputs which have quite *negligible experimental errors.* These include: the CP-violating parameter $|\epsilon|$; the $B_d-\bar{B}_d$ mass difference, Δm_d; the Δm_s bound mentioned above; [10] and the value of the sine of the Cabibbo angle, $\sin\theta_C \equiv \lambda$.

ii) There are a number of parameters which have, what one may call *controllable errors.* These include $|V_{cb}|$; the ratio $|V_{ub}|/|V_{cb}|$ and the $SU(3)$-

breaking parameter $\xi = f_{B_s}\sqrt{B_{B_s}}/f_{B_d}\sqrt{B_{B_d}}$.

iii) There are also associated theoretical parameters, whose errors are due to *theoretical uncertaintes* in determining appropriate hadronic matrix elements. These include B_K, the parameter which measures how different the matrix element $< K|(\bar{d}s)^2_{V-A}|\bar{K} >$ is from the vacuum saturation result ($B_K = 1$); and the analogous parameter for the B_d system, $B_d f^2_{B_d}$, which is connected to the matrix element $< B_d|(\bar{d}b)^2_{V-A}|\bar{B}_d >$.

The parameters with *controllable errors*, in principle, are the ones which can be improved using more experimental information. [a] In contrast, it is difficult to reduce, or even correctly estimate, the theoretical errors associated with parameters like B_K, because they are connected with the way one attempts to calculate the hadronic matrix elements. Although Soni [11] in this meeting has given 15% errors on B_K and $f_{B_d}\sqrt{B_{B_d}}$ [$B_K = 0.85\pm0.13$; $f_{B_d}\sqrt{B_{B_d}} = (230\pm30)MeV$], the systematic error on each of these quantities is much harder to pin down, since it depends on the uncertainties associated with going from a quenched to a fully unquenched calculation of the relevant matrix elements.

The fit of Stocchi shown in Fig. 1 uses the parameters detailed in Table 1. Although I find this set of parameters (and the fit!) perfectly reasonable, I believe that one should not take the 95% confidence limit region shown in Fig. 1 strictly as such. There is enough "theoretical error" in this whole business that the most prudent approach is to transform a "formal" 95% confidence limit region into an "effective" 68% limit region. Even after doing such an unorthodox thing, one cannot but be impressed by how well the CKM model fits the data!

As I mentioned earlier, the new result from LEP reported here by Willocq [10] on the $B_s - \bar{B}_s$ mass difference is of considerable importance for the final $\rho - \eta$ plane fit. Last year this parameter was bounded a bit more weakly ($\Delta m_s > 12.4$ ps^{-1}) so that the final $\rho - \eta$ plane fit permitted values of $\rho < 0$. [12] With the present bound, however, one cannot contemplate any longer negative values for ρ— a point emphasized by Deshpande [13] at this meeting. If $\rho > 0$, it follows that the CKM CP-violating phase γ cannot be as large as 90^0. In turn, this implies that one cannot any longer contemplate the superweak solution for the unitarity triangle angles α and β, in which $\sin 2\beta = \sin 2\alpha$. Indeed, the fit of the CKM parameters presented by Stocchi here leads to a best value for $\sin 2\beta$ and $\sin 2\alpha$ which are quite different from

[a] I will discuss explicitly below how one can effect this error reduction in the case of $|V_{cb}|$. Similar ideas can be brought to bear on $|V_{ub}|/|V_{cb}|$ and also on ξ.

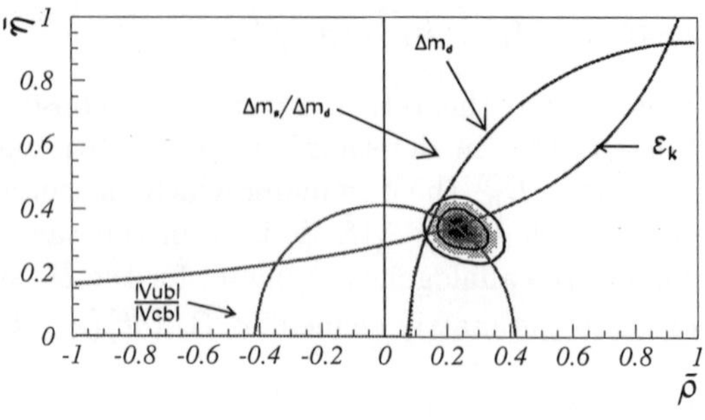

Figure 1. CKM fit presented in Ref [2].

Table 1. Input parameters for the fit in Fig. 1, from reference [2]

Parameter	Value
$\|\epsilon\|$	$(2.280 \pm 0.019) \times 10^{-3}$
$\|V_{ub}\|/\|V_{cb}\|$	$0.080 \pm 0.006 \pm 0.016$ Cornell
$\|V_{ub}\|/\|V_{cb}\|$	$0.104 \pm 0.012 \pm 0.015$ LEP
Δm_d	(0.472 ± 0.016) ps^{-1}
Δm_s	> 14.3 ps^{-1}
$\|V_{cb}\|$	$(40.4 \pm 1.9) \times 10^{-3}$
B_K	$0.86 \pm 0.06 \pm 0.08$
$f_{B_d}\sqrt{B_{B_d}}$	$(210 \pm 29 \pm 12)$MeV
ξ	$1.11 \pm 0.02 \pm 0.05$

each other: [2]

$$\sin 2\beta = 0.75 \pm 0.06 : \quad \sin 2\alpha = -0.22 \pm 0.24 \ . \tag{1}$$

I remark, that the 95% confidence limit on $\sin 2\beta$ which one infers from the CKM analysis,

$$0.62 \leq \sin 2\beta \leq 0.88, \tag{2}$$

is precisely in the range determined by present experiments. The value reported here for $\sin 2\beta$, obtained by ALEPH [8] [9]

$$\sin 2\beta = 0.93^{+0.64}_{-0.88}{}^{+0.36}_{-0.24} \ , \tag{3}$$

when combined with the CDF result [7] [$\sin 2\beta = 0.79^{+0.41}_{-0.44}$], leads to an average value

$$< \sin 2\beta >= 0.91 \pm 0.35 \ , \tag{4}$$

which is perfectly compatible with the range for $\sin 2\beta$ obtained indirectly from the CKM analysis. Furthermore, this experimental result, by itself, is already almost significant statistically.

It is possible to imagine improving considerably the allowed value for $\sin 2\beta$ and other unitarity triangle parameters by improving the *controllable errors* on $|V_{ub}|/|V_{cb}|$ and on the other parameters which enter in the CKM analysis. Let me illustrate how these improvements can come about by focusing specifically on the case of $|V_{cb}|$. If one could trust the parton model, and one knew the mass of the b-quark precisely, one could directly extract $|V_{cb}|$ from the semileptonic width for b-quarks to decay into final states containing c-quarks, given by the standard formula:

$$\Gamma(b \to c\ell\bar{\nu}_e) = |V_{cb}|^2 \frac{G_F^2}{192\pi^3} m_b^5 \ . \tag{5}$$

This formula, however, is of no use since m_b itself is not known precisly enough and, furthermore, there are corrections to the parton model!

What one does, in practice, is to use the Heavy Quark Effective Theory (HQET) to replace m_b by the mass of the B-meson. In so doing the rate (5) above is corrected by the addition of terms involving the matrix elements of the operators 0_i which enter in the operator produced expansion underlying the HQET. [14] To $0(1/m_b^2)$ one has two new operators contributing to the total width, which are characterized by two parameters λ_1, and λ_2 given by

$$\lambda_1 = \frac{1}{2M_B} < B|\bar{b}(iD)^2 b|B > \tag{6}$$

$$\lambda_2 = \frac{g_3}{2M_B} < B|\bar{b}\sigma_{\mu\nu}G^{\mu\nu}b|B > \ ,$$

where $G^{\mu\nu}$ is the gluon field strength tensor. In addition, to this order in the heavy quark expansion, [14] the Fermi momentum $\bar{\Lambda}$ enters in the formulas relating the mass of the b-quark to that of the B and B^* mesons:

$$M_B = m_b + \bar{\Lambda} - \frac{(\lambda_1 + 3\lambda_2)}{m_b} \tag{7}$$

$$M_B^* = m_b + \bar{\Lambda} - \frac{(\lambda_1 - \lambda_2)}{m_b} \ .$$

Using the physical masses for M_B and M_B^* yields directly a value for λ_2 [$\lambda_2 \simeq 0.12\ \mathrm{GeV}^2$].

Using the heavy quark expansion one can obtain a formula for the semileptonic rate $\Gamma(B \to X_c \ell \bar{\nu}_\ell)$ which involves M_B rather than m_b, at the price of some corrections involving λ_1, λ_2, and $\bar{\Lambda}$. One finds, including leading order QCD effects, [15] the formula:

$$\Gamma(B \to X_c \ell \bar{\nu}_\ell) = |V_{cb}|^2 \frac{G_F^2}{192\pi^3} M_B^5 \Phi(\frac{M_D}{M_B})[1 - 1.54\frac{\alpha_s}{\pi} + ...] \qquad (8)$$
$$\cdot \{1 - 1.65\frac{\bar{\Lambda}}{M_B} - 3.18\frac{\lambda_1}{M_B^2} + 0.02\frac{\lambda_2}{M_B^2} + ...\} .$$

Here $\Phi(M_D/M_B)$ is a phase space factor, while the first square bracket contains the $0(\alpha_s)$ perturbative QCD corrections. The terms in the curly brackets are the dominant corrections coming from the heavy quark expansion. If, besides λ_2, one knew λ_1 and $\bar{\Lambda}$ with good accuracy, one could determine $|V_{cb}|$ from the semileptonic width with a small $0(1/m_b^3)$ theory error. The parameters λ_1, λ_2, and $\bar{\Lambda}$ are, in principle, *controllable* since they can be inferred from properties of the lepton spectrum and of the hadronic mass spectrum in non-leptonic B-decays . [15] Specifically, moments of the lepton spectrum and/or the hadronic mass spectrum can be used to determine λ_1 and $\bar{\Lambda}$.

Unfortunately, as shown in Fig. 2, a preliminary analysis by the CLEO collaboration — presented here by Thaler [5] — give inconsistent results. The values for λ_1 and $\bar{\Lambda}$ determined from the leptonic spectrum do not seem to agree with those computed from an analysis of the moments of the hadronic mass spectrum. In my view, the latter values [$\lambda_1 = (-0.13 \pm 0.01 \pm 0.06)\ \mathrm{GeV}^2$; $\bar{\Lambda} = (0.33 \pm 0.02 \pm 0.08)\ \mathrm{GeV}$] are probably more trustworthy, since they are less sensitive to some of the necessary experimental cuts. Indeed, these values are quite close to the values reported by W. M. Zhang [16] at this meeting, coming from a B-bound state analysis using an explicit phenomenological wavefunction for the B-meson. At any rate, using these techniques, it is likely that one will be able to eventually reduce the theoretical error on $|V_{cb}|$ to about 1×10^{-3} (2.5%) and on $|V_{ub}|$ to about 4×10^{-4} (10%), with a corresponding reduction of the allowed region in the $\rho - \eta$ plane. As Misiak [17] discussed in BCP3, this same approach can be used in other contexts — in particular, to better constrain the dependence on the photon energy of the branching ratio for $B \to X_s\gamma$ above some minimum photon energy cut.

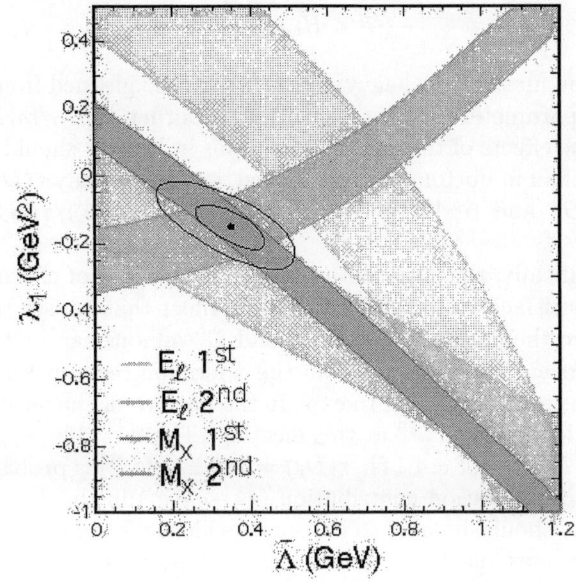

Figure 2. Preliminary analysis by the CLEO Collaboration of the moments of the hadronic mass spectrum (dark shaded) and moments of the leptonic energy spectrum (light shaded) to determine λ_1 and $\bar{\Lambda}$.

2 Lifetimes of Heavy Hadrons

There is reasonable theoretical understanding of the specific pattern of lifetimes of heavy hadrons containing c and b quarks. This topic was reviewed here by Bigi [18] and specific aspects were considered by Melic [19] and by Yang [20]. However, as the data on these lifetimes is now rather precise, [21] a puzzle has emerged. I want to discuss this briefly here.

One can understand the total decay width – and hence the lifetime – of a heavy hadron theoretically via the HQET. The total width is given by the discontinuity of the correlator of the weak Hamiltonian responsible for the decay with itself. This discontinuity is given by a sum of matrix elements of operators of increasing dimensonality, arising in an operator product expansion of the correlator. Schematically, one finds in this way the formula [18]

$$\Gamma_{\rm tot} \sim G_F^2 m_Q^5 \left[< H_Q|\bar{Q}Q|H_Q > + \frac{C_g}{m_Q^2} < H_Q|\bar{Q}\sigma_{\mu\nu}G^{\mu\nu}Q|H_Q > \right. \qquad (9)$$

$$+\frac{C_{4f}}{m_Q^3} < H_Q|\bar{Q}\Gamma q\bar{q}\Gamma Q|H_Q> +...\Big] \,,$$

where m_Q is the mass of the heavy quark. As can be gleaned from the above, the expansion parameter for the total width is of order Λ_{QCD}/m_Q. Hence one expects that the effects of the non-leading terms in Eq. (9) should be bigger in charm decays than in bottom decays. Indeed, experimentally $\tau(D^+)/\tau(D^o) \sim 2.5$ while the B^+ and B_d^o lifetimes are nearly equal [$\tau(B^+)/\tau(B_d^o) = 1.066 \pm 0.024$ [21]]

More specifically, the situation regarding the decays of charmed hadrons seems rather satisfactory. [18] The dominant effect that causes the large difference between the lifetimes of the D^+ and D^0 mesons can be traced to the Pauli interference [22] originating from the 4-fermion term in Eq. (9). These 4-fermion terms are also crucial to explain the pattern of charmed baryon lifetimes, as detailed by Melic [19] at this meeting. The fact that $\tau(D_s)/\tau(D^o)$ is near to 1.2 [$\tau(D^o) = 408 \pm 4.1$ fs; $\tau(D_s) = 486 \pm 15$ fs [21]] is probably evidence for some weak annihilation contribution ($c\bar{s} \to W^+$) in the D_s width. [18] If so, perhaps one should be able to see some evidence for glueball or η' decays of the D_s, processes like $D_s \to$ glueball $\ell\bar{\nu}_\ell$; $D_s \to \eta'\ell\bar{\nu}_\ell$.

Although the prediction that the lifetime differences in the B-sector should be small is well borne out by the data, there is a nagging problem with the Λ_b lifetime. [18] The Pauli interference effects suppress the lifetime of the Λ_b relative to that of the B_d-meson and one can write

$$\frac{\tau(\Lambda_b)}{\tau(B_d)} = 1 - \Delta \tag{10}$$

Bigi [18] estimates $\Delta \simeq 0.05$, while Melic [19] can push Δ to perhaps $\Delta \simeq 0.1$ by taking a very light b-quark mass. However, experimentally this ratio is closer to 0.8, with the latest results giving $\tau(\Lambda_b)/\tau(B_d) = 0.79 \pm 0.05$. [8] It is possible that this signals a real discrepancy between theory and experiment that needs explaining. However, in my view, given the fine agreement of all the other heavy hadron lifetimes with the predictions coming from HQET, perhaps one should not worry overmuch at this stage about a (2-3) σ discrepancy. Hopefully, future data from the Tevatron should help resolve this issue.

3 Charmless B-decays: Rates and CP-Asymmetries

At BCP3 there was lots of discussion of 2-body charmless B-decay. Data on these decays was presented by J. Smith [23] and J. Alexander [24], while various theoretical aspects were touched upon by H-Y Cheng [25]; C. D. Lu [26], H.-n. Li [27], M. Suzuki [28] and C. Bhattacharya. [29]

H.- Y. Cheng [25] in particular, summarized the recent considerable progress achieved in calculating within QCD the relative rates for the decays of B-meson into 2 pseudoscalar states $[B \to PP]$, 2 vector states $[B \to VV]$ and a vector and a pseudoscalar state $[B \to VP]$. The present treatment formalizes (and, in a sense, justifies and explains) the old diagrammatic approach to this problem of Ling Lie Chau. [30] The basic idea of these calculations is well known. One starts with an effective weak Hamiltonian given by a sum of operators

$$H_{\rm eff} = \sum_i C_i(\mu) 0_i(\mu) \ , \tag{11}$$

and one tries to estimate the matrix elements of these operators by using factorization. However, then one correct this procedure by including (some) non-factorizable pieces.

Technically, when one uses factorization, in the process of splitting up the operators as a product of currents the scale dependence of the matrix elements of $0_i(\mu)$ is not preserved:

$$< M_1 M_2 | 0_i(\mu) | B > = < M_1 | J_i | B > < M_2 | J_i | B > \ . \tag{12}$$

However, one can reorganize the Wilson coefficient expansion [31] to effectively recover μ-independent coefficients $C_i^{\rm eff}$ - at least to $0(\alpha_s)$. For Penguin operators these effects are rather large, amounting to about a 50% increase for $C_6^{\rm eff}$. [25] Non-factorizable contributions are incorporated as $1/N_c$ corrections. As Cheng [25] discussed, the recent advance comes from understanding that, depending on which operator 0_i one is considering, the correction factors have different N_c-factors associated with them . This has been justified in the heavy quark limit recently by Beneke, Buchalla, Neubert and Sachrajda. [32] In particular, for $(V-A)(V-A)$ operators $N_c^{\rm eff} \simeq 2$, while for $(V+A)(V-A)$ operators $N_c^{\rm eff} \simeq 6$.

H.- Y. Cheng [25] compared the theoretical predictions resulting from the above considerations with experimental results. The overall comparison is quite good, but there are two extant problems. These are:

i) It is difficult to get the branching ratio $BR(B \to KK) \simeq 0.25 BR(B \to K\pi)$ as is observed experimentally. Typically, one finds much larger $B \to \pi\pi$ branching fractions than are observed.

ii) It is very hard to push the $BR(B \to \eta' K)$ to the very large level observed $(BR \sim 80 \times 10^{-6})$ Typically, theoretically one can reach at most $BR(B \to \eta' K) \sim 50 \times 10^{-6}$. In fact, to get to values as large as this, it is necessary that $N_c^{\rm eff} = 2$ for $(V-A)(V-A)$ operators, as is suggested theoretically.

J. Smith [23] discussed a, theoretically less sophisticated, fit of the measured charmless B-decays. In this fit, which was carried out by Hou, Smith and Würtheim, [33] one simply considers the amplitude for the B-decays in question to be given by a Tree plus a Penguin contribution, neglecting throughout any strong rescattering phases. Further, these authors use (naive) factorization to calculate the relevant matrix elements. Since the Tree amplitudes depended on the CKM phase γ $[T = |T|e^{-i\gamma}]$, this fit determines this phase. Remarkably the fit discussed by Smith [23] determines γ quite accurately:

$$\gamma = (114^{+25}_{-21})^o \ . \tag{13}$$

However, it is difficult to judge the reliability of the approach. Furthermore, also in this case the branching fraction for the decay $B \to \eta' K$ is too low $[BR(B \to \eta' K) \sim 30 \times 10^{-6}]$.

The situation regarding *direct CP-asymmetries* is much less clear, both experimentally and theoretically, Experimentally, as Alexander [24] indicated, one is statistics limited so that $\delta A_{\text{CPviol.}}$ is typically of 0(0.2). For example, one has [24]

$$\begin{aligned} A_{\text{CPviol.}}(K^{\pm}\pi^{\mp}) &= -0.04 \pm 0.16 \\ A_{\text{CPviol.}}(K^{0}\pi^{\pm}) &= +0.18 \pm 0.24 \end{aligned} \tag{14}$$

Improvements will scale as $1/\sqrt{N}$ and one will need an integrated luminosity of $0(100 \text{ fb}^{-1})$ to get to $\delta A_{\text{CPviol.}} \simeq 0.04$!

Theoretically to be able to predict these direct CP-asymmetries one needs a good estimate of the strong rescattering phases. If one writes for the amplitudes of two charge -conjugate processes

$$T = T_1 + T_2 e^{i\delta_w} e^{i\delta_s}; \ \ \bar{T} = T_1 + T_2 e^{-i\delta_w} e^{i\delta_s} \ , \tag{15}$$

one sees that $A_{\text{CPviol.}}$ vanishes if there is no strong rescattering phase δ_s between the two amplitudes:

$$A_{\text{CPviol.}} = \frac{2r \sin\delta_w \sin\delta_s}{1 + r^2 + 2r\cos\delta_w \cos\delta_s} \ , \tag{16}$$

where $r = T_1/T_2$. For sizeable effects, in addition to having a non-negligible rescattering phase, one needs a relatively large weak CP-violating phase δ_w (which is probably OK in the CKM model) and a ratio $r \sim 1$. Since, typically, the two amplitudes involved are Penguin and Tree amplitudes, one needs these amplitudes to be comparable in size and to have a large rescattering phase between them — not too likely a possibility! Numerically, for example, if $r = 0.25, \delta_w \equiv \gamma = 60^o$ and $\delta_s = 30^o$, one obtains $A_{\text{CPviol.}} \simeq 0.1$.

At BCP3 three different approaches were discussed to try to estimate the rescattering phases δ_s which one might expect. All three approaches have some inherent difficulties, demonstrating how challenging really it is to have a reliable estimate for δ_s. C. D. Lu [26] used the old idea of Bander, Silverman and Soni [34] to extract a rescattering phase simply from the phase associated with the discontinuity of Penguin graphs. This phase depends on the momentum transfer carried by the gluon, $\delta_s(k^2)$. However, since the relevant k^2 are rather low, it is natural to question whether one can trust the discontinuity calculated in a pure quark picture to such values of k^2. In contrast, H.-n. Li [27] in his talk at BCP3 estimated δ_s as the rescattering phase which emerges in the Brodsky-Lepage bound state formalism [35] by means of factorization. In this case, the question is whether one can really use these techniques given the large energy release in the decay process $B \to M_1 M_2$.

Mahiko Suzuki [28] presented a more general discussion of the problem based on hadronic methods. He argued, I believe correctly, that it is really difficult to apply perturbative QCD ideas — even if one includes some resummation — for the process at hand, since the effective scale is only of order $k_{\rm eff} \sim \sqrt{\Lambda_{QCD} m_b} \sim 1.5$ GeV. However, Suzuki also pointed out that the rescattering phase δ_s is also difficult to estimate by hadronic methods. This is because the phase associated with, say, the $|K\pi>$ final state is not simply the phase associated with elastic $K\pi$ scattering. For the energies in question

$$|K\pi>=\sum_f e^{i\delta_f}|\alpha_f> \quad , \tag{17}$$

and most of the states that contribute in the above sum are inelastic (roughtly 80%). Given this fact, Suzuki tried to estimate the effective phase δ_s that emerges by using a random phase approximation. Using this approximation, he was able to correlate the resulting phase δ_s with the elasticity of the process, with the result depending on how favored or disfavored is the factorization of the amplitude. When factorization is not favoreda δ_s is bigger. However, if one really has a large rescattering phase δ_s there are also sizable distorsions of the amplitude — a result which Suzuki [28] points out was first understood by Fermi! Basically, not only is there a change in the imaginary part associated with the amplitude, but also the real part is affected:

$$T \to T\exp\left[\frac{P}{\pi}\int\frac{ds'\delta_s(s')}{s'-s}\right] \ . \tag{18}$$

These considerations make it unlikely that one will be able to really ever get a reliable theoretical prediction for a direct CP-asymmetry. However, this fact should not discourage experimentalists from looking for such CP-asymmetries.

4 Imputs from the Strange Quark Sector

In the last year, the new results on ϵ'/ϵ announced by KTeV [36] and by the NA48 Collaboration [37] have generated an enormous amount of interest. Not surprisingly, this subject was also a topic of considerable discussion at BCP3.

4.1 ϵ'/ϵ Results and their Interpretation

In BCP3 the experimental situation regarding ϵ'/ϵ was reviewed by A. Roodman [38] and by A. Nappi [39], while the theoretical aspets of these recent results were discussed by Bertolini, [40] Soni, [11] Masiero, [41] and Soldan. [42] In my view, the nicest aspect of the new KTEV/NA48 results is that they resolve the discrepancy that existed betwen the old Fermilab result, coming from the E731 experiment, [43] and the results of the old CERN experiment NA31. [44] The present results for ϵ'/ϵ:

$$\text{Re } \epsilon'/\epsilon = \begin{cases} (28.0 \pm 4.1) \times 10^{-4} & \text{Ref [35]} \\ (18.5 \pm 7.3) \times 10^{-4} & \text{Ref [36]} \end{cases} \tag{19}$$

when combined with the E731 and NA31 results lead to a world average value for this quantity:

$$< \text{Re } \epsilon'/\epsilon >= (21.2 \pm 4.6) \times 10^{-4} \ , \tag{20}$$

which clearly establishes the existence of $\Delta S = 1$ CP-violation. This result provides the first *direct* confirmation that $\sin\gamma \neq 0$ or, equivalently, that $\eta \neq 0$— something which could only be inferred from the CKM analysis we discussed earlier.

This said, however, it is difficult to extract from the present measurement of ϵ'/ϵ a precise value for η. This is because the value of ϵ'/ϵ, although proportional to η, is rather uncertain due to uncertainties in the calculation of the relevant hadronic matrix element. This point was emphasized by Bertolini [40] at this meeting. His argument can be illustrated by making use of an approximate formula for ϵ'/ϵ due to Buras and his collaborators. [45] As is well known, both matrix elements of gluonic Penguin operators and of electroweak Penguin operators contribute to the process $s \to d\bar{q}q$. Although the gluonic Penguin contributions are of $0(\alpha_s)$, and hence enhanced with respect to the electroweak Penguin contributions which are of $O(\alpha)$, they are suppressed by the $\Delta I = 1/2$ rule since they are proportional to A_2/A_0. [46] The electroweak Penguin contributions, on the other hand, although naturally small do not suffer from the $\Delta I = 1/2$ suppresion and, furthermore, are enhanced by a factor of m_t^2/M_W^2. [47] As a result, for ϵ'/ϵ both Penguin operator contributions are of a similar magnitude and, because they enter with opposite sign, they render the theoretical value for this quantity rather uncertain.

This can be appreciated quite nicely from the approximate formula for ϵ'/ϵ derived by Buras and his collaborators. [45] One has

$$\epsilon'/\epsilon \simeq 34\eta \left[B_6^{1/2} - 0.53B_8^{1/2}\right] \left(\frac{110MeV}{m_s(2GeV)}\right)^2 \times 10^{-4} . \tag{21}$$

Here B_6 and B_8 are the matrix elements of the gluonic Penguin operator and of the electroweak Penguin operator, respectively — normalized so that in the vacuum insertion approximation $B_6 = B_8 = 1$. Because, as we have seen, $\eta \simeq 0.3$, the above formula in this approximation gives $\epsilon'/\epsilon \sim 5 \times 10^{-4}$, which is much below the value measured by KTeV and NA48. Indeed, to get agreement with the present world average [cf Eq. (20)], one needs to stretch all the parameters in Buras's formula. Namely, one needs to maximize η [$\eta \simeq 0.4$]; decrease the value one assumes for $m_s(2GeV)$ [perhaps to m_s as low as $m_s(2GeV) \simeq 90$ MeV]; increase $B_6^{1/2}$ from unity — something which is not clear one can obtain in lattice calculations, [48] but which appears to be true in the chiral quark model [40]; and decrease $B_8^{1/2}$ — something which emerges naturally both in lattice and $1/N$ calculations. [49]

In this meeting, Bertolini [40] emphasized that what improves the *postdictions* of ϵ'/ϵ is to incorporate in the calculation the trend observed in the chiral quark model and in the $1/N$ approximation that $B_6^{1/2}/B_8^{1/2} \simeq 2$, and not unity as expected in the vacuum insertion approximation. However, this is just a phenomenological observation and one will not really trust a theoretical result for ϵ'/ϵ until the lattice results quantitatively arrive at a value for ϵ'/ϵ. Unfortunately, at the moment, there is considerable controversy on what this value might be. This was clear from Soni's talk [11] at BCP3.

Soni [11] argued forcefully that the extant lattice calculations for ϵ'/ϵ which use staggered fermions are not to be trusted, since they have trouble correctly incorporating the needed counterterms and cannot account well for chiral mixing. However, when one uses domain wall fermions, which have the correct chiral behaviour, one arrives at a result for ϵ'/ϵ which is difficult to believe: [11]

$$[\epsilon'/\epsilon]_{DWF} = (-120 \pm 60) \times 10^{-4} . \tag{22}$$

Not only is the sign of ϵ'/ϵ reversed from what one measures experimentally — due to a large negative contribution from the, so called, "eye diagrams" [11] — but the overall magnitude is also rather large.

Given the theoretical disarray concerning ϵ'/ϵ, it appears to me premature to try to invoke the presence of some new physics to "explain" the experimental value of ϵ'/ϵ. However, considering new physics with *correlated* predictions is interesting. For example, as discussed by Masiero [41] here, in supersymmetric models with a large δ_{RL} phase, not only does one get ϵ'/ϵ to be large, but one also predicts largish values for the electron dipole moment and the lepton flavor violating process $\mu \to e\gamma$. As Pakvasa [50] observed, such models also tend to give a rather large CP-violating contribution in Λ-decays.

4.2 Rare K-decays

In contrast to ϵ'/ϵ, as D'Ambrosio [51] discussed in BCP3, there are rare K-decays where the theoretical prediction are on much firmer ground. In particular, the "golden modes" $K_L \to \pi^0 \nu\bar{\nu}$ (whose branching ratio is proportional to η^2) and $K^+ \to \pi^+ \nu\bar{\nu}$ (whose branching ratio is proportional to $|V_{td}|^2$) have theoretical errors of order 1% and 5%, respectively. The Brookhaven experiment BNL E787 has observed one event corresponding to this latter process and one infers a branching ratio [52]

$$BR(K^+ \to \pi^+ \nu\bar{\nu}) = (1.5^{+3.5}_{-1.3}) \times 10^{-10} \ , \tag{23}$$

which is totally consistent with the expectations of the standard model $[(BR)_{SM} = (0.82 \pm 0.32) \times 10^{-10}]$. There are strong hopes [52] that the proposed BNL experiment E949 will be able to actually pin down a value for $|V_{td}|$, since they expect of order 10 events at the SM level of sensitivity.

Clearly, since the branching ratio for $K_L \to \pi^0 \nu\bar{\nu}$ is directly proportional to η^2 and it has a negligible theoretical error, a measurement of this process would have a significant impact on our knowledge of the CKM matrix. Unfortunately, the decay $K_L \to \pi^0 \nu\bar{\nu}$ is extremely challenging experimentally, since one expects a very small branching ratio in the SM $[(BR)_{SM} \simeq 3 \times 10^{-11}]$ and, further, one has an all neutral final state. Nevertheless, as discussed here by Hsiung [53] there are a number of proposed experiments in a planning stage, both at Brookhaven [KOPIO], Fermilab [KAMI] and KEK [E391]. However, to determine η at a significant level (with error of order $\delta\eta/\eta \sim 0.1$) one needs to measure the $K_L \to \pi^0 \nu\bar{\nu}$ branching ratio to an accuracy of order 20%. This is very hard indeed!

As Pakvasa [50] emphasized in his talk at BCP3, the theoretical predictions for CP-violation in Hyperon decays are likely to have less uncertainty than those for ϵ'/ϵ. However, in Hyperon decays, at least in the standard model, the expectations for CP-violating phenomena are very small and appear to be well below the present experimental reach. [54] Typically, one expects theoretically for Λ-decays a CP-violating contribution of order $A_{\text{CPviol.}}(\Lambda) \sim 10^{-5}$, while experimentally one can reach only a level of order $A_{\text{CPviol.}}(\Lambda)|_{exp} \sim 10^{-3} - 10^{-4}$.

5 Looking at the Future: Determing the Phases in the Unitarity Triangle

The most interesting challenge of the B-factories now entering into operation, and of future collider B experiments, is to try to pin down the angles in the unitarity triangle. As Trischuk [7] discussed at this meeting for the most favorable mode, involving the decay $B_d \to \psi K_s$, one can expect to reach the level $\delta \sin 2\beta \simeq 0.1$ at the B-factories, with an integrated luminosity of 50 fb^{-1}. To reach this same level of accuracy at the Tevatron, one probably will need about 2 fb^{-1} of integrated luminosity. These luminosity numbers are likely to be achieved in the next five years. Although the "uncertainty" $\delta \sin 2\beta \simeq 0.1$ is precisely that which obtains from present day CKM fits, [2] the results on $\sin 2\beta$ both at the B-factories and at the Tevatron will involve an *actual* measurement of a CP-violating asymmetry. Thus they are extremely important— even if they probably will not significantly improve the knowledge of $\sin 2\beta$ obtained indirectly through CKM fits.

In the same vein, it is also important to check that the value of $\sin 2\beta$ measured in a variety of different physical processes is in fact the same. Indeed, as remarked by a number of people at this meeting , [55, 56] this may well be the best way to look for physics beyond the standard model. For instance, if there were to be an extra $b \to s$ Penguin phase Φ_P, the process $B_d \to \psi K_s$ would still approximately measure $\sin 2\beta$, while the process $B_d \to \Phi K_s$ (which is Penguin dominated) would measure $\sin(2\beta + \Phi_P)$.

Extracting the two other angles in the unitarity triangle, α and γ, from experiment is likely to be much more challenging. This will require measuring many processes, as Sheldon Stone [57] emphasized in his nice overview at BCP3. Furthermore, one will need to use, at the same time, theory input rather judiciously. [13, 58] That this is so can be appreciated in a number of ways. For example, as is well known, [59] the process $B_d \to \pi^+\pi^-$ does not measure purely $\sin 2\alpha$ since there is likely significant Penguin pollution. In principle, one could imagine estimating these effects by studying $B_d \to \pi^o\pi^o$. [59] However, this decay is estimated to have a $BR \sim 10^{-7}$, which is too small to make this an effective means to control Penguin pollution. Deshpande [13] at this meeting suggested what may be a useful alternative. Namely, using a theoretical calculation to extract α from the value α_{meaus} gotten experimentally from the $B_d \to \pi^+\pi^-$ asymmetry. The model calculations he presented appeared rather encouraging.

Gronau [58] discussed, analogously, how theoretical input — in this case the $SU(3)$ transformation properties of the weak Hamiltonian — can be used to constrain γ for the $\Delta S = 1$ $B \to K\pi$ process. The piece of the decay amplitude which transforms as the $\overline{15}$ dimensional representation in the Tree amplitude essentially fixes the electroweak-Penguin amplitude

$$T_{EWP}(\overline{15}) = \delta_{EW} T_{\rm tree}(\overline{15}) \ , \tag{24}$$

with $\delta_{EW} = 0.65 \pm 0.15$ a calculable number in the standard model. Then using that $|K^o\pi^+ > +\sqrt{2}|K^+\pi^o >$ transforms as the 27 representation one can obtain useful interelations among $B \to K\pi$ process, from which one can extract γ. These techniques can be used also in other contexts and Gronau [58] estimated that they may lead to a determination of γ with an error $\delta\gamma = \pm 20^o$.

Obviously, with new data coming in from the B-factories, and soon also from the Tevatron, we should expect interesting news for BCP4!

Acknowledgements

I would like to thank both George Hou and Hai-Yang Cheng for their splendid hospitality in Taipei. This work was supported in part by the Department of Energy under Contract DE-FE03-91ER40662, Task C.

References

References

1. L. Wolfenstein, Phys. Rev. Lett. **51** (1983) 1945.
2. A. Stocchi, these Proceedings.
3. C. S. Kim, these Proceedings.
4. G. Eigen, these Proceedings.
5. J. Thaler, these Proceedings.
6. C. H. Jin, these Proceedings.
7. W. Trischuk, these Proceedings.
8. S. L. Wu, these Proceedings.
9. R. Forty, these Proceedings.
10. S. Willocq, these Proceedings.
11. A. Soni, these Proceedings.
12. See, for example, R. D. Peccei, hep-ph/9904456, to appear in the Proceedings the 17$^{\rm th}$ International Workshop on Weak Interactions and Neutrinos (WIN99), Cape Town, South Africa, 1999.
13. N. Deshpande, these Proceedings.

14. For a review, see for example, M. B. Wise, hep-ph/9805486, Lectures at the 1997 Les Houches Summer School ; M. Neubert, Phys. Rep. **245** (1994) 259.
15. A. F. Falk, M. Luke and M. J. Savage, Phys. Rev. **D49** (1994) 3367.
16. W. M. Zhang, these Proceedings.
17. M. Misiak, these Proceedings.
18. I. I. Bigi, these Proceedings.
19. B. Melic, these Proceedings.
20. K. C. Yang, these Proceedings.
21. F. Ukagawa, these Proceedings.
22. B. Guberina, S. Nussinov, R. D. Peccei and R. Rückl, Phys. Lett **B 89** (1979) 111.
23. J. Smith, these Proceedings.
24. J. Alexander, these Proceedings.
25. H.- Y. Cheng, these Proceedings.
26. C. D. Lu, these Proceedings.
27. H.- n. Li, these Proceedings.
28. M. Suzuki, these Proceedings.
29. G. Bhattacharya, these Proceedings.
30. L. L. Chau, Phys. Rept. **C95** (1983) 1; L. L. Chau and H.- Y. Cheng, Phys. Rev **D36** (1987) 137.
31. H.- Y. Cheng, H.- n. Li, K.- C. Yang, Phys. Rev. **D60** (1999) 094005.
32. M. Beneke, G. Buchalla, M. Neubert and C.T. Sachrajda, Phys. Rev. Lett. **83** (1999) 1914.
33. W.-S. Hou, J. G. Smith and F. Würtheim, hep-ex/9910014.
34. M. Bander, D. Silverman and A. Soni, Phys. Rev. Lett. **44** (1980) 7.
35. G. P. Lepage and S. J. Brodsky, Phys. Rev. Lett. **43** (1979) 545; Phys. Rev. **D22** (1980) 2157.
36. KTeV Collaboration, A. Alavi-Harati *et al.*, Phys. Rev. Lett. **83** (1999) 22.
37. NA48 Collaboration, V. Fanti *et al.*, Phys. Lett. **B465** (1999) 335.
38. A. Roodman, these Proceedings.
39. A. Nappi, these Proceedings.
40. S. Bertolini, these Proceedings.
41. A. Masiero, these Proceedings.
42. P. H. Soldan, these Proceedings.
43. E731 Collaboration, L. K. Gibbons *et al.*, Phys. Rev. Lett. **70** (1993) 1203.
44. NA31 Collaboration, G. D. Barr *et al.*, Phys. Lett. **B317** (1993) 233.
45. S. Bosch, A.J. Buras, M. Gorbahn, S. Jäger, M. Jamin, M.E. Lauten-

bacher and L. Silvestrini, Nucl. Phys. **B565** (2000) 3.

46. G. Buchalla, A. J. Buras and M. E. Lautenbacher, Rev. Mod. Phys. **68** (1996) 1125.
47. J. Flynn and L. Randall, Phys. Lett. **B216** (1989) 221; *ibid.* **B224** (1989) 221; Nucl. Phys. **B326** (1989) 3.
48. M. Ciuchini, E. Franco, L. Giusti, V. Lubicz and G. Martinelli, hep-ph/9910237, to be published in the Proceedings of KAON '99, Chicago, Illinois, 1999.
49. T. Hambye, G.O. Koehler, E.A. Paschos and P.H. Soldan Nucl. Phys. **B564** (2000) 391: see also, P. H. Soldan, these Proceedings.
50. S. Pakvasa. these Proceedings.
51. G. D' Ambrosio, these Proceedings.
52. S. Kettel, these Proceedings.
53. Y. B. Hsiung, these Proceedings.
54. K. B. Luk, these Proceedings
55. Y. Okada, these Proceedings.
56. X. G. He, these Proceedings.
57. S. Stone, these Proceedings.
58. M. Gronau, these Proceedings.
59. M. Gronau and D. London, Phys. Rev. Lett. **65** (1990) 3381.

List of Participants

Kazuo Abe	*KEK*	kazuo.abe@kek.jp
Koya Abe	*Tohoku University*	abek@awa.tohoku.ac.jp
Iris D. Abt	*Max-Planck Institut für Physik*	isa@mppmu.mpg.de
Hiroaki Aihara	*University of Tokyo*	aihara@phys.s.u-tokyo.ac.jp
Andrew Gerard Akeroyd	*KEK*	akeroyd@ccthmail.kek.jp
Jim Alexander	*Cornell University*	jima@lns.cornell.edu
Elie Aslanides	*Centre de Physique Particules de Marseille*	aslanides@cppm.in2p3.fr
David Asner	*University of California, Santa Barbara*	asner@mail.lns.cornell.edu
Seungwon Baek	*KEK*	swbaek@ccthmail.kek.jp
Stefano Bertolini	*INFN/SISSA*	stefano@he.sissa.it
Gautam Bhattacharyya	*Saha Institute of Nuclear Physics*	gb@tnp.saha.ernet.in
Ikaros Bigi	*University of Notre Dame*	bigi@undhep.hep.nd.edu
Dominique Boutigny	*LAPP, Annecy*	boutigny@in2p3.fr
Andrzej Bozek	*Institute of Nuclear Physics, Krakow*	bozek@bmail.kek.jp
Thomas E. Browder	*University of Hawaii*	teb@uhheph.phys.hawaii.edu
Chia-Hung Chang	*National Taiwan Normal University*	chchang@phy03.phy.ntnu.edu.tw
Darwin Chang	*National Tsing Hua University*	chang@phys.nthu.edu.tw
Paoti Chang	*National Taiwan University*	pchang@phys.ntu.edu.tw
Yuan Chao	*National Taiwan University*	john@hep1.phys.ntu.edu.tw
Junegone Chay	*Korea University*	chay@kupt4.korea.ac.kr
Augustine Chen	*National Central University*	t220042@twncu865.ncu.edu.tw
Ching-Ming Chen	*National Taiwan University*	jcvk4501@ms5.hinet.net
Chuan-Hung Chen	*National Cheng Kung University*	chchen@acer75.phys.ncku.edu.tw
Ju-Chin Chen	*National Taiwan University*	snyang4@phys.ntu.edu.tw

Kai-Feng Chen	*National Taiwan University*	kfjack@hep1.phys.ntu.edu.tw
Chia-Chin Cheng	*National Tsing Hua University*	g873310@phys.nthu.edu.tw
Chih-Hsiang Cheng	*Stanford University*	chcheng@hep.stanford.edu
Hai-Yang Cheng	*Academia Sinica*	phcheng@ccvax.sinica.edu.tw
Chi-Yee Cheung	*Academia Sinica*	phcheung@ccvax.sinica.edu.tw
Cheng-Wei Chiang	*Carnegie Mellon University*	chengwei@andrew.cmu.edu
Ruslan Chistov	*ITEP*	chistov@iris1.itep.ru
Ting-Wai Chiu	*National Taiwan University*	twchiu@phys.ntu.edu.tw
Chih-Lung Chou	*Academia Sinica*	choucl@phys.sinica.edu.tw
Chung-Hsien Chou	*National Tsing Hua University*	d833302@phys.nthu.edu.tw
Chun-Khiang Chua	*National Taiwan University*	ckchua@phys.ntu.edu.tw
Bok Keun Chung	*APCTP*	bkchung@nms.kyunghee.ac.kr
Tzau-Kang Chyi	*National Chiao Tung University*	u8227502@cc.nctu.edu.tw
David Coupal	*Stanford Linear Accelerator Center*	coupal@slac.stanford.edu
Giancarlo D'Ambrosio	*INFN-Sezione di Napoli*	Giancarlo.Dambrosio@na.infn.it
Nilendra G. Deshpande	*University of Oregon*	desh@oregon.uoregon.edu
Gerald Eigen	*University of Bergen*	eigen@asfys2.fi.uib.no
Gad Eilam	*Technion*	eilam@physics.technion.ac.il
Marco Fabbrichesi	*INFN/SISSA*	marco@he.sissa.it
Roger Forty	*CERN*	Roger.Forty@cern.ch
Chao-Qiang Geng	*National Tsing Hua University*	geng@phys.nthu.edu.tw
Ascelin Gordon	*University of Melbourne*	gordon@physics.unimelb.edu.au
Christoph Grab	*DESY/ ETH-Zürich*	grab@particle.phys.ethz.ch
Michael Gronau	*Technion*	gronau@physics.technion.ac.il
Je-An Gu	*National Taiwan University*	wyhwange@phys.ntu.edu.tw
Junji Haba	*KEK*	haba@mail.kek.jp
Hiroshi Hamasaki	*KEK*	hamasaki@bmail.kek.jp

Thomas Handler	*University of Tennessee*	thandler@utk.edu
Jorn Dines Hansen	*Niels Bohr Institute*	dines@nbi.dk
Masashi Hazumi	*Osaka University*	hazumi@champ.hep.sci.osaka-u.ac.jp
Xiao-Gang He	*National Taiwan University*	hexg@penguin.phys.ntu.edu.tw
Yi-Lin He	*National Tsing Hua University*	d887304@phys.nthu.edu.tw
JoAnne Hewett	*Stanford Linear Accelerator Center*	hewett@slac.stanford.edu
Takeo Higuchi	*University of Tokyo*	higuchi@beldaq1.phys.s.u-tokyo.ac.jp
Pei-Ming Ho	*National Taiwan University*	pmho@phys.ntu.edu.tw
Michihiro Hori	*Hiroshima University*	michi@theo3.phys.sci.hiroshima-u.ac.jp
Yoshimoto Hoshi	*Tohoku Gakuin University*	hoshi@tjcc.tohoku-gakuin.ac.jp
George Wei-Shu Hou	*National Taiwan University*	wshou@phys.ntu.edu.tw
Yee Bob Hsiung	*Fermilab*	hsiung@fnal.gov
Li-Yi Hsu	*National Taiwan University*	cdhu6@phys.ntu.edu.tw
Shih-Chieh Hsu	*National Taiwan University*	schsu@hep1.phys.ntu.edu.tw
Chien-Lan Hsueh	*National Taiwan University*	r7222048@ms.cc.ntu.edu.tw
Chao-shang Huang	*ITP, Beijing*	csh@itp.ac.cn
Hsuan-Cheng Huang	*National Taiwan University/BELLE*	hchuang@bmail.kek.jp
Richard E. Hughes	*Ohio State University*	hughes@mps.ohio-state.edu
Chien-Wen Hwang	*National Chiao Tung University*	hwcw@cc.nctu.edu.tw
Pauchy W-Y Hwang	*National Taiwan University*	wyhwang@phys.ntu.edu.tw
Toru Iijima	*KEK*	toru.iijima@kek.jp
Hitomi Ikeda	*KEK*	hitomi@bmail.kek.jp
Kenji Inami	*Nagoya University*	kenji@hepl.phys.nagoya-u.ac.jp
Marco Incagli	*INFN- Pisa*	incagli@pi.infn.it
Akimasa Ishikawa	*Nagoya University*	akimasa@hepl.phys.nagoya-u.ac.jp
Yoshihito Iwasaki	*KEK*	yoshihito.iwasaki@kek.jp

David Jackson	*Osaka University*	djackson@champ.hep.sci.osaka-u.ac.jp
Jia-Huag Jian	*National Taiwan University*	chiueh1@phys.ntu.edu.tw
Changhao Jin	*University of Melbourne*	jin@physics.unimelb.edu.au
Chung Kao	*University of Wisconsin*	kao@pheno.physics.wisc.edu
Yeong-Chuan Kao	*National Taiwan University*	yckao@phys.ntu.edu.tw
Hiroyuki Kawamura	*Hiroshima University*	kawamura@phys.sci.hiroshima-u.ac.jp
Steve Kettell	*Brookhaven National Lab*	kettell@bnl.gov
Yong-Yeon Keum	*Academia Sinica*	keum@ccthmail.kek.jp
Choong Sun Kim	*Yonsei University*	kim@kimcs.yonsei.ac.kr
Yeong- Gyun Kim	*YITP, Kyoto University*	ygkim@theory.yonsei.ac.kr
David Kirkby	*Stanford University*	kirkby@hep.stanford.edu
Yuichiro Kiyo	*Hiroshima University*	kiyo@theo.phys.sci.hiroshima-u.ac.jp
Pyungwon Ko	*KAIST*	pko@muon.kaist.ac.kr
Jürgen Körner	*University of Mainz*	Koerner@thep.physik.uni-mainz.de
Otto C.W. Kong	*Academia Sinica*	kongcw@phys.sinica.edu.tw
Emi Kou	*Nagoya University*	kou@eken.phys.nagoya-u.ac.jp
Takeshi Kurimoto	*Toyama University*	krmt@sci.toyama-u.ac.jp
Shin-ichi Kurokawa	*KEK*	shin-ichi.kurokawa@kek.jp
Hung-Liang Lai	*National Cheng-Kung University*	lai@acer75.phys.ncku.edu.tw
Wen-Ping Lai	*Academia Sinica*	wpl@phys.sinica.edu.tw
Hau-Bin Lee	*National Taiwan University*	d5222004@ms.cc.ntu.edu.tw
Jake Lee	*Yonsei University*	jilee@theory.yonsei.ac.kr
Ming-Chieh Lee	*National Taiwan University*	mclee@hep1.phys.ntu.edu.tw
Shih-Chang Lee	*Academia Sinica*	sclee@sinica.edu.tw
Jeng-Yuan Leou	*National Taiwan University*	hexg3@phys.ntu.edu.tw
Hsiang-nan Li	*National Cheng-Kung University*	hnli@mail.ncku.edu.tw
Huei-Shih Liao	*National Tsing Hua University*	d847301@phys.nthu.edu.tw

Chong-Chung Lih	*National Tsing Hua University*	cclih@phys.nthu.edu.tw
Jyh-Liong Lim	*National Chiao Tung University*	u8221805@cc.nctu.edu.tw
Guey-Lin Lin	*National Chiao Tung University*	glin@cc.nctu.edu.tw
Huey-Wen Lin	*National Taiwan University*	hwlin@hep1.phys.ntu.edu.tw
Shia-Yu Lin	*National Taiwan University*	d8222017@ms1.cc.ntu.edu.tw
Wen-Long Lin	*National Taiwan Normal University*	wenlin@phy03.phy.ntnu.edu.tw
Yu-Hsing Lin	*National Taiwan University*	yslin@hep1.phys.ntu.edu.tw
Chun Chu Liu	*National Tsing Hua University*	d877304@phys.nthu.edu.tw
Nigel S. Lockyer	*University of Pennsylvania*	lockyer@lockyer.hep.upenn.edu
Ian Low	*Carnegie Mellon University*	ilow@andrew.cmu.edu
Cai-Dian Lu	*Hiroshima University*	lucd@theo.phys.sci.hiroshima-u.ac.jp
Kam-Biu Luk	*UC Berkeley/LBNL*	k_luk@lbl.gov
Antonio Masiero	*SISSA-Trieste*	masiero@sissa.it
Takahiro Matsumoto	*Nagoya University*	takahiro@hepl.phys.nagoya-u.ac.jp
Blazenka Melic	*Rudjer Boskovic Institute, Zagreb*	melic@thphys.irb.hr
Mikolaj Misiak	*CERN*	Mikolaj.misiak@cern.ch
Kenkichi Miyabayashi	*Nara Women's University*	miyabaya@hepl.phys.nara-wu.ac.jp
Takuya Morozumi	*Hiroshima University*	morozumi@theo.phys.sci.hiroshima-u.ac.jp
Tatsuya Nakada	*CERN*	tatsuya.nakada@cern.ch
Eiichi Nakano	*Osaka City University*	nakano@hep.osaka-cu.ac.jp
Mikihiko Nakao	*KEK*	mikihiko.nakao@kek.jp
Aniello Nappi	*University of Perugia*	nappi@pg.infn.it
Osamu Nitoh	*Tokyo U. of Agriculture and Technology*	nitoh@cc.tuat.ac.jp
Seishi Noguchi	*Nara Women's University*	noguchi@hepl.phys.nara-wu.ac.jp
Takayoshi Ohshima	*Nagoya University*	ohshima@hepl.phys.nagoya-u.ac.jp
Takeshi Okabe	*Nagoya University*	okabet@hepl.phys.nagoya-u.ac.jp
Yasuhiro Okada	*KEK*	yasuhiro.okada@kek.jp

Brian J.E. O'Reilly	*University of Colorado*	irish@pizero.colorado.edu
Hitoshi Ozaki	*KEK*	hitoshi@bmail.kek.jp
Sandip Pakvasa	*University of Hawaii*	pakvasa@phys.hawaii.edu
Hwanbae Park	*Korea University*	hbpark@hep.korea.ac.kr
Roberto D. Peccei	*UCLA*	robertop@college.ucla.edu
Chai Ho Rim	*APCTP*	rim@apctp.kaist.ac.kr
Aaron Roodman	*SLAC*	roodman@SLAC.Stanford.EDU
Jiro Ryuko	*Osaka University*	ryuko@champ.hep.sci.osaka-u.ac.jp
Hiroyuki Sagawa	*KEK*	sagawa@bmail.kek.jp
Hiroshi Sakamoto	*Kyoto University*	sakamoto@scphys.kyoto-u.ac.jp
Ichiro Sanda	*Nagoya University*	sanda@eken.phys.nagoya-u.ac.jp
Max Sang	*Max Planck Institute, Munich*	Max.Sang@cern.ch
Sven Schaller	*Max-Planck-Institut für Physik*	schaller@mppmu.mpg.de
Steve Schrenk	*Virginia Tech/KEK BELLE*	schrenk@bmail.kek.jp
Iain Scott	*Royal Holloway & Bedford New College*	I.Scott@rhbnc.ac.uk
Bi-Ching Shi	*National Taiwan University*	r8222038@ms1.cc.ntu.edu.tw
Jian-Qing Shi	*National Taiwan University*	hexg1@phys.ntu.edu.tw
Hsien-Hung Shih	*Academia Sinica*	hhshih@phys.sinica.edu.tw
Boris Shwartz	*BINP, Novosibirsk*	shwartz@bmail.kek.jp
Rahul Sinha	*Institute of Mathematical Sciences, Madras*	sinha@imsc.ernet.in
Tomasz Skwarnicki	*Syracuse University*	tomasz@physics.syr.edu
James G. Smith	*University of Colorado*	jgsmith@pizero.colorado.edu
Peter Soldan	*University of Dortmund*	soldan@dilbert.physik.uni-dortmund.de
Amarjit Soni	*Brookhaven National Lab*	soni@bnl.gov
Olaf Steinkamp	*University of Zürich*	olaf.steinkamp@desy.de
Achille Stocchi	*LAL-Orsay*	achille.stocchi@cern.ch
Sheldon Stone	*Syracuse University*	stone@suhep.phy.syr.edu

Hirotaka Sugawara	*KEK*	sugawara@director.kek.jp
Akira Sugiyama	*Nagoya University*	sugiyama@hepl.phys.nagoya-u.ac.jp
Jun-ichi Suzuki	*KEK*	jsuzuki@bmail.kek.jp
Kazuhito Suzuki	*Chiba University*	kazuhito@bmail.kek.jp
Mahiko Suzuki	*UC Berkeley*	msuzuki@LBL.gov
Shiro Suzuki	*Nagoya Univeristy*	suzuki@hepl.phys.nagoya-u.ac.jp
Hiroyasu Tajima	*University of Tokyo*	tajima@phys.s.u-tokyo.ac.jp
Tamotsu Takahashi	*Osaka City University*	takahash@hep.osaka-cu.ac.jp
Jun-ichi Tanaka	*University of Tokyo*	jtanaka@tkybel1.phys.s.u-tokyo.ac.jp
Ping-Kun Teng	*Academia Sinica*	pkteng@sinica.edu.tw
Yoshiki Teramoto	*Osaka University*	teramoto@hep.osaka-cu.ac.jp
Kunihiko Terasaki	*YITP, Kyoto University*	terasaki@yukawa.kyoto-u.ac.jp
Jon J. Thaler	*University of Illinois*	jjt@postoffice.hep.uiuc.edu
Makoto Tomoto	*Nagoya University*	makoto@hepl.phys.nagoya-u.ac.jp
Josip Trampetic	*Rudjer Boskovic Institute, Zagreb*	josip@thphys.irb.hr
William Trischuk	*University of Toronto*	william@physics.utoronto.ca
Shang-Yuu Tsai	*National Taiwan University*	wyhwangp@phys.ntu.edu.tw
Benjamin Tseng	*Academia Sinica*	btseng@phys.sinica.edu.tw
Jie-Jun Tseng	*National Chiao Tung University*	u8627512@cc.nctu.edu.tw
Toru Tsuboyama	*KEK*	toru.tsuboyama@kek.jp
Sadaharu Uehara	*KEK*	uehara@post.kek.jp
Koji Ueno	*National Taiwan University*	ueno@hep1.phys.ntu.edu.tw
Kazumasa Ukai	*Nagoya University*	ukai@eken.phys.nagoya-u.ac.jp
Fumihiko Ukegawa	*University of Tsukuba*	ukegawa@hepaxp1.px.tsukuba.ac.jp
Shoji Uno	*KEK*	shoji.uno@kek.jp
Yutaka Ushiroda	*Kyoto university*	ushiroda@bmail.kek.jp
Pratibha Vikas	*University of Minnesota / CMS*	Pratibha.Vikas@cern.ch

Chin-Chi Wang	*National Taiwan University*	ccwang@hep1.phys.ntu.edu.tw
Chung-Hsiang Wang	*National Lien-Ho Institute of Technology*	chwang@mail.lctc.edu.tw
Min-Zu Wang	*National Taiwan University*	mwang@phys.ntu.edu.tw
Stephane Willocq	*University of Massachusetts, Amherst*	willocq@physics.umass.edu
Andreas Hans Wolf	*Syracuse University*	wolf@phy.syr.edu
Henry Tsz-king Wong	*Academia Sinica*	htwong@phys.sinica.edu.tw
Chung-Yi Wu	*National Taiwan University*	cywu@phys.ntu.edu.tw
Sau Lan Wu	*University of Wisconsin-Madison*	wu@wisconsin.cern.ch
Frank Würthwein	*MIT*	fkw@fnal.gov
Mitsuhiro Yamaga	*Tohoku University*	yamaga@awa.tohoku.ac.jp
Taku Yamanaka	*Osaka University*	taku@fnal.gov
Kwei-Chou Yang	*Chung-Yuan Christian University*	kcyang@cycu.edu.tw
Shin-Nan Yang	*National Taiwan University*	snyang@phys.ntu.edu.tw
Herng Yao	*National Taiwan Normal University*	yao@phy03.phy.ntnu.edu.tw
Yu-Ting Yeh	*National Taiwan University*	r7222049@ms.cc.ntu.edu.tw
Masashi Yokoyama	*University of Tokyo*	yokoyama@tkybel1.phys.s.u-tokyo.ac.jp
Tadashi Yoshikawa	*KEK*	yosikawa@acorn.kek.jp
Chan-Ping Yu	*National Taiwan University*	r7222035@ms.cc.ntu.edu.tw
Hoi- Lai Yu	*Academia Sinica*	phhlyu@ccvax.sinica.edu.tw
C.P. Yuan	*NCTS and Michigan State University*	cpyuan@phux1.phys.nthu.edu.tw
Wei-Min Zhang	*National Cheng-Kung University*	wzhang@mail.ncku.edu.tw
Marco Zito	*DAPNIA-Saclay*	zito@slac.stanford.edu

5 12 '99